REAL-TIME SIMULATION TECHNOLOGIES

Principles, Methodologies, and Applications

Computational Analysis, Synthesis, and Design of Dynamic Systems Series

Series Editor
Pieter J. Mosterman

MathWorks, Inc.
Natick, Massachusetts

McGill University
Montréal, Québec

Discrete-Event Modeling and Simulation: A Practitioner's Approach,
Gabriel A. Wainer

Discrete-Event Modeling and Simulation: Theory and Applications,
edited by Gabriel A. Wainer and Pieter J. Mosterman

Model-Based Design for Embedded Systems,
edited by Gabriela Nicolescu and Pieter J. Mosterman

Model-Based Testing for Embedded Systems,
edited by Justyna Zander, Ina Schieferdecker, and Pieter J. Mosterman

Multi-Agent Systems: Simulation & Applications,
edited by Adelinde M. Uhrmacher and Danny Weyns

Real-Time Simulation Technologies: Principles, Methodologies, and Applications,
edited by Katalin Popovici and Pieter J. Mosterman

Forthcoming Titles:

Computation for Humanity: Information Technology to Advance Society,
edited by Justyna Zander, Ina Schieferdecker, and Pieter J. Mosterman

REAL-TIME SIMULATION TECHNOLOGIES

Principles, Methodologies, and Applications

Edited by
Katalin Popovici
Pieter J. Mosterman

CRC Press
Taylor & Francis Group
Boca Raton London New York

CRC Press is an imprint of the
Taylor & Francis Group, an **informa** business

CRC Press
Taylor & Francis Group
6000 Broken Sound Parkway NW, Suite 300
Boca Raton, FL 33487-2742

First issued in paperback 2017

© 2013 by Taylor & Francis Group, LLC
CRC Press is an imprint of Taylor & Francis Group, an Informa business

No claim to original U.S. Government works

Version Date: 20120411

ISBN 13: 978-1-138-07755-3 (pbk)
ISBN 13: 978-1-4398-4665-0 (hbk)

Library of Congress Cataloging-in-Publication Data

Real-time simulation technologies : principles, methodologies, and applications / editors, Katalin Popovici and Pieter J. Mosterman.
    p. cm. -- (Computational analysis, synthesis, and design of dynamic systems series)
    Includes bibliographical references and index.
    ISBN 978-1-4398-4665-0 (hardback)
    1. Computer simulation. 2. Real-time data processing. I. Popovici, Katalin. II. Mosterman, Pieter J.

QA76.9.C65R385 2012
003'.3--dc23

2012011979

Visit the Taylor & Francis Web site at
http://www.taylorandfrancis.com

and the CRC Press Web site at
http://www.crcpress.com

Contents

Preface...ix
Editors..xi
Contributors xiii
Introduction...xvii

SECTION I Basic Simulation Technologies and Fundamentals

Chapter 1   Real-Time Simulation Using Hybrid Models.......................................3

Roy Crosbie

Chapter 2   Formalized Approach for the Design of Real-Time Distributed Computer Systems...35

Ming Zhang, Bernard Zeigler, and Xiaolin Hu

Chapter 3   Principles of DEVS Model Verification for Real-Time Embedded Applications 63

Hesham Saadawi, Gabriel A. Wainer, and Mohammad Moallemi

Chapter 4   Optimizing Discrete Modeling and Simulation for Real-Time Constraints with Metaprogramming..97

Luc Touraille, Jonathan Caux, and David Hill

Chapter 5   Modeling with UML and Its Real-Time Profiles 123

Emilia Farcas, Ingolf H. Krüger, and Massimiliano Menarini

Chapter 6   Modeling and Simulation of Timing Behavior with the Timing Definition Language.. 159

Josef Templ, Andreas Naderlinger, Patricia Derler, Peter Hintenaus, Wolfgang Pree, and Stefan Resmerita

SECTION II   Real-Time Simulation for System Design

Chapter 7 Progressive Simulation-Based Design for Networked
Real-Time Embedded Systems.. 181

Xiaolin Hu and Ehsan Azarnasab

Chapter 8 Validator Tool Suite: Filling the Gap between
Conventional Software-in-the-Loop and Hardware-in-the-Loop
Simulation Environments.. 199

*Stefan Resmerita, Patricia Derler, Wolfgang Pree,
and Kenneth Butts*

Chapter 9 Modern Methodology of Electric System Design Using
Rapid-Control Prototyping and Hardware-in-the-Loop 219

Jean Bélanger and Christian Dufour

Chapter 10 Modeling Multiprocessor Real-Time Systems at
Transaction Level 243

Giovanni Beltrame, Gabriela Nicolescu, and Luca Fossati

Chapter 11 Service-Based Simulation Framework for Performance
Estimation of Embedded Systems.. 259

Anders Sejer Tranberg-Hansen and Jan Madsen

Chapter 12 Consistency Management of UML Models 289

Emilia Farcas, Ingolf H. Krüger, and Massimiliano Menarini

SECTION III   Parallel and Distributed Real-Time Simulation

Chapter 13 Interactive Flight Control System Development and
Validation with Real-Time Simulation.. 331

Hugh H. T. Liu

Chapter 14 Test Bed for Evaluation of Power Grid Cyber-Infrastructure 349

David C. Bergman and David M. Nicol

Chapter 15 System Approach to Simulations for Training: Instruction,
Technology, and Process Engineering 371

Sae Schatz, Denise Nicholson, and Rhianon Dolletski

Chapter 16 Concurrent Simulation for Online Optimization of
Discrete Event Systems 389

Christos G. Cassandras and Christos G. Panayiotou

SECTION IV Tools and Applications

Chapter 17 Toward Accurate Simulation of Large-Scale Systems
via Time Dilation 419

James Edmondson and Douglas C. Schmidt

Chapter 18 Simulation for Operator Training in Production Machinery 439

Gerhard Rath

Chapter 19 Real-Time Simulation Platform for Controller Design,
Test, and Redesign.. 461

Savaş Şahin, Yalçın İşler, and Cüneyt Güzeliş

Chapter 20 Automotive Real-Time Simulation: Modeling
and Applications... 501

Johannes Scharpf, Robert Höpler, and Jeffrey Hillyard

Chapter 21 Specification and Simulation of Automotive Functionality
Using AUTOSAR.. 523

Marco Di Natale

Chapter 22 Modelica as a Platform for Real-Time Simulation...........................549

John J. Batteh, Michael M. Tiller, and Dietmar Winkler

Chapter 23 Real-Time Simulation of Physical Systems Using Simscape™ 581

Steve Miller and Jeff Wendlandt

Chapter 24 Systematic Derivation of Hybrid System Models for
Hydraulic Systems..599

Jeremy Hodgson, Rick Hyde, and Sanjiv Sharma

Index...623

Preface

In real-time simulation and optimization, events occur on a natural time scale. Such real-time behavior has become important, if not essential, for a broad spectrum of applications that include embedded system design, control of dynamic processes, online system adaptation, streaming data acquisition, data assimilation for decision making, and operator training.

Over the past decades, commercially available computing power has been on a steady pace to becoming increasingly powerful, affordable, and available. This, in turn, has contributed to the emergence of highly sophisticated simulation software applications that enable high-fidelity real-time simulation of dynamic systems. In such systems, the correctness of behavior depends not only on the logical results of the computations or mathematical algorithms but also on the natural time scale at which these results are produced. As a consequence, it is critically important to preserve the characteristics of system dynamics in a temporal sense.

The advances in computing power have given rise to the emergence of an overwhelming offering of applications where real-time simulation is involved, ranging from simply providing value in some form to having become an absolute necessity. One such example is the national power grid, which is currently in a transformational phase to adapt and support increasingly distributed energy generation in order to facilitate our society's desire to rely on renewable energy sources for a sustainable future. The envisioned corresponding "smart grid" cannot be realized without essential modifications to the United States grid infrastructure, and to this end, a virtual environment for testing the potential modifications is simply indispensible.*

As another example, in addition to working toward a sustainable future, virtual reality is recognized by the National Academy of Engineering† as one of the grand challenges of engineering for the twenty-first century. For successful virtual reality systems, it is imperative that humans perceive a simulator as if it were part of the natural world. To fully simulate such interaction between computers and humans across the various interface modalities, it is especially critical that sensations of sound, touch, and motion are produced in real time.

The broad area of cyber-physical systems‡ exemplifies another application domain of real-time simulation. As these systems represent computer-synthesized environments that integrate computation and physical processes, simulation environments must operate in real time if they are to encapsulate and capture cyber and physical characteristics of the systems in one unified framework.

Advanced design methods are rapidly becoming the key enabler of highly competitive industries, and real-time simulation performs a key role here as well. For example, in the aerospace industry, computational modeling and simulation help

---

* http://www.eweek.com/c/a/IT-Infrastructure/Smart-Grids-Growing-to-a-96-Billion-Market-by-2015-172497/
† http://www.engineeringchallenges.org/
‡ http://www.nitrd.gov/Pcast/reports/PCAST-NIT-FINAL.pdf

identify critical items and assist in feasibility studies and design space exploration.* The value and benefit of such simulations rely to a large extent on the quality of the mathematical models and how well they capture system behavior in the physical world. This necessitates real-time simulation of sophisticated models that represent systems ranging from, for example, electromechanical components to entire subsystem with complex embedded computation.

All these applications are testimony to the importance and timeliness of a collection of chapters dedicated to real-time simulation that comprises prominent and insightful achievements in the field. This book intends to facilitate, and compiles highly acclaimed real-time simulation material from a variety of different domains such as pure simulation algorithms, distributed systems, and real-time systems. The collection of contributions is intended to serve as the basis of a comprehensive and cohesive approach for real-time simulation. The synergy between the different application domains in concert with the foundational theories, methodologies, and technologies then sets the stage for enabling the next-generation real-time simulations. Each chapter contributes supreme results that have helped firmly establish their respective real-time simulation technologies, applications, and tools and that continue to remove barriers and expand the bounds of real-time simulation. The contributing authors excel in their expertise, and we are as much delighted as we are honored by their participation in the effort that lead to this book. We sincerely hope that the readers will find the indulgence of intellectual achievement as enjoyable and stimulating as we do.

In closing, we would like to express our genuine thankfulness and appreciation for all the time and effort that each of the authors has invested. Our truly pleasant collaboration has certainly helped make the completion of this project as easy as possible. In addition to the authors who contributed but also reviewed other chapters, we are gratefully indebted to external reviewers who took on the essential task of providing constructive feedback and helped realize the exceptional quality that we set out to achieve. Of course, none of this would have been possible without the continuous and wonderful support of the team at Taylor & Francis, especially our publisher, Nora Konopka, and the staff involved in the verification and production process. Many thanks to each of you!

MATLAB® is a registered trademark of The MathWorks, Inc. For product information, please contact:
The MathWorks, Inc.
3 Apple Hill Drive
Natick, MA 01760-2098 USA
Tel: 508 647 7000
Fax: 508-647-7001
E-mail: info@mathworks.com
Web: www.mathworks.com

---

* ten Dam, A. A., and J. Kos. 1999. "Real-time simulation of impact for the aerospace industry." Report no. NLR-TP-99289, National Aerospace Laboratory NLR, Amsterdam, the Netherlands.

Editors

Katalin Popovici received her engineer degree in computer science from the University of Oradea, Romania, in 2004 and her PhD in micro- and nanoelectronics from Grenoble Institute of Technology, France, in 2008. Between 2005 and 2008, she was a member of the SHAPES (Scalable Software Hardware Computing Architecture Platform for Embedded Systems) European research project, where she worked on hardware–software codesign. Currently, she is a senior software engineer at MathWorks in Natick, Massachusetts, where she works on partitioning and mapping capabilities from Simulink® models to embedded and real-time systems, with focus on code generation for multicore and heterogeneous architectures.

Dr. Popovici's research interests include system-level modeling and design of multiprocessor system-on-chip, programming models, and code generation for embedded applications. She often serves on international technical program committees and gives lectures on hardware–software codesign.

Pieter J. Mosterman is a senior research scientist at MathWorks in Natick, Massachusetts, where he works on computational modeling, simulation, and code generation technologies. He also holds an adjunct professor position in the School of Computer Science at McGill University. Prior to this, he was a research associate at the German Aerospace Center (DLR) in Oberpfaffenhofen. He received his PhD in electrical and computer engineering from Vanderbilt University in Nashville, Tennessee, and his MSc in electrical engineering from the University of Twente, the Netherlands. His primary research interests include computer-automated multiparadigm modeling (CAMPaM) with principal applications in design automation, training systems, and fault detection, isolation, and reconfiguration.

Dr. Mosterman designed the electronics laboratory simulator that was nominated for The Computerworld Smithsonian Award by Microsoft Corporation in 1994. In 2003, he was awarded the IMechE Donald Julius Groen Prize for his paper on the hybrid bond graph modeling and simulation environment HyBrSim. In 2009, he received the Distinguished Service Award of The Society for Modeling and Simulation International (SCS) for his services as the editor-in-chief of *SIMULATION: Transactions of SCS*. Dr. Mosterman was a guest editor for special issues on CAMPaM of *SIMULATION, IEEE Transactions on Control Systems Technology*, and *ACM Transactions on Modeling and Computer Simulation*. He has chaired over 30 scientific events, has served on over 80 international program committees, has published over a 100 peer-reviewed papers, and is an inventor with over 30 awarded patents.

Contributors

Ehsan Azarnasab
Department of Electrical and Computer
  Engineering
University of Utah
Salt Lake City, Utah

John J. Batteh
Modelon, Inc.
Ann Arbor, Michigan

Jean Bélanger
Opal-RT Technologies
Montréal, Québec, Canada

Giovanni Beltrame
École Polytechnique de Montréal
Montréal, Québec, Canada

David C. Bergman
University of Illinois at Urbana-Champaign
Champaign, Illinois

Kenneth Butts
Powertrain Control Department
Toyota Motor Engineering and
  Manufacturing North America
Ann Arbor, Michigan

Christos G. Cassandras
Division of Systems Engineering
Center for Information and Systems
  Engineering
Boston University
Brookline, Massachusetts

Jonathan Caux
Laboratoire d'Informatique,
  de Modélisation et d'Optimisation
  des Systèmes
Université Blaise Pascal, Clermont
  Université
Clermont-Ferrand, France

and

Unité Mixte de Recherche 6158,
  LIMOS (Laboratoire d'Informatique,
  de Modélisation et d'Optimisation
  des Systèmes)
Centre National de la Recherche
  Scientifique
Aubière, France

and

Integrative BioComputing
Rennes, France

Roy Crosbie
Department of Electrical and Computer
  Engineering
California State University
Chico, California

Patricia Derler
Department of Electrical Engineering
  and Computer Sciences
University of California
Berkeley, California

Marco Di Natale
TECIP Center
Scuola S. Anna
Pisa, Italy

Rhianon Dolletski
Institute for Simulation and Training
University of Central Florida
Orlando, Florida

Christian Dufour
Opal-RT Technologies
Montréal, Québec, Canada

James Edmondson
Vanderbilt University
Nashville, Tennessee

Emilia Farcas
University of California
San Diego, California

Luca Fossati
European Space Agency's ESTEC
Noordwijk, the Netherlands

Cüneyt Güzeliş
Department of Electronics and
    Communication Engineering
Faculty of Engineering and Computer
    Science
Izmir University of Economics
Izmir, Turkey

David Hill
Modélisation et d'Optimisation des
    Systèmes
Université Blaise Pascal, Clermont
    Université
Clermont-Ferrand, France

and

Unité Mixte de Recherche 6158,
    LIMOS
Centre National de la Recherche
    Scientifique
Aubière, France

Jeffrey Hillyard
TESIS DYNAware GmbH
München, Germany

Peter Hintenaus
C. Doppler Laboratory Embedded
    Software Systems
University of Salzburg
and
Chrona.com
Salzburg, Austria

Jeremy Hodgson
MBDA Ltd.
Hertfordshire, United Kingdom

Robert Höpler
Systems and Optimization
München, Germany

Xiaolin Hu
Department of Computer Science
Georgia State University
Atlanta, Georgia

Rick Hyde
The MathWorks, Inc.
Cambridge, United Kingdom

Yalçın İşler
Department of Biomedical Engineering
Faculty of Engineering and
    Architecture
Izmir Katip Çelebi University
İzmir, Turkey

Ingolf H. Krüger
University of California
San Diego, California

Hugh H. T. Liu
University of Toronto Institute for
    Aerospace Studies
Toronto, Ontario, Canada

Jan Madsen
Department of Informatics and
    Mathematical Modeling
Technical University of Denmark
Copenhagen, Denmark

Massimiliano Menarini
University of California
San Diego, California

Steve Miller
The MathWorks, Inc.
Ismaning, Germany

Mohammad Moallemi
Department of Systems and Computer
    Engineering
School of Computer Science
Carleton University
Ottawa, Ontario, Canada

Andreas Naderlinger
C. Doppler Laboratory Embedded
    Software Systems
University of Salzburg
Salzburg, Austria

Denise Nicholson
Institute for Simulation and
    Training
University of Central Florida
Orlando, Florida

David M. Nicol
University of Illinois at
    Urbana-Champaign
Champaign, Illinois

Gabriela Nicolescu
École Polytechnique de Montréal
Montréal, Québec, Canada

Christos G. Panayiotou
KIOS Research Center for Intelligent
    Systems and Networks
Department of Electrical and Computer
    Engineering
University of Cyprus
Nicosia, Cyprus

Wolfgang Pree
C. Doppler Laboratory Embedded
    Software Systems
University of Salzburg
and
Chrona.com
Salzburg, Austria

Gerhard Rath
Department Product Engineering,
University of Leoben
Leoben, Austria

Stefan Resmerita
C. Doppler Laboratory Embedded
    Software Systems
University of Salzburg
and
Chrona.com
Salzburg, Austria

Hesham Saadawi
School of Computer Science
Carleton University
Ottawa, Ontario, Canada

Savaş Şahin
Department of Electrical and
    Electronics Engineering
Faculty of Engineering and
    Architecture
Izmir Katip Çelebi University
İzmir, Turkey

Johannes Scharpf
Lehrstuhl für
    Verbrennungskraftmaschinen
Technische Universität München
München, Germany

Sae Schatz
MESH Solutions, LLC
Orlando, Florida

Douglas C. Schmidt
Vanderbilt University
Nashville, Tennessee

Sanjiv Sharma
Systems Department
Airbus Operations Ltd.
Filton, United Kingdom

Josef Templ
C. Doppler Laboratory Embedded
    Software Systems
University of Salzburg
Salzburg, Austria

Michael M. Tiller
Model-Based Systems Engineering
Dassault Systemes
Vélizy-Villacoublay, France

Luc Touraille
Laboratoire d'Informatique,
    de Modélisation et d'Optimisation
    des Systèmes
Université Blaise Pascal, Clermont
    Université
Clermont-Ferrand, France

and

Unité Mixte de Recherche 6158,
    LIMOS
Centre National de la Recherche
    Scientifique
Aubière, France

Anders Sejer Tranberg-Hansen
Department of Informatics and
    Mathematical Modeling
Technical University of Denmark
Copenhagen, Denmark

Gabriel A. Wainer
Department of Systems and Computer
    Engineering
School of Computer Science
Carleton University
Ottawa, Ontario, Canada

Jeff Wendlandt
The MathWorks, Inc.
Natick, Massachusetts

Dietmar Winkler
Telemark University College
Porsgrunn, Norway

Bernard Zeigler
Arizona Center for Integrative
    Modeling and Simulation
and
Department of Electrical and Computer
    Engineering
University of Arizona
Tucson, Arizona

Ming Zhang
Computer Science Department
Georgia State University
Atlanta, Georgia

Introduction

Desktop personal computer–based simulation has become the primary method for analyzing and studying the behavior of dynamic systems. Such simulation includes executing on a digital computer mathematical models for engineered systems that comprise physics and interact with humans. To address the temporal aspect of the system dynamics, events are to be simulated in a time frame in which they would naturally occur, which is known as real-time simulation. Real-time simulation thus captures how the behavior of a system unfolds as time passes, all with high fidelity in correspondence to natural time scales.

CONTENT

The utility of real-time simulation is evidenced by its popularity in many engineering domains, for example, in the design of complex systems such as aircraft and automobiles, flight simulators to train pilots, fluid dynamics, weather forecasting, and the design of robots, including their control algorithms. The importance of timing accuracy of these simulations strongly depends on the application but in general, a certain degree of timing accuracy is crucial in ensuring intended system performance. For example, engineered systems are increasingly designed in a virtual representation, which necessitates incrementally closing the gap between the virtual domain and the physical domain, so as to gradually reveal the detailed actual effects of design decisions. Real-time simulation then faithfully produces the temporal characteristics of the dynamic behavior of a system. While real-time simulation must account for the pertinent time constants in the system that is simulated, there is substantial variation of these between application domains. Moreover, depending on the application domain, some variation on response times can be accepted, whereas in other domains (e.g., feedback control), tighter time bounds are strictly necessary.

In addition, not only must the algorithm for simulation support real time, an implementation can only be achieved by merit of real-time capable platforms and protocols. These protocols are particularly challenging to develop when the simulation is executing on a distributed platform. Whereas aspects such as a distributed nature, the modalities of interface technologies, and quality of service characteristics such as latency are domain specific, the core real-time simulation theory and methodologies transcend these domains to form the foundation for a technology that supports a terrific spectrum of applications ranging from embedded systems to training systems.

This book contains a compilation of work from internationally renowned authors on fundamentals and basic techniques of real-time simulations for complex and diverse systems. It elaborates on practices of real-time simulation and addresses the main facets in the context of large application domains, including the current state of the art, important challenges, and successful trends.

As such, this book provides a basis for scholars who wish to study real-time simulation as a fundamental and foundational technology and to develop principles that are applicable across a broad variety of application domains while further developing the domain-specific refinement and idiosyncrasies. The collected material enables different levels of insight and understanding, ranging from attaining a cursory familiarization to developing a high level of expertise in a given domain. This is made possible thanks to a comprehensive introduction to concepts related to modeling and simulation of systems followed by detailed chapters on real-time simulation for system design, parallel and distributed simulations, industrial tools, and a large set of applications.

ORGANIZATION

This book is divided into four parts: Section I: Basic Simulation Technologies and Fundamentals, Section II: Real-Time Simulation for System Design, Section III: Parallel and Distributed Real-Time Simulation, and Section IV: Tools and Applications. The following presents an overview of each of the parts along with a brief introduction to the contents of each of the chapters.

SECTION I: BASIC SIMULATION TECHNOLOGIES AND FUNDAMENTALS

Section I establishes the fundamentals of real-time simulation and sets the stage for further exploring the opportunities that real-time simulation provides as discussed in the sections that follow.

Chapter 1 introduces real-time simulation of discrete-event behavior and of continuous-time behavior possibly interspersed with discontinuities. The chapter defines the discrete and continuous simulations as two different approaches to the modeling of dynamic systems with a hybrid model being one that includes elements of both discrete and continuous models. The relevant features of discrete and continuous models are addressed first before delving into the intricacies of real-time simulation in case of hybrid models.

Chapter 2 focuses on exploiting the Discrete Event System Specification (DEVS) as a formal approach for the design of a real-time distributed computer systems. The chapter demonstrates (1) how DEVS can serve as a fundamental system design tool for distributed real-time computer systems; (2) how DEVS is employed to validate designs of complex real-time computer systems, such as distributed virtual environments, distributed real-time peer-to-peer systems, and distributed simulation systems; and (3) how DEVS can be utilized as a fundamental technology for building a more advanced framework to support design of distributed real-time systems, such as cooperative robotic systems.

Chapter 3 proposes a new methodology to verify simulation models of real-time applications based on the DEVS formalism. This methodology defines a new class of Rational Time Advance DEVS (RTA-DEVS) and a transformation to obtain a timed automata (TA) that is behaviorally equivalent to RTA-DEVS. The resulting TA models are a subset of deterministic safety automata and can be used in the UPPAAL (or other similar) model checkers.

Chapter 4 presents a set of software engineering techniques for improving performance of real-time simulations through code generation and metaprogramming. It applies C++ template metaprogramming to DEVS in the simulation domain. The "metasimulator" employs specific static pieces of information of the models such as component names to evaluate certain parts of the simulation during template instantiation and generates a residual simulator specialized for a given model, where only the dynamic operations remain. Thanks to this, several improvements are obtained over more classical approaches that do not use metaprogramming.

Chapter 5 introduces basic modeling capabilities of the Unified Modeling Language (UML). UML provides a built-in extension mechanism through profiles, which allows tailoring the UML to a particular domain or target platform. One such extension, the profile for Modeling and Analysis of Real-Time and Embedded Systems (MARTE) is specifically directed toward real-time simulation. The MARTE profile is described followed by a discussion on remaining challenges on the road toward a systematic and effective model-driven engineering approach utilizing the UML.

Chapter 6 addresses the modeling and simulation of safety-critical embedded applications where timing requirements are specified in the Timing Definition Language (TDL), which supports the Logical Execution Time abstraction. It further describes the TDL constructs and sketches the seamless integration of TDL with two simulation environments: the MATLAB® and Simulink® products from MathWorks and the open-source Ptolemy framework from the University of California, Berkeley.

SECTION II: REAL-TIME SIMULATION FOR SYSTEM DESIGN

Section II focuses on the importance of real-time simulation in system design and how computational modeling and simulation help identify critical issues in the design while assisting in feasibility studies and design space exploration.

Chapter 7 presents a progressive simulation-based design methodology that uses fast and real-time simulations at different stages of the design process for networked real-time embedded systems. The methodology supports systematic transitions from simulation models to system realization. The methodology is applied to the development of a cognitive radio network that exploits the functionality of software-defined radio and includes experimental results.

Chapter 8 describes a new simulation environment called *Validator* for verification and validation of the real-time behavior of embedded systems. The corresponding tool suite offers advanced features for debugging the timing behavior of embedded systems and as such fills the timing gap between conventional hardware-in-the-loop and software-in-the-loop simulation environments. The simulation tool is used in the validation of an engine controller system.

Chapter 9 provides an introduction to and overview of real-time simulators after it first defines real-time simulation. Particular application focus is on electromagnetic transients, power systems modeling and simulation, and control prototyping techniques.

Chapter 10 presents a transaction-level technique for the modeling, simulation, and analysis of real-time applications that execute on a multiprocessor systems-on-chip architecture. The novelty of the technique is that it is based on an application-transparent emulation of the operating system primitives, including support for real-time operating system elements. The proposed methodology enables a quick evaluation of the real-time performance of an application with different design choices, including the study of system behavior as task deadlines become stricter or looser. The approach has been verified on a large set of multithreaded, mixed-workload (real-time and non-real-time) applications and benchmarks.

Chapter 11 introduces a compositional framework for system-level performance estimation of heterogeneous embedded systems. The framework is simulation based and allows performance estimation to be carried out throughout all design phases ranging from early functional to cycle accurate and bit true descriptions of the system implementation. The key strengths of the framework are the flexibility and refinement opportunities as well as the possibility of having components described at different levels of abstraction coexist and communicate within the same model instance.

Chapter 12 discusses various approaches to address the consistency checking problem in UML models used for real-time modeling. The chapter first identifies a set of core requirements to address the consistency problem of the UML and then analyzes the existing approaches in the literature and evaluates how they fulfill these requirements. The approach is to define an explicit ontology that captures the modeled target domain and to define a simple execution framework based on this target ontology.

SECTION III: PARALLEL AND DISTRIBUTED REAL-TIME SIMULATION

Section III focuses on how to apply parallel and distributed simulation techniques to enable and reduce simulation time of large-scale applications in an effort to harness the computing resources of parallel computers.

Chapter 13 describes the process of flight control system (FCS) development where simulation plays a critical role in interactive design, validation, and verification. After a number of simulation-based platforms are introduced, one such interactive FCS development and real-time simulation test bed is presented as a case study. The design case study uses a business jet aircraft model to illustrate the interactive FCS developments and is followed by a summary of challenges and lessons learned.

Chapter 14 documents an approach to developing virtual supervisory control and data acquisition (SCADA) models and validating the models using a local test bed. Moreover, it shows how other researchers can use the local test bed to validate their own technologies. First, the chapter gives a background on the network communications simulator Real-Time Immersive Network Simulation Environment, which provides the basis for the SCADA models. Then, it details the unique characteristics of the power grid, the primary protocols used in SCADA, and the main steps taken to develop the virtual test bed.

Chapter 15 provides a basic overview of distributed real-time Simulation-Based Training (SBT), which is a specific application area of parallel and distributed simulations. The chapter opens with a brief description of distributed real-time SBT and its development, after which it lists the recurrent challenges faced by such systems with respect to instructional best practices, technology, and use.

Chapter 16 presents a real-time simulation approach for the online optimization of Discrete Event Systems (DES). Key to the approach is an observer that collects information (e.g., event lifetimes) from the sample path of the DES. This information is then concurrently processed by real-time simulation modules that construct the sample paths of the system under different parameter or policy settings. From the constructed sample-path estimate, various performance metrics can be estimated and used to select, in real time, the best possible such setting.

SECTION IV: TOOLS AND APPLICATIONS

Section IV focuses on industrial real-time simulation tools and applications in many engineering domains, ranging from design of large-scale and complex systems such as automobiles, including their control algorithms, to operator training.

Chapter 17 discusses time dilation as a new technology for accurate simulation of large-scale systems. The time dilation mechanism in the simulation context attempts to correct clock drifts in a distributed system allowing for closer approximation of target behavior by simulated tests. A prototype of this time dilation technology called DieCast has been implemented by researchers at the University of California at San Diego. This chapter explores the benefits of this technology to date, summarizes what testers must consider when using DieCast, and describes future work necessary to mature time dilation techniques and tools for simulation of large-scale distributed systems.

Chapter 18 describes the development of and experience with a simulator to train replacing the rolls of a steel rolling mill. Because production is fully automated and operating continuously, the actual machine is not available for training. Instead, an operator-training simulator replaces the functionality of the actual machine by its simulation. To attain a high level of fidelity, the machine simulation executes with a replica of the actual electronic control system hardware in the loop.

Chapter 19 describes a real-time platform for controller design, test, and redesign. Emphasis is on recreating actual operating conditions for examining and analyzing the real-time performance of designed controllers while still having the opportunity to modify and tune the controller candidates based on their performance. The controller design, test, and redesign platform was developed to provide a high level of flexibility in choosing from various controller and plant models in the early stages of the controller design while creating conditions as close as possible to the physical world in the final design stage.

Chapter 20 describes important aspects of how automotive development benefits from real-time techniques for modeling and simulations and hardware-in-the-loop methods. Real-time simulation is discussed as a key enabler of earlier and more informed design tradeoffs and decisions. Prerequisites for and results of successful detailed real-time engine simulation are included.

Chapter 21 continues the specific automotive perspective by presenting specification and simulation of applications using the Automotive Open System Architecture (AUTOSAR). AUTOSAR was created to develop an open industry standard with a common software infrastructure based on standardized interfaces for software component specification and integration at different layers. The chapter provides an introduction to AUTOSAR as a language for system-level modeling and software component modeling.

Chapter 22 introduces the modeling language Modelica as a modeling platform to support real-time simulations of physical systems. A brief introduction to the Modelica language and its fundamental language features are given while highlighting its applicability to a wide range of engineering domains and key features that make it a convenient modeling platform to support real-time simulation. Selected simulations show the computational impact of symbolic model formulation and manipulation.

Chapter 23 presents real-time simulations of multidomain physical systems using Simscape™ of MathWorks to support model-based design. This chapter outlines the steps in moving from desktop to real-time simulation by illustrating how to find a combination of model complexity, solver choice, solver settings, and real-time target that permits execution in real time. These steps are formulated such that they apply to real-time simulation irrespective of which real-time hardware is used.

Chapter 24 develops a systematic approach to designing models of plant or actuator dynamics that are used to validate the performance of a controller. The approach maintains the inherent nonlinear behavior of a hydraulic circuit while being based on a simplified representation of fluid dynamics. The approach is demonstrated on a circuit consisting of a servo valve controlling the fluid pressure in a closed pipe.

Section I

Basic Simulation Technologies and Fundamentals

1 Real-Time Simulation Using Hybrid Models

Roy Crosbie

CONTENTS

1.1 Introduction: Discrete and Continuous Models 4
    1.1.1 Discrete Models 4
    1.1.2 Continuous Models 5
1.2 Discrete Modeling 5
    1.2.1 Queuing Models 5
    1.2.2 Digital System Models 6
    1.2.3 DEVS Formalism 6
    1.2.4 Time Management for Discrete Simulation 7
    1.2.5 Software for Discrete Simulation 8
    1.2.6 Example of a Discrete-Event Simulation 8
1.3 Continuous Modeling 8
    1.3.1 Nature of Continuous Models 8
    1.3.2 Time Management for Continuous Simulations 9
        1.3.2.1 Types of Numerical Integration Algorithm 9
        1.3.2.2 Fixed versus Variable Step 11
        1.3.2.3 Explicit versus Implicit 12
        1.3.2.4 Single-Step versus Multistep 12
        1.3.2.5 Variable-Order Algorithms and Stiff Systems 13
        1.3.2.6 Difference Equations 13
    1.3.3 Software for Continuous Simulation 13
    1.3.4 Example of a Continuous Simulation 15
1.4 Hybrid Models 15
    1.4.1 Time Management for Hybrid Simulations 16
    1.4.2 Software for Hybrid Simulations 16
    1.4.3 Examples of Hybrid Simulations 19
        1.4.3.1 Continuous Model with a Discrete Element 19
        1.4.3.2 Example of a Discrete-Based Hybrid Simulation 21
        1.4.3.3 Balanced Hybrid Simulation 22
1.5 Real-Time Hybrid Simulation 24
    1.5.1 Timing Issues in Real-Time Hybrid Simulation 24
    1.5.2 High-Speed Real-Time Hybrid Simulation 25

        1.5.3   Numerical Integration for High-Speed Real-Time Simulation 26
        1.5.4   HSRT Multirate Simulation 28
1.6   Concluding Remarks 30
References 31

1.1   INTRODUCTION: DISCRETE AND CONTINUOUS MODELS

The use of computers to simulate the behavior of physical or natural systems is as old as the history of computers itself. Adaptation of physical simulators, such as the original mechanical pilot training simulators, to *real-time* computer-based simulators was another early development in computer applications. Although the main focus here is on real-time simulation with hybrid models (explained below), the topic of simulation using hybrid models in general will be addressed first, with a more specific treatment of real-time applications introduced later.

The development of a simulation of an actual system involves the creation of a conceptual model of the *simuland* (the actual system to be simulated), which may be based on a set of rules, or a set of mathematical equations, or some other method of defining the state of the simuland and the way in which it changes with time. We use the term *actual system* rather than *real system* because sometimes the simuland does not exist and the use of the term *real* can be confusing. In the early days of simulation, a distinction quickly arose between two types of model for both real-time and non-real-time simulations. These were referred to as *discrete models* and *continuous models* and the associated processes as *discrete simulation* and *continuous simulation*. This led to the use of the terms *discrete system* and *continuous system* to characterize the actual systems that were being simulated even though it is often not the system that is discrete or continuous but the model that is used to represent it. The practice of referring to the systems themselves as discrete or continuous persists even though many actual systems can be represented by either discrete or continuous models. Traffic flow provides an example. Discrete models are often used in traffic studies in which each vehicle is represented and events such as arrival at an intersection or at a line of stationary traffic or a change of a stop light cause the system state to change. Study of traffic flow on a freeway, where the volume may be much greater is, however, often represented by a continuous model that treats the traffic as if it were a fluid flowing along the freeway using traffic flow rates as model variables.

Discrete and continuous simulations represent two different approaches to the modeling of dynamic systems. Both types of model require the system to be characterized by a system state that changes as time advances. It is the nature of this change that distinguishes the two approaches. A hybrid model is one that includes elements of both discrete and continuous models. Before dealing with hybrid models, we address the relevant features of discrete and continuous models.

1.1.1   Discrete Models

In a discrete model, it is assumed that the system state changes at certain discrete instants of time and remains unchanged between these times. These changes in the state of the system occur instantaneously as the result of an event that triggers them.

In a discrete model, time advances from event to event with appropriate updating of the system state at each event time. A prioritized future event queue is usually maintained as a central data structure into which new future events are entered as they are identified, and from which the next event is taken as the simulation clock advances to the next event time.

1.1.2 CONTINUOUS MODELS

In a continuous model, the state of the system is assumed to change in a continuous fashion as defined by the model differential equations, which relate the instantaneous rates of change of system variables to the current state of the system. In practice, the advance of time in the simulation is quasi-continuous because of the discrete time steps adopted by the numerical integration algorithms that are used to generate approximate solutions to the differential equations. In effect, even continuous simulations execute discretely. Switching phenomena are accommodated by allowing local discontinuities to occur in the state variables and their derivatives, in which case the model can be described as piecewise continuous and belongs to a class of hybrid systems discussed below. The details of time management in continuous simulations are also discussed in more detail below.

1.2 DISCRETE MODELING

As previously stated, a discrete model is one in which the state of the simuland is assumed to change only at specific instances in time. There are two major types of system that can be simulated using discrete models.

1.2.1 QUEUING MODELS

The first and probably most familiar type of discrete model is based on queuing models. In a typical queuing model, entities such as customers or parts arrive at service points representing operators or service units, which process them in turn. The waiting entities form queues, and both the arrival times of new entities and the service times of the servers are often generated from statistical distributions using random number generators. Changes in the state of the system caused by queue arrivals or departures, or a completion of service, are referred to as *events* and the time at which the event occurs is the *event time*.

A simulation based on a discrete model establishes an initial state of the system and a future event queue with event timings. The simulation then advances to the first of these event times, and the appropriate changes are made to the system state. These changes may generate changes in the entries in the event queue, including the identification of additional events, and the event queue is modified accordingly. Once the current system state is fully established, the simulation moves on to the next event and the process is repeated. This repeated sequence continues until the simulation satisfies some terminating condition. Models of this kind are used extensively to represent manufacturing plants, distribution networks, service facilities and systems, business operations, and many other applications.

Over time, three main approaches have been developed for simulations that use these models, known as the activity-based, event-based, and process-based methods. Historically, the term *activity-based simulation* was used to describe simulations in which time advances in small steps with checks for changes at each step. More recently, the term has also been used for Discrete Event System Specification (DEVS) simulations, where only components that are potentially active are evaluated (e.g., in fire spread applications, where only the cells modeling the fire front are evaluated) [1,2]. The original kind of activity-based simulation is inefficient and suitable only for simple applications. Event-based simulation in which time advances from event to event in a single software thread has been the basis of many popular discrete simulation languages, but, as parallel computing options increase, process-based simulation using parallel processors and multiple software threads has become the more popular approach. In this approach, the simulation is divided into processes that can be run in parallel [3]. Each process schedules its events in the correct order, but out-of-order execution can occur in which a process receives an event from another process with an event time that is earlier than its current time. Two strategies are used for dealing with this problem known as *conservative* and *optimistic synchronization*. Conservative synchronization requires a process to block its next impending event until it is certain that it is safe. Optimistic synchronization uses a roll-back approach when an out-of-order event occurs. This approach is discussed further in Section 1.2.4.

1.2.2 DIGITAL SYSTEM MODELS

As noted above, there are two distinct types of discrete model, the queuing models described in Section 1.2.1 and models that are used to represent systems that can be described by means of difference equations or z-transforms. These digital system models include digital electronic circuits and systems, and sampled-data and digital control systems in which the discrete subsystem is updated at typically equally spaced time intervals [4].

This type of discrete model consists of a set of difference equations that define the next state of the system in terms of its current and past states and its external inputs. If the model is defined, in its entirety or in part, by means of z-transforms, it is usually a straightforward task to convert the z-transform model into difference equation form.

Given a difference equation model, and once the initial state of the system is established, the simulation proceeds by advancing time incrementally by the specified time step and calculating the state of the system at each step. This method is widely used in the simulation of all kinds of digital electronic circuits, computer systems, and digital controllers. As mentioned earlier, a continuous simulation is itself a process of this kind.

1.2.3 DEVS FORMALISM

In 1976, Zeigler [1] introduced DEVS, a formalism based on General System Theory that provides a means of specifying a mathematical object called a system. The

DEVS conceptual framework consists of three basic objects: the *model*, the *simulator*, and the *experimental frame*. The model is a set of instructions that are intended to replicate system behavior, the simulator exercises these instructions to generate the behavior, and the experimental frame describes the way in which the model will be exercised by the simulator. The following description, taken from the introduction to a DEVS tutorial by Zeigler and Sargoujhian [5], draws an interesting comparison between discrete and continuous models:

> The Discrete Event System Specification (DEVS) formalism provides a means of specifying a mathematical object called a system. Basically, a system has a time base, inputs, states, and outputs, and functions for determining next states and outputs given current states and inputs. Discrete event systems represent certain constellations of such parameters just as continuous systems do. For example, the inputs in discrete event systems occur at arbitrarily spaced moments, while those in continuous systems are piecewise continuous functions of time.

Although its name, and the above description, suggests that DEVS is restricted to discrete-event models, it has also been extended to continuous models as described by Cellier and Kofman [6]. It is therefore capable of representing hybrid models. Several software products have been developed that are based on DEVS including simulation with continuous and real-time DEVS models [2].

1.2.4 TIME MANAGEMENT FOR DISCRETE SIMULATION

The management of time in a discrete simulation is based on knowledge of the times of future events, which are either known *a priori* or are maintained in a prioritized event queue with priorities that are determined by the time ordering of the events in the queue. It is not necessary to know the time of all events at the start of a simulation but in many cases, as long as the time of the next event is known, time management becomes fairly straightforward. The simulation will begin with a specification of the initial state of the system (length of various queues, status of the queue servers, etc.), and some future events can be determined using the random number generators to generate event times (arrival times, service times, etc.) for the various elements in the system. Time is then advanced to the next event time and appropriate changes made to the state of the system. A queue length may be increased or decreased by an arrival in the queue or by a completion of service; a unit may be seized or released by the simulation and so on. The simulation proceeds alternating between updating the system state as the result of an event and advancing time to the next event. The terminal event time is determined by the termination condition set by the programmer.

As mentioned above, it is becoming common to use a parallel-processing approach to accelerate execution times particularly for process-based discrete simulation software. Using this approach, time may advance in an unsynchronized way on different processors or when using different software threads. This can cause one part of the simulation to advance beyond the generation in another processor or thread of an event that affects it. Conservative scheduling can be used to delay execution of a process until it is safe to do so. Alternatively, optimistic scheduling includes techniques for time to be rolled back so that the effect of an event that has advanced too far in time on the state of the system can be determined. An example of this technique, developed by David Jefferson, is known as *Time Warp* [7]. Because of the need to facilitate roll back of time, these methods are probably not viable for most real-time applications.

1.2.5 SOFTWARE FOR DISCRETE SIMULATION

Many software products have been developed over the years aimed at supporting discrete-event simulation. Well-known examples include General Purpose Simulation Systems, Simscript, SIMAN/Arena, Simula, and Dymola/Modelica. The Simula language [8] is notable in that it was the first object-oriented, process-based, discrete simulation language having been first released in the 1960s. As stated above, there are also several software products based on the DEVS formalism.

Simulation of digital electronics, digital control, and sampled-data systems can be accomplished using digital hardware design software such as VHDL or Verilog, which combine a specification of the system with a simulator to test the design.

1.2.6 EXAMPLE OF A DISCRETE-EVENT SIMULATION

A simple example of a discrete-event simulation is described by Matloff and Davis [9]. It involves two machines that break down from time to time and must be repaired. In one variation, a single repair technician is available. Machine uptime and repair time are exponentially distributed with mean values set by the programmer. Events occur when a machine breaks down and when a repair is completed and a machine returns to service. If a second machine breaks down while the first is being repaired, it goes into a waiting queue until the repair technician is available. Clearly, this problem could be extended to include, for example, more machines, possibly with different uptimes and repair times, and different numbers and types of technicians.

1.3 CONTINUOUS MODELING

A continuous model is one in which the system state is assumed to vary in a continuous manner with no instantaneous changes in the values of system states or their derivatives.

1.3.1 NATURE OF CONTINUOUS MODELS

Continuous models normally consist of a mathematical description of the actual system by means of a combination of differential and algebraic equations. This type of mathematical model can be expressed in the form of first-order differential equations, each defining a system state variable, and additional equations that define auxiliary variables. The independent variable is time, producing a mathematical model for a continuously varying dynamic system. Some models may have additional independent variables (usually spatial dimensions), in which case the mathematical model is formulated as a set of partial differential equations. A simple example is the simulation of temperature changes along a thin metal rod as it is heated. The state variable is temperature, and it varies with both time and displacement along the rod.

The mathematical model is a partial differential equation with two independent variables, time and displacement.

Simulation using a continuous model involves the initialization of the values of system states followed by calculation of the initial values of the other algebraic variables. This initialization process establishes the initial values of all system variables and is followed by a repetitive process of advancing time in steps using an algorithm for the numerical solution of differential equations. The time steps may be of constant size (usually the case for real-time simulation) or of variable size. Variable-step routines typically include an estimation of the error generated in the current step and a strategy to change the step size to satisfy a user-supplied error tolerance. Such variable-step routines are not normally suitable for real-time simulation because their computation times vary from step to step.

1.3.2 TIME MANAGEMENT FOR CONTINUOUS SIMULATIONS

Continuous simulations are based on solutions of ordinary or partial differential equations for which time is a continuously changing variable with no instantaneous changes from one value of time to the next, in contrast to the behavior of a discrete simulation in which time jumps from one event time to the next. Computers are programmed to produce approximate solutions of the mathematical model using numerical approximation in which time does advance in small discrete steps. The way in which time advances is specified partly by the programmer and controlled largely by the algorithm chosen by the programmer to solve the equations in the mathematical model. There are many algorithms available for solving differential equations, and the choice of algorithm is one of the key decisions made by the user of a continuous simulation. It will be helpful at this stage to embark on a brief digression about different types and properties of numerical integration algorithms.

1.3.2.1 Types of Numerical Integration Algorithm

Assume a continuous mathematical model of the form

$$y' \equiv dy/dt = f(y, x, u) \tag{1.1a}$$

$$x = g(x, y, y', u) \tag{1.1b}$$

where y is a vector of states, y' is a vector of state derivatives, x is a vector of auxiliary variables (algebraic variables), u is a vector of inputs, and f and g are arbitrary but well-behaved functions.

Numerical integration algorithms normally calculate the next state of a system in terms of the current state and, in some cases, past states. Time is assumed to advance in finite steps leading to equations of the form

$$y(t+h) = \mathrm{fn}\big(y(t), x(t), u(t)\big) \tag{1.2}$$

where t is current time; h is the time increment for the current step; and y, x, and u are the system state, other system variables, and system inputs, respectively. In many cases, the algorithm can be expanded into a finite or infinite series that matches the Taylor series expansion for the first several terms.

The Taylor series can be expressed as

$$y(t+h) = y(t) + hy'(t) + (h^2/2!)y''(t) + (h^3/3!)y'''(t) + \ldots \qquad (1.3)$$

The simple Euler integration method, for example, solves Equations 1.1a and 1.1b by calculating

$$y(t+h) = y(t) + hy'(t) \qquad (1.4)$$

In other words, Euler integration simply extrapolates the states, y, by assuming that the derivatives of y remain constant throughout the step and equal to the value at the start of the step. The process is illustrated in Figure 1.1.

Assume a first-order system,

$$y'(t) = f(y,t) \qquad (1.5)$$

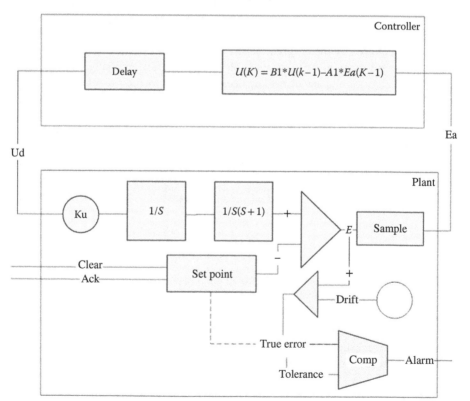

FIGURE 1.1   Euler integration.

Figure 1.1 shows part of the true solution for a state variable, $y(t)$. The initial value of y is y_0 at time $t = t(0)$. The initial value of the derivative, from Equation 1.5, is given by

$$y'(t_0) = f(y_0, t_0) \tag{1.6}$$

Applying Euler integration over a step of length h, we have

$$y(t_0 + h) = y(t_0) + hy'(t_0) \tag{1.7}$$

The derivatives at $(t_0 + h)$ can now be calculated from

$$y'(t_0 + h) = f(y_0 + h, t_0 + h) \tag{1.8}$$

This process can be repeated until the final time is reached.

This algorithm matches the first two terms in the Taylor series (up to the term in the first derivative, y') and is known as a first-order method. More complex methods are available that match the Taylor series to the terms in y'' (second-order methods), y''' (third-order methods), and so on. These Taylor-series-based methods attempt to approximate the solution over a single step using a polynomial approximation, linear for first order (Euler), a quadratic for second-order methods, and so on. An alternative approach, based on exponential approximations, will be introduced in Section 1.5.3. In general, a method of order n will have an approximation error dependent on the neglected terms of the Taylor series, which are the terms in h^{n+1} and above. This error is known as the truncation error. Additional errors can arise from the accumulation of round-off error, particularly when using lower-precision data formats and large numbers of steps. As we shall see, this can be of particular importance in some real-time simulations. There are many different numerical integration algorithms available. Runge–Kutta methods of different orders are widely used. Fourth-order Runge–Kutta methods are often chosen as providing a good trade-off between accuracy and computational complexity [10].

The two major factors that affect the accuracy of a numerical integration are order and step size. Generally speaking, errors decrease with decreasing step size and increasing order. In extreme cases, steps can be so short that round-off error can become an issue, particularly with lower-precision data formats. As step size increases, some algorithms can become unstable, even when the unstable behavior appears at a point at which errors were, up to that point, acceptable. In other words, acceptable stable solutions can make a rapid transition to unacceptable unstable solutions with very small increases in step size.

1.3.2.2 Fixed versus Variable Step

One of the choices faced by users of continuous simulation is whether to use an algorithm that advances time in equal increments (a fixed-step algorithm) or one that varies the size of time increments in order to satisfy user-supplied error tolerances.

Modern variable-step routines often use an approach in which estimated solutions of order n and $n - 1$ are generated at each step, and the error estimate is based on the difference between the two. The justification for this is that this difference approximates the value of the last term in the Taylor series before truncation. This is assumed to be a conservative estimate of the total truncation error. The default routine in MATLAB® and Simulink® [11] (ODE45) is, for example, a method of this type that compares fifth- and fourth-order approximations to the solution.

Fixed-step routines are simpler and are favored for real-time simulation because they result in a constant computation time. Variable-step routines are safer, popular, and often used as defaults in those simulation systems that offer a choice of algorithms.

1.3.2.3 Explicit versus Implicit

Algorithms that calculate the next state in terms of the current (and possibly earlier) states have right-hand sides that can be evaluated directly and are referred to as explicit algorithms. For some algorithms, however, Equation 1.2 is modified to include one or more terms in $(t + h)$ on the right-hand side:

$$y(t+h) = \mathrm{fn}\left\{y(t), y(t+h), x(t), x(t+h), u(t), u(t+h)\right\} \qquad (1.9)$$

which results in an implicit formula. This normally necessitates the solution of a set of simultaneous algebraic equations in each step. If the system of equations is linear, this can often be done by direct calculation, but nonlinear systems require an iterative solution procedure at each step. Implicit methods are generally unsuitable for real-time simulation because the amount of computation can vary from step to step without an *a priori* known upper bound. Moreover, it may prove necessary to provide external hardware-in-the-loop inputs ahead of time (i.e., $u(t + h)$ may be required at time t). Implicit methods can, however, offer improved stability and are often used for stiff systems (see Section 1.3.2.5).

1.3.2.4 Single-Step versus Multistep

The methods discussed so far are self-contained within each step and are called single-step methods. Some methods spread the calculation over more than one step and are called multistep methods. For example, the value of the system state at the next step may be determined in terms of the system state at the current and previous step times. Some algorithms require data from the past three or four steps. Equation 1.2 is now modified as follows:

$$y(t+h) = \mathrm{fn}\left\{y(t-k), x(t-h), u(t-h)\right\}, \quad \text{where} \quad k = 0, 1, 2, \ldots \qquad (1.10)$$

These methods are often cost-effective (in terms of the amount of computation necessary to achieve a given accuracy), but they do require special procedures to start them since past data is not available at the start of the simulation. These methods can also cause problems with hybrid and real-time simulations.

1.3.2.5 Variable-Order Algorithms and Stiff Systems

Many integration algorithms experience stability problems when applied to stiff systems. Stiff systems are systems that have a wide variation in the system dynamics, often described as having widely varying time constants or eigenvalues. These systems are characterized by containing both high-speed and low-speed dynamic properties, all of which must be captured by the method of solution. When the high-speed modes are active (i.e., variables are changing rapidly), short step sizes, commensurate with the time constants involved, are necessary to capture the rapid changes of state. The major problem with stiff systems arises when the fast modes are inactive (i.e., damped out) and the solution trajectories are smooth. Typically, longer steps would be used to cover the smooth trajectories, but in a stiff-system simulation, the dormant high-frequency modes can be stimulated by the integration algorithm causing instability unless the step size remains short, which can be very time-consuming.

This problem is often addressed by the use of special stable, stiff-system algorithms. One of the best known is the method of Gear [12], an implicit method based on backward differentiation, which adjusts the order as well as the step size of the algorithm to maintain both accuracy and stability while minimizing the amount of computation required. The original Gear DIFSUB routine was subsequently improved by Hindmarsh in the widely used GEAR routine [13]. Variable-order, variable-step methods of this kind are not normally suitable for time-critical real-time simulations.

1.3.2.6 Difference Equations

Whichever method is chosen to solve the differential equations in the mathematical model, the effect is to convert them into a set of approximately equivalent difference equations. It is these difference equations that are actually evaluated in each step of the simulation, and the way in which this process is implemented can have a significant bearing on the way the simulation is coded, particularly for real-time applications.

1.3.3 SOFTWARE FOR CONTINUOUS SIMULATION

There is a long history of languages for continuous simulation (traditionally referred to as continuous system simulation languages or CSSLs) dating back to the 1950s and 60s. Initially the software available to support continuous simulation developed from analog computer techniques. Analog computers used specially designed electronic amplifiers that were configured to perform mathematical operations such as integration, addition, multiplication, and function generation. These amplifiers were connected together by the programmer using patch cords to represent the differential equations to be solved. System constants and parameters were set on potentiometers (variable voltage dividers). The first continuous simulation systems for digital computers mimicked this approach. Names such as DAS (Digital Analog Simulator) and MIDAS (Modified Integration DAS) [14,15] were common. These systems required the programmer to produce a flow diagram showing the interconnection of components (adders, multipliers, integrators, etc.), which were similar to the physical components of an analog computer. In effect, the digital program simulated an analog computer simulation of the system! Since graphical input was not available, the flow diagrams were entered in tabular form showing how the various inputs and outputs were interconnected (similar to the Netlist for a modern electric circuit simulator such as SPICE). The expressiveness of this approach was limited by the available repertoire of components, and flow diagrams based only on these primitive elements rapidly became very large. With the increasing popularity of general-purpose high-level programming languages such as FORTRAN and Algol 60, a demand developed for simulation languages in which the program could be expressed in terms of program statements rather than flow diagrams. To distinguish between the two approaches, the names *block-structured* (for the flow-diagram method) and *statement-structured* (for the statement-based languages) were introduced. The emergence of statement-structured simulation languages as a replacement for the block-structured simulators was hailed as a great step forward.

The statement-structured languages include CSMP (Continuous System Modeling Program), DSL (Digital Simulation Language), ACSL (Advanced Continuous Simulation Language), CSSL4 (Continuous System Simulation Language 4), ESL (European Simulation Language), and, more recently, MATLAB and Simulink. An important development was the publication, in 1967, of a specification for a CSSL by Simulation Councils, Inc. (SCi) in their journal *SIMULATION* [16]. Although this specification was far from providing a complete standard, it did provide a template for the CSSLs that followed its publication. One of its main and enduring features was a simulation execution structure made up of three main regions designated the *Initial, Dynamic,* and *Terminal* regions. These names are fairly self-explanatory with calculations at time zero (or some other user-specified initial time) performed in the Initial region, the solution of the mathematical model with advancing time in the Dynamic region, and the terminating conditions and postrun calculations in the Terminal region.

Later, by virtue of the combination of greatly increased computer power and inexpensive graphical displays, a new generation of block-structured simulation software was developed. Modern continuous simulation software typically provides a graphical editor by means of which a flow diagram of the system can be defined. Unlike the original block-structured languages, modern simulation software offers a hierarchical approach in which primitive elements can be combined into more complex components and added to libraries. Extensive libraries are available from the software vendors, and users also have the capability of developing their own model libraries. Some products still provide the user with access to an underlying equivalent statement-structured form of the program, which experienced users can revise to meet special needs. Current versions of ESL and ACSL, for example, provide this feature.

With modern graphically based simulation languages, the user is able to build a block diagram of the simuland, often using simulation elements selected from toolboxes of common components and subsystems. The management of the simulation is implicit and controlled by the user through the use of interactive screens by which the parameters of the simulation (e.g., start time, terminating condition, integration algorithm and step size, output details, etc.) are selected. Some critics view the use of toolboxes with concern, claiming that this removes the details of the precise nature of the models used from the user and that this has the potential to generate convincing but inaccurate simulations. Used with care, however, and with adequate knowledge of the details of the model of each subsystem, this approach provides a very convenient way to develop simulations quickly.

1.3.4 EXAMPLE OF A CONTINUOUS SIMULATION

Consider the simple example of a small metal sphere that has been heated to a temperature T_0 at time $t = 0$ and is cooling down in air with an ambient temperature T_a. The differential equation that describes this process, at least approximately, is

$$dT/dt = K(T_a - T) \text{ given } T(t = 0) = T_a \text{ is the ambient temperature} \quad (1.11)$$

Note that we can calculate the initial value of $dT/dt = K^*(T_a - T_0)$. If $T_a < T_0$, then dT/dt will be negative, indicating cooling, and if $T_a > T_0$, then dT/dt is positive for heating. Using the Euler method, this would produce an estimate for $T(h)$ of

$$T(h) = T_0 + h * K * (T_a - T_0) \quad (1.12)$$

The derivative dT/dt at $T = h$ can then be calculated and the process repeated until the terminating condition is reached.

The process is the same for more complex integration algorithms or for larger systems of differential equations. In all cases, the calculation alternates between the determination of the current values of derivatives and auxiliary variables and the use of the integration algorithm to advance time by one step and calculate the corresponding values of the state variables. This alternation between establishing the current system state and advancing time mirrors the process used in discrete simulations in which the advance of time to the next event time alternates with the reevaluation of the system state.

1.4 HYBRID MODELS

In simple terms, a hybrid model is one that contains characteristics of both discrete and continuous models. There is extensive literature on the theory and analysis of hybrid systems (e.g., Ref. [17]).

For the purposes of developing simulations, a hybrid system can be viewed in some cases as a basically discrete model containing one or more states that change in a continuous fashion between events, and in others as a basically continuous model in which the states change continuously for most of the time but in which one or more of the states or their derivatives can occasionally change instantaneously. In yet other cases, the discrete and continuous processes may be balanced in an integrated model that is equally distributed between discrete and continuous components.

This distinction is important. Many currently available simulation software products provide comprehensive support for either discrete or continuous models with limited additional features that support the alternate approach. The balance between discrete and continuous elements in the model will, therefore, often dictate the choice of software to perform the simulation.

1.4.1 TIME MANAGEMENT FOR HYBRID SIMULATIONS

Regardless of the balance between discrete and continuous elements in a model, time management for hybrid simulations requires synchronization between the processing of events and the continuous advance of time for the continuous elements. In some cases, a period of continuous simulation, which may be of limited duration, can be triggered by a discrete event. In this case, time-management control will switch from the discrete to continuous modes with the duration of time steps determined by the integration algorithm. The simulation will remain in continuous mode until either the continuous simulation terminates (generating another event) or an event occurs in the discrete part of the model that impacts the continuous part. It can cause the continuous simulation to terminate, or it can cause a change of input or even a change to the continuous mathematical model in some way. Conversely, changes in the state of the continuous system can generate new discrete events, for example, when a continuous variable exceeds some critical value, such as a temperature or pressure limit.

When discrete events are occurring during a period in which a continuous process is active, it is important to manage the time advance of the continuous process so that the end of a step coincides with the time of an event. In this regard, it is helpful to use a distinction, first introduced by Cellier [18], between *time events* and *state events*.

A timed event is one for which the event time is known in advance. An example would be a clock signal in a sampled-data system. The numerical integration step size should, where possible, be controlled so as to coincide with the known timing of timed events.

A state event is one that is triggered by a condition that depends on the values of state or other dynamic variables in the simulation, such as a temperature that rises above a set point. The nature of state events is such that their event times are not normally known in advance, and special procedures are required to process them.

In non-real-time applications, synchronization with both types of event is often achieved by using a variable-step integration algorithm and additional synchronization procedures between the discrete and continuous simulations. In real-time applications, it may be necessary to reduce the step size of a fixed-step integration algorithm to minimize timing errors. In the case of state events, the discrete event is triggered by the continuous simulation, and it is again necessary either to use a variable-step algorithm to ensure accurate timing of the event or, particularly for real-time applications, to adopt a sufficiently short step size.

1.4.2 SOFTWARE FOR HYBRID SIMULATIONS

Several software packages that were originally developed specifically for either discrete or continuous simulation have subsequently added features that support hybrid simulations.

Early work on developing numerical integration routines to handle discontinuities was carried out by Hay, Crosbie, and Chaplin [19,20]. They adapted a conventional fourth/fifth-order variable-step routine by adding an interpolation stage that is activated whenever an end-of-step check shows that a state event occurred during the step. They used a succession of interpolations, starting with a linear interpolation between the start and end of the step to produce a first approximation to the event time. This is followed by a quadratic interpolation using values at the start, approximate event time, and end of step to produce a more accurate value for the event time. Further iterations with third- and higher-order interpolations are possible but not normally necessary to locate the event time with acceptable accuracy. This approach is very effective in non-real-time simulation and has been widely adopted, but it is not suitable for real-time simulation because of the variable and extended computation time whenever a state event occurs within a step.

This technique was later included in ESL, one of the first continuous simulation languages to introduce methods for processing discontinuities accurately and automatically. ESL was developed for the European Space Agency by Hay, Crosbie, and Pearce [21]. The way in which the discontinuities (events) are specified by the user is elegant and is based on the use of logical statements and a program construct known as a *when clause*. A when clause is a section of code that is executed when and only when (i.e., once only) a specified condition becomes true. More conventional conditional *if* statements can also be used to switch between different descriptions of a subsystem. Almost any discontinuous element can be described using *if* and *when* constructs.

A simple example of the use of the *when* clause is the code for a submodel for a T flip-flop. This device is assumed to toggle its single logical output whenever a trigger is applied to its input, which can be specified in ESL as

```
SUBMODEL TFLOP(LOGICAL:flag:=LOGICAL:trigger);
                INITIAL
                  flag:=false;
                DYNAMIC
                  when trigger then
                    flag:=not flag;
                  end_when;
              END TFLOP;
```

The first line defines a submodel named TFLOP that has a single logical output named *flag* and a single logical input named *trigger*. The *flag* is set to *false* in the INITIAL region. In the DYNAMIC region, which is executed at each time step, the *flag* is inverted when and only when the logical variable *trigger* changes from *false* to *true*. More statements could be inserted between *when* and *end_when*, and all of them would be executed whenever the *trigger* becomes *true*.

A more complex example, using both *if* and *when*, is illustrated by the following code template, which implements a method of defining a system that switches between three different states. A line preceded by a double hyphen (--) is a comment line.

```
INITIAL
--  Initialize state
    state := 1;
DYNAMIC
    when state=1 AND transition12 then
        state := 2;
    when state=1 AND transition13 then
        state := 3;
```

```
    when state=2 AND transition21 then
         state := 1;
    when state=2 AND transition23 then
         state := 3;
    when state=3 AND transition31 then
         state := 1;
    when state=3 AND transition32 then
         state := 2;
    end_when
-- Assign variables depending on state
    v1 :=if state=1 then <expression11>
             else_if state=2 then <expression12>
             else_if state=3 then <expression13>;
--Repeat for all variables
```

It is assumed that the simulation is initialized with the system in State 1. There may be additional initialization code in the INITIAL region. The DYNAMIC region uses *when* clauses to define the six possible transitions between the three states by means of the logical variables *transition12*, etc., which are normally all *false*. A state transition is indicated by setting the corresponding transition variable to *true*. The values of variables in a particular state are set using conditional *if* statements. Note that these are structured as assignments (:=) with conditional right-hand sides.

One of the strengths of ESL is that once a discontinuity has been defined in the code using *if* or *when* constructs (or by selecting a graphical element that is itself defined using such statements), it is automatically recognized and implicitly detected when the software runs with appropriate control of the step size. When models containing such discontinuities are simulated using a variable-step integration routine, and when the value of one of the *transition* variables changes from *false* to *true*, a discontinuity detection mechanism is invoked that adjusts step sizes in a way that ensures that the event time coincides with the end of an integration step, within a user-specified tolerance. This approach is not ideal for real-time simulation unless it can be guaranteed that the process will always reach a satisfactory conclusion within the available frame time. ESL does, however, support real-time simulation of these hybrid systems. The object-oriented modeling language Modelica [22] and the MathWorks product Simscape™ [23] also use the notion of *if* and *when* statements and corresponding semantics. Other languages, including ACSL and Simulink, have provided alternate ways of defining state events based on the definition of crossing detectors. Additional coding is usually necessary to link the specification of the event to its processing during execution.

Turning now to models that are basically discrete, some older discrete languages have added features for handling continuous elements. Simscript, for instance, features a continuous process that is illustrated by an example in the next section.

As stated earlier, when it was first developed in 1976, the DEVS formalism and the software products that implemented it were focused on discrete models. In the 1990s, continuous features were developed. We have already established that a numerical solution of differential equations is, in effect, a discrete process and is therefore accessible to representation using the DEVS formalism. It is necessary to capture the difference equations that arise from the continuous mathematical model combined with the selected numerical integration algorithm with DEVS. With a feature of this kind, a DEVS implementation is clearly capable of handling hybrid models. One example is PowerDEVS [24], a version of DEVS aimed specifically at power-electronic circuits and systems, but capable of supporting a range of hybrid simulations including real-time simulations. PowerDEVS is also interesting in that it uses a QSS (Quantized State System) [6,25] approach to continuous simulation. QSS deals with systems in which system states, as well as time, change in finite increments. Simulations using QSS thus involve discretization of both time and state variables.

1.4.3 EXAMPLES OF HYBRID SIMULATIONS

Understanding hybrid simulation techniques can be made easier by initially separating them into three groups: those that are basically continuous with discrete elements, those that are basically discrete with continuous elements, and those that are equally balanced. These distinctions can best be illustrated by simple examples.

1.4.3.1 Continuous Model with a Discrete Element

One of the most commonly quoted examples of a hybrid system in the literature is the bouncing ball. A bouncing ball simulation provides an example of a continuous model (dynamics of the motion of the ball) with discrete elements (impact with a hard surface). In its simplest form, the continuous model equates the acceleration of the ball to acceleration due to gravity ($d^2x/dt^2 = -g$ for $x > 0$) and assumes a hard impact in which the velocity of the ball changes instantaneously ($dx/dt^+ = -k*dx/dt$), where the coefficient of restitution k is a dissipation factor.

Another simple example, which will be developed later, is provided by the electrical circuit in Figure 1.2. An alternating current supply feeds a load consisting of a resistor and capacitor in parallel through a series resistance and a diode. The diode is represented by a very simple model of an ideal switch. If the voltage across it is positive, its resistance is assumed to be zero and the diode can be represented by a short circuit. If the voltage is negative, the resistance is infinite and the diode is represented by an open circuit. The voltage produced by the supply is $v = V\cos(wt)$. Assuming the capacitor is initially discharged, the initial value of v is equal to V and the diode will conduct. A simple mathematical model can be produced that recognizes two modes of operation of the circuit, the conducting mode and the nonconducting mode.

1.4.3.1.1 Conducting Mode

Assuming ideal components, the key differential equation is provided by the relationship between voltage and current for the capacitor, which can be expressed as

$$i_c = Cdv_c/dt \quad \text{or} \quad dv_c/dt = i_c/C \qquad (1.13)$$

where

$$i_c = i - i_R$$
$$i = (v - v_c)/R_L$$
$$i_R = v_c/R_L$$

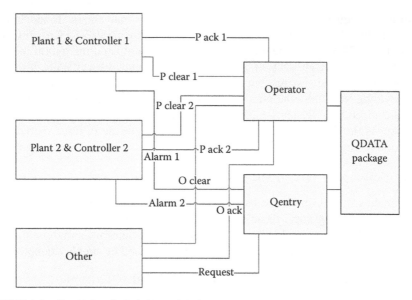

FIGURE 1.2 Example of a hybrid model: An electric circuit containing a switch.

The sequence of calculation is as follows:

The value of the state v_c is known at the beginning of a time step, either as an initial condition or the result of the calculation from the previous step. Given the value of v_c at the current time t, it is possible to calculate the values of i_R, i, and i_c also at time t (in addition to v, which is defined for all time in the model). Given i_c, it is then possible to calculate the current value of the derivative dv_c/dt. This completes the computation of the state of the system at the given time instant. This is the point at which the algorithm chosen by the user to solve the differential equation takes over. Time is advanced by one time step, and the value of v_c at time $t + h$, where h is the time increment, is calculated using the integration algorithm. A simple example would be to use Euler integration, which simply assumes that the value of dv_c/dt at t is maintained constant over the next time increment so that the new value of v_c is calculated as

$$v_c(t + h) = v_c(t) + hv_c'(t), \quad \text{where} \quad v_c' = dv_c/dt \qquad (1.14)$$

The process is then repeated starting at $t = t + h$ with the corresponding value of v_c and is continued until the system mode changes to nonconducting or the terminating condition for the simulation is reached.

1.4.3.1.2 Nonconducting Mode

In the nonconducting mode, the RC load is disconnected from the supply. The governing equations are now

$$dv_c/dt = i_c/C \qquad (1.15)$$

where

$$i_c = -i_R$$
$$i_R = v_c/R_L$$

Given v_c at the beginning of a time step, i_R and i_c can be calculated followed by dv_c/dt. The numerical integration algorithm can then be used to calculate v_c at the end of the next step.

A key question now is how to make the decision to switch between modes. As was discussed earlier, the answer to this question can be quite different for real-time and non-real-time simulations. When real-time execution is not required, a variable-step algorithm, preferably one that is combined with a discontinuity detection scheme, could be most appropriate. For real time, it may be necessary to use a fixed-step method, in which case the length of the step is constrained by the need to limit timing errors. This question is discussed further in Section 1.5.1.

1.4.3.2   Example of a Discrete-Based Hybrid Simulation

Discrete models are widely used to represent industrial processes. Many of these models deal with the progress of materials or components as they move through a manufacturing process. Objects are held in queues waiting for equipment to become available to perform the next stage in the manufacturing process. Service times are often generated randomly using an appropriate distribution. In some cases, it is necessary to represent the process more accurately, and this may involve the use of a continuous model of a particular part of the total manufacturing process. Fayek [26] describes an application involving the movement of metal ingots into and out of a furnace.

A steel plant has a soaking pit furnace which is being used to heat up steel ingots. The interarrival time of the ingots is determined to be exponentially distributed with a mean of 1.5 hours. If there is an available soaking pit when an ingot arrives, it is immediately put into the furnace. Otherwise it is put into a warming pit where it is assumed to retain its initial temperature until a soaking pit is available.

Fayek describes both totally discrete and combined continuous–discrete (or hybrid) models to address this problem. In the discrete version, the time to heat an ingot in the furnace is uniformly distributed between 4 and 8 hours. In the hybrid version, it is determined by solving a differential equation representing the change in temperature of the ingot:

$$dh/dt = (H - h_i) * c_i \tag{1.15}$$

where h_i is the 1000°F temperature of the ith ingot, H is the furnace temperature (assumed to be 1500°F), and c_i is the heating time coefficient of the ith ingot, which is equal to $(0.07 + x)$, where x is normally distributed with a mean of 0.05 and a standard deviation of 0.01.

Ingots are to be heated to a final temperature that is normally distributed between 800 and 1000°F.

In a further variation of the simulation, the furnace temperature is assumed variable. It is normally heating up toward a given final temperature and is also reduced when cold ingots are put into the furnace. This adds a second differential equation for the furnace temperature as follows:

$$dH/dt = (2500 - H) * 0.05 \qquad (1.16)$$

Introducing a cold ingot is assumed to produce an instantaneous fall in the furnace temperature depending on the difference between the furnace and ingot temperatures divided by the number of ingots in the furnace. This hybrid simulation was implemented in a version of Simscript V that supports continuous features. The simulation ran successfully, and the ability to model the dynamics of the heating and cooling effects provided a more valid representation of the entire process.

1.4.3.3 Balanced Hybrid Simulation

Many of the documented applications involving hybrid simulations are of the above kinds, either mainly discrete or mainly continuous. An example of a more balanced hybrid simulation is provided by a study performed for the European Space Agency by the author and colleagues using ESL [27]. It involves an artificially created simulation that was intended to demonstrate the ability of ESL to perform simulations that combined a balance of continuous and queuing elements. This is not generally a straightforward task with most simulation software systems. It contains continuous and both types of discrete elements discussed earlier (queues and sampled data).

The continuous model (derived from an ACSL example) is illustrated in Figure 1.3. It contains a subsystem with a transfer function $1/s(s + 1)$. The output, X, of this system is compared to a set point X_c to form an error signal E. A sampled version of E is passed to a digital controller, which generates a control signal U. The control is a linear combination of the current error, $E(n)$, the previous error, $E(n - 1)$, and the previous control, $U(n - 1)$. The difference equation describing the controller is

$$U(n) = B1 * U(n-1) + A0 * E(n) - A1 * E(n-1) \qquad (1.17)$$

The constants $A0$, $A1$, and $B1$ depend upon the equivalent lead and lag time constants of the controller (T_{lead} and T_{lag}) and the sampling period T_s. A delay is added to the updating of U to represent the computation delay in the digital controller.

This basic example was extended (see Figure 1.4) by making the assumption that the actual set point, unknown to the controller, is drifting (represented by an

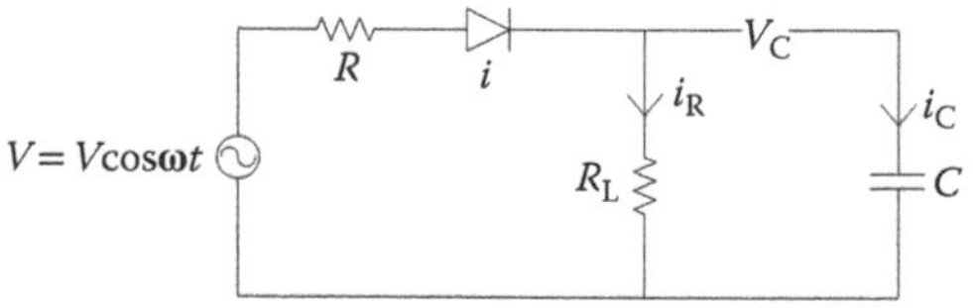

FIGURE 1.3 Continuous part of combined continuous–discrete benchmark.

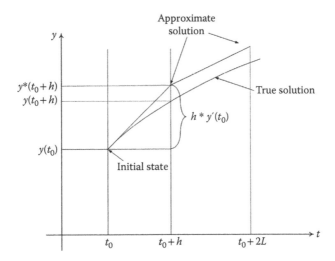

FIGURE 1.4   Combined continuous–discrete benchmark.

additional sinusoidal variation). It is assumed that a mechanism exists for indicating when the true error from the actual set point exceeds a given threshold by generating a logical alarm signal. This alarm message is transmitted over a communication link to an operator console where it is queued with other messages until the operator is available to take corrective action. A correction message is then sent back to the system, which has the effect of resetting the set point. The system also has the ability to determine when the alarm condition clears spontaneously, in which case it sends a cancellation message, which will cancel the original alarm if it is received before the alarm has been acted upon. This system incorporates the following features of combined continuous/discrete systems:

- State events generated from within a continuous subsystem
- Synchronous A/D sampling of continuous variables
- Updating of continuous variables through a D/A converter
- Components described by difference equations/z-transforms
- Digital computation delays
- Queuing of events awaiting processing
- Cancellation of queued events
- A server (operator) responding to queued events in sequence

The entire simulation was coded in ESL using ESL primitives. ESL is a traditional CSSL with discrete features. It is to be expected that more recent hybrid simulation software products that are aimed specifically at hybrid systems would facilitate the development of simulations of this kind.

One of the outcomes of this study was a list of recommended enhancements to the ESL language to make it more capable of supporting balanced hybrid models with features similar to those in the benchmark. These included

- Add support for discrete variable types and a notation for difference equations
- Support the use of z-transforms
- Improve processing of time events
- Introduce an EVENT block for defining state and time events
- Support queues so that they can be easily specified and manipulated by the user
- Provide a greater range of random number distributions
- Provide support for user-specified statistical outputs
- Provide support for producing results in the form of event lists

These changes were aimed at producing a more balanced hybrid simulation capability. Most of them remain unimplemented.

1.5   REAL-TIME HYBRID SIMULATION

Most real-time simulations are characterized by the necessity to incorporate some physical component in the simulation. This can be hardware, software, or a human or humans in the loop. In all cases, it is necessary for the simulation to be synchronous with a real-time clock to ensure correct timing of the interactions between the simulation and the external agent.

1.5.1   Timing Issues in Real-Time Hybrid Simulation

Challenges face the potential user of hybrid simulation, whether real time or not. The choice of software to support the development of the simulation can be critical, and using it to properly represent both the continuous and discrete features of the model and their interaction requires great care. Precise time management, particularly the accurate interchange of timing control between discrete and continuous processes, is essential. At this point, it is necessary to draw a distinction between integration step size and frame time in real-time simulations. The integration step size is precisely what it says, the increment of time by which the current time is advanced in a single application of the selected numerical integration algorithm. In real-time simulation, it is normally constant. The frame time is the normally constant time interval between data transfers to external elements such as hardware in the loop or other simulations, and a frame may consist of several integration time steps. It can be regarded as a real-time equivalent of the communication interval in standard CSSL terminology. The required frame time can dictate the choice of integration algorithm, especially for higher-speed, real-time simulations with the need for very short frame times.

Taken together, these factors pose a formidable challenge to the designer of hybrid real-time simulations. We now consider some of the specific issues that arise in their creation.

The basic requirement for a real-time simulation is that all computation within a frame should be completed within the fixed frame time. In some applications, this requirement can be relaxed, allowing occasional data drops (perhaps caused by communication delays, for example, with hardware-in-the-loop), but in general, the hard real-time requirement must be observed.

The detailed nature of real-time hybrid simulations depends to some extent on the length of the acceptable frame time, which can vary, in different types of applications, from seconds to microseconds. For longer frame times, the ability of the processor or processors to perform the necessary calculations in the available time is less often an issue, but time synchronization between different parts of the simulation, particularly in those cases where the simulation is distributed over different systems, is key.

Distributed real-time simulations often involve different software components to be integrated into a joint simulation. The question of interoperability between software elements then becomes of major concern. Special software environments have been developed to support this kind of interoperability. One of the most widely used is the high-level architecture (HLA) [28]. This architecture is widely used, particularly in military applications, where different simulations running on different platforms in different geographical locations are involved. An HLA simulation is known as a *federation* and its components as *federates*. Data that is passed between different federates using a publish-and-subscribe technique is time stamped to ensure that time synchronization is maintained using a real-time interface.

Wainer, Kim, et al. [29,30] have described some interesting DEVS-based real-time hybrid simulations, which make use of an interface that supports combinations of model components that use different DEVS-based software. A simulation of a robot that negotiates obstacles uses PowerDEVS to simulate the robot and ECD++ (another DEVS-based software) to simulate the interaction with the obstacles.

A number of commercial systems are available that support real-time hybrid simulations including MathWorks xPC Target™ [31] and Real-Time Windows Target™ [32], the combination of National Instruments (NI) LabView with MathWorks Simulink using the Simulation Interface Toolkit from NI [33], RT-Lab from OPAL-RT [34], the rtX simulator from Applied Dynamics International [35], and the Real-Time Digital Simulator (RTDS), a simulator that specifically targets electrical power systems, from RTDS Technologies [36].

1.5.2 HIGH-SPEED REAL-TIME HYBRID SIMULATION

Particular problems arise with those applications that involve short intervals between repeated discrete events. This is true for instance of real-time simulations of modern power-electronic systems. Electric power can be delivered in many different forms. The most obvious distinction is between alternating current (ac) and direct current (dc). Although line frequencies of 50 or 60 Hz are commonly used for domestic supply, other frequencies may be used. For example, 400 Hz is a common choice for many power distribution systems, in aircraft for example. Higher frequencies require smaller transformers, filters, and other related equipment but involve higher losses over longer distances. Many pieces of equipment (such as computers, phones, radios, televisions, for example) require dc supplies and commonly incorporate a converter from ac to dc in the line cord. The dc voltages vary for different applications and for both ac and dc supplies, there are different limits on the amount of allowable harmonic distortion (unwanted ac frequencies).

Converters convert electric power both ways between ac and dc. Converters involve a significant amount of switching whether they are producing dc from an ac waveform or vice versa. The switching is often controlled by pulse-width modulation controllers that vary the intervals between switching events while keeping the average frequency constant. In modern converters, the switching frequencies are set as high as possible to minimize harmonic distortion and to limit the size of associated components. For higher power applications (such as the power system for an electric ship), switching frequencies may be in the kilohertz range. For small appliances, frequencies may be measured in tens or even hundreds of kilohertz. For a 10 kHz switching frequency, the average interval between switching events is 100 μs. In a non-real-time simulation, a variable-step integration algorithm with discontinuity detection would probably be used to minimize errors in the timing of the switching event. In a real-time simulation, using a fixed-step integration algorithm, timing errors are minimized by using sufficiently short steps. If, for example, the maximum allowable timing error is 2% of the average period of the controller switching waveform, then the step size should be no longer than 2 μs. The worst case arises when the actual switch time is immediately after the end of a step. In this case, the integration step will proceed using the equations that were valid before the switching event for the entire step, and the corresponding simulated switching event will not occur until the end of the step. In this case, the timing error is equal to the integration step length. Techniques aimed at compensating for switch-timing errors in real-time simulations are discussed in Section 1.5.3.

Note that real-time simulations that require frame times of less than about 10–20 μs require special processing systems because conventional real-time simulation systems are not capable of reliably maintaining such short frames. Digital signal processors (DSPs) [37] and field-programmable gate arrays (FPGAs) [38] have been used for this purpose. The DSPs or FPGAs need not be more powerful (in a MFLOPS sense) than the conventional system. The key is that the processor that executes the critical simulation code is not vulnerable to interruption by the operating system. A dedicated core or cores in a multicore processor could also be employed in this manner. The RTDS power-system simulator uses both DSPs and FPGA-based processors and offers high-speed units capable of 2-μs frame times.

1.5.3 NUMERICAL INTEGRATION FOR HIGH-SPEED REAL-TIME SIMULATION

Methods to address the need for short frame times in high-speed real-time (HSRT) simulation can be divided into the use of simple algorithms that minimize the amount of computation and the use of techniques that compensate for midstep switching events.

As discussed earlier, the simplest integration algorithm is provided by the Euler method, which only requires that the derivatives of system variables at the start of a step (all of which are known) are multiplied by the fixed-step length to provide an estimate of the change in the value of each variable during the step. This simple extrapolation has the advantage of simplicity. It is not widely used in general in continuous simulations because of its larger truncation errors, but in cases where the step size is constrained for other reasons, it can be effective. It does, however, present stability concerns. Euler integration is the simplest of a class of single-step methods of different orders, known as Runge–Kutta methods, which produce an approximation to the average rate of change of a state variable during a step. Stability margins for integration algorithms can generally be expressed in terms of the eigenvalues of the system (which can be viewed as corresponding roughly to inverse time constants). Euler integration becomes unstable if the step size exceeds twice the shortest time constant (more strictly $2/\lambda$, where λ is the largest eigenvalue). This situation can arise at a point at which slightly shorter (stable) step lengths produce acceptably accurate results, so the transition from acceptable to totally unusable results can be immediate. For nonlinear systems with eigenvalues that can change from step to step, this can be a serious problem. It is consequently advisable for users to be fully aware of the dynamics of their systems and of the potential for numerical instability. Higher-order Runge–Kutta methods have slightly improved stability behavior, but the margin is still, in general, no more than about $2.8/\lambda$ (an improvement of only about 40% over the Euler method). One of the advantages of using variable-step algorithms is that they usually ensure that the step length remains in the stable region, but their use is not normally an option in real-time simulation.

Another option for real-time simulation is a fixed-length, explicit, multistep method such as Adams–Bashforth. This approach has been advocated by Robert M. Howe [39], who has a long history of innovative contributions to real-time simulation methodology. Howe has also developed techniques specifically for switched linear circuits, such as those found in power-electronic applications, that compensate for the fact that switching instants occur during a step [40]. Howe's approach involves splitting a step into substeps separated by switching events and calculating analytic solutions to the differential equations using precomputed coefficients. These techniques require additional computation but can also reduce errors significantly. They permit the use of longer steps for the same accuracy at a cost of more computation per step, and they represent a serious option for HSRT simulation involving linear differential equation models.

De Kelper et al. [41] has also developed techniques for minimizing the effect of switching errors in real-time simulations of electrical power systems that adapt the methods for non-real-time variable-step integration described in Section 1.4.2 to fixed-step methods, particularly for real-time applications. De Kelper has adapted the method to real-time fixed-step simulations in two ways. The first is to limit the interpolation to a single linear process, and the second is to advance the solution from the switching point by the same fixed-step length but to again use linear interpolation to produce a solution at the end of the original step. In other words, the first step goes from t to $t + h$ and is interpolated to find a solution at $t + d$, which is the time at which the linear interpolation locates the switching event. The second step goes from $t + d$ to $t + d + h$, and a linear interpolation is used to produce a solution at $t + h$.

Bednar and Crosbie [42] has developed an efficient integration method for linear models that builds on a state-transition approach [43]. The state-transition approach is based on the solution to the vector differential equation

$$dx/dt = Ax + Bu \qquad (1.18)$$

Assuming a sample and hold input,

$$u(t) = u(kT) \quad \text{for} \quad kT <= t < (k+1)T \tag{1.19}$$

then the approximate solution is

$$x[(k+1)T] = \Phi * x(kT) + \Gamma * u(kT) \quad \text{for} \quad k = 0, 1, \ldots \tag{1.20}$$

where Φ is the transition matrix.

This approach provides a class of fixed-step, explicit methods of varying complexity and accuracy depending on how many terms of two infinite series are used. A version using three and two terms, respectively, and known as the ST(3,2) method has proved effective. In a number of tests, it compared favorably with other methods for accuracy, stability, and computation time. The method can also be applied to some nonlinear systems of equations [44].

1.5.4 HSRT MULTIRATE SIMULATION

Most applications that require HSRT simulation involve components with lower bandwidth dynamics that do not require the short frame-time that HSRT methods are designed to deliver. In these cases, it is better to design a multirate simulation in which the system is separated into segments that are simulated using different frame rates. This means that those segments that do not depend on HSRT techniques can be implemented on conventional processors. A multirate benchmark has been devised to support the study of high-speed, real-time, multirate simulations.

Multirate simulation has been a recognized simulation technique for many years (e.g., Ref. [45]). It has often been used to accelerate the execution of large simulations. The basic idea is that it is not necessary to update the dynamics of the slower parts of a system at the same rates as are required for the faster parts. It is a simple idea that has stood the test of time when applied carefully and conservatively. It is particularly suitable for HSRT applications for which it is possible to execute the high-speed segments on the special high-speed processors and the slower segments on more conventional computer systems.

It is not, however, without its problems. Step sizes for the different segments must be chosen with care. It is also necessary to distinguish between integration step size and communication interval. The segments in a multirate simulation communicate with each other at communication intervals that are normally related by integer multiples. Within a communication interval, the simulation can proceed using a single integration step, equal in length to the communication interval, or by using several integration steps, which together advance the simulation by the duration of one communication interval. These integration steps can be of either fixed or variable length as long as synchronism with the communication interval is maintained.

A multirate simulation is effectively a sampled-data system, and issues can arise with the adequacy of the data passed between segments, particularly when a faster segment has to wait several communication intervals before it receives an update from a slower segment. The input from the slow segment can be kept constant between updates (zero-order hold), but this introduces an effective time delay of approximately half the slower frame time. Extrapolation techniques (first-, second-, or even fractional-order hold) can be used to compensate. Another question is how data is passed from a faster to a slower segment given that the faster segment will produce several data points between data transfers to the slower one. Some method of averaging the fast outputs over each slow frame is required. There is a danger of aliasing if rapidly changing outputs from a fast segment are not filtered or averaged in some way. In some cases, this can be achieved by carefully selecting the segment boundaries so that the outputs from the fast segment do not have high-frequency components. In other cases, it may be necessary to add antialiasing filters. If sufficient care is not taken in selecting communication intervals and data interchange processes, multirate simulations can become inaccurate or even unstable.

A multirate benchmark simulation, based on a simulation of an unmanned underwater vehicle (UUV), was developed in cooperation with the University of South Carolina and the University of Glasgow [46,47]. The UUV uses a battery bank as its energy source feeding an ac motor drive through a dc to ac converter. The drive powers the vessel, which is modeled as a six-degree-of-freedom platform with control surfaces. The system is divided into subsystems that are simulated with different frame rates: (1) converter and switching controller (fast), (2) feedback controller (slow medium), (3) motor drive (fast medium), and (4) the vessel, battery, and interface graphics (slow). The UUV system is illustrated in Figure 1.5. Typical simulation frame rates used in this simulation are

Battery, ship, 3D graphics: 100 mS
Converter, switch controller: 2–5 μS
Motor: 50–100 μS
Feedback controller: 1 mS

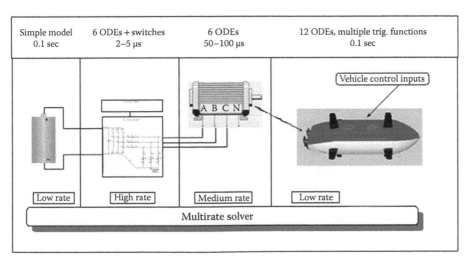

FIGURE 1.5 Simplified representation of unmanned underwater vehicle model (vessel inputs and outputs and controls are omitted).

The simulation has been implemented using the Virtual Test Bed (VTB), developed by the University of South Carolina [48] in both non-real-time and real-time versions. As an indication of the increase in speed afforded by the use of multirate techniques, implementations of this simulation on a Microsoft Windows system running in non–real time showed a speedup of three orders of magnitude from versions that used the fastest rate for the entire system, with no significant change in the accuracy of the results. The VTB provides a flexible simulation environment with 2D and 3D graphics for output and user controls. It supports multidisciplinary simulation with both mechanical and electrical components and has a library containing a large range of electrical and mechanical components.

The real-time version replaced the UUV drive system (motor and propulsion drive shaft) with hardware. It was not feasible to reproduce the actual physical components in the laboratory, so a small low-power motor was used to represent the actual motor. The load on the drive shaft was represented by a generator coupled to the motor output shaft. The output of the generator was connected to a programmable resistive load that was controlled by the simulation. This configuration supported an equivalent real-time simulation of the UUV system that could support the development of high-speed, real-time, multirate simulation techniques.

1.6   CONCLUDING REMARKS

In this chapter, we have considered relevant aspects of discrete and continuous as well as hybrid simulations in both real-time and non-real-time contexts. The main focus, however, is on real-time simulation with hybrid models. We have seen that the distinction between discrete and continuous simulation is, in a sense, a distinction without a difference since a continuous simulation is itself essentially a discrete process. The key to hybrid simulations lies in the approach to time management and the synchronization of events between different parts of the simulation.

First we consider the non-real-time requirements. In general, discrete simulations usually advance in unequal time steps from event to event, and continuous simulations are also more likely to advance in unequal time steps using variable-step algorithms. Both discrete and continuous simulations can use fixed time steps. They are commonly used for simulating discrete time systems such as digital controllers and are a feature of the older activity-based discrete simulation approach. Where possible, however, the use of variable time steps can deliver significant improvements in performance.

For real-time simulation, the important factor is the available computer speed. If the computer is capable of executing code significantly faster than the related real events, then it can be allowed to idle while real events catch up. In fact, although it is often stated that variable-step integration algorithms are unsuitable for real-time continuous simulations, their use, particularly in conjunction with a fixed frame time, is not unfeasible as long as the longest possible execution time for the calculations within a frame can be guaranteed to be less than the time available.

For real-time hybrid simulations, time management must be capable of advancing the continuous simulations in fixed time steps (or frames) toward the time of the next discrete event. Synchronizing the fixed increment of time applied to the continuous simulation with the variable increments between events in the discrete simulation is a key factor in the time-management protocol. In the UUV simulation described above, the continuous simulation time steps are sufficiently short so that the timing errors that result from imposing a fixed time step are assumed to be acceptable. The methods described by Howe [39,40] and De Kelper et al. [41] help to compensate for these errors. An alternative, more widely used for non-real-time simulations, would be to use the methods that locate the time at which imminent state events occur and adjust the step size to coincide with the event. In real time, however, this introduces a double step in which time first advances to the time of the state event and then continues to the end of a normal step. This can only be done if the available processor power can accommodate the double step plus the overhead associated with locating the state event. The situation can be further complicated by multirate techniques, which require the event handler to keep track of the different sequences of fixed increment events for the different rates as well as the discrete events for each multirate segment.

Developing real-time hybrid simulations is in many cases a challenging undertaking with today's available simulation software products. The picture is most promising for applications that are dominantly discrete with limited continuous elements or dominantly continuous with discrete elements restricted to switching events. Adding a need for more complete and balanced discrete/continuous features similar to the example described above, or for high-speed, or for multirate simulation significantly increases the degree of difficulty and is very likely to involve detailed programming of algorithms and time management. With the increasing interest in all kinds of hybrid simulations, there are growing and exciting opportunities for simulation software developers to meet this need.

REFERENCES

1. Zeigler, B. P. 1976. *Theory of Modeling and Simulation*. 1st ed. New York: Wiley Interscience.
2. Wainer, G. A., and P. J. Mosterman, eds. 2011. *Discrete-Event Modeling and Simulation: Theory and Applications*. Boca Raton, FL: CRC Press.
3. Fujimoto, R. M. 2003. "Advanced tutorials: Parallel Simulation: Distributed Simulation Systems." In *Proceedings of the 2003 Winter Simulation Conference*, edited by S. Chick, P. J. Sánchez, D. Ferrin, and D. J. Morrice, New Orleans, LA, December 7–10, 2003, pp. 124–134.
4. Olukotun, K., M. Heinrich, and D. Ofelt. 1998. "Digital System Simulation: Methodologies and Examples." In *Proceedings of 35th Design Automation Conference, DAC '98*, San Francisco, CA, June 15–19, 2011, pp. 658–63.
5. Zeigler, B. P., and H. S. Sarghoujian. 2003. "Introduction to DEVS Modeling & Simulation with JAVA: Developing Component-based Simulation Models." http://www.acims.arizona.edu/PUBLICATIONS/publications.shtml#devsHLA.
6. Cellier, F. E., and E. Kofman. 2006. *Continuous System Simulation*. New York: Springer Verlag.
7. Jefferson, D. et al. 1987. "The Time-Warp Operating System." In *Proceedings of 11th Symposium on Operating Systems Principles*. New York: ACM.
8. Birtwistle, G. M., O.-J. Dahl, B. Myhrhaug, and K. Nygaard. 1973. *SIMULA Begin*. Chartwell-Bratt, Bromley, England 1979.

9. Matloff, N. "Introduction to Discrete-Event Simulation and the SimPy Language," http://heather.cs.ucdavis.edu/~matloff/156/PLN/DESimIntro.pdf.

10. Ralston, A., and P. Rabinowitz. 2001. *A First Course in Numerical Analysis*. 2nd ed. Mineola, NY: Dover Publications.

11. Klee, H., and R. Allen. 2011. *Simulation of Dynamic Systems with MATLAB and Simulink*. Boca Raton, FL: CRC Press Inc.

12. Gear, C. W. 1962. *Numerical Initial Value Problems in the Numerical Solution of ODEs*. New York: Wiley.

13. Byrne, G. D., A. C. Hindmarsh, K. R. Jackson, and H. G. Brown. 1977. "A Comparison of Two ODE Codes: GEAR and EPISODE." *Computers & Chemical Engineering* 1 (2): 133–47.

14. Clancy J. J., and M. S. Fineberg. 1965. "Digital Simulation Languages: A Critique and a Guide." In *Proceedings of the November 30–December 1, 1965, Fall Joint Computer Conference*, 23–6. Montvale, NJ: AFIPS Press.

15. Brennan R. D., and R. N Linebarger. 1964. "A Survey of Digital Simulation: Digital Analog Simulator Programs." *SIMULATION* 3 (6): 22–36.

16. SCi Software Committee. 1967. "The SCi Continuous System Simulation Language (CSSL)." *SIMULATION* 9 (6): 281–303.

17. Goebel, R., R. G. Sanfelice, and A. R. Teel. 2009. "Hybrid Dynamical Systems." *IEEE Control Systems Magazine* 29 (2): 28–93.

18. Cellier, F. E. 1979. *Combined Continuous/Discrete System Simulation by use of Digital Computers*, Doctoral dissertation, ETH Zurich.

19. Hay, J. L., R. E. Crosbie, and R. I. Chaplin. 1974. "Integration Subroutines for Systems with Discontinuities." *Computer Journal* 17 (3).

20. Crosbie, R. E., and J. L. Hay. 1974. "Digital Techniques for the Simulation of Discontinuities." In *Proceedings of Summer Simulation Conference*, Houston, TX, July 9–11, 1974, pp. 87–91.

21. Crosbie, R. E., S. Javey, J. L. Hay, and J. G. Pearce. 1985. "ESL—A New Continuous System Simulation Language." *SIMULATION* 44 (5): 242–46.

22. Modelica Association. March 2010. "Modelica: A Unified Object-Oriented Language for Physical Systems Modeling: Language Specification, Version 3.2." https://www.modelica.org/documents/ModelicaSpec32.pdf.

23. MathWorks. September 2009. *Simscape User's Guide*. Natick, MA: MathWorks.

24. Bergero, F., and E. Kofman. 2010. "PowerDEVS: A Tool for Hybrid System Modeling and Real-Time Simulation." *SIMULATION* 87 (1–2): 113–32.

25. Kofman, E., F. E. Cellier, and G. Migoni. "Continuous System Simulation and Control." In *Discrete-Event Modeling and Simulation: Theory and Applications*, edited by G. A. Wainer and P. J. Mosterman, 75–107. Boca Raton, FL: CRC Press.

26. Fayek, A-M. 1987. *Introduction to Combined Discrete-Continuous Simulation Using PC SIMSCRIPT II.5*. La Jolla, CA: CACI.

27. Crosbie, R.E. August 1993. "Combined Continuous-Discrete Simulation." ISIM Internal Report ref \estec\extlab\tsk4_fin.001, ISIM International Simulation Ltd, 161 Claremont Rd. Salford M6 8PA, UK.

28. Kuhl, F., R. Weatherly, and J. Dahmann. 1999. *Creating Computer Simulation Systems: An Introduction to the High-Level Architecture*. Upper Saddle River, NJ: Prentice Hall PTR.

29. Moallemi, M., and G. Wainer. 2010. "Designing an Interface for Real-Time and Embedded DEVS." In *Proceedings of Symposium on Theory of Modeling and Simulation (DEVS10)*, Orlando, FL, April 11–15, 2010, SCS.

30. Cho, S. M., and T. G. Kim. 1998. "Real-Time DEVS Simulation: Concurrent, Time-Selective Execution of Combined RT-DEVS Model and Interactive Environment." In *Proceedings of 1998 Summer Computer Simulation Conference*, Reno, NV, July 19–22, 1998, SCS.

31. Ledin, J., M. Dickens, and J. Sharp. 2003. "Single Modeling Environment for Constructing High-Fidelity Plant and Controller Models." In *Proceedings of 2003 AIAA Modeling and Simulation Technologies Conference*, Austin, TX, August 21–24, 2003.
32. MathWorks. *Real-Time Windows Target™*. Natick, MA: MathWorks. http://www.mathworks.com/products/rtwt/
33. Isen, F. W. 2009. *DSP for MATLAB and LabVIEW: Synthesis Lectures on Signal Processing*. San Rafael, CA: Morgan and Claypool.
34. Venne, P., J-N. Paquin, and J. Belanger. "The What, Where and Why of Real-Time Simulation." http://www.opal-rt.com/technical-document/what-where-and-why-real-time-simulation.
35. Applied Dynamics International. http://www.adi.com/products_sim_tar_rtx.htm.
36. Forsyth, P., and R. Kuffel. 2007. "Utility Applications of a RTDS Simulator." In *Proceedings of Power Engineering Conference (IPEC 2007)*, 112–17, December 3–6. Piscataway, NJ: IEEE Press.
37. Crosbie, R. E., J. J. Zenor, D. Word, and N. G. Hingorani. October–December 2004. "Fast Real-Time DSP Simulations for On-Line Testing of Hardware and Software." *Modeling & Simulation* 3 (4).
38. Word, D., J. J. Zenor, and R. Powelson. 2008. "Using FPGAs for Ultra-High-Speed Real-Time Simulation." In *Proceedings of Conference on Grand Challenges in Modeling and Simulation*, Edinburgh, Scotland, June 16–19, 2008.
39. Howe, R. M. 1985. "Transfer Function and Characteristic Root Errors for Fixed-Step Integration Algorithms." *Transactions of the Society for Computer Simulation*, 2 (4): 293–300.
40. Howe, R. M. 2010. "Improving Accuracy and Speed in Real-Time Simulation of Electric Circuits." *International Journal of Modeling, Simulation, and Scientific Computing* 1 (1): 47–83.
41. De Kelper, B., L. A. Dessaint, K. Al-Haddad, and H. Nakra. 2002. "A Comprehensive Approach to Fixed-Step Simulation of Switched Circuits." *IEEE Transactions on Power Electronics* 17 (2): 216–24.
42. Bednar, R., and R. E. Crosbie. 2007. "Stability of Multi-Rate Simulation Algorithms." In *Proceedings of the Summer Computer Simulation Multiconference*, San Diego, CA, July 16–19, 2007, SCS San Diego.
43. Ogata, K., 1987. *Discrete-Time Control Systems*, State Space Analysis. Englewood Cliffs, NJ: Prentice Hall.
44. Bednar, R., and R. E. Crosbie. 2009. "Solution of Non-Linear Differential Equations Using State-Transition Methods." In *Proceedings of Conference on Grand Challenges in Modeling and Simulation*, Istanbul, Turkey, July 13–16, 2009, SCS.
45. Palusinski, O. A. 1985. "Simulation of Dynamic Systems Using Multi-Rate Integration Techniques." *Transactions of SCS* 2 (4): 257–73.
46. Zenor, J. J., R. Bednar, and S. Bhalerao. 2008. "Multi-Party, Multi-Rate Simulation of an Unmanned Underwater Vehicle." In *Proceedings of Conference on Grand Challenges in Modeling and Simulation*, Edinburgh, Scotland, June 16–19, 2008.
47. Zenor J. J., D. J. Murray-Smith, E. W. McGookin, and R. E. Crosbie. 2009. "Development of a Multi-Rate Simulation Model of an Unmanned Underwater Vehicle for Real-Time Applications." In *Mathmod Conference*, Vienna, Austria, February 11–13, 2009.
48. Dougal, R. A. 2005. "Design Tools for Electric Ship Systems." In *Proceedings of IEEE Electric Ship Technologies Symposium*, Philadelphia, PA, July 25–27, 2005, pp. 8–11.

2 Formalized Approach for the Design of Real-Time Distributed Computer Systems

Ming Zhang, Bernard Zeigler, and Xiaolin Hu

CONTENTS

2.1 Introduction 35
2.2 Formal Approaches to the Design of Real-Time Distributed
Computer Systems 37
2.3 DEVS as a Model-Based Design Formalization Tool 40
    2.3.1 Model-Based System Design 40
    2.3.2 DEVS and RT-DEVS 40
    2.3.3 DEVS as a Formalized Aid to System Design 44
2.4 DEVS-Based Formal Design Approaches for Real-Time Distributed
Computing Systems 45
    2.4.1 Design Aid for Real-Time Distributed VE 45
    2.4.2 Design Aid and Verification for Real-Time Distributed Systems 49
    2.4.3 DEVS Approach to the Design of Distributed Real-Time
        Cooperative Robotic Systems 56
2.5 Summary 59
References 59

2.1 INTRODUCTION

Real-time distributed computer systems combine virtual environment (VE)–based systems, real-time peer-to-peer (P2P)–based systems, quality-of-service (QoS)–aware distributed computer systems, and hybrid systems (the systems that incorporate both real-time and non-real-time behaviors). Efficient system design for such systems requires a sound framework that can facilitate effective design modeling and simulation (M&S) as well as support for flexible testing of design alternatives. Traditional approaches to system design do not use formal system specification languages or formalisms and are based on accumulated domain knowledge. Thus, system design, in many cases, is separated from system implementation, system testing, and system validation. With the rapid advances of distributed real-time computer systems, the complexity of such systems is increasingly challenging traditional nonformalized design approaches. Therefore, formal design methods have been widely used for system design in recent years, and such efforts have been verified by many researchers as suitable for large-scale and complex systems. Among different formal approaches, it is worth noting that model-based system design has proved to be one of the most efficient methods to address the key concerns in complex system design [1–4]. Indeed, the model-based design approach uses a formal language to describe system design models, and such design models are then simulated to predict the performance of the system in real-world scenarios. In particular, the model-based formal design using Discrete Event System Specification (DEVS) [5–7] differentiates it from most other approaches because of its use of a unique integrative framework that can address most of the issues faced in complex system design, and in particular, distributed real-time computer systems.

The DEVS formalism has become one of the most important components of M&S theory. It provides a conceptual framework and an associated computational approach for solving methodological problems in M&S communities. This computational approach is based on the mathematical theory of systems including the hierarchy of system specifications and specification morphisms. It manipulates a framework of elements to determine their logical relationships. As a formal system specification language, DEVS formalism and its associated M&S environment offers a very useful toolset for helping with complex system design and verification. Therefore, DEVS can be applied for precisely modeling the system components and their interactions and is therefore ideal for the design and evaluation of real-time dynamic software or embedded systems. Moreover, the recent advance of distributed DEVS opens up new potential for more efficient system design for real-time distributed systems including QoS-aware systems.

System design is a broad topic that has been addressed by many researchers [8,9]. Most recently, the advance of M&S techniques has greatly advanced system design by providing a more effective and reliable tool for designing large-scale and complex real-time distributed computer systems. In this chapter, we analyze and demonstrate how DEVS as a formalized approach can aid system design in a very flexible and efficient manner. We hold that DEVS, real-time DEVS (RT-DEVS), and distributed DEVS will continue to be among the most powerful tools for aiding complex system design, and in particular, real-time distributed computer systems including VE- and P2P-based systems.

The main contributions of this chapter are to

1. Demonstrate how DEVS can be used as a fundamental system design tool for distributed real-time computer systems.
2. Demonstrate how DEVS is used for design validation for complex real-time computer systems, such as distributed VEs, distributed real-time P2P systems, and distributed simulation systems.
3. Demonstrate how DEVS can be used as a fundamental technology for building a more advanced framework to support design of distributed real-time systems, such as cooperative robotic systems.

The rest of this chapter is organized as follows: Section 2.2 introduces existing formal approaches for design of real-time distributed computer systems. Section 2.3 focuses on a DEVS-based formal design approach including a brief review of model-based system design, as well as DEVS foundational theories and how it works as a sound tool to aid system design. Section 2.4 discusses DEVS-based formal design approaches for real-time distributed computing systems with several case studies. Finally, Section 2.5 concludes this chapter.

2.2 FORMAL APPROACHES TO THE DESIGN OF REAL-TIME DISTRIBUTED COMPUTER SYSTEMS

With the rapid advances of real-time distributed computer systems, the need for more efficient system design is called on to reduce the design cost and design life cycle. Moreover, the traditional design approaches cannot meet the requirement of today's large-scale and complex real-time distributed computer systems because traditional nonformalized design approaches are error-prone, costly, and difficult to validate. Therefore, formalized system design approaches have been widely used in many areas for more efficient design of real-time distributed computer systems.

As a matter of fact, formalized system design has been proposed and clarified by some earlier researchers that can be traced back as far as the 1990s. For instance, Yau et al. [8] presented a research paper in 1990 that foresaw the best approaches for designing distributed computer systems. Their work classified the design approaches into three categories: dataflow oriented, communication oriented, and object oriented. A dataflow-oriented design approach basically uses structured analysis/structured design [9,10] as the fundamental design method, which is based on the functional decomposition of the system. Therefore, the concurrency related issues are not directly handled in the design phase, which results in inefficient design for distributed real-time-based computer systems.

To overcome this problem, communication-oriented design methods are developed. However, because of the focus on concurrency naturally the functional aspect of the system is a secondary concern, which results in limitation of the communication-oriented approach to the design of communication intensive applications with simple data structures and computations within the system. Based on this classification, STATEMATE [11], various types of Petri nets [12–14], SREM [15], and Raddle [16] are representatives of communication-oriented design approaches for distributed computer systems. Among these approaches, Petri net–based methods are well known in terms of behavior modeling and considered by many authors to be the best in this category. The key idea of a Petri net is to use "places," "transition," and a number of tokens to represent a system model. As such, the dynamic behavior of the system's concurrent processes can be specified in terms of communication and synchronization among processes.

It is worth mentioning that some Petri net–based approaches have been aware of the limitation of the pure communication-oriented classic Petri net method, and, therefore, some "object" based architectural features were added to the Petri net concept. For example, a Net-based and Object-based Architectural Model (NOAM) for architectural modeling and prototyping of real-time distributed computer systems has been proposed [17]. NOAM supports progressive system decomposition and refinement to construct a hierarchically structured system architectural model that is modular, scalable, and executable. However, to understand and build a NOAM-based system model is a nontrivial task.

In contrast, an object-orient design approach can overcome the shortcomings of both dataflow-oriented and communication-oriented approaches. Most of today's real-time distributed computer systems are built upon object-oriented concepts and programming languages; therefore, the object-oriented system design approach naturally fits the modern concept of system designs. Basically, object-oriented design can take full advantage of an object-oriented methodology to use a formal design specification language and tools in one unified domain. This approach can address both structural decomposition and communication in a distributed computer system. Indeed, many of today's formal design tools are based on an object-oriented philosophy. However, there are major concerns with respect to synchronization and parallelism support as well as effective modeling tools (that can take full advantage of object-oriented methodology).

Timed automata as a formalism also shows its capability for modeling distributed real-time systems [18]. It employs timed automata for modeling a communication protocol with a main focus on verification of desired/undesired states in time-critical distributed applications. Other work [19] studied a formal design approach for middleware for distributed real-time embedded (DRE) systems. Considering the growing complexity of the middleware-based architecture, this work relied on timed automata to meet the need for detailed modeling of low-level middleware mechanisms. Meanwhile, evaluation of such models through model checking tools can greatly reduce the error at an earlier design phase. This work proposed a model-driven approach for the design of middleware-based DRE systems in which timed automata play the key role in formal system specification and model verification. Indeed, the use of timed automata has been a well-known key ingredient for designing real-time distributed computer systems. Application of timed automata for designing real-time distributed computer systems has been increasingly extended to new areas, including design of QoS-aware distributed real-time systems. As an example, recent work [20] used timed automata in a model-based approach for modeling and verification of both system behavior and QoS requirements in a distributed wireless network environment. In this research, timed automata were used for behavior modeling and model verification.

Unified Modeling Language (UML), as an object-oriented formal method, has also been widely adopted for designing computer systems including real-time distributed systems. The feasibility of using real-time UML as a formal design tool for distributed real-time computer systems has been discussed in Selic [21]. UML holds promise in the real-time domain, not only because it incorporates most of the basic modeling abstractions that the domain relies on but also because the flexibility of its stereotype mechanism allows these general abstractions to be specialized to the desired degree of accuracy. Other work by Gomaa and Menascé [22] investigated the component interaction patterns of a large-scale distributed system using UML as a formal specification language for performance modeling and system design.

Both the synchronous and asynchronous communication mechanisms were studied by UML-based modeling. Moreover, a domain-specific formal semantic definition for a subset of the UML 2.0 component model was introduced, and an integrated sequence of design steps was then discussed [23]. Using UML for QoS-aware real-time distributed systems has also been investigated by many researchers. For instance, Bordbar et al. [24] discuss the design of a distributed system by using UML as the formal language for the specification. This work uses UML to model the computational viewpoint of the system with a focus on the description of QoS within that viewpoint. The resulting UML model can then serve as a template via which specific distributed system designs can be constructed.

Also known as an object-oriented formal system specification language, DEVS is playing an increasingly important role in aiding the design of complex real-time computer systems including real-time embedded systems, distributed real-time P2P systems, distributed VEs, and many more. The key difference between DEVS and aforementioned other object-oriented formal design approaches (such as those based on Petri nets and timed automata) lies in

- The system design models can be directly simulated in one integrative environment that includes model specifications, simulators, and "experimental frames" (under which models are evaluated using appropriate simulators).
- Because of their composability, the design model components are reusable through a model repository due to the separation of models from their simulators.
- The model and simulator structure are natively hierarchical, which makes it easy to map them onto today's complex computer systems.

To compare DEVS with aforementioned other well-known formal system design methods, we study how they are fundamentally different. For instance, in the approach using Petri nets, the real-time system design can be formalized and represented using a Petri net graph, through which the system behavior can be abstractly visualized and tested. The malfunctioning or undesired system behavior can thus be identified and analyzed. However, while complex hierarchical systems actually represent the majority of today's real-time distributed computer systems, it is not straightforward to model such systems with Petri nets. How an individual system component affects the overall system design outcome is difficult to analyze. Moreover, how to obtain an optimal design from multiple design alternatives is also beyond the basic capability of a Petri net–based approach. In comparison, these issues are addressed fundamentally in a DEVS-based approach. Furthermore, another significant advantage of a DEVS-based approach is that the simulation models can be integrated with a real-time distributed computer system to form a virtual simulation environment. This environment then can combine both the system's component models and real system components. This unique feature is not present in other formal methods, and it will be further demonstrated in Section 2.4.3. In the following sections, after the fundamental DEVS theories are introduced, we will demonstrate how DEVS can effectively aid system design in many different scenarios.

2.3 DEVS AS A MODEL-BASED DESIGN FORMALIZATION TOOL

As we have introduced, DEVS separates system models from their simulators, thus it natively supports a formal and model-based system design approach. In this section, we will discuss some background of DEVS as an effective model-based system design tool. To set the stage for the discussion of DEVS, we review model-based system design.

2.3.1 MODEL-BASED SYSTEM DESIGN

With the rapid advance of today's computer systems, a formalized system design process has been assuming an increasingly important role, in particular, for the design of distributed real-time computer systems. Traditionally, the design of such systems is independent from their implementation, test, and design validation. Thus, the overall design practice is time-consuming and error-prone. Model-based design is one of the modern approaches for complex system design. It is based on design models that attempt to capture the key system design parameters, and generally it requires the use of formal system specification languages, such as timed automata and DEVS. For instance, Schulz and Rozenblit [1] proposed a novel system codesign method using a formal specification language, such as DEVS, for aiding the design of embedded systems. Hu and Zeigler [2] proposed a model-based design framework for dynamic distributed real-time systems that introduces the concept of "model continuity," which it leverages extensively. Indeed, embedded systems and distributed real-time systems are not the only areas in which a model-based formal design approach have been applied. The design of today's VEs is also affected by the concept of a model-based approach, and the feasibility and effectiveness of using a model-based approach for designing VE systems have been investigated. For example, VR-WISE (a Virtual Reality modeling framework) has been employed as a tool for behavior specification and development [3]. It is actually a model-based design method, which makes the design of VEs more intuitive. One of the key advantages of VR-WISE is that it requires less virtual reality background and can be used by a broader audience [3]. Similarly, a simulation-based design methodology using composable models has been presented [4] that uses a tightly integrated design environment for the development of both the form and behavior of the system components.

Indeed, a model-based design approach opens a new direction toward solving complex distributed real-time computer systems. The application areas have been expanded to VEs, real-time P2P networks, and QoS-aware distributed systems, just to name a few. Compared to traditional approaches, model-based design involves not only building accurate design models but also simulating the design model in predefined experimental frames (EFs), through which design defects can be easily captured in the early design phases. Moreover, the optimal system design can be easily discovered by simulating models of design alternatives. Meanwhile, model validation becomes easy because of the use of a formal specification language.

2.3.2 DEVS AND RT-DEVS

DEVS [5] is a mathematical formalism originally developed for specifying discrete event systems. DEVS is a well-known theoretical approach to M&S and has attracted many researchers for reliable and efficient system M&S. The key idea of DEVS is to use "atomic" models to express individual component's behavior and to use "coupled" models to represent the interactions between the components in a system. The DEVS modeling framework fundamentally supports the reusability of individual models and also provides an efficient methodology for hierarchical model construction, which in turn can provide maximum flexibility for system M&S. As a pioneering formal M&S methodology, DEVS provides a concrete simulation theoretical foundation, which promotes fully object-oriented M&S techniques for solving today's complex M&S problems.

As such, the standard and basic DEVS formalism consists of two formalisms, one for atomic models (AMs) and one for coupled models. The atomic DEVS is expressed as follows:

$$AM = \langle X, S, Y, \delta_{int}, \delta_{ext}, \lambda, ta \rangle$$

X: set of external input events,
S: set of sequential states,
Y: set of outputs,
δ_{int}: $S \to S$: internal transition function,
δ_{ext}: $Q*X^b \to S$: external transition function,

where

$Q = \{(s, e) \mid s \in S, 0 \le e \le ta(s)\}$
λ: $S \to Y^b$: output function,
ta: time advance function.

As a matter of fact, the AM is a building block for a more complex coupled model, which defines a new model constructed by connecting basic model components. Two major activities involved in defining a coupled model are specifying its component models and defining the couplings that create the desired communication networks. Therefore, a DEVS-coupled model is defined as follows:

$$DN = \langle X, Y, \{M_i\}, \{I_i\}, \{Z_{ij}\} \rangle$$

where

X: set of external input events;
Y: a set of outputs;
D: a set of components names;
    for each i in D,
    M_i is a component model
    I_i is the set of influences for i
        for each j in I_i,
        $Z_{i,j}$ is the i-to-j output translation function.

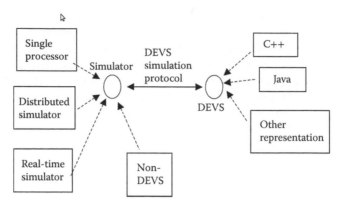

FIGURE 2.1 Discrete Event System Specification (DEVS) modeling and simulation frame-work. (From Zhang, M., *Toward a Flexible and Reconfigurable Distributed Simulation: A New Approach to Distributed DEVS*, PhD dissertation, Electrical and Computer Engineering Department, University of Arizona, Spring 2007. With permission.)

The DEVS M&S framework is very different from traditional module and func-tion-based ones. It provides a very flexible and scalable M&S foundation by separat-ing models and simulators. Figure 2.1 shows how DEVS model components interact with DEVS and non-DEVS simulators through the DEVS simulation protocol. We can also see that DEVS models interact with each other through DEVS simulators. The separation of models from simulators is a key aspect in DEVS, which is criti-cal for scalable simulation and middleware-supported distributed simulation such as those using CORBA (Common Object Request Broker Architecture), HLA (high-level architecture), and MPI (Message Passing Interface).

The advantages of such a framework are obvious because model development is in fact not affected by underlying computational resources for executing the model. Therefore, models maintain their reusability and can be stored or retrieved from a model repository. The same model system can be executed in different ways using different DEVS simulation protocols. In such a setting, commonly used middleware technologies for parallel and distributed computing could be easily applied on sepa-rately developed DEVS models. Therefore, within the DEVS framework, model com-ponents can be easily migrated from single processor to multiprocessor and vice versa.

If we have a closer look at the DEVS-based modeling framework, we find that it is based on a hierarchical model construction technique as shown in Figure 2.2. For example, a coupled model is obtained by adding a coupling specification to a set of AMs. This coupled model can then be used as a component in a larger system with new components. A hierarchical coupled model can be constructed level by level by adding a set of model components (either atomic or coupled) as well as coupling information among these components. Consequently, a reusable model repository for developers is created. The DEVS-based modeling framework also supports a model component as a "black box," where the internals of the model are hidden and only the behavior of it is exposed through its input/output ports.

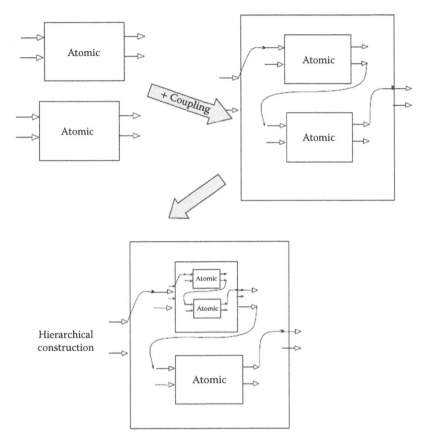

FIGURE 2.2 Coupled modules formed via coupling and their use as components. (From Zhang, M., *Toward a Flexible and Reconfigurable Distributed Simulation: A New Approach to Distributed DEVS*, PhD dissertation, Electrical and Computer Engineering Department, University of Arizona, Spring 2007. With permission.)

One interesting aspect of DEVS formalism is that a coupled DEVS model can be expressed as an equivalent basic model (or AM) and so DEVS models adhere to closure under coupling. Such an equivalent basic model derived from a coupled model can then be employed in a larger coupled model. Therefore, DEVS formalism provides a robust composition framework that supports closure under coupling and hierarchical construction.

RT-DEVS extends the aforementioned basic DEVS for real-time M&S applications. It differentiates from basic DEVS in that an activity set A and its associated mapping function ψ are added to the existing atomic DEVS to provide the real-time interaction capabilities for DEVS models in their environments. Thus, different model components in a real-time system can be encapsulated uniformly by this new formalism. Indeed, RT-DEVS formalism aims to solve the discrete event-based system problems because of the real-time requirement and is able to specify real-time distributed systems as DEVS models. The RT-DEVS formalism is defined by Hong and Kim [26], and an atomic RT-DEVS model can be expressed as follows:

$$\text{Real-time atomic model} = \langle X, S, Y, \delta_{int}, \delta_{ext}, \lambda, \text{ta}, A, \psi \rangle$$

X: set of external input events,
S: set of sequential states,
Y: set of outputs,
$\delta_{int} : S \rightarrow S$: internal transition function,
$\delta_{ext} : Q*X^b \rightarrow S$: external transition function

where

$Q = \{(s, e) \mid s \in S, 0 \leq e \leq \text{ta}(s)\}$
X^b is a set of bags over elements in X,
$\lambda : S \rightarrow Y^b$: output function,
Y^b is a set of bags over elements in Y,
ta: time advance function (advances in real time),
A: set of activities with the constraints,
$\Psi : S \rightarrow A$: an activity mapping function.

As a real-time extension of standard DEVS, RT-DEVS establishes the basis for using DEVS for system design when real-time concerns are key to a successful design. As a matter of fact, RT-DEVS has been adopted by many researchers as a fundamental method to investigate the hard-to-predict behavior of complex real-time distributed systems. Compared to other real-time formal specification languages, RT-DEVS extends all the advantages of standard DEVS, which makes the system design practices more efficient and reliable.

2.3.3   DEVS AS A FORMALIZED AID TO SYSTEM DESIGN

As discussed earlier, in recent years, many formal languages have been used to aid in system design, including UML [27], timed automata [28], Petri nets, [29] and statecharts [30]. However, to a large extent these formal languages have not been very suitable for the design of large-scale and complex systems because of the lack of effective model-driven engines (simulators), reusable model repositories, flexible model composition, etc.

DEVS [5] distinguished itself from these formalized system design approaches in that it can provide a more efficient system design modeling framework in which design alternatives can be easily constructed and simulated, as shown in Figure 2.3. Indeed, DEVS model-based design has been used by many researchers for solving complex system design problems. The key differences between DEVS and other formal approaches is that DEVS supports reusable hierarchical model development, and the DEVS design model can be simulated in one unified framework to

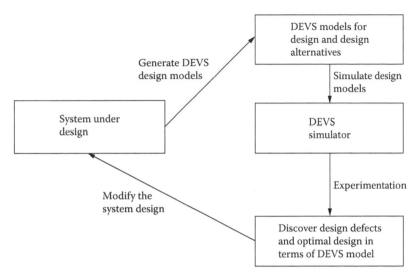

FIGURE 2.3 Discrete Event System Specification (DEVS) model-based approach for system design.

quickly discover system design problems. In other words, DEVS can make it much easier to identify design problems, validate system design, and discover optimal system design.

2.4 DEVS-BASED FORMAL DESIGN APPROACHES FOR REAL-TIME DISTRIBUTED COMPUTING SYSTEMS

In this section, we will present how DEVS and RT-DEVS are used for aiding the design of real-time distributed computer systems. Several different types of distributed systems are involved, and we will see that the DEVS approach is not affected by particular system types. As such, it has great potential to be applied to any of today's distributed real-time systems when other formal approaches are limited in their functionality.

2.4.1 DESIGN AID FOR REAL-TIME DISTRIBUTED VE

Employing a VE is one of the leading computer techniques that has many application areas. Compared to other computer techniques, it can essentially provide a very attractive and user-friendly human–computer interaction platform by which the end users can play with the computer systems in a "like-real" fashion. Indeed, VE has been widely used in training, gaming, E-learning, and many other related areas [31–33]. It is worth mentioning that modern VE design is a complex process considering the many different components and their interactions in the system. Moreover, the design of modern VE systems involves many different novel computer techniques including object-oriented design, object-oriented programming languages, scripting languages, HTML- and XML-based web technology, and three-dimensional (3D) graphic rendering, just to name a few. In particular, the techniques for real-time 3D graphic rendering and distributing VE components in real time are playing important roles in today's VE applications. Therefore, how to effectively apply the 3D technique into a distributed real-time computing environment brings a lot of challenges that are not faced in traditional VE approaches.

Indeed, designing an effective real-time distributed VE is a complex practice that involves using many different system design approaches. Compared to common software components, the interconnected and interacting virtual objects in a VE system must be precisely defined in terms of both visual effects and the behavior specifications. The real-time performance in a distributed computing environment, in fact, increases the complexity for specifying and implementing the visual object–based virtual world. This holds particularly true for traditional design techniques for VE, as they do not separate the design of visual objects and their behaviors, and moreover, they make little use of formalized methods as integrative design approaches. The problems brought about by nonformalized VE design are significant and include among many others: nonreusability of components, difficulty in behavior validation, and time-consuming implementation. Therefore, nonformalized VE design is not suitable for designing a large-scale and/or a complex VE system, in particular when it is distributed and will operate in real time. Instead, an integrative design platform to address the usability, performance, load balancing, etc. is required. Meanwhile, requirement specifications can also be an issue in the nonformalized VE design. Note that although substantial research is available on using formalized design practices for VE design [34–36], this work mainly focuses on the behavior design of the visual components/objects.

Traditionally, VE system design separates visual objects design from their behavior implementation. Visual objects design is directly manipulated by some visual modeling tools such as 3ds Max [37], AC3D [38], etc. However, the design of the behavior and implementation for a visual object is distinctly more difficult than the design of the visual object appearance because of the complexity of the behaviors of visual objects and their interactions. Furthermore, the real-time performance of the visual objects in a distributed environment is relatively difficult to predict in the design phase, which increases the necessity of a formalized design approach. Indeed, many formalized designs have been investigated by researchers to achieve a more efficient VE system design. These formalized design approaches use system specification languages to ease the design. For example, state-transition diagrams [34], FlowNet [35], HyNet [36], Tufts [39], Petri nets [29], and statecharts [30] have all been used for this purpose.

The drawbacks of these formal methods for the design of a complex VE system were introduced in Section 2.3.3 and have been gradually recognized in recent years. For example, FlowNet uses a graphical notation to specify the behavior of components, and when the specified VE system becomes large and complex, the resulting "FlowNet" is very difficult to understand and manage. Moreover, there exists no mapping of the specification created by these formalisms to the implemented visual objects, which makes design validation and optimization difficult to achieve. As we have discussed earlier, the DEVS-based M&S framework is a model-based system approach that is able to provide an integrated platform for aiding VE design. In the following paragraph, we discuss a practical example of using DEVS as an efficient tool for aiding the design of a web-based 3D VE system. We will see how a DEVS-based system design model can be easily realized in a simulation software environment, and how design validation and design optimization can be achieved through simulating a DEVS design model.

In recent research [32,33], a DEVS model-based approach for aiding the design of a 3D VE system, a virtual hospital, was discussed. This "virtual hospital" is a web-based distributed E-learning VE involving the design and implementation of many visual objects (such as "patient," "x-ray scanner," "touch sensor," etc.) and system components (such as web components, database, user interface, etc.). The visual objects in the system are specified and designed using an X3D tool, while the interactions between these visual objects are defined and implemented using a scripting language. The behavior validation of the visual objects and the visual objects optimization are two key aspects in the design of this real-time distributed VE system. The authors used DEVS as the basic formalized tool to model the key visual objects and their interactions to validate their behavior specification and to find an optimal design. In fact, the behavior validation was conducted by running the DEVS model under a specified EF. This use of DEVS allows for the validation visual components' behaviors quickly at an earlier stage of the system design. Thus, for example, some redundant or inefficient visual components can be removed in the final system implementation without losing any necessary system's functionalities.

As shown in Figure 2.4, the key visual objects are realized in a DEVS integrated software framework, which is able to show models and their interaction using DEVS formalism. Here, the behavior of individual visual objects is defined using DEVS

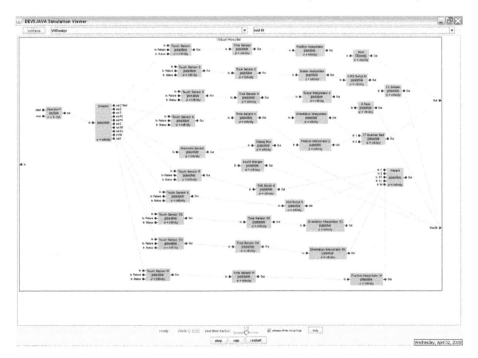

FIGURE 2.4 DEVSJAVA representation of a virtual hospital. (From Boukerche, A. et al., *Concurrency & Computation: Practice & Experience*, 21, 1422, 2009. With permission.)

AM formalism, while the interaction of these individual visual object models are described by DEVS-coupled model specifications, or more specifically, the interactions are modeled by constructing messaging channels through individual input and output ports of the model.

As stated previously, each DEVS model component has strictly defined "states," input and output "event" ports, as well as how the model responds to internal and external events. As an example, the "Touch Sensor" visual object is described as an atomic DEVS as follows [24]:

```
X={"In"}
Y={"Out"}
S={"passive", "Touched"}
δ_int{"passive", "Touched"} = {"passive"};
δ_ext{"In",{"passive", 0 < e < infinite}} = {"Touched"};
λ{"Touched"} = {"Out"};
ta("passive") = infinite;
ta("Touched") = 0;
```

All other individual AMs are described similarly as the above "touch sensor" model. In terms of the behavior validation of each individual visual object, the representing DEVS model is tested under DEVSJAVA with different injected initial states. For instance, the "patient" visual object is modeled as an atomic DEVS "patient" model, and simulating this DEVS model can validate whether the behavior of the "patient" component defined in X3D is correct or not. Through testing and validating all these individual visual object DEVS models, any design error (in terms of visual objects' behavior) can be discovered and eradicated easily. As a next step, the design inefficiency can also be identified by simulating a modeled subsystem of the virtual hospital. As an example, the authors examined the image animation of the "CT Scanner PC" component by simulating a corresponding DEVS design model (a coupled DEVS model in DEVS representation). As shown in Figures 2.5 and 2.6, a sequential operation steps of system components and their driving scripts

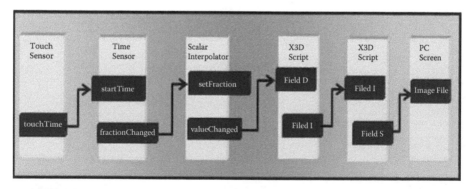

FIGURE 2.5 Image animation of CT scanner PC in the original X3D design. (From Boukerche, A. et al., *Concurrency & Computation: Practice & Experience*, 21, 1422, 2009. With permission.)

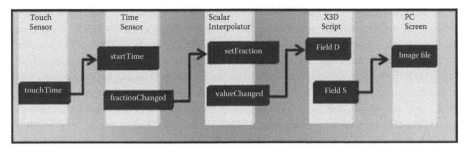

| Touch Sensor | Time Sensor | Scalar Interpolator | X3D Script | PC Screen |

FIGURE 2.6 Image animation of CT scanner PC in the improved X3D design. (From Boukerche, A. et al., *Concurrency & Computation: Practice & Experience*, 21, 1422, 2009. With permission.)

(X3D Script) were presented, in which "X3D Scripts" take the data from "Scalar Interpolator" to generate the image file used in "PC Screen" (a visual component in the VE system being designed). Through mapping such operations to the DEVS model and then simulating them, it is discovered that the two "X3D Script" can be replaced by a new one as a better design. This is achieved by simulating the representing DEVS models to find a more efficient design alternative. Here, we can see that DEVS works smoothly as a guiding tool for design optimization. It is worth mentioning that DEVS models are easy to build, and simulation of DEVS models can help predict the performance of the designed VE system. Therefore, DEVS can not only be used as a design validation tool but also as a guiding tool for design optimization.

2.4.2 DESIGN AID AND VERIFICATION FOR REAL-TIME DISTRIBUTED SYSTEMS

Indeed, VE-based systems can benefit from DEVS in terms of design validation and optimization. Real-time distributed computer systems, including real-time distributed simulation systems and QoS-aware distributed real-time computing systems, can also take advantage of using DEVS as the fundamental system design aiding tool. Indeed, the capability of precisely defining system components and their interaction makes DEVS a unique approach for effectively designing real-time distributed computer systems.

Real-time-based formal system specification languages have been widely studied, including UML-RT [27], timed automata [28], and others. For instance, UML-RT is quite suitable for the high-level formal description of a system (software, hardware, and the embedded system level), whereas timed automata can be used to precisely model the components of a system. As an extension of standard DEVS, RT-DEVS opens a wide area of specifying real-time computer systems, including distributed real-time systems and QoS-aware distributed systems. Compared to UML-RT and timed automata, RT-DEVS inherits the basic DEVS methodology and can also provide specification of real-time parameters, which can greatly help the design of a real-time distributed system in a unified platform (for example, the DEVSJAVA software platform).

As a well-known distributed simulation standard, HLA [40] has been widely adopted in the simulation community. Real-time interface (RTI) is the core in a HLA-based distributed simulation system. Therefore, the study of HLA-RTI has been attracting many researchers for a better RTI design. As a matter of fact, a real-time RTI (RT-RTI) system generally requires the participating simulation components to be deterministic, which exactly suits the DEVS formalism-based approach because DEVS and RT-DEVS are ideal for expressing the deterministic model behaviors to satisfy the requirements of designing a RT-RTI system. There exists very limited research that uses DEVS to aid a novel HLA-RTI design. In the following example, we will show how RT-DEVS is used to predict the key system design factors that can affect the overall performance of a RT-RTI system.

Other work [41] proposed a novel HLA-RTI design and used DEVS to validate their proposed real-time HLA/RTI [42] system design. RT-DEVS was used to specify the key RTI components, and the key performance parameters were captured by simulating the corresponding DEVS RTI models. As shown in Figure 2.7, the RTI design was realized as a DEVSJAVA simulation model, which represented the proposed new RTI design. The individual AMs were specified by RT-DEVS formalism. The overall real-time HLA/RTI system was modeled basically in three layers: task generation and task director (the "taskGen" and "director" models), HLA/RTI services (the six key HLA/RTI services are modeled as "DDM," "DM," "FM," "OM," "ObjM," and "TM," respectively), and the thread pools (five regular pools and one

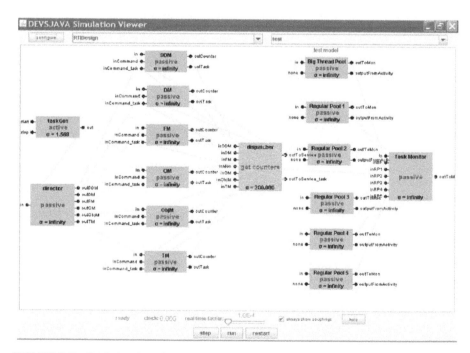

FIGURE 2.7   Real-time interface (RTI) design models in DEVSJAVA. (From Boukerche, A. et al., *Simulation*, 84, 231, 2008. With permission.)

bigger pool). Thus, the dynamic behaviors of a designed HLA/RTI system can then be simulated by DEVSJAVA simulators.

For instance, the "service queue" AM is described with RT-DEVS as [41]

```
X={"input tasks", "output counter", "output task"}
Y={"counter", "task"}
S={"passive", "outCounter", "outTask"}
δint{"passive", "outCounter", "outTask"}={"passive"}
δext{"output counter",{"passive",0<e<infinite}}={"outCounter"};
δext{"output task",{"passive", 0<e<infinite}}={"outTask"};
λ{"outCounter"}={"counter"}; λ{"outTask"}={"task"};
ta("outCounter")=0 ; ta("outTask")=0 ; ta("passive")=infinite
A={}
Ψ={}
```

and "Thread Pool" is described with RT-DEVS [41] as follows:

```
X={"input tasks"}
Y={"message"}
S={"passive", "busy"}
δint{"passive", "busy"}={"passive"}
δext{"input tasks",{"passive", 0<e<infinite}}={"busy"};
λ{"busy"}={"message"};
ta("busy")=constant value ; ta("passive")=infinite
A={"create and run task thread"}
Ψ{"busy"}={"create and run task thread"}.
```

As shown in the above RT-DEVS formalism, each model component (AM) has strictly defined "states," input and output "events," as well as how the model responds to internal and external events (transition functions). Different from standard DEVS, the real-time activity of the model is also defined using an activity set and an activity mapping function. Therefore, the overall system behaviors can be modeled and simulated in an integrated framework. Given that the key aspect of RT-RTI is its real-time performance (or, say, "task satisfactory rate"), the performance of a proposed novel RT-RTI core system must be evaluated and validated. In this research, the authors effectively used RT-DEVS as the design validation tool.

Figure 2.8 shows simulation on the DEVSJAVA platform of both the RT-DEVS models of the original RT-RTI and proposed novel RT-RTI designs. Indeed, the DEVSJAVA models in Figure 2.7 captured the key RT-RTI components and their real-time interactions so that the key performance parameters can be obtained. Consequently, the simulation results can be compared to the experimental results obtained from the executions in a physical RT-RTI environment. In particular, the key concern here is how dynamic thread pool management in the new RT-RTI design can benefit the performance of the overall RT-RTI system in terms of the tasks that are served to meet their deadlines. The experimental results, obtained from both the real RT-RTI execution and the DEVS model simulation, validated that the proposed

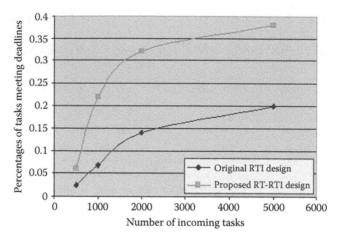

FIGURE 2.8 Discrete Event System Specification (DEVS) simulation result for real-time interface (RTI) design. (From Boukerche, A. et al., *Simulation*, 84, 231, 2008. With permission.)

novel RT-RTI design outperformed the original RTI system in terms of providing a better task serving rate in a real-time distributed simulation scenario. The DEVS simulation result, shown in Figure 2.8, provides a solid and consistent result that supports the proposed new RT-RTI design. In this example, we saw that the DEVS component-based design approach is able to discover, predict, and validate the key design concerns for a real-time distributed simulation system.

In other work [43], RT-DEVS has been used to validate a design of a QoS-aware distributed real-time system. The design of such systems is a new area to explore with DEVS and the research created and implemented DEVS design models for both a tree-based (hierarchical) and a flat service management systems. These DEVS design models were then simulated in a real-time fashion, and the DEVS simulation results conform very well with the experimental results obtained in an actual P2P-based distributed cluster environment. Meanwhile, the "states" and "state transitions" of each component in the designed system were also validated by using a DEVS-based unit test method. It is worth mentioning that a QoS-aware system has many time-sensitive real-time parameters that must be considered during the design process, but how these parameters affect the overall real-time performance of the final designed system is generally very difficult to evaluate or predict. DEVS and RT-DEVS indeed provide a powerful system tool to facilitate the evaluation of such complex design processes in a quick and accurate way.

Another example demonstrates how RT-DEVS was used as a formalization technique toward aiding the algorithm design and validation of a P2P-based distributed system. In particular, a novel load-balancing algorithm in a P2P-based network was designed in which both dynamic QoS and service migration play key roles in terms of affecting the overall system performance [44]. The proposed load-balancing algorithm was based on a genetic algorithm [45], and the design of this algorithm was validated by a DEVS-based formal approach. The algorithm was modeled in a DEVSJAVA environment, which represents the key components and the behaviors of the algorithm in detail. Thus, the proposed algorithm can be simulated in terms of how it satisfies the user requirement in a dynamic distributed service-oriented system. The concept of DEVS "variable structure" was used to fulfill the requirement of modeling the variations of components in a dynamic system, which is actually one of the most important features for analyzing a service migration scheme in the proposed algorithm design.

The DEVSJAVA integrated environment provides the capability of visualizing the hierarchical system structure, which helps build the design models quickly. In terms of experimental evaluations, the proposed load-balancing algorithm (based on a genetic algorithm) was compared with two other systems: one only using task scheduling and one using no load-balancing algorithm. The proposed algorithm used a load-balancing manager that has dynamic QoS parameters implemented, and these parameters can be updated using a genetic algorithm. The three experimental distributed systems were also modeled in DEVSJAVA as shown in Figure 2.9, in which the computing services on hosted machines were modeled as "service nodes" and the multiple threads on one service were modeled as "node-i-thread-i" models. There is also a "task generation" model to generate tasks to the modeled system as well as a "MigrationManager" for task migrations. Two sets of experimental scenarios were created. In the first scenario, tasks were fed to the system using a fixed interval of 100 ms, whereas in the second scenario, the tasks were sent to the system using a different feeding rate so that the adaptability of the system could be evaluated. For

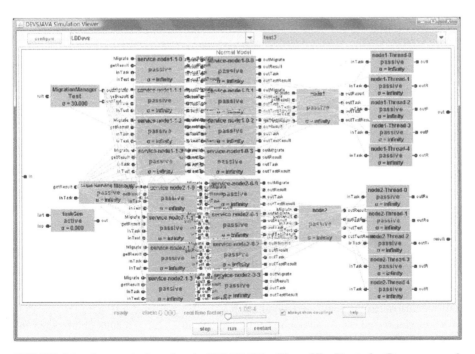

FIGURE 2.9 System design view in DEVSJAVA. (From Xie, H. et al., *Concurrency & Computation: Practice & Experience*. Vol. 22, P1223–1239, 2010. With permission.)

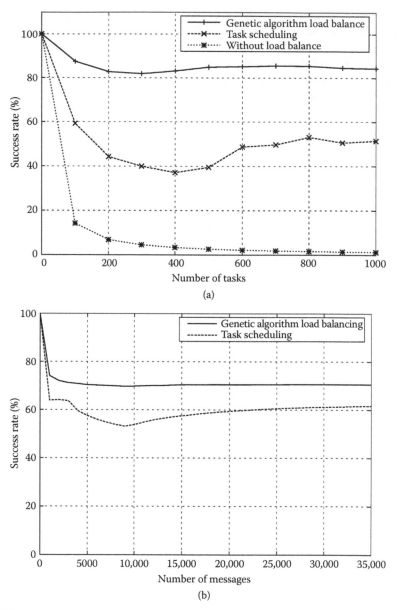

FIGURE 2.10   Simulation result obtained for the first scenario (a) in a real-world experiment; (b) in a Discrete Event System Specification (DEVS) simulation. (From Xie, H. et al., *Concurrency & Computation: Practice & Experience*. Vol. 22, P1223–1239, 2010. With permission.)

both scenarios, experiments were conducted in both a practical P2P network and the DEVS model-based EF. As shown in Figures 2.10 and 2.11, the results obtained from DEVS models conforms well with the results captured in the practical P2P network. We can see that the DEVS simulations predicted higher task success rates when compared to the experiments in practice (conducted in real P2P-enabled computer

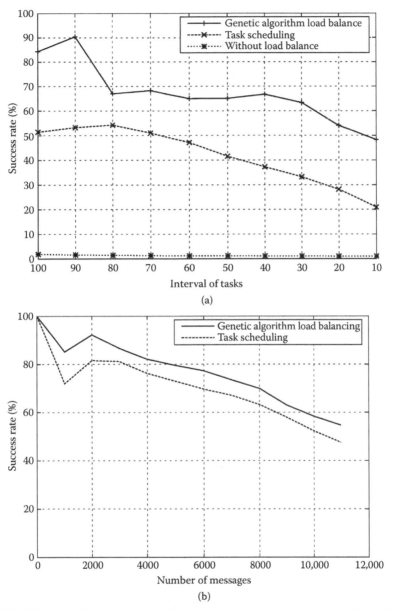

FIGURE 2.11 Simulation result obtained for the second scenario (a) in a real-world experiment; (b) in a Discrete Event System Specification (DEVS) simulation. (From Xie, H. et al., *Concurrency & Computation: Practice & Experience.* Vol. 22, P1223–1239, 2010. With permission.)

networks) in both scenarios. This is because of the different experimental settings, and the key concern here is to validate the differences in the algorithms, not to exactly predict the performance of the algorithms in practice. Indeed, the experimental results imply that DEVS can provide accurate prediction of the performance of complex load-balancing algorithms for dynamic P2P-based distributed real-time systems. In the case where practical experiments are difficult to capture, DEVS and RT-DEVS can still provide effective prediction of the system performance using easy-to-build simulation-based models.

2.4.3 DEVS Approach to the Design of Distributed Real-Time Cooperative Robotic Systems

In the previous discussion, we demonstrated some examples that use DEVS as a formal design validation tool to assist the design of complex real-time distributed systems. In this section, we will demonstrate how the DEVS formal approach can provide an integrative M&S environment to aid the design of distributed real-time cooperative robotic systems. In this subsection, we show how a physical robot can interact with a DEVS-based robot model in a simulation-based VE [46,47].

The DEVS approach presented in this section is based on an architecture proposed and implemented by one of the authors (Hu), which employs the following four steps for the design and testing of a cooperative robotic system [46]:

1. Centralized simulation to analyze the system under test within a model of the environment linked by abstract sensor/actuator interfaces on a single computer.
2. Distributed simulation, in which models are deployed to the real network that the system will be executed on and simulated by distributed real-time simulators.
3. Hardware-in-the-loop simulation, in which the environment model is simulated by a DEVS real-time simulator on one computer, whereas the control model under test is executed by a DEVS real-time execution engine on the real hardware.
4. Deployed execution, in which a real-time execution engine executes the control model that interacts with the physical environment through the earlier sensor/actuator interfaces that have been appropriately instantiated.

As the technology of robotic systems advances rapidly, systematic development methods and integrative development environments play increasingly important roles in handling the complexity of these systems. Compared to the conventional approach, the system architecture of Hu and Zeigler [46] supports robot-in-the-loop type of simulation, which can effectively overcome the difficulties of traditional approaches when the experimental environment confines the use of physical robots. One of the key ideas of this approach is to constitute an incremental study and measurement process for cooperative robotic systems combined with a simulation-based VE to support systematic analysis and measurement of large-scale cooperative robotic systems. Another highlight in this approach lies in the support of dynamic system reconfiguration in a real-time simulation of cooperative robotic systems. To enable this feature, "variable structure" DEVS [48] is used. This capability indeed makes dealing with the evolving cooperative robotic systems much easier. The concept of "experimental frame" is explored to form an integrative environment for studying the behaviors of large-scale cooperative robotic systems. In other words, the system architecture can be treated with a hybrid approach that allows real robots as well as virtual ones to be studied together in a simulation-based VE. This simulation-based VE was originally developed in the context of mobile robot applications; however, it can also be applied to other robotic applications such as object transportation and material handling. In the sequel, we demonstrate how physical robots can interact with the VE to form a cooperative real-time robotic system.

In general, robotic systems can be viewed as a particular form of real-time systems that monitor and respond to, or control, an external environment. This environment is connected to the computer system through sensors, actuators, and other input–output interfaces [6]. A robotic system from this point of view consists of sensors, actuators, and the decision-making unit. Thus, a cooperative robotic system is composed of a collection of robots that communicate with one another and interact with an environment. This view of robotic systems suggests a basic architecture for the simulation-based VE that we developed: an environment model and a collection of robot models, which include decision-making models, sensors, and actuators. The environment model represents the physical environment within which a robotic system will be operated. It forms a VE for the robots and may include virtual obstacles, virtual robots, or any other entities that are useful for simulation-based study. The robot model represents the control software that governs the decision making and communication of a robot. It also includes sensor and actuator interfaces to support interactions between the robot and its environment. In the approach by Hu, the "model continuity" methodology is used, which clearly separates the decision-making unit of a robot (which is modeled as a DEVS atomic or coupled model) from the sensors and actuators (which are modeled as DEVS `abstractActivities`). The decision-making model defines the control logic while the sensor/actuator `abstractActivities` represent the sensors or actuators, including their behaviors, interfaces, and properties of uncertainty and inaccuracy. DEVS-based model couplings are added between sensor/actuator `abstractActivities` and the environment model; therefore, messages can be passed between the decision-making model and the environment model through sensor/actuator `abstractActivities`.

It is worth noting that such separation between the decision-making model of a robot and its sensor/actuator interfaces can bring several advantages: First, it separates the decision making of a robot from hardware interaction. This makes it easier for the designer to focus on the decision-making model, which is the main design interest. Second, the existence of a sensor/actuator interface layer makes it possible for the decision-making model to interact with different types of sensors/actuators as long as the interface functions between them remain the same. Thus depending on different experimental and study objectives, a decision-making model can be equipped with different types of sensors and actuators. As a matter of fact, in this cooperative robotic system, models of sensors/actuators (modeled as `abstractActivities`, also referred to as virtual sensors/actuators hereafter) are developed to simulate the behavior of physical sensors/actuators. Meanwhile, physical sensor/actuator interfaces (implemented as `RTActivities`) are developed to drive the physical sensor/ actuators of a robot. A virtual sensor/actuator and its corresponding physical sensor/actuator interface share the same interface functions with the decision-making model. During simulation, a decision-making model uses virtual sensors/actuators to interact with a VE; during operation, the same decision-making model, which

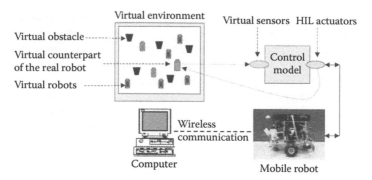

FIGURE 2.12   Robot-in-the-loop simulation. (From Hu, X., and Zeigler B. P., *Integrated Computer-Aided Engineering*, 12, 353, 2005. With permission.)

resides on a real robot, uses physical sensor/actuator interfaces to interact with a physical environment.

Figure 2.12 shows an example of the aforementioned cooperative robotic system, in which one physical mobile robot operates in a VE. In this example, the mobile robot uses its virtual sensors to get sensory input from the VE and uses its real motor interface to move the robot. As a result, the physical robot moves in a physical space based on the sensory input from a VE. Within this VE, the robot "sees" virtual obstacles that are simulated by computers and makes decisions based on those inputs. Meanwhile, virtual robots (robot models) can also be added into the environment so the physical robot can sense them and communicate/coordinate with them. This capability of robot-in-the-loop simulation brings simulation-based study one step closer to the physical world. Furthermore, for large-scale cooperative robotic systems that include hundreds of robots, it makes it possible to conduct system-wide tests and measurements without waiting for all physical robots to be available. In this latter case, the robots not yet physically available can be provided by the simulation-based VE.

Figure 2.13 illustrates how EFs, models/systems, and simulation methods can play together to carry out simulation-based measurement. This process includes three lines of development and integration: the "models/system" that will be tested and measured, the "EFs" that specify the measurement, and the "methods" that are employed to carry out the measurement. The process starts from the system specification that is formulated as DEVS-expressible formalisms. The system specification is further divided into two specifications: the design specification that guides the development of models/system and the measurement specification that guides the development of EFs. Three methods, corresponding to three stages of study, are used to carry out the measurement incrementally. These methods are conventional simulation, VE-based simulation, and real-time execution. Similarly, three types of EFs exist: EF for model-world study, EF for virtual-world study, and EF for physical-world study. Techniques are under development to derive EF development from the measurement specification in automated or semiautomated ways.

As a summary, the simulation-based VE supports a powerful hybrid approach that allows physical robots as well as virtual ones to be studied together. As the

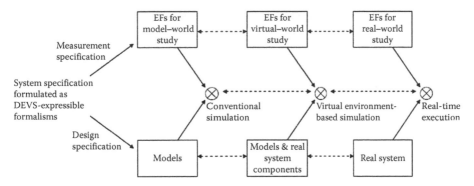

FIGURE 2.13 Experimental frames (EFs), models, and study methods. (From Hu, X., and Zeigler B. P., *Integrated Computer-Aided Engineering*, 12, 353, 2005. With permission.)

technology of robotic systems advances rapidly, systematic development methods and integrative development environments play an increasingly important role in handling the complexity of these systems. We can clearly see that DEVS is key to making it possible to build such a hybrid simulation framework, which can then aid the design of cooperation robotic systems more easily and effectively.

2.5   SUMMARY

In this chapter, we demonstrated that DEVS and RT-DEVS can serve as an efficient system design framework for many of today's complex real-time distributed systems. Compared to other formalized approaches, DEVS shows many advantages with its unique features, which include fast hierarchical model construction/reconstruction, model reusability, and ease of model validation. Furthermore, the flexible M&S tools and environments that DEVS enables can help in discovering optimal system design quickly as well as in validating existing designs.

Furthermore, we also expect that distributed DEVS can aid the design of distributed real-time systems in a more accurate and effective method. This is because distributed DEVS models can better represent physical systems. Thus, simulating distributed DEVS design models can provide a more accurate picture to identify the design defects and capture the key design parameters that must be effectively manipulated.

REFERENCES

1. Schulz, S., J. W. Rozenblit, M. Mrva, and K. Buchenrieder. 1998. "Model-Based Codesign." *IEEE Computer* 31 (8): 60–7.
2. Hu, X., and B. P. Zeigler. 2005. "Model Continuity in the Design of Dynamic Distributed Real-Time Systems." *IEEE Transactions on Systems, Man and Cybernetics—Part A: Systems and Humans* 35 (6): 867–78.
3. Pellens, B., F. Kleinermann, O. D. Troyer, and W. Bille. 2006. "Model-Based Design of Virtual Environment Behavior." In *Proceedings of the 12th International Conference on Virtual Systems and Multimedia 2006*, October 18–20, Xi'an, China, LNCS 4270, pp. 29–39.
4. Paredis, C. J. J., A. Diaz-Calderon, R. Sinha, and P. K. Khosla. 2001. "Composable Models for Simulation-Based Design." *Engineering with Computers* 17: 112–28.

5. Zeigler, B. P., T. G. Kim, and H. Praehofer. 2000. *Theory of Modeling and Simulation.* 2nd ed. New York: Academic Press.
6. Shaw, S. C. 2001. *Real-Time Systems and Software.* New York: John Wiley & Sons.
7. Wainer, G. A., and P. J. Mosterman. 2010. *Discrete-Event Modeling and Simulation: Theory and Applications.* Boca Raton, FL: CRC Press.
8. Yau, Stephen S., Xiaoping, Jia, and Doo-Hwan, Bae. 1990. "Trends in Software Design for Distributed Computing Systems." In *Proceedings of Second IEEE Workshop on Future Trends of Distributed Computing Systems*, Cairo, Egypt, September 30–October 2, 1990, pp. 154–60.
9. Sommerville, I. 1989. *Software Engineering.* 3rd ed. Wokingham, U.K.: Addison-Wesley Publishing Co.
10. Marco, D. 1918. *Structured Analysis and System Specification.* New York: Yourdon Press.
11. Harel, D., H. Lachover, A. Naamad, A. Pnueli, M. Politi, R. Sherman, A. Shtull-Trauring, and M. Trakhtenbrot. 1990. "STATEMATE: A Working Environment for the Development of Complex Reactive Systems." *IEEE Transactions on Software Engineering* 16 (4): 406–14.
12. Yau, S. S., and M. U. Caglayan. 1983. "Distributed Software System Design Representation Using Modified Petri Nets." *IEEE Transactions on Software Engineering* SE-9 (6): 733–45.
13. Ramamoorthy, C. V., Y. Yaw, and W. T. Tsai. 1988. "Synthesis Rules for Cyclic Interaction among Processes in Concurrent Systems." In *Proceedings of the 12th Annual International Computer Software & Applications Conference*, Chicago, 1988, pp. 497–504.
14. Yau, S. S., and C.-R Chou. September 1988. "Control Flow Analysis of Distributed Computing System Software Using Structured Petri Net Model." In *Proceedings: Workshop on the Future Trends of Distributed Computing Systems in the 1990s*, Hongkong, September 14–16, 1988, pp. 174–83.
15. Alford, M. 1985. "SREM at the Age of Eight: The Distributed Computing Design System." *IEEE Computer* 18 (4): 36–45.
16. Evangelist, M., V. Y. Shen, I. R. Forman, and M. Graf. 1988. "Using Raddle to Design Distributed Systems." In *Proceedings of the 10th International Conference on Software Engineering*, Singapore, April 11–15, 1988, pp. 102–11.
17. Yi Deng, Shengkai Lu, and Michael Evangelist. 1997. "A Formal Approach for Architectural Modeling and Prototyping of Distributed Real-Time Systems." In *Proceedings of the Thirtieth Annual Hawaii International Conference on System Sciences*, Maui, Hawaii, January 7–10, 1997, pp. 481–90.
18. Jan, K., and H. Zdenek. "Verifying Real-Time Properties of Can Bus by Timed Automata." http://dce.felk.cvut.cz/hanzalek/publications/Hanzalek04a.pdf., published 2004 online, accessed on July 2010.
19. Subramonian, V., C. Gill, C. Sanchez, and H. Sipma. "Composable Timed Automata Models for Real-Time Embedded Systems Middleware." http://www.docin.com/p-87686529.html, published on 2005, accessed on July 2010.
20. Bordbar, B., R. Anane, and K. Okano. 2006 "A Timed Automata Approach to QoS Resolution." *International Journal of Simulation* 7 (1): 46–54.
21. Selic, B. 1999. "Turning Clockwise: Using UML in the Real-Time Domain." *ACM Communications* 42 (10): 46–54.
22. Gomaa, H., and D. A. Menascé. 2000. "Design and Performance Modeling of Component Interconnection Patterns for Distributed Software Architectures." In *Proceedings of the ACM WOSP 2000*, Ottawa, ON, Canada, pp. 117–26.
23. Giese, H., M. Tichy, S. Burmester. W. Schafer, and S. Flake. "Towards the Compositional Verification of Real-Time UML Designs." In *Proceedings of the 9th European Software Engineering Conference*, Helsinki, Finland, September 1–5, 2003, pp. 38–47.

24. Bordbar, B., J. Derrick, and G. Waters. 2002. "Using UML to Specify QoS Constraints in ODP." *Computer Networks* 40: 279–304.
25. Zhang, M. Spring 2007. *Toward a Flexible and Reconfigurable Distributed Simulation: A New Approach to Distributed DEVS.* PhD dissertation, Electrical and Computer Engineering Department, University of Arizona.
26. Hong, J. S., and T. G. Kim. 1997. "Real-time Discrete Event System Specification Formalism for Seamless Real-Time Software Development." *Discrete Event Dynamic Systems: Theory and Applications* 7: 355–75.
27. UML 2.0 Superstructure Specification. 2003 August. Available from OMG Adopted Specification.
28. Alur, R., and D. L. Dill. 1994. "A Theory of Timed Automata." *Theoretical Computer Science* 126: 183–235.
29. van Bilion, W. R. 1988. "Extending Petri-Nets for Specifying Man-Machine Dialogue." *International Journal of Man-Machine Studies* 28: 437–55.
30. Harel, D. 1987. "Statecharts: A Visual Formalism for Complex Systems." *Science of Computer Programming* 8: 231–74.
31. Mosterman, P. J., J. O. Campbell, A. J. Brodersen, and J. R. Bourne. 1996. "Design and Implementation of an Electronics Laboratory Simulator." *IEEE Transactions on Education* 39 (3): 309–13.
32. Hamidi, A., A. Boukerche, L. Ahmad, and M. Zhang. 2008. "Supporting Web-Based E-Learning Through Collaborative Virtual Environments for Radiotherapy Treatment: A Formal Design." In *Proceedings of the IEEE International Conference on Virtual Environments, Human-Computer Interfaces, and Measurement Systems*, Istanbul, Turkey, July 14–16, pp. 51–6.
33. Boukerche, A., A. Hamidi, and M. Zhang. August 2009. "Design of a Virtual Environment Aided by a Model Based Formal Approach Using DEVS." *Concurrency & Computation: Practice & Experience* 21 (11): 1422–36.
34. Hoare, C. A. R. 1978. "Communicating Sequential Processes." *Communications of the ACM* 28 (8): 666–77.
35. Smith, S., and D. Duke. 1999. "Virtual Environments as Hybrid Systems." In *Proceedings of the Eurographics UK 17th Annual Conference*, pp. 113–28.
36. Wieting, R. 1996. "Hybrid High-Level Nets." In *Proceedings of the 1996 Winter Simulation Conference*, Coronado, CA, USA, December 8–11, 1996, pp. 848–55.
37. AutoDesk 3ds Max, http://usa.autodesk.com/adsk/servlet/index?siteID=123112&id=5659302.
38. AC3D, http://www.inivis.com/.
39. Jacob, R. J. K. 1996. "A Visual Language for Non-WIMP User Interfaces." In *Proceedings IEEE Symposium on Visual Languages*, Boulder, Colorado, USA, September 3–6, 1996, pp. 231–38.
40. Dahmann, J. S., R. M. Fujimoto, and R. M. Weatherly. "The Department of Defense High Level Architecture." In *Proceedings of the 1997 Winter Simulation Conference*, Atlanta, GA, USA, December 7–10, 1997.
41. Boukerche, A., M. Zhang, and A. Shadid. 2008. "DEVS Approach to Real Time RTI Design for Large-Scale Distributed Simulation Systems." *Simulation* 84: 231.
42. IEEE Standard 1516-2000. 2000 September. "IEEE Standard for Modeling and Simulation (M&S) High Level Architecture (HLA)—Framework and Rules."
43. Xie, H., A. Boukerche, M. Zhang, and B. P. Zeigler. 2008. "Design of a QoS-Aware Service Composition and Management System in Peer-to-Peer Network Aided by DEVS." In *Proceedings of IEEE International Symposium on Distributed Simulation and Real Time Application*, Vancouver, BC, Canada, October 27–29, 2008.

44. Xie, H., A. Boukerche, and M. Zhang. 2010. "A Formalized Approach Formalization Designing a P2P-Based Dynamic Load Balancing Scheme." *Concurrency & Computation: Practice & Experience* 22: 1223–39.
45. Melanie, M., 1996. *An Introduction to Genetic Algorithms*. Cambridge, MA: MIT Press.
46. Hu, X., and B. P. Zeigler. 2005. "A Simulation-Based Virtual Environment to Study Cooperative Robotic Systems." *Integrated Computer-Aided Engineering* 12 (4): 353–67.
47. Hu, X. Fall 2003. "A Simulation-Based Software Development Methodology for Distributed Real-Time Systems." PhD dissertation, Electrical and Computer Engineering Department, University of Arizona.
48. Hu, X., B. P. Zeigler, and S. Mittal. 2005. "Variable Structure in DEVS Component-Based Modeling and Simulation." *Simulation* 81 (2): 91–102.

3 Principles of DEVS Model Verification for Real-Time Embedded Applications

Hesham Saadawi, Gabriel A. Wainer,
and Mohammad Moallemi

CONTENTS

3.1 Introduction 63
3.2 Background 65
    3.2.1 Difficulties of DEVS Formal Verification 68
    3.2.2 Rational Time-Advance DEVS 69
3.3 DEVS Verification Methodology 71
3.4 Case Study: Controller for an E-Puck Robotic Application 80
    3.4.1 DEVS Model Specification 81
    3.4.2 Implementation on the ECD++ Toolkit 84
    3.4.3 Executing the Models 88
    3.4.4 Verifying the Model 89
3.5 Conclusions 93
References 93

3.1 INTRODUCTION

Embedded real-time (RT) software systems are increasingly used in mission critical applications, where a failure of the system to deliver its function can be catastrophic. Currently existing RT engineering methodologies use modeling as a method to study and evaluate different system designs before building the target application. Having a system model enables the verification of system properties and functionality before building the actual system. In this way, deployed systems would have a very high reliability, as the formal verification permits detecting systems errors at the early stages of the design. To apply such methodologies for embedded control systems, a designer must abstract the physical system to be controlled and build a model for it. This model can then be combined with a model of the proposed controller design for study and evaluation.

In general, different techniques are used to reason about these models and gain confidence in the correctness of a design. Informal methods usually rely on extensive testing of the systems based on system specification. These techniques can reveal errors but cannot prove nonexistence of errors. Instead, formal techniques can prove the correctness of a design. Unfortunately, formal approaches are usually constrained in their applications, as they do not scale up well and they require the user to have expert knowledge in applying formal techniques. Another drawback of applying formal techniques is that they must be applied to an abstract model of the target system to be practical. However, in doing so, what is being verified is not the final executable system. Even if the abstract model is correct, there is a risk that some errors creep into the implementation through the manual process of implementing specifications into executable code [1].

A different approach considers using modeling and simulation (M&S) to gain confidence in the model correctness. M&S of RT systems also enables testing much like testing a physical system, even for cases where physical testing may be too costly or impossible to achieve [2]. If the models used for M&S are formal, their correctness is verifiable and a designer can also observe the system evolution and its inner workings. Another advantage of executable models is that they can be deployed to the target platform, thus providing the opportunity to use the controller model not only for simulations but also as the actual code executing in the target hardware. The advantage of this methodology is that the verified model is itself the final implementation executing in RT. This avoids any new errors that would appear during transformation of the verified models into an implementation, thus guaranteeing high degree of system correctness and reliability.

In the following sections, we introduce a new methodology proposed to verify simulation models based on the Discrete Event Systems Specification (DEVS) formalism [3]. The reason for introducing a new methodology for DEVS verification is that most existing such methods are limited to a constrained set of DEVS subclasses. This prevents the verification of a wide range of existing DEVS models and forces the modeler to use less-expressive subclasses. In addition, these DEVS subclasses require special verification tools that may not add much value over standard verification tools for timed automata (TA) [4,5]. The value of these special verification tools is questionable, as most verification algorithms used for restricted DEVS subclasses rely on the same timed model-checking algorithm used for the verification of TA. In that sense, these algorithms have the same time and space complexity as those of TA model-checking algorithms, and thus, DEVS verification tools do not provide any advantages over TA verification. On the contrary, verification tools for TA are widespread, and they usually contain many performance optimizations.

For these reasons, we define a new class of rational time-advance DEVS called RTA-DEVS that is close to classic DEVS in semantics and expressive power (enabling the verification of most existing DEVS models [6]). Then, we define a transformation to obtain a TA that is behaviorally equivalent to RTA-DEVS [7]. The advantage of doing so is that many classic DEVS models would satisfy the semantics of RTA-DEVS models, and they could be simulated with any DEVS simulator, but they can be transformed to TA to validate the desired properties formally. RTA-DEVS has followed Finite and Deterministic DEVS, called FD-DEVS [8], in restricting the time advance function to nonnegative rational numbers but also relaxed the restriction of FD-DEVS on external transition functions. This makes RTA-DEVS closer to general DEVS and adds expressiveness. However, RTA-DEVS still restricts the elapsed time in a state used in the external transition function to be a nonnegative rational number. This restriction translates to having nonnegative rational constants in guards in the transformed TA model and ensures termination of the reachability analysis algorithms implemented in UPPAAL [9]. As per the theory of TA presented in Ref. [4], irrational constants in TA guards render reachability analysis undecidable (as proved in Ref. [5]).

3.2 BACKGROUND

DEVS was originally defined in the 1970s as a mechanism for specifying discrete event models specification [3]. It is based on dynamic systems theory, and it allows one to define hierarchical modular models. A system modeled with DEVS is described as a composite of submodels, each of them being behavioral (atomic) or structural (coupled). Each model is defined by a time base, inputs, states, outputs, and functions to compute the next states and outputs. A DEVS atomic model is formally described by

$$M = \langle X, S, Y, \delta_{int}, \delta_{ext}, \lambda, \mathrm{ta} \rangle$$

where

X is the set of external inputs.
Y is the set of external outputs.
S is the set of system states.
$\delta_{int}: S \to S$ is the internal transition function.
$\delta_{ext}: Q \times X \to S$, where $Q = \{(s,e) \mid s \in S, 0 \le e \le \mathrm{ta}(s), e \in \mathfrak{R}_0^{+\infty}\}$ is the external transition function (where e is the time elapsed since the last transition, a positive real value).
$\lambda: S \to Y \cup \emptyset$ is the output function.
$\mathrm{ta}: S \to \mathfrak{R}_0^{+\infty}$ is the time advance function, which maps each state to a positive real number.

A DEVS atomic model is the most basic DEVS component. The behavior of a DEVS model is defined by transition functions in atomic components. An atomic model M can be affected by external input events X and can generate output events Y. The state set S represents the state variables of the model. The internal transition function δ_{int} and the external transition function δ_{ext} compute the next state of the model. When an external event arrives at elapsed time e (which is less than or equal to ta(s) specified by the time advance function), a new state s' is computed by the external transition function. Otherwise, if ta(s) finishes without input interruption, the new state s' is computed by the internal transition function. In this case, an output specified by the output function λ can be produced based on the state s. After a state transition, a new ta(s') is computed, and the elapsed time e is set to zero.

A DEVS coupled model is composed of several atomic or coupled submodels. The property of closure under coupling allows atomic and coupled models to be integrated to form a model hierarchy. Coupled models are formally defined as follows:

$$CM = <X, Y, D, M_d, \text{EIC}, \text{EOC}, \text{IC}, \text{Select}>$$

where

X is the set of input ports and values.
Y is the set of output ports and values.
D is the set of the component names (an index of submodels).
EIC is the set of External Input Couplings, which connects the input events of the coupled model itself to one or more of the input events of its components.
EOC is the set of External Output Couplings, which connects the output events of the components to the output events of the coupled model itself.
IC is the set of Internal Couplings, which connects the output events of the components to the input events of other components.
Select: $2^D \rightarrow D$ is a tie-breaking function, which defines how to select an event from a set of simultaneous events.

CD++ [10] allows defining models following these specifications. The tool is built as a hierarchy of models, and each of the models is related to a simulation entity. CD++ includes a graphical specification language, based on DEVS Graphs [11], to enhance interaction with stakeholders during system specification while having the advantage of allowing the modeler to think about the problem in a more abstract way. DEVS graphs can be formally defined as [12]

$$GGAD = <X_M, S, Y_M, \delta_{int}, \delta_{ext}, \lambda, \text{ta}>$$

where

$X_M = \{(p,v)| p \in IPorts, v \in X_p\}$ is the set of input ports.
$Y_M = \{(p,v)| p \in OPorts, v \in Y_p\}$ is the set of output ports.
$S = B \times P(V)$ represents the states of the model, where
    $B = \{b \mid b \in Bubbles\}$ is the set of model states.
    $V = \{(v,n) \mid v \in Variables, n \in R_0\}$ represents the intermediate state variables of the model and their values.
$\delta_{int}, \delta_{ext}, \lambda$, and ta have the same meaning as in traditional DEVS models.

CD++ uses this formal notation to define atomic models, as seen in Figure 3.1. A unique identifier defines each model, which can be completely specified using a graphical specification based on the formal definition above. That is, states are represented by bubbles including an identifier and a state lifetime, state variables can be associated with the transitions, and there are two types of transitions: external and internal.

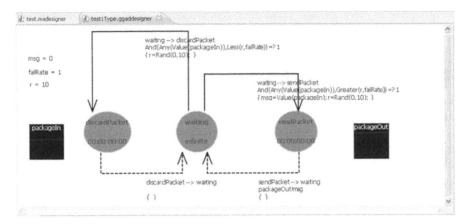

FIGURE 3.1   An atomic model defined as a DEVS graph.

The DEVS graph in Figure 3.1 shows a simple component for a packet routing model. As we can see, there are three states in this DEVS atomic model: *waiting, discardPacket*, and *sendPacket*. The model uses two input/output ports (corresponding to the X_M and Y_M sets in the formal specification): *packageIn* and *packetOut*. Three variables are defined for this model and initialized: *msg, failRate*, and *r*. Internal transitions are shown with dashed arrow lines. The internal transition *sendPacket* → *waiting* uses the output function, which is defined to send the value of the variable *msg* to output port *packetOut*. The external transitions are shown with solid arrow lines, with a condition that would enable that transition only if it is evaluated to true, and an expression to update some of the model variables (when needed). For instance, the transition from *waiting* to *sendPacket* is activated when a packet is received on *packageIn* (*Any(Value(packageIn, 1))*) and the *failRate* is greater than a random value *r* (*Greater(r, failRate)*). In that case, the model changes to state *sendPacket* and also assigns the value of the packet to the *msg* intermediate variable. It also computes a new random value.

Each DEVS graph is translated into an analytical definition that the runtime engines use to execute. The internal transitions employ the following syntax:

```
int: source destination [outport!value]* ( { (action;)* } )
```

Here, *source* and *destination* represent the initial and final states associated with the execution of the transition function. As the output function should also execute before the internal transition, an *output* value can be associated with the internal transition. One or more *actions* can be triggered during the execution of the transition (changing the values of state variables). External transitions are defined as follows:

```
ext : source destination ( { (action;)* } )? expression
```

In this case, when the expression is true (which includes inputs arriving from input ports), the model will change from state *source* to state *destination*, while also executing one or more actions. These notations are generated as a direct translation from the graph represented in Figure 3.1.

3.2.1 DIFFICULTIES OF DEVS FORMAL VERIFICATION

DEVS formal definitions for atomic models are the most generic DEVS [3]. The first difficulty in DEVS formal verification, is that model-checking techniques are only decidable for finite state systems. In case of infinite-state systems, irrelevant details must be abstracted to obtain a finite state system before applying model-checking techniques. Another difficulty is the nondeterminism in DEVS behavior. This can be caused by stochastic behavior in a DEVS model due to the use of a probabilistic function in the definition of the external transition function δ_{ext} or time advance function δ_{int} [13]. Another major difficulty in applying automatic formal verification techniques such as model checking to DEVS models is that the DEVS time advance function can take values of irrational real numbers. These values cannot be represented in a finite reachability graph that is used in model-checking algorithms, and thus, the algorithm will not be able to terminate, hence rendering the verification problem undecidable.

Several techniques have been introduced to overcome these problems and provide reasonable approximation to DEVS while enabling formal verification. As will be discussed in the following paragraphs, the techniques range from formal model checking of restricted classes of DEVS, the generation of test traces from DEVS models for simulation testing, the specification of high-level system requirements in TA and verifying DEVS model against those requirements, and introducing clock constructs to DEVS to conform with TA.

One approach, called *real-time DEVS* (RT-DEVS) introduces a time advance function that maps each state to a range with maximum and minimum time values and introduces an activity associated with every system state [14]. This work also introduced an RT-DEVS *executive* that executes these models in RT. RT-DEVS was also used to design RT controllers as shown in Ref. [15] for a train-gate system. Further work on verifying RT-DEVS was introduced in Refs. [16,17], where the authors relied on TA as used by UPPAAL and defined (although not formally proved) a transformation method from RT-DEVS to UPPAAL. This transformation allows weak synchronization between components of the TA model as RT-DEVS semantics uses weak synchronization.

Other approaches use a limited version of DEVS that can be verified. For instance, a method based on Finite and Deterministic DEVS (FD-DEVS) was introduced [8] where the time advance function maps states into rational numbers and the external transition function cannot use the elapsed time value. The verification relies on reachability analysis, similar to TA algorithms. FD-DEVS is limited, thus it may not fit some applications that require the full expressiveness of DEVS. Likewise, although reachability analysis algorithms have been defined (and verification is possible), there are no tools available that implement these algorithms.

To avoid these limitations, other approaches tried to map DEVS models to TA [18]. The conversion method mapped a DEVS model through its components and its simulator. The approach suggests trace equivalence as the basis for parallel DEVS and TA model equivalence. This work did not consider some DEVS features that may not map to TA, such as irrational values in DEVS transition functions. Moreover, some limitations also exist for relying on trace equivalence between DEVS and TA, as we will show in Section 3.3. A similar approach presented in Ref. [19] uses TA to specify the high-level system requirements, after which these requirements are modeled as a DEVS model. The system requirements are then verified through simulation of the DEVS model.

The work by Hernandez and Giambiasi [20] showed that the verification of general DEVS models through reachability analysis is undecidable. The authors based their deduction on building a DEVS simulation Turing machine. Since the halting problem in Turing machines is undecidable (i.e., with analysis only, we cannot know in which state a Turing machine would be), they concluded that this is also true for DEVS models. In other words, we cannot recognize if we have reached a particular state starting from an initial state, and consequently, reachability analysis for general DEVS is impossible. Based on this result, reachability analysis may be possible only for restricted classes of DEVS. This result was based on introducing state variables with infinite number of values into the DEVS formalism. Therefore, limiting the number of states of a DEVS model is mandatory for decidable reachability. Hence, further work [21] introduced a new class of DEVS called time-constrained DEVS (TC-DEVS), which expanded the definition of DEVS atomic models with multiple clocks incremented independently of other clocks. Classic DEVS atomic models can be seen as having only one clock that keeps track of the elapsed time in a state and is reset on each transition. TC-DEVS also added clock constraints similar to TA (to function as guards on external and internal transitions). However, it is different from UPPAAL TA in that it allows clock constraints in state invariants to include clock differences. TC-DEVS is then transformed into an UPPAAL TA model. This work, however, did not include a transformation of TC-DEVS state invariants to UPPAAL TA when the model has invariants with clock differences, as this is not allowed in UPPAAL TA.

For large and more complex DEVS models, where formal verification is not feasible, testing would be the only choice. Techniques have been presented to generate testing sequences from model specifications that can then be applied against the model implementation to verify the conformance of the implementation to specifications [22,23].

3.2.2 Rational Time-Advance DEVS

RTA-DEVS was proposed to provide the system modeler with a formalism that is expressive and sufficient to model complex systems behavior, while being verifiable by formal model-checking techniques. RTA-DEVS is a subclass of DEVS that has removed the main difficulties of the formal model verification discussed in Section 3.2.1; yet, it is sufficiently powerful to model complex system behavior.

As in classical DEVS, we must define RTA-DEVS atomic models. The main difference is that RTA-DEVS employs a different definition for the time advance function, ta, and for the external transition function, δ_{ext}. The Atomic Rational Time-Advance is defined as follows:

$$AM_{TC} = <X, Y, S, \delta_{int}, \delta_{ext}, \lambda, ta>$$

where

    X is the set of external inputs.
    Y is the set of external outputs.
    S is the set of system states.
    $\delta_{int}: S \rightarrow S$ is the internal transition function (as in classic DEVS).
    $\delta_{ext}: T \times X \rightarrow S$ with $T = \{(s,e)/s \; 0 \leq e \leq ta(s), \; e \in Q_{0,+\infty}\}$ is the external transition function (e is the time elapsed since the last transition, which takes a *positive rational value*).
    $\lambda: S \rightarrow Y \cup \emptyset$ is the output function.
    ta: $S \rightarrow Q_{0,+\infty}$ is the time advance function that maps each state to a *positive rational number.*

Coupled RTA-DEVS models are defined as in classic DEVS, as discussed in Section 3.2.

A coupled RTA-DEVS model M can be simulated with an equivalent atomic RTA-DEVS model, whose behavior is defined as follows:

$$M = \langle X, Y, S, s_0, \delta_{ext}, \delta_{int}, \lambda, ta \rangle$$

where

    X and Y are the input and output event sets, respectively. X is the set of all input events accepted and Y is the set of all output events generated by coupled model M.
    $S = X_{i \in D} \; V_i$ is the model state, expressed as the Cartesian product of all component states, where V_i is the total state for component i, $V_i = \{(s_i, t_{ei})| \; s_i \in S_i, \; t_{ei} \in [0, ta(s_i)]\}$. Here, t_{ei} denotes the elapsed time in state s_i of component i, and S_i is the set of states of component i.
    $s_0 = X_{i \in D} \; v_{0i}$ is the initial system state, with $v_{0i} = (s_{0i}, 0)$ the initial state of component $i \in D$.
    ta: $S \rightarrow T$ is the time advance function. It is calculated for the global state $s \in S$ of the coupled model as the minimum time remaining for any state among all components, formally:
        ta$(s) = \min\{(ta(s_j) - t_{ei}) \; | \; i \in D\}$ where $s = (...(s_i, t_{ei}),...)$ is the global total state of the coupled model at some point in time, s_i is the state of component i, and t_{ei} is elapsed time in that state.
    $\delta_{ext}: X \times V \rightarrow S$ is the external transition function for the coupled model, where V is total state of the coupled model, $V = \{(s, t_e)| \; s \in S, \; t_e \in [0, ta(s)]\}$.
    $\delta_{int}: S \rightarrow S$ is the internal transition function of the coupled model.
    $\lambda: S \rightarrow Y$ is the output function of the coupled model.

3.3 DEVS VERIFICATION METHODOLOGY

In this section, we introduce our methodology to transform RTA-DEVS models into TA models. The resulting TA models are a subset of deterministic safety automata (which can be used in UPPAAL or other similar model checkers). The transformation methodology can be summarized as follows:

1. Define a clock variable for each atomic RTA-DEVS model (i.e., x).
2. Replace every state in RTA-DEVS with a corresponding one in TA (i.e., L_1 for source s_1 and L_2 for destination s_2).
3. Model the RTA-DEVS internal transition from s_1 to s_2 as a TA as follows:
   a. For the RTA-DEVS source state s_1, define a TA source state L_1. For the RTA-DEVS destination state s_2, define a TA destination state L_2.
   b. Reset the clock variable on the entry to each state ($x: = 0$).
   c. Put an invariant in the source state derived from the time advance function for that state, that is, $x < ta(s_1)$.
   d. Optionally, define a transition with a guard. This guard should be the complement of the invariant in the source state, that is, $x \geq ta(s_1)$.
   e. Define an action for each output function defined.
4. The RTA-DEVS external transition is modeled in TA with the following items:
   a. A source state and some destination state(s), that is, L_1 for source s_1 and L_2 for destination s_2.
   b. A clock reset on the entry into each state.
   c. An invariant in the source state that corresponds to the time advance function for that state, that is, $x < ta(s_1)$.
   d. For the external transition(s) with guards of clock constraints, these constraints should be disjoint to obtain a deterministic TA model.
   e. The action label on TA transitions for each RTA-DEVS input event to source state s_1.

By applying the above-mentioned steps, we obtain a TA model that executes every transition defined in the RTA-DEVS model under study. As already known, the RTA-DEVS behavior is completely defined by its transition functions, which defines all transitions in the RTA-DEVS model. Thus, the resulting TA model executes the RTA-DEVS.

However, TA models cannot have irrational constant values in guards or state invariants. This implies that for any DEVS model containing a state lifetime of irrational values, it will not be possible to directly apply the transformation shown in Table 3.1. In this case, the irrational values would have to be approximated to the nearest rational value according to a choice by the modeler, based on the required precision for the equivalent RTA-DEVS model. In doing so, the transformation should take into account the following rules. These rules avoid building invalid RTA-DEVS or TA models that contain time-action locks (that prevent the model execution progress) or loops where execution progresses infinitely without allowing time to advance [24].

Rule 1: When approximating an irrational value triggering an internal transition that is coupled with an external transition, the choice of approximation value should be consistent for all constants using this irrational number.

TABLE 3.1

Transformation of Rational Time-Advance DEVS to Behaviorally Equivalent Time Automata

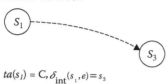

RTA-DEVS   Equivalent TA

Internal transition

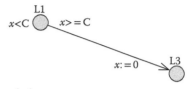

$ta(s_1) = C, \delta_{\text{int}}(s_1, e) = s_3$

clock x;

External transition

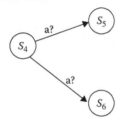

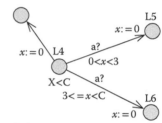

$ta(s_4) = C, \delta_{ext}(s_4, a, e) = s_5, 0 < e < 3$
$\delta_{ext}(s_4, a, e) = s_6, 3 \leq e < ta(s_4)$

clock x;

Formally, assume we have the following defined in a DEVS coupled model as shown in Figure 3.2:

$$\delta^A_{\text{int}}(S_i, C_{\text{irr}}) = S_j, \lambda^A(S_i) = a, \text{ta}^A(S_i) = C_{\text{irr}}$$

$$\delta^B_{\text{ext}}(S_k, e, a) = (S_l, 0) \quad C_{\text{irr}} \leq e \prec \infty$$

$$\delta^B_{\text{ext}}(S_k, e, a) = (S_m, 0) \quad 0 \prec e \prec C_{\text{irr}}$$

It should be approximated in RTA-DEVS as follows:

$$\delta^A_{\text{int}}(S_i, C_r) = S_j, \lambda^A(S_i) = a, \text{ta}^A(S_i) = C_r$$

$$\delta^B_{\text{ext}}(S_k, e, a) = (S_l, 0) \quad C_r \leq e \prec \infty$$

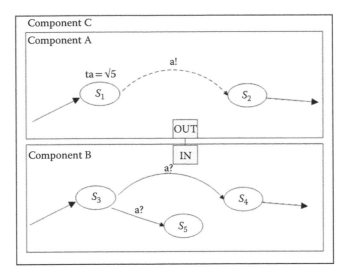

FIGURE 3.2 A coupled DEVS model.

$$\delta^B{}_{ext}(S_k,e,a) = (S_m,0) \quad 0 \prec e \prec C_r$$

where

C_{irr} is an irrational real number.

C_r is a rational real number.

δ^A_{int}, λ^A, and ta^A are functions defined for component A.

Rule 2: When approximating an irrational value for elapsed time in the definition of the external transition function, the choice of the approximation value should be consistent for all constants using this irrational number. Formally, assume we have the following DEVS definition of an external transition function in a model similar to the one shown in Figure 3.3:

$$\delta_{ext}(S_i,e,a) = (S_j,0) \quad C_{irr} \leq e \prec \infty$$

$$\delta_{ext}(S_i,e,a) = (S_k,0) \quad 0 \prec e \prec C_{irr}$$

It should be approximated in the RTA-DEVS model with the following form to avoid creating action locks:

$$\delta_{ext}(S_i,e,a) = (S_j,0) \quad C_r \leq e \prec \infty$$

$$\delta_{ext}(S_i,e,a) = (S_k,0) \quad 0 \prec e \prec C_r$$

The second rule is to avoid action locks that may happen if we define the external transition function with conditions on its transitions where there is a gap in time

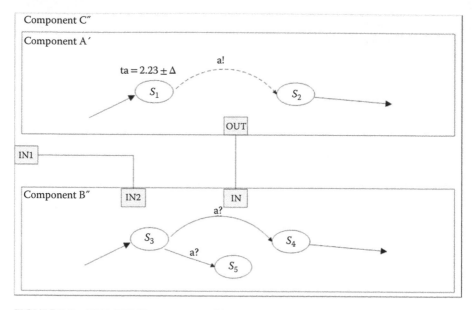

FIGURE 3.3   RTA-DEVS component with external input.

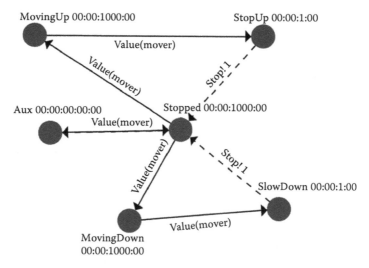

FIGURE 3.4   Elevator RTA-DEVS model.

(where the function is not defined). Another possibility is to have an approximated external transition function in which conditions on different transitions overlap in time, thus creating nondeterminism that is not in the original DEVS model.

To further clarify the method, we show a simple example representing the behavior of an elevator [6]. The RTA-DEVS model in Figure 3.4 models the movement of the elevator.

TABLE 3.2

Elevator Model States

Elevator State	State Description
Stopped	The elevator speed is zero and it is stopped at one of the floors.
StopUp	The elevator is moving up and preparing to stop. It decelerates.
MovingUp	The elevator is moving up with a constant speed.
SlowDown	The elevator is moving down and preparing to stop. It decelerates.
MovingDown	The elevator is moving down with a constant speed.
Aux	An auxiliary state to allow output from the elevator model when internal transition occurs: Aux → Stopped.

As shown, the elevator can be in one of the five states (listed in Table 3.2).

In this model, the elevator starts in the *Stopped* state and waits for the controller commands to move to satisfy a button request from the user. The decisions for the proper direction and the start and stop of movement are all taken by the controller. The states of the elevator are represented by circles in the figure. External transitions are enabled when the function *Value(mover)* evaluates to true. This function is defined as in Table 3.3 for the different transitions shown in Figure 3.4.

Likewise, the behavior of the internal transitions are defined as in Table 3.4.

By following the transformation steps summarized in Table 3.1, we can construct the equivalent TA model as shown in Figure 3.5. This model is constructed to be behaviorally equivalent to the DEVS model of Figure 3.4. This equivalence is essential to ensure that any properties that we must check in the DEVS model are preserved in the constructed TA model.

By applying the methodology we identified in Section 3.3, we go through the following steps to obtain the TA model in Figure 3.5.

- Define a clock variable for each atomic RTA-DEVS model. This results in variable x.
- Replace every state in RTA-DEVS with a corresponding one in TA. A location is created for each state in DEVS with the same name as is shown in the TA model.
- Model the RTA-DEVS internal transition as a TA as follows.
  a. A source state L_1 and a destination state L_2: *SlowingDown* and *StopUp* states in Figure 3.5 *represent source states of SlowDown and StopUp states depicted in* Figure 3.4 Reset the clock variable on the entry into each state ($x = 0$).
  b. Put an invariant in the source state derived from the time advance function for that state. The invariant at both states *SlowingDown* and *StopUp* is $x < 1$.

TABLE 3.3
Elevator External Transitions

Transition	Function Definition	Expression
Stopped → MovingUp	Value(mover)?2	Mover = = 2
MovingUp → StopUp	Value(mover)?0	Mover = = 0
Stopped → Aux	Value(mover)?0	Mover = = 0
MovingDown → SlowDown	Value(mover)?0	Mover = = 0
SlowDown → Stopped	stop!1	Stop = = 1
StopUp → Stopped	stop!1	Stop = = 1
Stopped → MovingDown	Value(mover)?1	Mover = = 1

TABLE 3.4
Elevator Internal Transitions

Transition	Action Definition	Outport!value
SlowingDown → Stopped		stop!1
StopUp → Stopped		stop!1

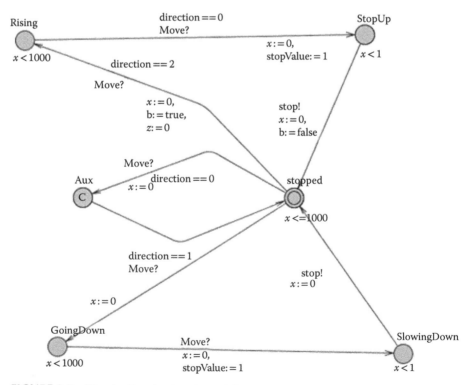

FIGURE 3.5   Elevator timed automata model.

    c.   Optionally, define a transition with a guard. This *guard* should be the complement to the invariant in the source state. None are defined in this model.

    d.   Define an action for each output function that is defined. This corresponds to the two actions *stop!1* in the model.

- The RTA-DEVS external transition is modeled in TA with the following items.

    a.   A source state and some destination state(s). Transitions are defined in the TA model that correspond to the DEVS model.

    b.   A clock reset on the entry into each state.

    c.   An invariant in the source state that corresponds to the time advance function for that state. This corresponds to the three occurrences of the invariant $x < 1000$.

    d.   For the external transition(s) with guards of clock constraints, these constraints should be disjoint to obtain a deterministic TA model. For example, in the elevator model, direction $= =0$, direction $= = 1$, and direction $= = 2$.

    e.   For each event on external transition of the RTA-DEVS model, place a synchronization channel on the corresponding TA transition. For example, the *move?* and *stop!* channels in the TA model of Figure 3.5 represent external events of *mover(value)* and *stop!1* in the elevator model of Figure 3.4.

It is important to preserve the equivalence properties also when we map any verification results obtained from the TA model back to the DEVS model. To ensure this equivalence, the transformation from DEVS to TA is done based on the notion of bisimulation equivalence [7]. This equivalence ensures that for each state in DEVS, there is a corresponding one in TA and vice versa. It also ensures that for each transition in DEVS, there would be a corresponding equivalent one in TA and vice versa. Once we have a TA model that is behaviorally equivalent to the DEVS model, any property we wish to verify in the DEVS model can be verified in the TA model, and verification results would apply directly to the DEVS model.

The DEVS Elevator-Controller is shown in Figure 3.6. By applying the transformation steps discussed in the beginning of Section 3.3, we obtain the TA model as shown in Figure 3.7. In this transformation, we represented DEVS states with lifetime of zero as committed locations in the TA model. Examples of these are states Aux1, Stopping, and Moving. Committed locations of TA prevent time to elapse in them and hence serve our purpose for this transformation.

To apply the UPPAAL model checker on this elevator system, the elevator system must be represented as a closed system that allows UPPAAL to explore all its transitions and states. To do so, we define a simple *environment* model that represents a user requesting the services of the elevator, as shown in Figure 3.8. In this model, the third floor button is pressed after 5 time units. This causes the Elevator-Controller to receive the floor value and then send the corresponding command to the elevator model to reach the third floor. The environment model then simulates different user requests.

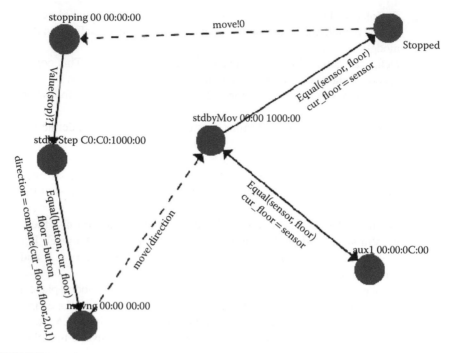

FIGURE 3.6 Elevator controller model as a DEVS graph.

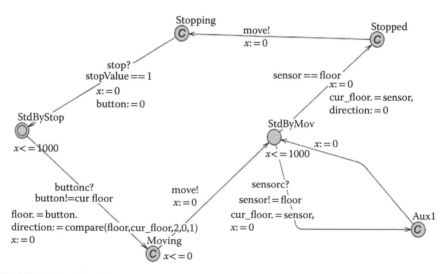

FIGURE 3.7 Timed automata controller model in UPPAAL.

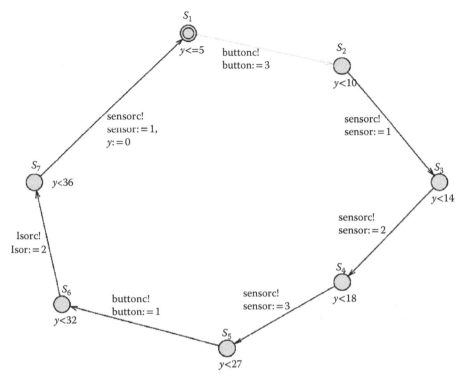

FIGURE 3.8 Environment inputs (button and sensor).

The system composed of the Elevator, the Elevator-Controller, and the Environment can be checked using UPPAAL to verify certain properties about the system. For instance, some of the important properties would be as follows:

a. Does the DEVS model progress? Can we detect deadlock conditions?
b. If no deadlocks are found, is it always guaranteed whenever a user pushes a floor button that the elevator would reach that floor (i.e., the normal operation of the elevator system)?
c. If the elevator eventually reaches the floor, is there a guaranteed upper bound between the request and the arrival of the elevator?

For the first question, we applied UPPAAL to our model to check for any deadlocks that may be present in the elevator. To check for that failure, we had formulated a simple query, expressed in computational tree logic (CTL) [25–27] as follows:

```
A[] not deadlock
```

After running the checker, it shows that this property is satisfied, that is, there is no deadlock, as shown in Figure 3.9.

The property (b) is an example of the system *liveness*, in which we are interested to check if by pressing a certain floor button, the elevator would *eventually* reach

```
UPPAAL version 4.0.6 (rev. 2986), March 2007 -- server.
A[] not deadlock
Property is satisfied.
```

FIGURE 3.9   Elevator verification results in UPPAAL.

that floor. For example, if the user presses the third-floor button, the elevator should *eventually* reach the third floor. This property is expressed in CTL as follows:

```
button == 3 --> ElevatorController.cur_floor == 3
```

This states that whenever a user input for the third-floor button occurs, the *cur_floor* variable in the *ElevatorController* would eventually reach that floor. This property was also satisfied in UPPAAL model checker for the given model.

To check the third property (c), that is, whether the elevator would reach the requested third floor within some bounded time, we extend the model for bounded time checking by adding the Boolean variable b and a global clock z as shown in the elevator model in Figure 3.5. The variable b would be set to true for the time when the elevator starts traveling up until it reaches the *Stopped* state again. Therefore, by checking the accumulated time while b is true, it would provide us the property we must check. Then, the property can be expressed with the following query:

```
A[] ( b imply z < 27 ) which is satisfied.
```

However, the query

```
A[] ( b imply z < 26 ) is not satisfied.
```

This shows that the elevator would reach the third floor after requested to go there after no less than 26 time units, but is guaranteed to be there after 27 time units or more.

3.4   CASE STUDY: CONTROLLER FOR AN E-PUCK ROBOTIC APPLICATION

In this section, we present a case study where we use DEVS to build a model of a controller for an E-puck robot and later the same model is used as an actual controller. The E-puck (shown in Figure 3.10) is a desktop-size mobile robot with a wide range of possibilities (signal processing, autonomous control, embedded programming, etc.).

The E-puck contains various sensors covering different modalities: (i) eight infrared (IR) proximity sensors placed around the body measure the closeness of obstacles, (ii) a 3D accelerometer provides the acceleration vector of the E-puck, (iii) three microphones to capture sound, and (iv) a color CMOS (Complementary Metal Oxide Semiconductor) camera with a resolution of 640 × 480 pixels. It also includes the following actuators: (i) two stepper motors, making it capable of moving forward and backward, and spinning in both directions; (ii) a speaker, connected to an audio codec; (iii) eight red light-emitting diodes (LED) placed all around the top; and (iv) a red front LED placed beside the camera.

FIGURE 3.10   E-puck robot.

In the following sections, we introduce a DEVS model for a simple controller for the E-puck, the corresponding implementation in CD++, and the formal verification of different properties of the model through the transformation introduced in Section 3.4 combined with the use of the UPPAL model checker.

3.4.1   DEVS MODEL SPECIFICATION

The controller is designed to steer the robot in a field while avoiding obstacles. We have defined a DEVS model with an atomic component (*epuck0*) that imitates the behavior of the controller, shown in Figure 3.11. There are eight input ports (*InIR0, … InIR7*), each of them modeling the connection to one proximity sensor. The input ports periodically receive the distances to the obstacles from the sensors. There are also two output ports: *OutMotor*, which transfers the output commands to the motors, and *OutLED*, to turn on/off the LEDs.

The controller can command the following actions based on the inputs received from the sensors: *move forward, turn 45 degrees left, turn 45 degrees right, turn 90 degrees left, turn 90 degrees right, turn 180 degrees*, and *stop*. Initially, the robot starts moving forward while receiving the periodic inputs from proximity sensors and analyzing them. As soon as it detects an obstacle, it performs one of the turning actions based on the position of the obstacle. The robot continues turning until it finds an empty space ahead. The controller also uses LEDs to signal the action that is being performed. For example, if the robot is moving forward, the front LED (*led0*) turns on and if it is turning 45 degrees to the left, *led7* turns on. Figure 3.12 illustrates two sample imaginary scenarios in which obstacles block the robot's path.

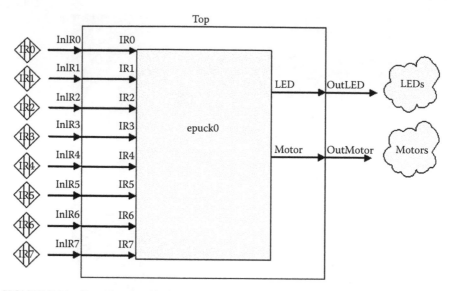

FIGURE 3.11   E-puck controller DEVS model hierarchy.

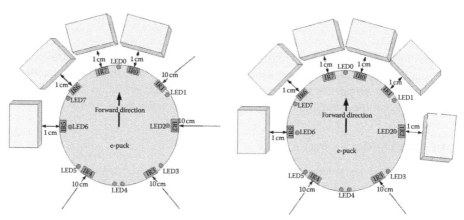

FIGURE 3.12   (a) Scenario 1: front-left blocked. (b) Scenario 2: front completely blocked.

The DEVS formal specification of the epuck0 atomic component is as follows:

$$AM = < X, S, Y, \delta_{ext}, \delta_{int}, \delta, ta >$$

where

X = {(IR0, R), (IR1, R), (IR2, R), (IR3, R), (IR4, R), (IR5, R), (IR6, R), (IR7, R)}.
S = {move forward, turn 45° left, turn 45° right, turn 90° left, turn 90° right, turn 180°, stop, prepare move forward, prepare turn 45° left, prepare turn 45° right, prepare turn 90° left, prepare turn 90° right, prepare turn 180°, prepare stop}.

$Y = \{(\text{LED}, (100, 0, 10, 20, \dots, 70, 1, 11, 21, \dots, 71)), (\text{Motor}, (0, 1, \dots, 6))\}$.

δ_{ext} = If there is an obstacle trigger the proper state change based on Figure 3.13.

δ_{int} = Change the state based on Figure 3.13.

λ = Generate appropriate outputs to the robot based on Figure 3.13.

ta = move forward → ∞; turn 45° left → 100 milliseconds; turn 45° right → 100 milliseconds; turn 90° left → 200 milliseconds; turn 90° right → 200 milliseconds; turn 180° → 400 milliseconds; stop → ∞; prepare move forward → 0 second; prepare turn 45° left → 0 second; prepare turn 45° right → 0 second; prepare turn 90° left → 0 second; prepare turn 90° right → 0 second; prepare turn 180° → 0 second; prepare stop → 0 second.

Table 3.5 summarizes the integer outputs of the DEVS model and their associated actions to be performed in the robot hardware. The driver interface programmed by the user transforms the numeric values to actions in the robot.

Figure 3.13 illustrates an abstract state diagram of the *epuck0* atomic component. The DEVS graph state diagram summarizes the behavior of a DEVS atomic component by representing the states, transitions, inputs, outputs, and state durations graphically. As we can see, initially, the robot moves forward and if no obstacle is detected from *IR0*, *IR1*, *IR6*, and *IR7* (the four sensors scanning the front direction, as seen in Figure 3.12), it continues moving forward. As soon as an obstacle is detected, the value of the sensor *IR6* is examined. If this sensor shows no obstacle, the left corner of the robot is open resulting in a 45° turn toward the left. Otherwise,

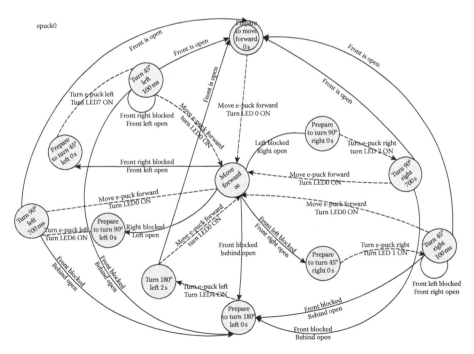

FIGURE 3.13 epuck0 atomic component state diagram.

TABLE 3.5
DEVS Output Mapping Table

Port Name	Port Value	Hardware Command	Comment
OutLED	100	Turn all LEDs off	
	0, 10, 20 ... 70	Turn LED off	The most significant digit is the number of LEDs to be turned off
	1, 11, 21 ... 71	Turn LED on	The most significant digit is the number of LEDs to be turned on
OutMotor	0	Set horizontal and rotational speed to 0	Stop
	1	Set horizontal speed to 0.5	Move forward
	2	Set rotational speed to 1	Turn 45° left
	3	Set rotational speed to −1	Turn 45° right
	4	Set rotational speed to −1	Turn 90° right
	5	Set rotational speed to 1	Turn 90° left
	6	Set rotational speed to 1	Turn 180°

it checks *IR1*, and if there is space, the robot turns 45° to the right. If both *IR1* and *IR6* are blocked, the controller examines *IR2*; if there is space, the robot performs a 90° turn to the left. The same occurs with *IR2*. If all sensors are blocked, the robot tries turning to the opposite direction (180°).

3.4.2   Implementation on the ECD++ Toolkit

To program a DEVS model on ECD++ [28,10], three main components are necessary.

1. A model file in which the model hierarchy, model components, input and output ports of each component and input/output couplings are declared. The model file is passed to the ECD++ executable file as a runtime argument, and the latter instantiates the model components based on the declarations in the model file.
2. Source files of the model components. For each atomic component, a C++ class is defined, and the external and internal transitions and the output function are programmed as methods of this class.
3. A driver interface. A driver function is overridden by the user for each input or output port at the top level of the model hierarchy that is connected to a hardware counterpart.

The ECD++ model file also contains information about the period of the input drivers and the duration of the states for each atomic component. In this example, we have tuned the input period of the IR sensors to 50 milliseconds. Therefore, for every 50 milliseconds, the external transition of the *epuck0* atomic component is invoked, and based on the updated values, the next action is decided. The following is the ECD++ model file of the e-puck robot controller model.

```
1    [top]
2    components : epuck0@epuck
3    out : outmotor outled
4    in : inir0 inir1 inir2 inir3 inir4 inir5 inir6 inir7
5    link : inir0 ir0@epuck0
6    link : inir1 ir1@epuck0
7    link : inir2 ir2@epuck0
8    link : inir3 ir3@epuck0
9    link : inir4 ir4@epuck0
10   link : inir5 ir5@epuck0
11   link : inir6 ir6@epuck0
12   link : inir7 ir7@epuck0
13   link : motor@epuck0 outmotor
14   link : led@epuck0 outled
15   inir0 : 00:00:00:100
16   inir1 : 00:00:00:100
17   inir2 : 00:00:00:100
18   inir3 : 00:00:00:100
19   inir4 : 00:00:00:100
20   inir5 : 00:00:00:100
21   inir6 : 00:00:00:100
22   inir7 : 00:00:00:100
23
24   [epuck0]
25   preparationTime : 00:00:00:000
26   turn45Time : 00:00:00:100
27   turn90Time : 00:00:00:700
28   turn180Time : 00:00:02:000
```

Line 1 defines the *top* coupled component and line 2 declares its components. Lines 3 and 4 declare the output and input ports within the *top* coupled component, respectively. Lines 6–14 define the internal couplings. Lines 15–22 declare the periods of each input port. Lines 24–27 declare the duration of states within the *epuck0* component.

The external function performs the state transitions based on the DEVS graph diagram presented in Section 3.4.1. The following is the source code of the external transition function of *epuck0* atomic component.

```
1  if(state!=Mov_Fwd && IR0>0.04 &&  IR7>0.04 && IR1>0.02 &&
IR6>0.02){
2  }else if((state==Mov_Fwd)&&(IR0<0.05 || IR1< 0.02) &&
IR6>0.04){
3    state = Pre_Trn_45_Lft;
4    holdIn( Atomic::active, preparationTime );
5  }else if((state==Trn_45_Lft)&&(IR0<0.05 || IR1<0.02) &&
IR6>0.04){
```

```
6     state = Trn_45_Lft;
7     holdIn( Atomic::active, turn45Time);
8   }else if((state == Mov_Fwd)&& (IR6< 0.02 || IR7< 0.05) &&
IR1> 0.04){
9     state = Pre_Trn_45_Rgt;
10    holdIn( Atomic::active, preparationTime);
11  }else if((state==Trn_45_Rgt)&& (IR6< 0.02 || IR7< 0.05) &&
IR1> 0.04){
12    state = Trn_45_Rgt;
13    holdIn( Atomic::active, turn45Time);
14  }else if(state == Mov_Fwd && IR[0]< 0.05 && IR[7]< 0.05 &&
IR[2] > 0.04){
15    state = Pre_Trn_90_Lft;
16    holdIn( Atomic::active, preparationTime);
17  }else if(state == Mov_Fwd && IR[0]< 0.05 && IR[7]< 0.05 &&
IR[5] > 0.04){
18    state = Pre_Trn_90_Rgt;
19    holdIn(Atomic::active, preparationTime);
20  }else if(state!=Trn_180&&IR[0]<0.05&&IR[7]<0.05&&IR[2]<0.05
&&IR[5]<0.05){
21    state = Pre_Trn_180;
22    holdIn( Atomic::active, preparationTime);
23  }
```

Line 1 shows the case when *moving forward* is the current state and there is no obstacle ahead. Line 2 manages the case when *IR0* or *IR1* (right side of the robot) is obstructed. In that case, the state of the robot is changed to *prepare turn 45° left* (line 3), and in line 4, the time duration of this state is set. The other cases and the state changes are also indicated in the above-mentioned code snippet. The internal transition function and the output function are similar. For instance, the following code snippet shows a part of the internal transition function:

```
1  Model &epuck::internalFunction( const InternalMessage & )
2  {
3        switch (state){
4          case Pre_Mov_Fwd:
5          case Trn_45_Lft:
6          case Trn_45_Rgt:
7          case Trn_90_Lft:
8          case Trn_90_Rgt:
9          case Trn_180:
10             state = Mov_Fwd ;
11             passivate();
12             break;
13
14          case Pre_Trn_45_Lft:
15             state = Trn_45_Lft ;
16             holdIn( Atomic::active, turn45Time );
17             break;
18 ...
```

Lines 4–9 show a part of the internal transition for the states *prepare move forward*, *turn 45° left*, *turn 45° right*, *turn 90° left*, *turn 90° right*, and *turn 180°*, after which the model continues to *move forward* (line 10). Line 14 shows the case for *prepare turn 45° left* state, after which the component transfers to *turn 45° left* state.

The following code snippet shows a part of the ECD++ output function (λ) implementation:

```
1  Model &epuck::outputFunction( const InternalMessage &msg )
2  {
3        switch (state){
4            case Pre_Mov_Fwd:
5                sendOutput( msg.time(), led, 100) ;//Turn all
Leds off
6                sendOutput( msg.time(), motor, 1) ;//Moving
Forward
7                sendOutput( msg.time(), led, 1) ;//Turn Led 0 on
8                break;
9
10           case Trn_45_Lft:
11               sendOutput( msg.time(), led, 70) ;//Turn Led 7
off
12               sendOutput( msg.time(), motor, 1) ;//Moving
Forward
13               sendOutput( msg.time(), led, 1) ;//Turn Led 0 on
14               break;
15 …
```

In this case, line 4 handles the outputs of state *prepare move forward* in which three different outputs are generated. Line 5 is the output command to turn off all LEDs. Line 6 shows the moving forward command sent to the *motor* port and line 7 is the command to turn *led0* on. These outputs are then decoded and converted by the corresponding drivers. Lines 9–13 show the outputs for the *turn 45° left* state where the *led7* is turned off first, then the motors are instructed to move forward and *led0* is turned on afterwards.

The following code snippet shows the driver interface function for the *OutMotor* output port of the *top* coupled component. The outputs generated in the output function for the *motor* port of the *epuck0* atomic component are inputted to this function as an integer argument and the respective hardware command is spawned here.

```
1  bool OutMotor::pDriver(Value &value)
2  {
3      switch((int)value){
4          case 0: //Stop
5              playerc_position2d_set_cmd_vel(position2d, 0, 0,
0, 1);
6              break;
7          case 1: //Moving Forward
8              playerc_position2d_set_cmd_vel(position2d, 1, 0,
0, 1);
```

```
9               break;
10          case 3: //Turn 45 deg. Right
11          case 4: //Turn 90 deg. Right
12              playerc_position2d_set_cmd_vel(position2d,
0,0,-10,1);
13              break;
14          case 2: //Turn 45 deg. Left
15          case 5: //Turn 90 deg. Left
16          case 6: //Turn 180 deg. (turn from left)
17              playerc_position2d_set_cmd_vel(position2d, 0, 0,
10,1);
18              break;
19  };
20  }
```

Lines 4–6 show the case of the stop command, which is encoded with value 0. Line 5 is the command to stop the robot. Lines 7–9 handle the moving forward output command and lines 10–13 manage the right turning commands. For both 45° and 90° turn actions, the robot starts spinning to the right, while the calibrated duration of the respective state accomplishes the desired degree of spinning. A more accurate approach to perform the turning actions would measure the spinning angle constantly and stop when the desired angle is reached.

3.4.3 Executing the Models

The e-puck model was first tested using virtual-time simulation mode. We designed a virtual space with obstacles and ran the simulation with inputs supplied from an event file, in which two series of inputs to the sensors are defined. Figure 3.14

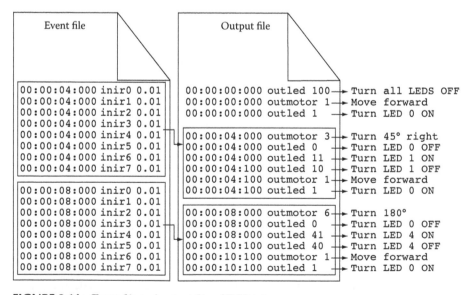

FIGURE 3.14 Event file and output file of ECD++.

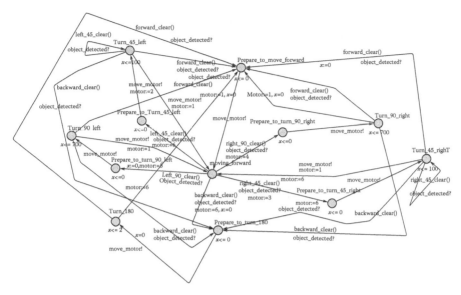

FIGURE 3.15 Robot controller timed automata model.

shows the contents of the event file and output file of the ECD++ and the action associated with the outputs of the controller model. The event file is structured in the format of "time, input port and value" and in this example it consists of two series of events representing the two scenarios shown in Figure 3.12. The first three lines of the output file are the initial outputs of the model that move the robot forward. After 4 seconds of simulation, the first series of inputs is injected into the model, which results in the second series of outputs. The latter spins the robot 45° to the right and performs the appropriate LED commands and after 100 milliseconds moves the robot forward again with the appropriate LED commands. A similar scenario happens for the second series of inputs at time 8 of simulation in which the robot turns 180°.

After verifying the behavior of the model in various scenarios, we tested the model using the actual e-puck robot. The model was executed in RT mode in which the model interacts with the target platform (in this case, the robot hardware). The same behavior was observed and the robot found its way through the obstacles.*

3.4.4 Verifying the Model

To obtain a TA model that is behaviorally equivalent to the DEVS model shown in Figure 3.13, we followed the procedure discussed in Sections 3.3 and 3.4. The equivalent TA is shown in Figure 3.15. In this model, the Boolean conditions of

---

* The results can be seen at http://youtube.com/arslab.

the DEVS state model were defined in functions in the TA model as shown in the following code snippet.

```
1  bool forward_clear(){return ir0>4 && ir7 >4 && ir1>2 &&
ir6>2;}
2  bool left_45_clear(){return (ir0<5 || ir1<2) && ir6>4;}
3  bool right_45_clear(){return (ir6>2 || ir7 <5) && ir1>4;}
4  bool backward_clear(){return ir0<5 && ir7 <5 && ir2<5 &&
ir5<5;}
5  bool right_90_clear(){return ir0<5 && ir7 <5 && ir5>4;}
6  bool left_90_clear(){return ir0<5 && ir7 <5 && ir2>4;}
```

In this TA model, Boolean functions constitute guards on the transitions that evaluate to true whenever the sensor values satisfy the condition given in the DEVS model. While the DEVS model of the robot-controller was tested and simulated with the real robot moving in a specific environment, to verify the robot-controller model, we built a closed system where this model interacts with other models representing the motor and the environment in which the robot travels.

Figure 3.16 shows the TA model of the motor. This model represents the motor states, starting in *Preparing_To_Move_Forward*. It then synchronizes the motor model with the controller model through the *move_motor* channel. The motor model shows six states that the motor can visit depending on the value of the shared variable *motor*, which is updated by the Robot Controller model.

In Figure 3.17, a simple model of the environment in which the robot may travel is shown. This model represents an environment that looks like a contoured closed layout. This layout was modeled in TA by the values assigned to different sensors on the robot *ir0, ir1, ... , ir7*. These values are also shown in the event file used to test the robot DEVS model.

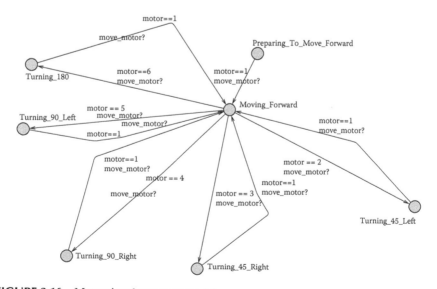

FIGURE 3.16   Motor timed automata model.

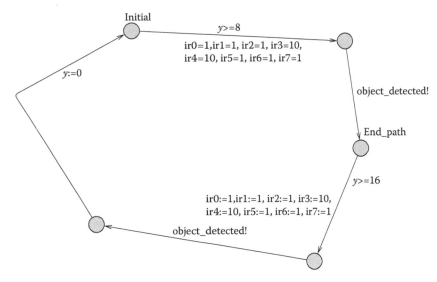

FIGURE 3.17 Environment timed automata model.

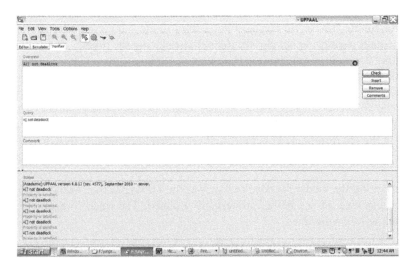

FIGURE 3.18 UPPAAL verification for deadlocks.

The verification of the system composed of Robot-Controller, Motor, and Environment in the UPPAAL tool revealed that it is free of deadlocks as shown in Figure 3.18. This ensures the controller is always able to successfully guide the robot through the given layout.

More complex layouts can also be modeled in TA to verify the system for more complicated behavior. For example, the TA environment model could be constructed to randomly assign values in reasonable range to the sensors. This would model generating arbitrary shaped obstacles around the robot. The verification for deadlock would then be executed to reveal if a deadlock is possible at any particular shape

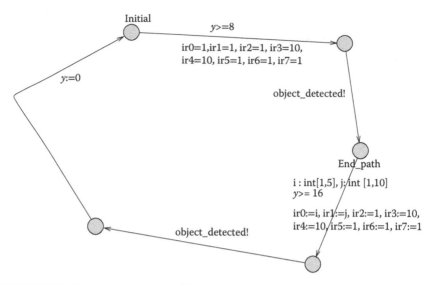

FIGURE 3.19 Environment model with random sensor inputs.

facing the robot. This would reveal either a fault in the robot-controller or the controller being too simple to handle irregular shapes facing the robot.

Another example of verification that we can explore with verification includes different kinds of environments for the robot controller. In this example, we would attempt to check if the controller may enter into a deadlock and stop progress. To do so, we modify the environment model to randomly generate different sensor readings that simulate the robot approach to the situation where there is variety in the environment. This modified environment model is shown in Figure 3.19. In this model, sensor *ir0* obtains a value in the range of [1,5], *ir1* a value in [1,10].

By checking the model for deadlock again, the UPPAAL tool would verify this property and give a trace if this property is violated. The trace comprises values shown in the following lines.

```
1   motor = 1
2   ir0 = 5
3   ir1 = 1
4   ir2 = 1
5   ir3 = 10
6   ir4 = 10
7   ir5 = 1
8   ir6 = 1
9   ir7 = 1
```

This shows that the current design for the robot controller would not handle this combination of sensor values. In this case, either the designer reevaluates the design to address the shortcoming or it may be stated as an assumption on the simple robot controller that this shortcoming is acceptable.

3.5 CONCLUSIONS

Mixing formal software verification and validation together with simulation techniques provides a strong methodology for simulation models verification and validation models. Because it is amenable to RT simulation, this methodology constitutes a significant contribution to the field of embedded software engineering. Simulation models must be validated against actual system properties to make sure the modeler captured the essence of the system under study. Formal verification provides this assurance without the need to run exhaustive simulations and manually analyze simulation results. Simulation models also must be verified against any errors that may have been introduced during model building. Errors such as infinite transitions in a bounded time, which result in an illegitimate DEVS model, are difficult to debug manually and thus good candidates for formal verification.

Embedded systems design must model both the physical system under control and the controller. These models hold more value if they are written in a formalism that can be simulated, such as is the case for DEVS. This allows the designer to simulate the system, change design, and simulate again to reach a correct and optimal design. Formal verification helps verify absence of defects in the system. Once the system is proven to be free of defects, the controller DEVS model, being verified within the complete system, is deployed as the executable controller by running on an embedded DEVS simulator. This eliminates any transformation between the verified model and its implementation, thus avoiding potential defects from creeping into the final implementation because of necessary transformations.

REFERENCES

1. Merz, S., and N. Navet. 2008. *Modeling and Verification of Real-Time Systems: Formalisms and Software Tools*. Hoboken, NJ: John Wiley & Sons, Inc.
2. Wainer, G., E. Glinsky, and P. MacSween. 2005. "A Model-Driven Technique for Development of Embedded Systems based on the DEVS Formalism." In *Model-Driven Software Development. Volume II of Research and Practice in Software Engineering*, edited by S. Beydeda and V. Gruhn. Berlin, Heidelberg: Springer-Verlag.
3. Zeigler, B. P., H. Praehofer, and T. G. Kim. 2000. *Theory of Modeling and Simulation*. 2nd ed. New York: Academic Press.
4. Alur, R., and D. Dill. 1994. "Theory of Timed Automata." *Theoretical Computer Science* 126: 183–235.
5. Miller, J. 2000. "Decidability and Complexity Results for Timed Automata and Semilinear Hybrid Automata." In *Hybrid Systems: Computation and Control*, LNCS 1790, pp. 296–309, London, UK: Springer-Verlag.
6. Saadawi, H., and G. Wainer. 2009. "Verification of Real-Time DEVS Models." In *Proceedings of DEVS Symposium 2009*, pp. 143:1–143:8, San Diego, CA: Society for Computer Simulation International.
7. Saadawi, H., and G. Wainer. 2010. "Rational Time-Advance DEVS (RTA-DEVS)." In *Proceedings of DEVS Symposium 2010*, Orlando, FL, pp. 143:1–143:8, San Diego, CA: Society for Computer Simulation International.

8. Hwang, M-H., and B. P. Zeigler. July 2009. "Reachability Graph of Finite and Deterministic DEVS Networks." *IEEE Transactions on Automation Science and Engineering* 6 (3): 468–78.

9. Behrmann, G., A. David, and K. Larsen. 2004. "A Tutorial on Uppaal." In *Proceedings of the 4th International School on Formal Methods for the Design of Computer, Communication, and Software Systems*, LNCS, 3185, pp. 200–237, Springer-Verlag.

10. Wainer. G. 2009. *Discrete-Event Modeling and Simulation: A Practitioner's Approach.* Boca Raton, FL: CRC Press, Taylor & Francis.

11. Zeigler, B. P., H. Song, T. Kim, and H. Praehofer. 1995. "DEVS Framework for Modelling, Simulation, Analysis, and Design of Hybrid Systems." In *Proceedings of HSAC*, Ithaca, NY, LNCS 999, pp. 529–551, London, UK: Springer-Verlag .

12. Christen, G., A. Dobniewski, and G. Wainer. 2004. "Modeling State-Based DEVS Models in CD++." In *Proceedings of MGA, Advanced Simulation Technologies Conference*, Arlington, VA. San Diego, CA: Society for Computer Simulation International.

13. Castro, R., E. Kofman, and G. Wainer. 2010. "A Formal Framework for Stochastic DEVS Modeling and Simulation." *Simulation, Transactions of the SCS 86* (10): 587–611.

14. Hong, J. S., H. S. Song, T. G. Kim, and K. H. Park. 1997. "A Real-Time Discrete Event System Specification Formalism for Seamless Real-Time Software Development." *Discrete Event Dynamic Systems* 7 (4): 355–75.

15. Song, H. S., and T. G. Kim. February 2005. "Application of Real-Time DEVS to Analysis of Safety-Critical Embedded Control Systems: Railroad Crossing Control Example." *Simulation* 81 (2): 119–36.

16. Furfaro, A., and L. Nigro. 2008. "Embedded Control Systems Design Based on RT-DEVS and Temporal Analysis using UPPAAL." In *Computer Science and Information Technology*, IMCSIT 2008, October 20–22, 2008, pp. 601–08, Polish Information Processing Society.

17. Furfaro, A., and L. Nigro. June 2009. "A Development Methodology for Embedded Systems Based on RT-DEVS." *Innovations in Systems and Software Engineering 5*: 117–27.

18. Han, S., and K. Huang. 2007. "Equivalent Semantic Translation from Parallel DEVS Models to Time Automata". In *Proceedings of 7th International Conference on Computational Science, ICCS,* pp. 1246–1253, Berlin, Heidelberg: Springer-Verlag.

19. Giambiasi, N., J.-L. Paillet, and F. Châne. 2003. "Simulation and Verification II: From Timed Automata to DEVS Models." In *Proceedings of the 35th Winter Simulation Conference (WSC '03)*, New Orleans, LA, pp. 923–931, San Diego, CA: Society for Computer Simulation International.

20. Hernandez, A., and N. Giambiasi. 2005. pp. 923–931, San Diego, CA: Society for Computer Simulation International. "State Reachability for DEVS Models." In *Proceedings of Argentine Symposium on Software Engineering*, Buenos Aires, Argentina, pp. 251–265.

21. Dacharry, H., and N. Giambiasi. 2007. "Formal Verification Approaches for DEVS." In *Proceedings of the 2007 Summer Computer Simulation Conference (SCSC)*, San Diego, CA, pp. 312–319, San Diego, CA: Society for Computer Simulation International.

22. Hong, K. J., and T. G. Kim. 2005. "Timed I/O Test Sequences for Discrete Event Model Verification." In *AIS 2004*, Jeju, Korea, pp. 275–84, LNAI 3397.

23. Labiche, Y., and G. Wainer. 2005. "Towards the Verification and Validation of DEVS Models." In *Proceedings of the 1st Open International Conference on Modeling & Simulation*, Clermont-Ferrand, France, pp. 295–305.

24. Bowman, H., and R. Gomez. 2006. *Concurrency Theory: Calculi and Automata for Modelling Untimed and Timed Concurrent Systems.* London: Springer-Verlag.

25. Emerson, E. A., and E. M. Clarke. 1982. "Using Branching-Time Temporal Logic to Synthesize Synchronization Skeletons." *Science of Computer Programming* 2 (3): 241–66.
26. Alur, R., C. Courcoubetis, and D. L. Dill. 1990. "Model-Checking for Real-Time Systems." In *5th Symposium on Logic in Computer Science (LICS'90)*, pp. 414–25, IEEE.
27. Henzinger, T. A. 1994. "Symbolic Model Checking for Real-Time Systems." *Information and Computation* 111: 193–244.
28. Moallemi, M., and G. Wainer. 2010. "Designing an Interface for Real-Time and Embedded DEVS." In *Proceedings of 2010 Symposium on Theory of Modeling and Simulation*, TMS/DEVS'10, Orlando, FL.*Transition,* pp. 137.1–137.8, San Diego, CA: Society for Computer Simulation International.

4 Optimizing Discrete Modeling and Simulation for Real-Time Constraints with Metaprogramming

Luc Touraille, Jonathan Caux, and David Hill

CONTENTS

4.1 Introduction 97
4.2 Code Generation and Metaprogramming for Optimizing Real-Time
Simulation...98
    4.2.1 Concepts and Definitions...99
    4.2.2 Text Generation..100
        4.2.2.1 Data Generation 100
        4.2.2.2 Instructions Generation..101
    4.2.3 Domain-Specific Languages...101
        4.2.3.1 Embedded DSLs 103
    4.2.4 Multistage Programming Languages 105
        4.2.4.1 C++ Template Metaprogramming 106
    4.2.5 Partial Evaluation 108
    4.2.6 Comparison of the Different Approaches......................................111
4.3 Application to Simulation: DEVS-Metasimulator....................................112
    4.3.1 DEVS/RT-DEVS...112
    4.3.2 Application of Metaprogramming to DEVS Simulation 115
4.4 Conclusion 118
References...118

4.1 INTRODUCTION

Real-time simulation is characterized by the strong constraints that must be respected regarding the time spent simulating an activity. Usually we can run simulation faster than real time, and we synchronize with virtual time with techniques that are now mastered by many engineers [1]. Sometimes, the simulation can execute at a higher rate than the actual system evolves; in this case, all that is necessary is delaying the simulation so that virtual time stays synchronized to the expected wall clock real time. However, when the simulation involves heavy computation or the simulated activities are very short-lived, meeting the deadlines may imply optimizing certain parts of the simulation to reduce its execution time. A simulation run that is slower than real time can occur when dealing with detailed physical models. For instance, we experienced this when simulating the details of particles for physical studies in nuclear medicine [2,3]. In these works, we faced modeling cases where it was impossible to meet the real-time constraints. Medical scanning experiments that last 20–30 minutes can be simulated in a few seconds with analytical models. However, the results obtained can be very imprecise for small tumors (up to 100% error for tumors lesser than 1 cm in diameter). To be more precise, we must build 3D spatial stochastic models, which run for more than 9 hours on our current computer, and many replicates are necessary to reduce the error (less than 10% error achieved in 3 days of computing using the European Computing Grid) [4].

Even in simpler cases, achieving real-time simulation can be complicated. We have met such a case when dealing with fire spreading simulations following detailed physical experiments where physicists wanted to meet strong requirements in terms of both performance and precision. When a fixed spatial precision is set for the predictions obtained with fire spreading modeling, efficient models and high-performance computations are required [5]. The efficiency of discrete event simulators is now receiving significant attention [6–9].

In this chapter, we present a set of software engineering techniques for improving performance through code generation and metaprogramming for greater runtime efficiency. The gain obtained through such techniques enables reconsidering some simulation applications where it would be a distinct benefit to have precise human-in-the-loop simulations that are currently too slow to be considered for real-time applications. There are a number of optimization techniques that can be applied; in Section 4.2, we focus on what can be achieved with code generation and metaprogramming. After the introduction of several metaprogramming techniques, we propose a comparison to show the different benefits they provide regarding performance and design. Section 4.3 applies C++ Template Metaprogramming (TMP) to the Discrete Event System Specification (DEVS) in the simulation domain. We then close with some conclusions in Section 4.4.

4.2   CODE GENERATION AND METAPROGRAMMING FOR OPTIMIZING REAL-TIME SIMULATION

An intuitive and (most of the time) easily verifiable notion is that the more generic a program is, the less efficient it is. A solution that is especially tailored to answer a specific problem or small set of problems has a good chance of being more effective than a solution solving a bigger set of problems. For example, a piece of code computing the result of an equation will perform better than a full-fledged equation solver. However, this classic trade-off between abstraction/genericity and performance can be overcome to attain the best of both worlds by using metaprogramming and particularly dedicated program generation.

4.2.1 CONCEPTS AND DEFINITIONS

The simplest definition of a metaprogram is "a program that creates or manipulates a program." This definition encompasses many concepts: source code generation, compilation, reflection, string evaluation, and so on. There is currently no broadly accepted taxonomy of metaprogramming systems. Two interesting propositions can be found in Refs. [10,11], where the authors identify concepts and relationships to characterize metaprograms. When dealing with performance, the most useful aspect of metaprogramming is program generation. Sheard [10] provides the following definition for a program generator:

> A program generator (a meta-program) solves a particular problem by constructing another program (an object program) that solves the problem at hand. Usually, the generated (object) program is "specialized" for the particular problem and uses fewer resources than a general purpose, non-generator solution.

The definition by Damaševičius and Štuikys [11] is more general and simply states that

> Software generation is an automated process of creation of a target system from a high-level specification.

These definitions are nicely encompassed by the concept of *multistage programming* [12], that is, decomposing the execution of a program into several steps. As Figure 4.1 shows through some examples, any kind of program execution can be seen as a sequence of steps. In particular, execution can be performed in a single stage (interpretation as in PHP program execution), two stages (e.g., compilation + execution), or more (e.g., code generation + compilation + execution).

Adding a new stage to the interpretation approach means reducing the abstraction level to finally produce a code optimized for performing the operations described in the source program. Optionally, each stage can be parameterized by an input other than the source specification. For example, a compiler produces a machine code corresponding to the input source code and specific and optimized for a specific target machine.

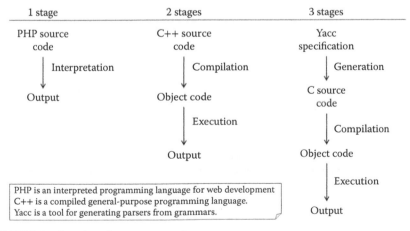

FIGURE 4.1   Samples of n-stage executions.

In Sections 4.2.2 to 4.2.5, we will present several metaprogramming techniques that allow the generation of specialized and optimized artifacts without sacrificing genericity and abstraction. We adopt a pragmatic point of view, so that emphasis will be on practical examples rather than on theoretical aspects.

4.2.2 TEXT GENERATION

Metaprogramming can be as simple as writing strings into a file. For example, the following C code is a metaprogram that generates a C file containing the code for printing the string passed to the metaprogram:

```
#include <stdio.h>

int main( int argc, char * argv[] )
{
    FILE * output;
    output = fopen( argv[1], "w" );

    fprintf( output, "#include <stdio.h>\n\n" );
    fprintf( output, "int main(void)\n" );
    fprintf( output, "{\n" );
    fprintf( output, " printf(\"%s\");\n", argv[2] );
    fprintf( output, "}" );

    return 0;

}
```

CODE 1   C metaprogram generating C source code printing the string given in argument.

Of course, this contrived example adds little value compared to writing the object program directly. However, this technique becomes quite valuable when the object code is too tedious to write by hand or when the metaprogram performs computations on its arguments before generation as all the computations that are performed beforehand need not be executed in the object program. Compared to other metaprogramming techniques, a major advantage of source code generation is that the output is readable by a human, thereby helping debugging. However, the string approach presented here has the drawback that the validity of the object program cannot be enforced at the meta level.

We can make a distinction between two types of generation: data generation and instructions generation.

4.2.2.1 Data Generation

The aim of data generation is to compute a set of values before execution so that they are already loaded in memory when the program starts. For example, some programs use a lookup table containing the sine of many values instead of computing the sine each time it is needed. Most of the time, this lookup table is populated at start-up; however, a more efficient approach is to generate the code for initializing the lookup table directly with the correct values. This technique can also be used to accelerate stochastic simulation by having a metaprogram generate the pseudorandom numbers instead of generating the stream during execution. The metaprogram creates a file containing the initialization of an array with the generated numbers. This file can then be compiled and each simulation that must use the random stream only has to link to the object file obtained. In these programs, the time that it takes to obtain a pseudorandom number is significantly decreased (in previous work [13], we observed a factor of 5). This approach fits with the large memories we have even on personal computers since it favors execution time at the expense of additional object file space. When the amount of pseudorandom numbers is too large to be stored in memory, memory mapping techniques can be used [14,15].

4.2.2.2 Instructions Generation

With data generation, we mostly generate values. It is nevertheless metaprogramming, because we also generate the code to store these values in a data structure accessible at runtime. However, this is not the nominal use of metaprogramming. A metaprogram occasionally generates data, but mainly generates instructions. We can consider the metaprogram as a metasolution to a given problem. Instead of solving the problem, the program generates a solution to the problem. A metaprogram is often more generic than a regular program, and it can create solutions specialized for the inputs it receives. For example, in the work by Missaoui, Hill, and Perret [16], a metaprogram was used to generate programs for computing every reverse translation of oligopeptides containing a given set of amino acids. The metaprogram identified the set of amino acids to be considered as the input and generated a nonrecursive pile handling program with a big number of nested loops, used to compute the backtranslations of any oligopeptide (composed of the amino acids specified at generation). The resulting programs would have been difficult, if not impossible, to write manually without sacrificing some efficiency.

Both generation approaches have been used in the real-time context for decades. In the work by Auer, Kemppainen, Okkonen, and Seppänen [17], a code generator is presented to implement full tasks for real-time systems. The code is produced from a structured data/control flow description following the real-time structured analysis/ structured design (RT-SA/SD) methodology [18] and the program transformation paradigm (which is now a classical concept of the model-driven paradigm). Petri net programs and PL/M-x86 code were generated, and the resulting Petri nets were used for dynamic analysis, helping the discovery of significant errors hidden in the RT-SA/SD specifications. In the work by Webster, Levy, Harley, Woodward, Naidoo, Westhuizen, and Meyer [19], we find the generation of parallel real-time control code. A prototype of a controller hardware platform with the design and implementation of the associated software showed efficient real-time response with customized code generation for a set of particular processors. Parallelism was used to optimize real-time responses and to increase controller throughput.

4.2.3 Domain-Specific Languages

Domain-specific languages (DSLs) are programming or modeling languages dedicated to a particular domain. They contrast with general-purpose programming languages (GPLs) such as C, Java, and LISP and general-purpose modeling languages such as

UML, which can be used in any domain. A DSL usually defines a textual or graphical syntax for manipulating the domain entities. Specifications written in the DSL are compiled either directly to machine code or more often to source code in a GPL.

The advantages of using DSLs are numerous. The most obvious is that they provide abstractions that are at the appropriate level for the user, hence expediting development of models and applications. However, the same effect can be more or less achieved in a GPL through libraries. More interesting are the opportunities that DSLs bring regarding verification and optimization. Indeed, since the language is domain-dependent, the tools for manipulating specifications can themselves be domain-dependent. This means that verification about the program can be performed at design time using domain knowledge. Similarly, the DSL compiler can perform several domain-specific optimizations [20]. For the sake of illustration, let us consider electronic circuit simulation. Using a traditional library,* the user would manipulate classes and functions. For example, there could be classes for representing each type of logical gate. When compiling this program, all the GPL compiler sees is classes and functions, and therefore, it is only able to perform generic optimizations such as loop unrolling and inlining. Now, if the same program were written in a DSL for circuit simulation (such as VHSIC Hardware Description Language (VHDL) [21]), the DSL compiler could perform more specific optimizations. For example, the DSL compiler could be aware that the combination of two NOT gates is equivalent to no gates at all and consequently avoid generating the code for these two gates in the object program.

DSLs have been used in simulation for a long time. In a previous work [22,23], we surveyed some of the most prominent simulation DSLs, which are briefly discussed in the current chapter. Back in the 1960s, IBM introduced GPSS (Global Purpose Simulation System [24]), a language for discrete event simulation with both a textual syntax and a graphical syntax, which could be compiled to machine code or to interpretable pseudocode. In the late 1970s, SLAM (Simulation Language for Alternative Modeling) [25] was introduced. It made it possible to use several modeling approaches (process, event, and continuous [26]), possibly combined together. A few years later, SIMAN (SIMulation ANalysis [27]) was released. It allowed modeling discrete, continuous, or hybrid systems and was used in several modeling and simulation software such as ARENA, which is still evolving and widely used. We can also cite the QNAP2 (Queuing Network Analysis Package 2 [28]) language, based on the theory of queuing networks. The trend in the past few years has been to make languages even more specialized. Indeed, the notion of specificity is rather subjective. After all, even though the above-mentioned languages seem more specific than a GPL such as FORTRAN or C, they can also be seen as much more generic than a DSL for flight simulation, for instance. The more specific a DSL is, the easier it is to use it and to perform verification and optimization on the model. However, it also greatly limits its scope. As a consequence, the cost of developing the DSL should not be higher than the outcome obtained by using it. Fortunately, there are more and more tools available to ease their development. One of the most active areas in this regard is model-driven engineering (MDE) [29,30]. Indeed, the MDE approach and the tools developed to

---

* "Traditional library" as opposed to "active library," a notion that is presented later in this section.

support it enable, among other things, the creation of new DSLs in a very short time. DSLs are supported by providing metamodels and model transformations to develop abstract syntaxes, concrete syntaxes, code generators, graphical editors, and so on.

Several DSLs have been developed for real-time programming; some of them target a broad range of real-time systems (e.g., Hume [31], or the graphical modeling language used in Simulink® [32]), while others are more specialized and are peculiar to a given business such as avionics [33] or robotics [34]. All these DSLs make it easier to write real-time applications by providing high-level constructs, for example, for describing time constraints in an operation. Moreover, these languages most of the time exhibit characteristics that are difficult to obtain with GPL, such as determinacy or bounded time/space. They can also outperform GPLs by applying domain-specific optimizations. For example, Simulink can minimize a certain type of loops in models, thereby removing the need to solve them at runtime with a computationally expensive loop solver.

4.2.3.1 Embedded DSLs

The development of a DSL from scratch can be quite cumbersome. Tools are available to facilitate the process, such as Lex and Yacc [35] or MDE-related tools, but it is still not a seamless experience. Moreover, the DSL puts some burden on the user who must learn an entirely new syntax and use some specific tools to generate or execute its code. Finally, it sometimes happens that a DSL must contain a GPL as a sublanguage to provide the user with complete flexibility on some part of a program (these are sometimes called *hybrid DSLs*).

An interesting approach to overcome these issues is the concept of embedded domain-specific languages (EDSLs), sometimes called *internal DSLs* [36]. These DSLs are embedded in a host language (usually a GPL), meaning that they are defined using the host constructs and syntax. This implies that every expression in the DSL must be a valid expression in the GPL. Even though this seems to be an unacceptable restriction, the gains obtained by using an EDSL are quite attractive:

- The syntax is the same as the host GPL, and therefore, the user can focus on learning the DSL semantics.
- The GPL can be used in synergy with the DSL to have both the practical expressivity of the DSL and the theoretical expressivity of the GPL.
- It is not necessary to write a custom lexer, parser, and code generator because all of these are provided for the GPL and can be exploited at no additional charge for the DSL, including the generic optimizations performed by the GPL optimizer.

Fowler identified several techniques for writing DSLs in object-oriented languages and more broadly in any procedural language [36]. These techniques aim at providing a *fluent interface* to a library, meaning an API designed to flow like natural language: the code using the library should be readable as if it were written in a language of its own. Method chaining is one of the techniques that can be used to provide a fluent interface. Code 2 hereafter shows a sample of code for operating a robot that uses a fluent interface. The interface of the "Robot" class is designed so that the code using it can be read as if using a language specific to robot manipulation.

```
Robot wall_e;

wall_e.
    right( 60 ).
    forward().
        length( 100 ).
        speed( 20 ).
    left( 20 ).
    deployArm();
```

CODE 2   Sample use of a fluent interface for robot manipulation.

The syntax of the DSL can be even richer in languages that support operator over-loading such as C++ and C# and/or operator definition such as Scala and F#. In the first case, the DSL designer can reuse existing operators in his/her own language to provide a more usable syntax, while in the second case, the designer can even create his/her own keywords to extend the original GPL syntax.

However, as enjoyable as providing a language-like interface to a library can be, it is not acceptable in the case at hand if it unfavorably affects performance. In gen-eral, providing a fluent interface implies additional runtime computations, notably to keep track of the context. Does it mean that EDSLs should not be employed when execution time is a concern? No, not if we couple the notion of EDSL with the one of *active library*. Veldhuizen and Gannon [37], who coined the term in 1998, provide the following definition:

> Unlike traditional libraries which are passive collections of functions and objects, Active Libraries may generate components, specialize algorithms, optimize code [and] configure and tune themselves for a target machine [...].

In other words, an active library can provide the same generic abstractions as a traditional one while at the same time providing maximum efficiency, thanks to code generation and optimization, achievable through metaprogramming. We can see an active library as a library generator that creates libraries tailored for the user-specific needs, depending on its parameterization.

Therefore, an EDSL implemented as an active library will be able to perform the same operations the compiler executes when dealing with external DSL. Ahead of execution, the library can perform abstract syntax tree rewriting, evalua-tion of constant expressions, domain-specific optimizations, and so on, effectively resulting in a code specialized for the user's program. The languages providing the most powerful facilities for implementing "active EDSLs" are functional lan-guages such as dialects of the LISP family, which provide macros for extending the syntax and quasiquoting/unquoting* for generating code, and languages with very flexible syntax such as Ruby or Scala. However, these languages are not as widely used as other languages such as Java or C, and their strength lies in aspects other than performance. On the contrary, the C++ language is one of the most used languages and exhibits some of the best performance. And the good news for

---

* These terms will be defined in Section 4.2.4.

this now "old" language is that it supports active libraries, EDSLs, and metaprogramming, through the use of the template mechanism, as will be discussed in Section 4.2.4.

A pioneer in the area of C++ active EDSLs is Blitz++ [38], a scientific computing library that uses metaprogramming to optimize numerical computations at compile time, making it possible to equal and sometimes exceed the speed of FORTRAN. Since then, frameworks have been developed to facilitate the writing of C++ EDSL. The most accomplished one is probably Boost.Proto [39], a C++ EDSL for defining C++ EDSLs. This active library makes it extremely easy to define grammars, evaluate or transform expression trees, and so on, all in C++ and with most of the computation happening at compile time. It has been successfully exploited in several domains such as parsing and output generation (Boost.Spirit), functional programming (Boost.Phoenix), or parallel programming (Skell BE [40]).

In the real-time domain, most EDSLs are based on functional languages, which are more amenable to static analysis than imperative languages and easier to adapt into DSLs. A good example is Atom, a Haskell-based EDSL [41] that provides several guarantees such as deterministic execution time and memory consumption, notably by performing task scheduling and thread synchronization at compile time. A short overview of the available DSLs for real-time systems modeling can be found in Ref. [42].

4.2.4 MULTISTAGE PROGRAMMING LANGUAGES

Multistage programming languages (MSPLs) are languages including constructs for concisely building object programs that are guaranteed to be syntactically correct [12].

In traditional programming languages, program generation is performed through either strings or data types manipulation. The first approach, presented at the beginning of this section, has the major drawback that there is no way of proving that the resulting program will be correct. After all, `"obj.meth(arg)"` and `"î#%2$µ"` are both strings, but the first one may have a meaning in the object program while the second will probably be an erroneous piece of code. This issue can be solved by using data types to represent the object program (e.g., manipulating instances of classes such as Class, Method, Expression, and Variable), but doing so makes the metaprogram much less readable and concise.

To overcome these problems, MSPLs contain annotations to indicate which pieces of code belong to the object program. These pieces of code will not be evaluated during execution and will simply be "forwarded" to the object program. They must nevertheless be valid code, hence providing syntactic verification of the object program at the meta level. This is the core of the LISP macros system, where the quote operator is used to annotate object code. More interestingly, dialects of the LISP family usually include a backquote (quasiquote in Scheme) operator. A backquoted expression, as a quoted expression, will be included in the object program, but it can also contain unquoted expressions that will be evaluated before being included.* Backquoted expressions can be nested to provide an arbitrary number of stages. The LISP code

---

* A similar feature, command substitution, can be found in Unix shells. Incidentally, the character used to denote commands that must be substituted is also the backquote.

below provides an example of a power function that generates the code for computing x^n for a fixed n. The pow function takes two parameters, the base x and the exponent n, and is recursively defined ($x^0 = 1$, $x^n = x * x^{n-1}$). However, since the multiplication is backquoted, it will not be performed immediately. Instead, it will be included in the generated code. The two operands being unquoted (through the comma operator), they will be replaced by the result of their evaluation. Eventually, the pow function will generate unrolled code without recursive calls nor test for base case.

```
(define (pow x n)
    (if (= n 0)
        1
        `(* ,x ,(pow x (- n 1)))
    )
)

(pow 'var 4)

;Output: (* var (* var (* var (* var 1))))
```

CODE 3 LISP Code generating the code computing x^n.

This backquote feature, even though powerful, can create some issues regarding symbol binding. In particular, instead of being bound at the metaprogram level, they are bound during evaluation (in the object program), which is often not the desired behavior. The MetaML language [43] has been developed, among other reasons, to solve these problems.

Functional languages are well suited for metaprogramming, thanks to their homoiconicity ("data is code"), and have been successfully used to develop real-time applications. For instance, in the late 1980s, the expert system G2 was developed in Common LISP and heavily relied on macros to generate efficient code [44]. Thanks to that, this expert system was used for many soft real-time applications, including space shuttle monitoring and ecosystem control (Biosphere II).

However, functional languages are far from being mainstream and are not very well suited to computation-intensive simulations. More classic choices in this domain are C, C++, and Java. In these languages, an arbitrary number of stages would be difficult to achieve, although one of them still has some multistaging capability.

4.2.4.1 C++ Template Metaprogramming

C++, the well-known multiparadigm language, includes a feature for performing generic programming: templates. As a reminder, a class (resp. function) template is a parameterized model from which classes (resp. functions) will be generated during compilation, more precisely during a phase called template instantiation. This feature is very useful for writing generic classes (resp. functions) that can operate on any type or a set of types meeting some criteria. Moreover, templates proved to be more powerful than what was originally thought when they were introduced in the language. In 1994, Unruh found out that they could be used to perform numerical computations such as computing prime numbers, and Veldhuizen later established that templates were Turing complete [45]. These discoveries gave birth to a metaprogramming technique called C++ TMP [46].

Veldhuizen described C++ TMP as a kind of partial evaluation (see Section 4.2.5); however, it does not really correspond to the accepted definition. A more appropriate way to look at it would be to consider C++ with templates as a two and a half stage programming language: one stage and a half for template instantiation and compilation (which are in this case two indivisible steps), and one stage for execution.

C++ TMP exhibits several characteristics very close to functional programming, namely, lack of mutable variables, extensive use of recursion, and pattern matching through (partial) specialization. As an example, Code 4 hereafter presents the equivalent in C++ TMP of the LISP code shown above. The class template pow is parameterized by an exponent N. It contains a member function apply that takes as parameter a base x and multiplies it by x^{N-1}. To do so, the template is instantiated with the decremented exponent, and the apply member function of this instantiation is invoked. The base case is handled through template specialization for N = 0. When the compiler encounters pow<4>::apply, it successively instantiates pow<3>::apply, pow<2>::apply, pow<1>::apply, and finally pow<0>::apply, which matches the specialization, hence stopping the recursion. Since the functions are small and simple, the compiler can inline them and remove the function calls, effectively resulting in a generated code that only performs successive multiplications.

```
template <unsigned int N>
struct pow
{
    static double apply(double x)
    {
        return x * pow<N-1>::apply(x);
    }
};

template <>
struct pow<0>
{
    static double apply(double x)
    {
        return 1;
    }
};
[...]
double res = pow<4>::apply(var);
// the generated assembly code will be equivalent to
// var * var * var * var;
```

CODE 4 C++ template metaprogram for computing x^n.

Writing metaprograms with C++ TMP is not as seamless an experience as it is in LISP or MetaML. The syntax is somewhat awkward since this usage of templates was not anticipated by the C++ Standards Committee. Fortunately, several libraries have been developed to smooth the process. The most significant ones are the MetaProgramming Library (MPL) and Fusion, two libraries belonging to the Boost repository. Boost.MPL is the compile-time equivalent of the C++ Standard Library. It provides several sequences, algorithms, and metafunctions for manipulating types during compilation. Boost.Fusion makes the link between compile-time metaprogramming and runtime programming, by providing a set of heterogeneous containers (tuples) that can hold elements with arbitrary types and several functions and algorithms operating on these containers either at compile time (type manipulation) or at runtime (value manipulation).

An application of C++ TMP to modeling and simulation is provided in Section 4.3.

4.2.5 PARTIAL EVALUATION

The last metaprogramming technique presented in this section is known as *partial evaluation* [47]. Partial evaluation is an automated form of multistage programming technique where the programmer is not required to invest any additional effort to obtain the benefit of program specialization.

In mathematics and computer science, *partial application* is the technique of transforming a function with several parameters to another function with a smaller arity where some of the parameters have been fixed to a given value. Partial evaluation is quite similar, except that it applies to programs instead of mathematical functions. A partial evaluator is an algorithm that takes as input a source program and some of its inputs, and generates a *residual* or *specialized* program. When run with the rest of the inputs, the residual program generates the same output as the original program. However, in the meantime, the partial evaluator had the opportunity to evaluate every part of the original program that depended on the provided inputs. Consequently, the residual program performs fewer operations than the source program, hence exhibiting better performance. Formally, we consider a program as a function of static and dynamic data given as input, which produces some output:

$$P : \left(I_{\text{static}}, I_{\text{dynamic}} \right) \xrightarrow{\text{yields}} O$$

Given this definition, a partial evaluator can be defined as follows:

$$PE : \left(P, I_{\text{static}} \right) \xrightarrow{\text{yields}} P*$$

such that

$$P* : \left(I_{\text{dynamic}} \right) \xrightarrow{\text{yields}} O$$

As an example, in Code 5, C code of the binary search algorithm is shown. The algorithm recursively searches a given value in an ordered array and returns the index of the value (or −1 if it is not found).

```c
int search( int * array, int value, int size )
{
    return binary_search( array, value, 0, size-1 );
}

int binary_search( int * array, int value, int firstIndex, int lastIndex )
{
    if ( lastIndex < firstIndex ) return -1;    // value not found

    int middleIndex = firstIndex + ( lastIndex - firstIndex ) / 2;

    if ( array[ middleIndex ] == value )          // value found
    {
        return middleIndex;
    }
    else if ( array[ middleIndex ] > value )      // value is before
    {
        return binary_search( array, value, firstIndex, middleIndex - 1 );
    }
    else                                          // value is after
    {
        return binary_search( array, value, middleIndex + 1, lastIndex );
    }
}
```

CODE 5 Binary search in C.

The search function takes three parameters: the ordered array, the searched value, and the size of the array. Assuming we have a means to partially evaluate this function, Code 6 shows the residual function we would obtain with size = 3.

```c
int search3( int * array, int value )
{
    if ( array[ 1 ] == value )
    {
        return 1;
    }
    else if ( array[ 1 ] > value )
    {
        if ( array[ 0 ] == value )
            return 0;
        else
            return -1;
    }
    else
    {
        if ( array[ 2 ] == value )
            return 2;
        else
            return -1;
    }
}
```

CODE 6 Binary search residual function for a given size (3).

In this residual function, partial evaluation eliminated all recursive calls, the computation of the middle index, and the test for the "value not found" base case. Most of the time, specializing a function or a program will yield a code that is both smaller and faster since many instructions will be eliminated. However, since partial evaluation performs operations such as loop unrolling and function inlining, it can sometimes lead to code bloat. This phenomenon, called overspecialization, should be avoided in the context of embedded programming. In addition to improving performance, the specialization of functions or programs can also be used to verify assertions about inputs before execution. The assertions will be checked by the partial evaluator, which will abort the residual program production if one of them is not verified.

At this point, it should be obvious that partial evaluation shares the same goal as other multistage programming techniques, namely, to produce an optimized version of a program, specialized for some data. The main difference is that the process of specialization is assigned to a partial evaluator, that is, a program that will automatically perform the generation. The partial evaluator is in charge of determining which pieces of data are *static* and hence can be exploited to perform ahead-of-time computations and which ones are *dynamic* (not known before runtime).

A partial evaluator operates in two main phases. First of all, it must annotate the input program to discriminate between *eliminable* and *residual* instructions. This step, called *binding-time analysis*, must ensure that the annotated program will be correct with respect to two requirements: congruence (every element marked as eliminable must be eliminable for any possible static input) and termination (for any static input, the specializer processing the annotated program must terminate). The second step actually performs the specialization by using several techniques, the most prominent ones being symbolic computation, function calls unfolding (inlining), and program point specialization (duplication of program parts with different specializations).

There are not many partial evaluators available yet. Most of the existing ones target declarative programming (notably functional and logic programming), where programs can be easily manipulated. However, there also exist partial evaluators for some imperative programming languages such as Pascal and C. Regarding more recent languages, it is interesting to mention a work that aims at providing a partial evaluator for the Common Intermediate Language [48]. This Common Intermediate Language is the pseudocode of the Microsoft .NET framework to which many high-level programming languages are compiled (C#, VB.NET, F#...).

Partial evaluation can sometimes be combined with other metaprogramming techniques. In the work by Herrmann and Langhammer [49], the authors apply both multistage programming (see Section 4.2.4) and partial evaluation to the interpretation of an image-processing DSL. Their interpreter first simplifies the original source code by partially evaluating it with some static input (the size of the image), then generates either bytecode or native code using staging annotations, before eventually executing the resulting program with the dynamic input (the image to be processed). Even though all these steps are performed at runtime, the improvement in execution time as compared with classical interpretation is extremely good, up to 100×.

In yet other work [50], partial evaluation is applied to real-time programs written for the Maruti real-time operating system to obtain deterministic execution times. Indeed, to be reusable, programs often must use features such as recursion or loop with nonbounded bounds that hinder the analysis of the program, notably the estimation of the execution time. The authors show that by partially evaluating these programs for some input, it is possible to remove the stochasticity inherent in these features and thereby obtain deterministic and predictable execution times.

4.2.6 COMPARISON OF THE DIFFERENT APPROACHES

Table 4.1 presents a comparison of the metaprogramming techniques described previously. The criteria retained are as follows:

- Generation speed: How fast the metaprogram generates the object program.
- Syntactic correctness of the object program: Whether the syntactic validity of the object program is enforced at the metaprogram level (✓) or not (×).
- Ease of development/Ease of use: How easy it is to create and use metaprograms. It is important to draw the distinction between these two activities, especially in the case of DSLs and partial evaluation. Indeed, most of the time, the people developing the tools and the ones using it will not be the same, and the work they have to provide will be very different. For instance, developing a partial evaluator is a distinctly difficult task, while developing an application and partially evaluating it is almost the same as writing a "classical" application, since partial evaluation is mostly automatic. Regarding text generation and multistage programming, we considered the case of ad-hoc programs developed with a particular aim in mind, not libraries. Consequently, it is irrelevant to differentiate between development and use.
- Use of common tools: Whether the user can use widely available tools such as a compiler or an interpreter for a common language (C/C++, LISP...) (✓) or whether tools particular to the approach must be considered (×).
- Domain-specific optimizations: Potential of the metaprogram for performing optimizations specific to the domain at hand (e.g., scheduling tasks ahead of runtime).
- Non-domain-specific optimizations: Potential of the metaprogram for performing generic optimizations (e.g., unrolling a loop with a constant number of iterations). Only the optimizations performed by the metaprogram are considered, not the optimizations applied on the object program at a later stage (such as optimization during the compilation of generated source code).

Scores range from ★ to ★★★★★. Each criterion has been expressed so that a high score denotes an asset.

As can be seen from Table 4.1, there is no one-size-fits-all solution. Instead, each approach has its strengths and weaknesses, and selecting one must be done on a case-by-case basis, depending on the necessary features.

TABLE 4.1

Comparison of Metaprogramming Approaches

	Text Generation	Domain-Specific Languages		Multistage Programming	Partial Evaluation
		External DSLs	Embedded DSLs		
Generation Speed	★★★★★	★★★☆☆	★☆☆☆☆ to ★★★☆☆ (1)	★☆☆☆☆ to ★★★☆☆ (1)	★★☆☆☆
Syntactic correctness of the object program	✗	✗	✓	✓	✓
Ease of development	★★★☆☆	★★★★☆	★★☆☆☆	★★★☆☆	★☆☆☆☆
Ease of use		★★★★★	★★★★☆ to ★★★★★ (2)		★★★★☆
Use of common tools	✓	✗	✓	✓	✗
Domain-specific optimizations	★★★★★	★★★★☆	★★★☆☆	★★★★☆	★☆☆☆☆
Non-domain-specific optimizations	★★★★☆	★★★☆☆	★★☆☆☆	★★★☆☆	★★★★★

Note: (1) = depending on the language and the compiler used; (2) = depending on the host language, the potential syntaxes can be more or less constrained.

4.3 APPLICATION TO SIMULATION: DEVS-METASIMULATOR

In Section 4.2, we introduced several metaprogramming techniques for improving program performance. Hereafter, we present an application of one of these techniques, namely C++ TMP, to the simulation domain, more particularly to the Discrete EVent System specification (DEVS) formalism [51,6].

4.3.1 DEVS/RT-DEVS

The DEVS formalism was proposed by Zeigler in 1976. It establishes sound mathematical bases for modeling and simulation through three concepts: atomic models, coupled models, and abstract simulators. A DEVS atomic model is an entity holding a state and evolving through internal state changes and responses to external stimuli.

After undergoing an internal event, the model generates an event that is sent to the "outside world." Formally, a classic DEVS atomic model is defined by a 7-tuple:

$$M = \langle X, Y, S, \text{ta}, \delta_{\text{int}}, \delta_{\text{ext}}, \lambda \rangle$$

where

> X is the set of inputs. Usually, X is decomposed into several sets representing model *ports*.
>
> Y is the set of outputs. It is defined similarly to X, using ports.
>
> S is the state set, containing all the possible characterizations of the system.
>
> ta is the time advance function. It determines how long the system stays in a given state before undergoing an internal transition and moving to another state.
>
> δ_{int} is the internal state transition function. It defines how the system evolves in an autonomous way.
>
> δ_{ext} is the external state transition function. It defines how the system reacts when stimulated by an external event.
>
> λ is the output function. Invoked after each internal transition, it determines the events generated by the system, functions of the state it was in.

Atomic models can be combined to form coupled models. A DEVS coupled model is a hierarchical structure composed of atomic and/or other coupled models. The components are organized in a graph-like structure where output ports of components are linked to input ports of other components. As such, output events generated by some components become input events to others. Formally, a classic DEVS coupled model is characterized by a structure

$$N = \langle X, Y, D, \{M_d \mid d \in D\}, \text{EIC}, \text{IC}, \text{EOC}, \text{Select} \rangle$$

where

> X is the set of input ports and values, as in atomic models.
>
> Y is the set of output ports and values, as in atomic models.
>
> D is the set of component names.
>
> For each d in D, M_d is a DEVS model (coupled or atomic).
>
> EIC is the external input coupling; it connects inputs of the coupled model N to component inputs.
>
> EOC is the external output coupling; it connects component outputs to N outputs.
>
> IC is the internal coupling; it connects component outputs to component inputs.
>
> Select is the tie-breaking function. Classic DEVS is fundamentally sequential. Therefore, when two models are supposed to undergo an internal transition at the same time—and consequently generate two simultaneous events—only one of them must be activated. This arbitrage is performed through the Select function.

The operational semantic of these models have been defined through abstract simulators, which represent algorithms that correctly simulate DEVS models [51].

Many extensions have been proposed to increase the scope of DEVS: fuzzy-DEVS, dynamic structure DEVS, and so on. The real-time DEVS (RT-DEVS) [52] extension is particularly interesting in the context of this book. This formalism adapts classic DEVS to add real-time aspects and provides a framework for modeling systems and simulating them in real time so that they can interact with the physical world. As opposed to classic DEVS, an RT-DEVS simulator uses a real-time clock instead of a virtual one and hence must be deployed on a real-time operating system. The main idea is to fill the time between internal events with actual activities instead of virtually jumping through time.

The definition of RT-DEVS models is only slightly different than Classic DEVS models. Coupled models are specified in the same way, except for the lack of a Select function (which makes no sense in a real-time context, since in the physical world, events can occur simultaneously and are not "sequentialized"). Atomic models only differ in the addition of *activities* and the use of intervals to specify the time advance. An atomic RT-DEVS model is defined as follows:

$$\mathrm{RTAM} = \langle X, Y, S, \mathrm{ti}, \delta_{\mathrm{int}}, \delta_{\mathrm{ext}}, \lambda, \psi, A \rangle$$

where

X, Y, S, δ_{int}, δ_{ext}, and λ are the same as in Classic DEVS.
A is a set of activities. An activity is a function, associated with a state, which must be fully executed before the state of the model can change.
ψ is the activity mapping function, which associates each state with an activity.
ti is the interval time advance function. It specifies, for each state, the estimated interval of time in which the associated activity is expected to be completed.

As for Classic DEVS, abstract simulators have been defined to characterize what must be done to execute RT-DEVS models. These simulators deal with concurrent execution of activities, simultaneous events, and so on. However, some timing discrepancies can appear during simulation. Indeed, simulators must execute not only the activities specified by the model but also the simulation operations necessary to make the model evolve during execution. This implies that, in addition to activities, the simulator must execute internal and external transition functions, output functions, event routing, and so on. Because of this overhead, timing errors can arise. More specifically, activities can start and finish later than they should. Algorithmic approaches have been proposed to compensate for these discrepancies [52]. However, these solutions only work when the duration of activities is larger than the simulation overhead. When activities are short, RT-DEVS models can run too slow to meet the constraints defined by the interval time advance function. Consequently, it is important to reduce the overhead as much as possible. Hereafter, we will explain how metaprogramming can improve the efficiency of DEVS simulators as well as other characteristics such as correctness.

4.3.2 Application of Metaprogramming to DEVS Simulation

DEVS models can be decomposed into structural specifications (ports, model composition, and couplings) and behavioral specifications (transition functions and output function). Most of the time, the structure will not change during execution (one notable exception are dynamic structure DEVS models). Consequently, the simulation of DEVS models appears to be a good candidate for metaprogramming optimizations. Previous work [53] presents a DEVS simulator, implemented as an active library, which specializes itself for the model provided using C++ TMP. The formalism handled by the simulator presented is Classic DEVS and the same ideas can be applied to DEVS extensions—such as RT-DEVS—or to other formalisms.

Table 4.2 classifies elements of DEVS models with respect to the earliest stage at which they are accessible. For example, the names of components in coupled models are static pieces of information (known at compile time) and can therefore be exploited in the metaprogram.

The "metasimulator" uses the static information to evaluate some parts of the simulation during template instantiation and generates a residual simulator specialized for the given model, where only the dynamic operations remain. Thanks to this, several improvements are obtained over more classical approaches that do not use metaprogramming.

A first set of enhancements concerns model verification. Specifically, the metaprogram can check many assertions about the model provided by the library user. This enables testing the correctness of the model before execution even begins.

TABLE 4.2

Classification of the Different Elements of Atomic and Coupled DEVS Models

Static (Compile Time)	Dynamic (Runtime)
Atomic Model	
Input ports names and types	Values on output ports
Output ports names and types	Values on input ports
	Values on state variables
	Functions depending on these values:
	Internal and external transition functions
	Output function
	Time advance function
Coupled Model	
Input ports names and types	Values on input ports
Output ports names and types	Values on output ports
Components names and types	Components instances
Couplings	
Select function	

As a result, some errors are immediately caught and stop the generation of the object program (through compiler errors). The following is a list of the verifications that are performed by the metaprogram:

- Name checking. Names are made available at the compilation stage by making them types instead of strings. As a consequence, the metaprogram can check that the names used throughout the model are all correct.
- Identifier uniqueness. In addition to the preceding point, the metaprogram can also verify that each identifier is unique in its scope. For example, a model shall not have two ports with the same name. If this were the case, compilation would abort—generation of the object program would fail. Concretely, this is achieved by using static asserts, that is, assertions that are verified at compile time.
- Detection of incorrect couplings. The metaprogram is able to detect several kinds of invalid connections between coupled model components:
  - Port-type incompatibility. If the types of two connected ports are incompatible (i.e., the output event of the source component cannot be converted into an input event of the destination component), the metaprogram issues an error and stops its execution.
  - Direct feedback loops. The DEVS formalism forbids algebraic loops, meaning cycles of output-to-input connections without delay. This constraint cannot be enforced by the metaprogram in the general case since it depends on time advance functions, which are not always statically evaluable. However, direct feedback loops (connection of an output port of a component to an input port of the same component) can be detected based solely on the coupling information. Consequently, they can be discovered and rejected by the metaprogram.
  - Simultaneous events on a component. In Classic DEVS, a component must not receive simultaneous events on its input ports. Once again, this can be enforced by the metaprogram using the coupling information. This constraint no longer holds in RT-DEVS.

In addition to these forms of verification, the use of metaprogramming also improves simulation efficiency by making the metaprogram perform actual simulation operations that are usually performed at runtime and by removing some of the overhead usually associated with the use of generic high-level libraries. The following is the list of operations performed by the metaprogram that have an impact on the residual simulator performances:

- Virtual calls dispatch. Genericity is often achieved through inheritance and polymorphism. However, polymorphism comes at a price: since method calls must be bound at runtime, the compiler is deprived of several optimization opportunities, such as inlining. Moreover, the dynamic dispatch most of the time implies additional computation (e.g., a lookup in a virtual method table) generating some overhead. By using a technique called static polymorphism, it is possible to effectively achieve polymorphism at the metaprogram level and remove any necessity for it in the object program.

- Event routing. Events exchanged between components involve two pieces of information: the value carried by the event (dynamic) and the source or destination port of the event (static). Since the source of the event is available at the metaprogramming stage, as well as the coupling information, it is possible for the metaprogram to route the events from their source to their destination at compile time. The result in the residual simulator is that there is no longer a need for any tree structure. Instead, all components are on the same level and are directly connected to one another. During execution, events pass directly from one component to another without going up and down a hierarchy of coupled model handlers.

- Input events filtering. Atomic models usually have different reactions to different stimuli because the change in state depends on the port receiving the event. Consequently, external transition functions commonly make branching based on the input port involved. As discussed earlier, the routing of an event between models is performed by the metaprogram. Therefore, the identifier of the port receiving an event is statically known, which allows the metaprogram to filter events and dispatch to the correct behavior at compile time.

- Removal of certain superfluous operations. On the basis of the structure of a model, the metaprogram is able to remove some simulation operations that are not necessary. For example, if an atomic model has no output ports, there is no need to invoke its output function.

- Simultaneous internal events scheduling. When several components must undergo an internal event at the same time, the Select function is used to determine which one will be activated first. By representing this function with a data structure, it is possible for the metaprogram to perform the tie breaking. This optimization would not be applicable in RT-DEVS since there is no Select function in this extension of DEVS.

In a previous work [53], we tested this metaprogramming technique on a sample DEVS model, and the residual simulator was compared to other software: a program developed with another library supporting simulation of any DEVS model and a piece of software developed specifically for the model considered. The results showed that the residual program generated by the metaprogram was four times faster than the generic library (written in Java). We have also performed a comparison between the residual program and the several versions of C++ software especially crafted and optimized for the model considered. We observed an overhead only in one of these versions. This is explained by the fact that the "handwritten" simulator used some optimizations based on knowledge about the model that was not accessible to the metaprogram (in the same way that a compiler is not able to perform optimizations that depend on the domain of the software developed). However, the generation of the residual program is fully automated, and such a tool can be used by many modelers, whereas very few of them can implement efficient and optimized C++ code.

4.4 CONCLUSION

In this chapter, we introduced several metaprogramming techniques for improving program performance that can be used in the development of real-time software [54,55]. The optimization of the real-time simulation code can include low-level optimizations such as loop unrolling and recoding critical parts in assembly language or even in microcode. In this chapter, we considered optimization at a much higher level that we did not find in the current state of the art in real-time simulation. We discussed different code generation techniques (generation of data and instructions) and also presented the input of DSLs, programming or modeling languages dedicated to a particular domain. MSPLs were also discussed; they include constructs for concisely building object programs guaranteed to be syntactically correct. This class of programming languages includes C++, through the use of TMP. Finally, we offered some insights on partial evaluation, a powerful and automatic method for specializing programs to obtain better performance such as shorter execution time and lower memory footprint. Each of these approaches has already been successfully applied to real-time programming, as can be seen in the numerous references provided. We also proposed a comparison of all the techniques described. A table will help in selecting the most appropriate technique given a set of requirements, since no technique is universally optimal. To provide a more detailed case study using a specific technique, we presented how C++ TMP can be used to develop a DEVS metasimulator exhibiting significant improvements over traditional simulators.

The next avenue we want to investigate is applying metaprogramming techniques to multicore programming. A more powerful architecture must be considered when a simulation runs slower than real time. The increase in raw computing performance is now linked to parallel architectures. Among these architectures, the GP-GPU (General-Purpose Graphical Processing Unit) technology is an easy, inexpensive, and efficient way to allow processing of large data on a personal computer. Much currently available literature has already revealed that it is possible to use the computational capability of GPUs to implement real-time simulations of complex models [56–58]. However, to make the most of it, the development work is still substantial, mainly because of the slow memory accesses that often force the developer to significantly modify the sequential implementation. The new generation of GP-GPU and the template facilities of the latest release of the NVIDIA C/C++ compiler will help in making up for many of these problems. Together with the increase of computational power and the additional fast local memory available on the GPU cards, it will allow developers to implement much more efficient real-time simulations for various domains.

REFERENCES

1. Laplante, P. A. 1992. *Real-Time Systems Design and Analysis, An Engineer's Handbook.* 2nd ed., 361 pp. Hoboken, NJ: John Wiley & Sons.
2. Lazaro, D., Z. El Bitar, V. Breton, D. Hill, and I. Buvat. 2005. "Fully 3D Monte Carlo Reconstruction in SPECT: A Feasibility Study." *Physics in Medicine & Biology* 50: 3739–54.

3. El Bitar, Z., D. Lazaro, V. Breton, D. Hill, and I. Buvat. 2006. "Fully 3D Monte Carlo Image Reconstruction in SPECT Using Functional Regions." *Nuclear Instruments and Methods in Physics Research* 569: 399–403.

4. Reuillon, R., D. Hill, Z. El Bitar, and V. Breton. 2008. "Rigorous Distribution of Stochastic Simulations Using the DistMe Toolkit." *IEEE Transactions on Nuclear Science* 55 (1): 595–60.

5. Innocenti, E., X. Silvani, A. Muzy, and D. Hill. 2009. "A Software Framework for Fine Grain Parallelization of Cellular Models with OpenMP: Application to Fire Spread." *Environmental Modelling & Software* 24: 819–31.

6. Wainer, G. A., and P. J. Mosterman. 2011. *Discrete-Event Modeling and Simulation: Theory and Applications*, 534 pp. Boca Raton, FL: CRC Press.

7. Lee, W. B., and T. G. Kim. 2003. "Simulation Speedup for DEVS Models by Composition-Based Compilation." In *Summer Computer Simulation Conference*, Montreal, Canada, pp. 395–400.

8. Hu, X., and B. P. Zeigler. 2004. "A High Performance Simulation Engine for Large-Scale Cellular DEVS Models." In *High Performance Computing Symposium (HPC'04), Advanced Simulation Technologies Conference (ASTC)*, Arlington.

9. Muzy, A., and J. J. Nutaro. 2005. "Algorithms for Efficient Implementation of the DEVS & DSDEVS Abstract Simulators." In *First Open International Conference on Modeling and Simulation (OICMS)*, Clermont-Ferrand, France, pp. 273–80.

10. Sheard, T. 2001. "Accomplishments and Research Challenges in Meta-Programming." In *Proceedings of the 2nd International Workshop on Semantics, Applications, and Implementation of Program Generation (SAIG'2001)*, Florence, Italy, LNCS 2196, pp. 2–44.

11. Damaševičius, R., and V. Štuikys. 2008. "Taxonomy of the Fundamental Concepts of Metaprogramming." *Information Technology and Control, Kaunas, Technologija* 37 (2): 124–32.

12. Taha, W. 1999. *Multi-Stage Programming: Its Theory and Applications*. PhD dissertation, Oregon Graduate Institute of Science and Technology, 171 pp.

13. Hill, D., and A. Roche. 2002. "Benchmark of the Unrolling of Pseudorandom Numbers Generators." In *14th European Simulation Symposium*, Dresden, Germany, October 23–26, 2002, pp. 119–29.

14. Hill, D. 2002. "Object-Oriented Modelling and Post-Genomic Biology Programming Analogies." In *Proceedings of Artificial Intelligence, Simulation and Planning (AIS 2002)*, Lisbon, April 7–10, 2002, pp. 329–34.

15. Hill, D. 2002. "URNG: A Portable Optimisation Technique for Every Software Application Requiring Pseudorandom Numbers." *Simulation Modelling Practice and Theory* 11: 643–54.

16. Missaoui, M., D. Hill, and P. Perret. 2008. "A Comparison of Algorithms for a Complete Backtranslation of Oligopeptides." *Special Issue, International Journal of Computational Biology and Drug Design* 1 (1): 26–38.

17. Auer, A., P. Kemppainen, A. Okkonen, and V. Seppänen. 1988. "Automated Code Generation of Embedded Real-Time Systems." *Microprocessing and Microprogramming* 24 (1–5): 51–5.

18. Ward, P. T., and S. J. Mellor. 1985. *Structured Development for Real-Time Systems*, Vol. I: ISBN 9780138547875, 156 pp.; Vol. II: ISBN 9780138547950, 136 pp.; Vol. III: ISBN 9780138548032, 202 pp. Upper Saddle River, NJ: Prentice Hall.

19. Webster, M. R., D. C. Levy, R. G. Harley, D. R. Woodward, L. Naidoo, M. V. D. Westhuizen, and B. S. Meyer. 1993. "Predictable Parallel Real-Time Code Generation." *Control Engineering Practice* 1 (3): 449–55.

20. Lengauer, C. 2004. "Program Optimization in the Domain of High-Performance Parallelism." In *Domain-Specific Program Generation. International Seminar*, Dagstuhl Castle, Germany, LNCS 3016, pp. 73–91.

21. Ashenden, P. J. 2001. *The Designer's Guide to VHDL*. 2nd ed., 759 pp. San Francisco, CA: Morgan Kaufmann.
22. Hill, D. 1993. *Analyse Orientée-Objets et Modélisation par Simulation*. 362 pp. Reading, MA: Addison-Wesley.
23. Hill, D. 1996. *Object-Oriented Analysis and Simulation*. 291 pp. Boston, MA: Addison-Wesley Longman.
24. Gordon, G. M. 1962. "A General Purpose Systems Simulator." *IBM Systems Journal* 1 (1): 18–32.
25. O'Reilly, J. J., and K. C. Nordlund. 1987. "Introduction to SLAM II and SLAMSYSTEM." In *Proceedings of the 21st Winter Simulation Conference*, Washington, DC, pp. 178–83.
26. Crosbie, R. 2012. "Real-Time Simulation Using Hybrid Models." In *Real-Time Simulation Technologies: Principles, Methodologies, and Applications*, edited by K. Popovici, and P. J. Mosterman. Boca Raton, FL: CRC Press.
27. Pegden, C. D., R. E. Shannon, and R. P. Sadowski. 1995. *Introduction to Simulation Using SIMAN*. 2nd ed., 600 pp. New York: McGraw-Hill Companies.
28. Potier, D. 1983. "New User's Introduction to QNAP2." INRIA Technical Report No. 40.
29. Schmidt, D. C. 2006. "Model-Driven Engineering: Guest Editor's Introduction." *IEEE Computer Society* 39 (2): 25–31.
30. Gronback, R. C. 2009. *Eclipse Modeling Project: A Domain-Specific Language (DSL) Toolkit*. 736 pp. Upper Saddle River, NJ: Addison-Wesley Professional.
31. Hammond, K., and G. Michaelson. 2003. "Hume: A Domain-Specific Language for Real-Time Embedded Systems." In *Proceedings of the 2nd International Conference on Generative Programming and Component Engineering*, Erfurt, Germany, LNCS 2830, pp. 37–56.
32. Dabney, J. B., and T. L. Harman. 2003. *Mastering Simulink®*, 400 pp. Upper Saddle River, NJ: Prentice Hall.
33. Gamatié, A., C. Brunette, R. Delamare, T. Gautier, and J.-P. Talpin. 2006. "A Modeling Paradigm for Integrated Modular Avionics Design." In *Proceedings of the 32nd EUROMICRO Conference on Software Engineering and Advanced Applications*, Cavtat, Dubrovnik, pp. 134–43.
34. Peterson, J., P. Hudak, and C. Elliot. 1999. "Lambda in Motion: Controlling Robots with Haskell." In *Proceedings of the 1st International Workshop on Practical Aspects of Declarative Languages*, San Antonia, TX, LNCS 1551, pp. 91–105.
35. Levine, J. R., T. Mason, and D. Brown. 1992. *Lex & Yacc*. 2nd ed., 388 pp. Sebastopol, CA: O'Reilly & Associates.
36. Fowler, M. 2010. *Domain-Specific Languages*. 640 pp. Boston: Addison-Wesley Professional. ISBN 9780321712943.
37. Veldhuizen, T. L., and D. Gannon. 1998. "Active Libraries: Rethinking the Roles of Compilers and Libraries." In *Proceedings of the 1998 SIAM Workshop on Object Oriented Methods for Interoperable Scientific and Engineering Computing*, New York, pp. 286–95.
38. Veldhuizen, T. L. 1998. "Arrays in Blitz++." In *Proceedings of the 2nd International Scientific Computing in Object Oriented Parallel Environments (ISCOPE'98)*, LNCS 1505, pp. 223–30. Santa Fe: Springer.
39. Niebler, E. 2007. "Proto: A Compiler Construction Toolkit for DSELs." In *Proceedings of the 2007 Symposium on Library-Centric Software Design*, Montreal, Canada, pp. 42–51.
40. Saidani, T., J. Falcou, C. Tadonki, L. Lacassagne, and D. Etiemble. 2009. "Algorithmic Skeletons within an Embedded Domain Specific Language for the CELL Processor." In *Proceedings of the 18th International Conference on Parallel Architectures and Compilation Techniques*, Raleigh, pp. 67–76.

41. Hawkins, T. 2008. "Controlling Hybrid Vehicles with Haskell." *Presentation at the 2008 Commercial Users of Functional Programming Conference*, Victoria, Canada.
42. Peterson, R. S. 2010. *Testing real-time requirements for integrated systems*. Master's thesis, University of Twente. Retrieved from http://essay.utwente.nl/59656/ (last accessed April 2011).
43. Taha, W., and T. Sheard. 1997. "Multi-Stage Programming with Explicit Annotations." In *Proceedings of the ACM-SIGPLAN Symposium on Partial Evaluation and Semantic Based Program Manipulations PEPM'97*, Amsterdam, pp. 203–17.
44. Allard, J. R., and L. B. Hawkinson. 1991. "Real-Time Programming in Common LISP." *Communications of the ACM – Special Issue on LISP* 34 (9): 64–9.
45. Veldhuizen, T.L. 2003. *C++ Templates are Turing Complete*. Technical report, Indiana University.
46. Abrahams, D., and A. Gurtovoy. 2004. *C++ Template Metaprogramming: Concepts, Tools, and Techniques from Boost and Beyond*. 400 pp. Boston: Addison-Wesley Professional. ISBN 9780321227256.
47. Jones, N. D. 1996. "An Introduction to Partial Evaluation." *ACM Computing Surveys* 28 (3): 480–503.
48. Chepovsky, A. M., A. V. Klimov, Y. A. Klimov, A. S. Mischlenko, S. A. Romanenko, and S.Y. Skorobogatov. 2003. "Partial Evaluation for Common Intermediate Language." In M. Broy and A. V. Zamulin (Eds.), *Perspectives of System Informatics*, LNCS 2890, pp. 171–7.
49. Herrmann, C. A., and T. Langhammer. 2006. "Combining Partial Evaluation and Staged Interpretation in the Implementation of Domain-Specific Languages." *Science of Computer Programming* 62 (1): 47–65.
50. Nirkhe, V., and W. Pugh. 1993. "A Partial Evaluator for the Maruti Hard Real-Time System." *Real-Time Systems* 5 (1): 13–30.
51. Zeigler, B. P., H. Praehofer, and T. G. Kim. 2000. *Theory of Modeling and Simulation: Integrating Discrete Event and Continuous Complex Dynamic Systems*. 2nd ed., 510 pp. San Diego, CA: Academic Press.
52. Hong, J. S., H. S. Song, T. G. Kim, and K. H. Park. 1997. "A Real-Time Discrete Event System Specification Formalism for Seamless Real-Time Software Development." *Discrete Event Dynamic Systems* 7 (4): 355–75.
53. Touraille, L., M. K. Traoré, and D. Hill. 2010. "Enhancing DEVS Simulation through Template Metaprogramming: DEVS-MetaSimulator." In *Proceedings of the 2010 Summer Computer Simulation Conference*, Ottawa, Canada, pp. 394–402.
54. Allworth, S. T. 1981. *Introduction to Real-Time Software Design*. 128 pp. Macmillan. ISBN 9780333271353.
55. Ganssle, J. G. 1992. *The Art of Programming Embedded Systems*. 279 pp. San Diego, CA: Academic Press.
56. Keller, M., and A. Kolb. 2009. "Real-Time Simulation of Time-of-Flight Sensors." *Simulation Modelling Practice and Theory* 17 (5): 967–78.
57. Ritschel, T., M. Ihrke, J.-R. Frisvad, J. Coppens, K. Myszkowski, and H.-P. Seidel. 2009. "Temporal Glare: Real-Time Dynamic Simulation of the Scattering in the Human Eye." *Computer Graphics Forum* 28 (2): 183–92.
58. Wu, E., Y. Liu, and X. Liu. 2005. "An Improved Study of Real-Time Fluid Simulation on GPU." *Computer Animation and Virtual Worlds* 15 (3–4): 139–46.

5 Modeling with UML and Its Real-Time Profiles

Emilia Farcas, Ingolf H. Krüger,
and Massimiliano Menarini

CONTENTS

5.1 Introduction 123
    5.1.1 Challenges for MBE in Embedded Systems............................ 124
    5.1.2 MBE Process 125
    5.1.3 Outline 127
5.2 Requirements for Modeling Languages 127
5.3 Modeling with UML... 129
    5.3.1 Metamodeling Architecture.. 129
    5.3.2 Modeling with UML... 131
        5.3.2.1 Requirements Modeling with UML 131
        5.3.2.2 Modeling Logical and Technical Architectures
                with UML 133
        5.3.2.3 UML Basics by Example... 134
    5.3.3 UML Extension Mechanism.. 139
    5.3.4 UML Behavioral Semantics 140
5.4 UML Extensions for Real Time 142
    5.4.1 Overview of UML Profiles... 143
    5.4.2 Modeling and Analysis of Real-Time and Embedded Systems 145
        5.4.2.1 MARTE Basics 145
        5.4.2.2 MARTE Semantics 149
        5.4.2.3 MARTE Example 150
5.5 Discussion and Open Issues... 152
5.6 Summary and Outlook 155
Acknowledgments.. 156
References... 156

5.1 INTRODUCTION

Embedded systems, such as automotive, avionics, communications, and defense systems, are notoriously difficult to design and develop because of increasing complexity, heterogeneous requirements (combining, e.g., noncritical infotainment and hard real-time engine control), distributed infrastructure, distributed development and outsourcing, time to market, lifecycle management, and cost, to name just a few.

Model-based engineering (MBE) [1] greatly contributes to the development of dependable systems with models playing an important role in domain analysis, requirements elicitation, architecture specification, analysis, documentation, and evolution. Nevertheless, while modeling is accepted as a key engineering activity especially in all engineering disciplines *outside* of software engineering, it is often seen as standing in the way of "the code," and if it is used, it is often only for forward engineering (i.e., creating code from specifications), leaving activities such as diagnosis and failure management to be addressed separately.

The Unified Modeling Language™ (UML®) [2,3] is a general-purpose modeling language—a standard managed by the Object Management Group® (OMG®)—widely used in both academia and industry. It is a family of graphical notations underpinned by a single metamodel [4]. UML can be used at different levels of the development process, especially for requirement modeling and functional design, resulting in specification for behavior, structure, and quality of service (QoS) properties. There exist several UML compliant modeling tools that support code generation to C/C++, Java, Ada, different Real-Time Operating Systems (RTOSs), and Common Object Request Broker Architecture (CORBA®). Moreover, through its profile mechanism, UML can be tailored for various domains or different target platforms.

In this chapter, we present the modeling capabilities of UML and its extensions for real-time systems. We are not providing an introductory UML tutorial, but instead we assume that the reader is familiar with most basic UML notations. Furthermore, we discuss requirements for modeling languages, which helps us pinpoint how and to what degree UML can be used for modeling aspects of real-time systems. This also allows us to identify open issues, such as model consistency, that remain to be addressed to provide a systematic modeling methodology.

In the following paragraphs, we discuss the challenges in embedded systems and how models can be used in various steps in the development process.

5.1.1 Challenges for MBE in Embedded Systems

Embedded systems are often developed by integrating components that have been designed and implemented by different teams, often specialized in different disciplines such as mechanical, electronics, and software engineering. As the system behavior emerges from the interplay of multiple distributed components, a key challenge is the *correct* integration of all these components. System integration is often performed in vertical design chains such as in automotive and avionics, and the development chain typically involves several tools that are not integrated, as explained below.

For example, embedded systems in automobiles are developed as a joint engineering process between an original equipment manufacturer (OEM) and suppliers. The car manufacturer or OEM traditionally specifies the requirements of the various electronic control units (ECUs) for the car. Each ECU is produced by (external) companies known as suppliers. The OEM then integrates the supplier's components into the vehicle. Achieving independent component development by suppliers necessitates precise and expressive requirements and interface specifications, as well as an integration framework that addresses quality requirements that crosscut multiple components. In practice, requirements are seldom expressed precisely enough, and system integration has to consider the effects of interactions among distributed components. Therefore, projects often use joint iterative development between the OEM and suppliers to ensure corrective feedback cycles.

Furthermore, the specification and implementation of diagnostic functions, for example, is often implemented ad hoc based on a stream of documents exchanged between the OEM and the suppliers. In the automotive domain, model-based approaches leveraging tools such as MATLAB® [5] and Simulink® [6] from MathWorks and ASCET® [7] from ETAS are used to model control functions and generate implementations for different platforms. However, in practice, there is no formal model of diagnosis that is exchanged between parties. Consequently, it is impossible to validate the design and anticipate problems in putting together the various components in the later phases of the development. The lack of an integrated diagnostic model also limits the reuse of diagnostics across models and generations of cars. Redeveloping diagnostic functions, obviously, leads to higher cost in all areas of requirements engineering, design, development, maintenance, verification, and validation.

Models and MBE hold promise for overcoming these exemplar challenges. Models should serve as a common interface between requirements and architecture specification—using models is the only systematic way to ensure that parties can communicate across all development phases from requirements to acceptance tests. The ultimate goal of MBE is that engineers will spend most of their time modeling the system under consideration and then generate code for a specific target platform. This goal is already supported by various tools (including MATLAB and Simulink), but the models often do not include all aspects of the system, as explained earlier. When automatic code generation is not feasible at the level of the entire system, there is still significant benefit if modeling is used for requirements gathering and architecture verification before deploying the actual system. Various terms are used in the literature to denote the use of models in the development process (e.g., Model-Driven Architecture® [8], model-based design [9], model-driven engineering [10], and model-integrated computing [11,12]). We use the general term MBE as a superset for all model-based approaches.

In the past decade, significant advances in the area of model specification, transformation, analysis, and synthesis have brought the vision of MBE within reach. Challenges for a comprehensive methodology include providing modeling techniques that result in a consistent, integrated specification; models expressive enough to support both generic and domain-specific aspects of the system; model reusability and integration; model execution; and seamless tool suites.

5.1.2 MBE Process

In general, Fowler [13] identified three ways of using models (specifically, models in UML): sketch, blueprint, and programming language. In sketching, models are used to communicate some aspects of the system. As blueprints, the models should be sufficiently complete as to allow straightforward implementation. By using a modeling language as a programming language, the system is specified at the level of models, and then tools are used to automatically generate the code for the target platform.

In the case of embedded systems, sketches are useful for requirements elicitation because each requirement addresses only a partial view of the system and, thus, a model for an individual requirement is by definition an incomplete specification. Moreover, in the early phases of system engineering, rarely all requirements, let alone their interplay, are understood. On the one hand, sketches as an early analysis tool are generally not hampered by lack of precision in the modeling language proper. To use a model as a blueprint or a programming language, on the other hand, requires the precise specification of the platform and the environment in which the system will operate. Blueprints for embedded systems can support system verification before deploying the actual system, and such verification should include timing requirements, resource usage, and other characteristics of embedded systems. Using a model as a programming language requires powerful tools for code synthesis. Sometimes, code generated from models (e.g., MATLAB and Simulink) is manually adjusted to meet the timing constraints, in which case the link between the code and the model is lost. Moreover, topics such as system reconfiguration and diagnosis are difficult to address. Therefore, a comprehensive MBE approach should involve round-trip engineering.

Figure 5.1 shows a simplified view of the activities and models involved in an MBE process. We use the terms *logical architecture* and *technical architecture* as described in Pretschner et al. [14]. The *logical architecture* is the decomposition of the system into functional components independent of the platform that will execute them, whereas the *technical architecture* specifies the representation of the logical components in terms of platform-specific entities. A logical architecture can be mapped to multiple technical architectures, each capturing all aspects of a particular deployment. Various other terms are used in the literature to distinguish between logical and technical system aspects. For example, OMG's Model-Driven Architecture® (MDA®) [8] distinguishes between a platform-independent model (PIM) and a platform-specific model (PSM). Furthermore, the Department of Defense Architecture Framework [15] distinguishes between operational views and systems views to separate the logical and technical architectures. We maintain the use of logical and technical architecture to encompass both these standards.

The modeling process involves creating and refining abstractions in a series of steps. Starting with requirements, we first construct requirements models. Next, we construct logical architecture models, which show how the system achieves the

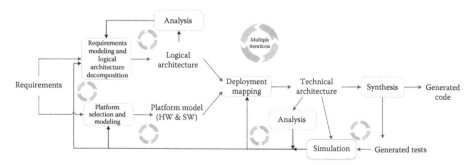

FIGURE 5.1 Simplified model-based engineering process.

functionality described in the requirements models. At the same time, we use the requirements analysis to select and model a deployment and execution platform. Next, we map the logical models onto the platform, thereby obtaining a technical architecture.

Analysis and simulation can be performed at various levels. Logical models can be analyzed for consistency and correctness, whereas technical models can be analyzed for realizability of the deployed system (e.g., schedulability analysis).

The modeling process involves feedback loops, and multiple iterations are possible in each step, within sequences of steps and within the entire process from the beginning. For example, analyzing the technical architecture can reveal hidden requirements and, therefore, may lead to reiterating the process from the initial phase of requirements modeling. We emphasize the need for an iterative process in alignment with the spiral [16] model of agile development methodologies, where requirements often resolve to partial specifications and refinements at one stage can trigger iterations at some earlier stages. An iterative process also accommodates architectural spiking, which is defined as taking a partial set of requirements/use cases and generating a system architecture and implementation based on them, then adding more and more use cases over subsequent rounds. Architectural spikes help in identifying the most critical parts of the system and implementing them first to verify that they can be addressed as expected and, therefore, to mitigate the risks associated with them.

5.1.3 OUTLINE

In the preceding paragraphs, we presented the modeling process in general and various ways of using models, with the implications in the case of embedded systems. We also discussed the current state of the art and challenges in MBE.

The remainder of the chapter is structured as follows. In Section 5.2, we identify requirements for a comprehensive modeling language for embedded systems. This allows us to better position UML's capabilities in response to this requirements spectrum. In Section 5.3, we present an overview of UML, its metamodeling architecture, extension mechanism, and behavior semantics. We also show how UML can be used for modeling real-time systems, by means of a simplified example from the Bay Area Rapid Transit (BART) [17] system. In Section 5.4, we provide an overview of UML profiles relevant for real-time systems and we cover more detailed aspects from the UML profile for Modeling and Analysis of Real-Time and Embedded systems (MARTE) [18]. In Section 5.5, we discuss how the UML and MARTE together meet the requirements presented in Section 5.2.

5.2 REQUIREMENTS FOR MODELING LANGUAGES

In the following, we present a set of requirements that modeling languages should meet for a comprehensive methodology of real-time systems. We have identified these requirements based on our experience in several automotive projects with industry partners. The list is not intended to be exhaustive, as further requirements have been presented elsewhere (e.g., Gerard et al. [19]). We focus our presentation on the specification capabilities of a modeling language and not on the entire end-to-end MBE approach nor on the tools necessary to implement it.

- *Consistency.* A modeling language should allow grouping of requirements, structure, behavior, and analysis in a single, integrated system model. Therefore, the language should allow consistency checking for models expressed in different notations, developed in different design iterations, or models that are part of different views/slices of the same system.
- *Traceability.* Requirements should be mappable to a precise specification of the system and from there to implementation while the mapping should be kept current during the system evolution. Traceability also applies to models at different levels of abstraction, enabling conformance checking for refinement operations.
- *Realizability.* Models often represent partial specifications that are refined in successive iterations in the development cycle. Models also represent different views on the system. The underlying question is whether the models allow a system to be constructed such that all requirements are fulfilled. At the very least, we would like to know which requirements stand in the way of realizability.
- *Distribution and integration.* System behavior emerges as the interplay of the functionality provided by subsystems, often developed independently by different suppliers. Thus, models should be capable of expressing concurrency and synchronization. Furthermore, since the OEM is responsible for the integration of subsystems, modeling should support overarching system specification, addressing the integration requirements as well as concerns that cut across the individual components such as resource optimization across the integrated system.
- *Interdisciplinary domains.* Embedded systems design involves multiple domains such as mechanical, electronics, and software. The system components are often designed at different stages in the development process, by different teams, using different tools and languages. A common modeling language should ease integration and tradeoff analysis, and it should reduce the need for *disruptive* feedback iteration cycles.
- *Nonfunctional properties.* A modeling language should allow specifying nonfunctional properties (such as performance, reliability, and power consumption) associated with behaviors, refinement relationships, deployment models, etc. Moreover, the set of nonfunctional properties should not be predefined and the language should support the specification of application-specific properties.
- *Resource models.* Embedded systems interact with the physical world and are constrained by the resources provided by the hardware and software platforms. Therefore, a specification should support modeling of platforms and resources, as well as allocation and optimization of resources to meet the functional and nonfunctional requirements.
- *Timing.* Time plays a critical role in real-time systems and, therefore, a modeling notation should express timing requirements in various temporal

models: (i) causal models, which are concerned only with the order of activities, (ii) synchronous models, which use the concept of simultaneity of events at discrete time instants, (iii) real-time scheduled models, which take physical durations and the timing of activities as influenced by central processing unit (CPU) speed, scheduler, utilization, etc., into account, and (iv) logical time models (e.g., Giotto [20–22] and TDL [23,24]), which consider that activities take a fixed logical amount of time, assuming that the platform can execute all activities to meet their constraints.

*   *Heterogeneous models of computation and communication.* Real-time systems are often embedded systems that control physical processes, which are often represented in terms of mathematical models. A modeling specification should support continuous behaviors, discrete event-based or time-based behaviors, or combinations thereof.

In the following sections, we present an overview of modeling with UML and its profiles. In Section 5.5, we revisit the requirements identified above and identify open issues in UML.

5.3 MODELING WITH UML

UML is a family of graphical notations that support modeling structural (i.e., static) and behavioral (i.e., dynamic) views of a system. Among others, the structural view includes class and component diagrams, while the behavior view includes sequence and state machine diagrams. UML provides 14 types of diagrams, though some are used more often and more prominently than others—a guide to the key aspects of the UML can be found in Fowler [13]. In this chapter, we focus on some of the commonly used diagrams.

UML was created to unify many object-oriented graphical modeling notations that became popular in the early 1990s. UML appeared in 1997 and underwent several changes from one version to another, the most radical ones taking place with the transition to UML 2. The most recent version is UML 2.3, whose specification consists of the UML Superstructure [3] defining the notation and semantics for diagrams and the UML Infrastructure [2] defining the language on which the superstructure is based.

In this section, we discuss the metamodeling architecture, the basic modeling capabilities of UML, the extension mechanisms, and behavioral semantics of UML. Then, in Section 5.4, we present an overview of the profiles for real-time systems and go into the details of the MARTE profile. In the end, we discuss the advantages and open issues in UML with respect to the requirements for modeling languages identified in Section 5.2.

5.3.1 Metamodeling Architecture

All graphical notations in UML are backed by a single metamodel. The notation (e.g., a class diagram notation) is the graphical syntax of the language. The metamodel defines the concepts of the language—the abstract syntax. Therefore, the UML metamodel defines the language elements and the relationship between them in the different UML graphical notations (e.g., sequence diagrams and class diagrams). A modeler who uses UML diagrams only as sketches is typically not concerned that much with the metamodel. However, for blueprints and especially for using UML as a programming language [13], the metamodel is very important.

The metalayer hierarchy for any language generally has three layers: (i) the metamodel, or the language specification, (ii) the model, and (iii) objects of the model. The metamodel defines how model elements in a model are instantiated. This layered structure can be applied recursively, such that the same layer that is a model instantiated from a metamodel can be seen as a metamodel of another model at the next lower level of instantiation.

The OMG has developed a modeling language (similar to a class diagram) called the Meta Object Facility (MOF™) [25], which is used to specify metamodels, such as the UML metamodel. From the perspective of MOF, the UML metamodel is seen as a user model that is based on the MOF metamodel. Therefore, MOF is commonly referred to as a meta-metamodel.

Figure 5.2 shows the four-layer metamodeling architecture used by the OMG: meta-metamodel (layer M3), metamodel (layer M2), model (layer M1), and the run-time system (layer M0). The top layer (M3) provides the meta-metamodel (MOF) that is used to build metamodels. The UML metamodel (layer M2) is an instance of the MOF. Layer M1 models are user models written in UML, and these models are instances of the UML metamodel. The M0 layer contains the runtime instances of the model instances defined in layer M1. The layers are numbered from M0 upward, but the architecture is not restricted to four layers. In general, more layers can be used by applying the recursive pattern as explained before.

Figure 5.2 also shows an example, where the meta-metaclass Class is defined as part of MOF. Then, UML defines the metaclasses Class and Attribute as part of the UML metamodel. Every model element in UML is an instance of exactly one model element in MOF. The UML Class is instantiated in a user model in a class called Train, with an attribute called Speed. Layer M0 contains an instance of a Train.

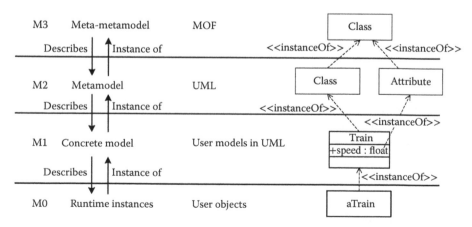

FIGURE 5.2 Metamodeling architecture.

The UML Infrastructure [2] defines a common package Core such that model elements are shared between UML and MOF. UML is defined as a model based on MOF used as metamodel; however, both UML and MOF depend on Core. Therefore, Core can be seen as the architectural kernel, and the Infrastructure defines the foundations for both M2 and M3 layers. The Superstructure [3] extends and customizes the Infrastructure to provide the modeling elements of UML.

5.3.2 MODELING WITH UML

In this section, we start with presenting the UML diagrams briefly, then we comment how they can be used for modeling requirements and architectures, and in the end we present a modeling example with the basic capabilities.

Figure 5.3 shows the types of diagrams provided by UML for modeling structure and behavior. A UML model consists of elements such as packages, classes, and associations. UML diagrams are graphical representations of a UML model. Examples of using UML diagrams for real-time systems are available in Douglass [26–28].

Structure diagrams show the static structure of the system at different abstraction levels and how system elements relate to each other. Class diagrams show system Classifiers such as classes and interfaces, their attributes, and relationships between them. Object diagrams show instances of Classifiers and instances of associations between them. Composite structure diagrams show the internal structure of a Classifier. Component diagrams show logical or physical components and dependencies between them via required and provided interfaces. Package diagrams show packages (i.e., namespaces used to group together elements that are related) and dependencies between them. Deployment diagrams show the assignment of software artifacts to execution nodes. A component may be implemented by one or more artifacts.

Behavior diagrams show the dynamic behavior of the system objects over time. Use case diagrams show a set of actions that some actors perform. Activity diagrams show sequences and conditions for coordinating lower-level behaviors. State machine diagrams model the behavior of a part of the system through finite state transitions. Sequence diagrams show the messages exchanged between entities. Communication diagrams show the interaction between entities, where the sequence of messages is given as a sequence numbering scheme. Interaction overview diagrams are similar to activity diagrams but focus on the overview of the control flow. Timing diagrams show interactions and conditions along a linear time axis.

Some diagram types blur the boundary between structure and behavior. For instance, sequence and object diagrams display aspects of both. Figure 5.3 classifies the diagrams according to their dominant trait.

5.3.2.1 Requirements Modeling with UML

For requirements engineering, modelers first create domain models to describe the existing system for which the software should be built, capturing domain entities and their structural and behavioral relationships in a systematic way. Domain models cover stakeholders, human actors that interact with the system, hardware devices, and the environment in which the system will operate. UML class diagrams can

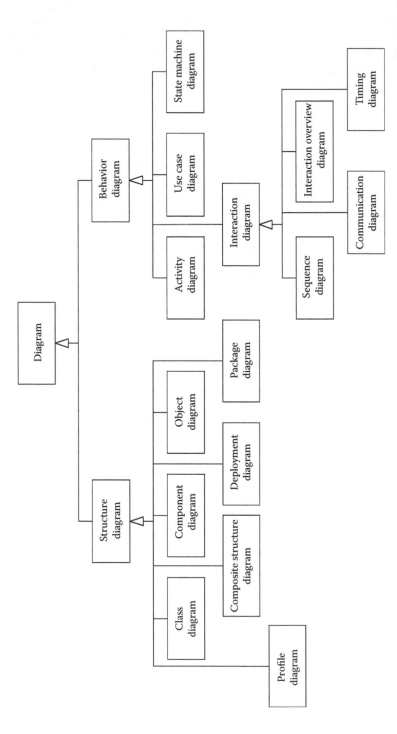

FIGURE 5.3 Unified Modeling Language structure and behavior diagrams. (From Object Management Group, "Unified Modeling Language (OMG UML), Superstructure, Version 2.3," formal/2010-05-05, OMG, 2010. With permission.)

be used for specifying structural domain model aspects, showing the relationships between system entities. UML class diagrams can also be used to specify business rules implicitly through class composition and multiplicity constraints, as well as explicitly through pre- and post-conditions expressed in Object Constraint Language (OCL) [29]. Such domain models can help in defining the questions for stakeholders and uncovering hidden requirements. Models also help in defining the boundaries between the target system and its environment.

UML use case diagrams identify actors and the capabilities of the system. Use cases are useful also in eliciting end-to-end timing and other QoS requirements, often expressed as constraints of the form "when the actor does X, the system should respond within Y ms."

The next level of detail beyond use cases is often a set of scenarios that depict interactions between the actors of the use case when using that specific capability of the system. Scenarios are modeled as sequence diagrams, capturing the operations performed by the system, the message protocol of interaction between the system and its actors, and constraints. UML activity and communication diagrams (previously called collaboration diagrams) can also be used to show how actors collaborate to perform tasks.

5.3.2.2 Modeling Logical and Technical Architectures with UML

After identifying requirements, architecture and design models show how the system is structured and how its internal entities behave to achieve system goals. All diagrams from Figure 5.3 can be used at this stage. In general, a type of structural or behavioral diagram can be used both for the logical and technical architecture. The models relevant to the logical architecture focus on capabilities and their mapping to logical entities, whereas the models relevant to the technical architecture focus on deployment entities. A platform model adds details such as middleware, operating system, network, and resources. The technical architecture then expresses the mapping between the logical architecture and the platform. In particular, package and deployment diagrams are used for the technical architecture. Furthermore, UML profiles such as MARTE [18] support the modeling of software and hardware resources in a standard way (see Section 5.4).

UML is the language choice of MDA, where both PIM and PSM are expressed as UML models. Model transformation from PIM to PSM is a core concern of MDA, and a PIM has to contain sufficient detail for a tool to generate a PSM. In MDA, the PSM contains the same information as an implementation, but in the form of a UML model instead of code. In general, UML is often used for modeling the logical architecture and other languages can be used for the technical architecture. However, if the transformation is not performed automatically or if it cannot be formalized, then the traceability between the logical and technical architecture is lost.

As concurrency and timing constraints are of particular concern in real-time systems, in the following, we discuss how UML supports them. Concurrency can be modeled in communication diagrams via the sequencing mechanism for events. Messages with the same sequence numbers (e.g., two messages numbered 3a and 3b) can be simultaneously triggered, provided that all messages with lower sequence numbers have been successfully triggered. Furthermore, concurrency can be expressed in sequence diagrams via the interaction operator PAR, which expresses parallel execution of a set of operations.

The Simple Time package defined in the UML Superstructure [3] provides basic support to represent time and durations and to define timing constraints. A time event denotes an absolute or relative point in time (relative to the occurrence of other events) when the event occurs. A time event is specified by an expression, which may reference observations. A time observation denotes a time instant to be observed during execution when a model element is entered or exited. A duration observation denotes an interval of time. For example, a time constraint can be associated with the reception of a message. In UML, timing constraints can be used in sequence diagrams or state machine diagrams. Simple Time enables triggering a transition in a state machine when a specific point in time has reached or after a certain amount of time has passed. However, the simple model of time does not account for multiple clocks or phenomena such as clock drifts, which occur in distributed systems, leaving more sophisticated models of time to be provided by profiles. In Section 5.4.2, we discuss in more detail how timing constraints are supported in the MARTE profile.

UML also introduces timing diagrams to show the effects of message/event interactions between entities over time. Timing diagrams are useful to depict different states of entities over time, especially in systems with continuous behavior relative to time, but sequence diagrams are more useful in actually modeling explicit time constraints.

5.3.2.3   UML Basics by Example

To show the modeling capabilities of UML, we use a simplified example of the BART system, particularly the part of the train system that controls speed and acceleration of the trains. BART is the commuter rail train system in the San Francisco Bay area. A full description of the case study is beyond the scope of this chapter, so we will exemplify some of the UML diagrams that can be used for modeling such a system—use case, class, sequence, and state machine diagrams. We will revisit the example in Section 5.4.2.3 when discussing additional capabilities introduced in MARTE and the issue of consistency in UML models.

The BART system automatically controls over 50 trains, most of them consisting of 10 cars. Tracks are unidirectional and sections of the track network are shared by trains of different lines. A track is partitioned into track segments, which may be bounded by gates. A gate can be viewed as a traffic light, establishing the right-of-way where tracks join at switches. Figure 5.4 depicts a domain model for the BART track system, showing in a UML class diagram the relationships between physical entities such as train, track, and gate. Such models facilitate establishing a common language for eliciting requirements from domain experts. Typically, specifying relationships and multiplicity constraints on a domain model leads to further discussions with the stakeholders to clarify the domain. For example, gates are not necessarily associated with switches, but can be used just to control the traffic flow.

Other work [17] describes the Advanced Automatic Train Control (AATC) system, which controls the train movement for BART. One important AATC requirement is to optimize train speeds and the spacing between the trains to increase throughput on the congested parts of the network, while constantly ensuring train safety. The specification strictly defines certain safety conditions that must never be violated, such as "a train must never enter a segment closed by a gate," or "the

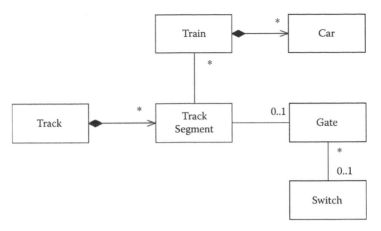

FIGURE 5.4 Class diagram: domain model for the Bay Area Rapid Transit tracks.

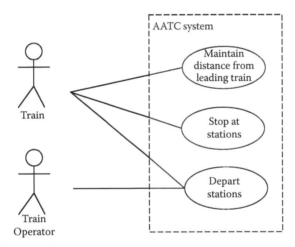

FIGURE 5.5 Use case diagram.

distance between trains must always exceed the safe stopping distance of the following train under any circumstances."

The system is controlled automatically. Onboard operators have limited responsibility: they signal the system when the platforms are clear, so a train can depart a station and they can operate the trains manually when a problem arises. Use case diagrams are useful in identifying the system boundaries (the control system that must be designed) and the external actors that interact with the system. Typically in UML, actors are human actors that use an application, but in embedded systems actors can be external physical resources such as devices and sensors. Nevertheless, actors represent logical roles, so a physical resource could play several roles in UML models. Figure 5.5 depicts a simple use case diagram for BART. Actors that interact with the AATC system are the Train and the Train Operator and so they are part of the system environment. The use cases depict the high-level goals of the system without details on how these goals are accomplished.

AATC consists of computers at train stations, a radio communications network that links the stations with the trains, and two AATC controllers on board of each train—the two controllers are at the front and back of the train. A track is not a loop. Thus, at the end of the line, the front and back controllers exchange roles and the train moves in the other direction. Each station controls a local part of the track network. Stations communicate with the neighboring stations using land-based network links. Trains receive acceleration and brake commands from the station computers via the radio communication network. The train AATC controller (from the lead car) is responsible for operating the brakes and motors of all cars in the train. The radio network has the capability of providing ranging information (from wayside radios to train radios and back) that allows the system to track train positions.

The system operates in half a second cycles. In each cycle, the station control computer receives train information, computes commands for all trains under its control, and forwards these commands to the train controllers. Figure 5.6 shows a sequence diagram depicting the interactions between three roles called Train, Station AATC, and Train Controller. Note that the Station AATC system obtains the status information directly from the Train by using the radio network, not from the Train Controller.

The sequence diagram features interaction frames, introduced in UML 2.0. A frame provides the boundary of a diagram and a place to show the diagram label (e.g., "Control Train Speed" in Figure 5.6). Frames also allow specifying combined fragments with operators and guards. Common examples of operators are LOOP for repetitive sequences, ALT for mutually exclusive fragments, and PAR for parallel execution of fragments. Figure 5.6 uses a LOOP operator to show that the

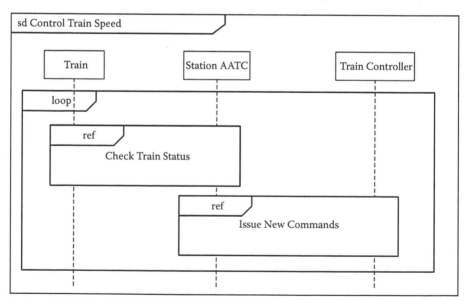

FIGURE 5.6 Sequence diagram that composes two other sequence diagrams.

system repeats the sequence of checking the train position and issuing new commands. Another operator is REF, which creates a reference to an interaction specified in another diagram. This REF operator allows composing primitive sequence diagrams into complex sequence diagrams. The expressiveness of UML 2 increased with the addition of these operators, which are borrowed from Message Sequence Charts [30,31].

Figure 5.7 depicts a simplified Check Train Status sequence diagram as referenced in Figure 5.6. The Train sends status information regarding its speed, acceleration, and range. The Station AATC system computes the train position from the status information and updates its Environmental Model. Status messages and commands are time-stamped in the so-called Message Origination Time Tag (MOTT). When a Train sends status information to a station, it attaches the time it sends the message as a MOTT. When the Station AATC estimates the train position, it attaches the original MOTT to the estimate. Furthermore, when the Station AATC sends a command, it again attaches the original MOTT, and the Train Controller checks the MOTT before executing the command. The station's control algorithm takes the MOTT, track information, and train status into account to compute new commands that never violate the safety conditions. To ensure this, each station computer is attached to an independent safety control computer that validates all computed commands for conformance with the safety conditions.

The actors in sequence diagrams (e.g., Train, Station AATC, etc.) are logical roles—in modeling the interactions, we concentrate on specific use cases and abstract from any concrete deployment architectures. In essence, a role shows *part* of the behavior the system displays during execution. What concrete deployment entity *plays* this role is left for a later modeling stage. The natural modeling entities

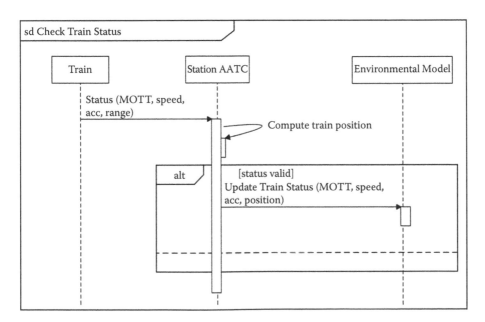

FIGURE 5.7   Sequence diagram for Check Train Status.

for roles in the UML are Classifiers—with the understanding that multiple roles may be aggregated into a single Classifier. The roles related to computing commands and safety are omitted from Figure 5.7, as they are relevant for another sequence diagram, called Issue New Commands, shown later in this chapter in Figure 5.17. The roles visible in a sequence diagram are a subset of the roles of the entire system.

Figure 5.8 shows a simplified domain model with the roles mentioned so far. We use the notation of a class diagram without the multiplicities—for a role domain model, we are interested in the roles that communicate and the links between them. The same diagram can be seen as a simplified Communication diagram, showing the communication links without the messages being exchanged. The role domain model is part of the logical architecture, as roles are logical entities that are later mapped onto physical components to define the technical architecture. A component can play several logical roles.

If a train does not receive a valid command within 2 s of the time stamp contained in the MOTT accompanying the status, it goes into emergency braking. Figure 5.9 shows the behavior of the Train Controller as a state-machine diagram with two states for normal operation and emergency mode.

In state-machine diagrams, we show states as boxes with rounded corners. Arrows denote state transitions. Labels on arrows indicate (i) the trigger (such as a message received), (ii) a guard (a condition that must be true for the transition to be taken) in square brackets, separated from (iii) the action (to be performed when the transition is taken)

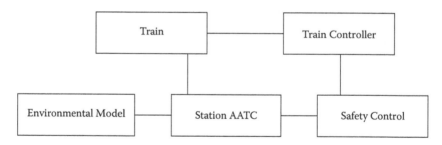

FIGURE 5.8 Role domain model as a class diagram.

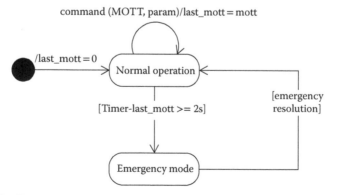

FIGURE 5.9 State-machine diagram for Train Controller.

by a "/." Actions include assignments to state variables and the sending of messages. All three pasts of a transition are optional. A solid circle indicates the initial "pseudo" state.

This example shows a frequently used pattern in modeling time with the basic capabilities of the UML: time is represented as an explicit parameter in messages exchanged among actors and these actors then perform explicit time arithmetic to determine transition triggers.

5.3.3   UML EXTENSION MECHANISM

The UML standard supports two types of extension mechanisms: lightweight extension through profiles and first-class extension through MOF. The profile mechanism allows UML metaclasses to be specialized for specific domains or different target platforms. In profiles, it is not possible to modify existing metamodels or to insert new metaclasses. Thus, in profiles it is impossible to remove constraints that apply to the UML metamodel, but it is possible to add new constraints that are specific to the profile. In contrast to profiles, in first-class extensibility, there are no restrictions on what changes can be made to a metamodel as MOF enables adding new metaclasses, removing existing classes, and changing relationships. In other words, the profile's extension defines a new dialect of UML, whereas the first-class extension defines a new language *related to* the UML.

Stereotypes, tagged values, and constraints are the main extension mechanisms available in a profile. A profile extends a reference metamodel such that the specialized semantics do not contradict the semantics of the metamodel. As such, the reference model is considered "read only." Stereotypes allow creating new model elements (not new metamodels), new constructs specific to a particular domain or platform. As such, a stereotype extends an existing metaclass and uses the same graphical notation as a class, with the keyword «stereotype» shown before or above the name of the stereotype. When the stereotype is applied to a model element, the name of the stereotype is given between «». A metaclass is extended by a stereotype by using a special kind of association relationship called an extension, which supports flexible addition/removal of stereotypes to classes. The notation for an extension is an arrow pointing to the extended class with the arrowhead as a solid triangle.

For example, for real-time systems, the MARTE profile defines the stereotype «RtUnit» to denote a real-time processing unit as shown in Figure 5.10. The left-hand

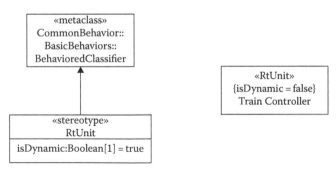

FIGURE 5.10   Stereotype example.

part of the figure shows the stereotype definition (the stereotype extends the metaclass BehavioredClassifier) and the right-hand part of the figure shows how the stereotype is applied to the model element Train Controller. A stereotype definition must be consistent with the abstract syntax and semantics of UML but can adapt the concrete syntax of UML to the domain. A stereotype can use an icon instead of a UML diagram element as, for example, when defining a clock element.

The properties of a stereotype are called tag definitions. When a stereotype is applied to a model element, the values of its properties are called tagged values. Thus, tagged values are defined as tag–value pairs, where the tag represents the property and the value represents the value of the property. In previous versions of UML, tagged values allowed defining additional properties for any type of model element. Starting with UML 2, a tagged value can be represented only as an attribute on a stereotype. For example, Figure 5.10 shows the Boolean attribute isDynamic defined for the stereotype «RtUnit». If this attribute is true, the real-time unit dynamically creates the schedulable resource required to execute its services [18]. When applying the stereotype, tagged values can be shown in the class compartment under the stereotype name as shown in Figure 5.10 for Train Controller. However, tagged values may also be shown in a comment attached to the stereotype.

A profile consists of a package that contains one or more related extension mechanisms. Profile diagrams (see Figure 5.3) allow defining custom stereotypes, tagged values, and constraints. Constraints allow extending the semantics of the UML metamodel by adding new rules. Constraints can be specified in OCL (not shown here for reasons of brevity).

Compared to pure stereotyping, the advantage of using the UML profile mechanism is that UML's meta- and meta-metamodels provide a shared semantic and syntactic foundation across all profiles.

5.3.4   UML BEHAVIORAL SEMANTICS

UML semantics is a topic that ignites fierce discussions in the modeling community. A common argument of UML critics is that UML has no behavioral semantics. While this was true for the first version of the language, OMG introduced an action-based semantics into the UML version 2.0 standard. Because UML supports expressing behavior in different specialized languages (i.e., state machines and activity diagrams), the semantics defined in the standard uses generic and fine-grained elements, which are composed to support all these sublanguages. UML is also intended to be used in different domains that have diverse execution requirements. Therefore, the UML standard defines variation points for which the documentation states alternative behaviors or leaves explicitly unspecified the behavior, calling for profiles to choose the most appropriate set of behaviors for the intended application domain.

Shortcomings of UML semantics are an important topic of discussion in the modeling community. Some scientists [32] point out that, even if a semantic model exists in the standard, it is not adequate to many usage scenarios. A first criticism is that the semantics is not defined formally, in terms of pure math of other formal languages. Consequently, it is impossible to prove that UML semantics is consistently defined or to develop verification tools that check behavioral properties of UML models. A second criticism is that the standard documents do not contain a chapter that coherently describes UML semantics. In fact, semantics information is spread across the standard and discussed together with the different UML notations. This makes it difficult for the reader to grasp UML semantics and thus paves the way for inconsistencies in the definition and misinterpretation by users. Additional critics complain about the very simple time model, which uses a centralized unique clock and must be completely replaced in many domains. Finally, the variation points that make UML applicable to any domain are subject to criticism as they force adopters to create semantic variations for each such domain.

In the remainder of this section, we describe some of the key elements of UML semantics and how they connect to structural elements of UML. We do not present the semantics of specific UML behavioral notations (e.g., interaction diagrams) but focus on the key elements that are used in the UML standard to construct the semantics of such notations. In our description, we highlight some variation points that are left open for profiles to specify. In Section 5.4.2.2, we discuss how the MARTE profile uses those variation points to create a behavioral semantics amenable to real-time embedded systems.

The first step for understanding UML behavioral semantics is defining how behavior is specified and which elements participate in an instance of such behavior. UML specifies behavior by defining flows of actions. These flows are always attached to a structural element, a Classifier. UML distinguishes between two types of objects with behavior: active and passive. Active objects, on the one hand, are the source of behavior. When created, they execute their actions independently of any other object. Passive objects, on the other hand, execute their actions in reaction to requests from other objects. Active objects subsume familiar programming concepts such as threads of execution and processes. The UML specification uses this more generic representation so that it can include behavior of entities that are not necessarily programs (e.g., people).

Figure 5.11 depicts the subset of the UML metamodel that defines the Behavior class and its connection to Behaviored Classifiers. Behavior is the superclass that

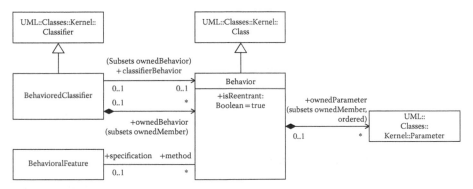

FIGURE 5.11 Subset of common behavior from package BasicBehavior. (From Object Management Group, "Unified Modeling Language (OMG UML), Superstructure, Version 2.3," formal/2010-05-05, OMG, 2010. With permission.)

represents all behaviors in the UML. Therefore, all the UML notations that describe behavior inherit this basic definition and enrich it by defining more specific types of behavior and their descriptions. Behaviors are associated to BehavioredClassifier. Because the UML Classifiers define sets of instances, the figure shows that behaviors are contained in instances of model elements. This is expressed by the use of a composition relation (black diamond) in the ownedBehavior association of Figure 5.11. The figure also shows that instances can be associated with a classifier Behavior, which represents the behavior that active objects start executing when created.

The UML standard specifies the behavior performed by Behavior objects in terms of actions and action flows. Actions are elementary operations that, given sets of inputs, produce sets of outputs and optionally modify the value of some structural elements. Multiple actions can be composed into flows to specify more complex operations. Flows specify networks of actions and, as such, identify dependencies between subsequent actions and how outputs feed into subsequent inputs. To support different types of systems requiring diverse semantic choices, UML supports variation points in the action model. One such variation point is the type of dependencies that flows can define. UML has two types of flow dependency: control flow, which starts the dependent action only when the previous one ends, and object flow, which starts dependent actions as soon as all their inputs are available. A second variation point arises by the fact that UML does not mandate when the elements of the action output sets become available. Different decisions on when to make the output values available create very different semantics. For example, UML could support semantics that assume synchronous reactions.

Messages and events are the final building blocks that form the base of the UML behavioral semantics. Messages support communication between different instances of UML elements. For example, different message patterns support synchronous calls, asynchronous calls, and signal broadcast. Messages are received in message pools and presented to the receiving instances when they are ready to receive. What happens to an incoming message when the pool is full and in which order messages in the pool are presented to the receiving instance are semantic variation points. Reception of messages is an example of an event. UML events represent the occurrence of a generic state or condition in the UML model instance. Types of events include call events, expression change events, and time events. Some events act as triggers, in which case a Behavior starts executing when the event occurs. Other events just capture a state. For example, a send signal does not even initiate a new behavior whereas a receive signal may.

A very good summary of the UML 2.0 semantics is provided by Selic [33]. We also suggest the interested reader to consult the standard documents [2,3], in particular, to understand how the UML behavior notations (Sequence Diagrams, State Machines Diagrams, Activity Diagrams, etc.) use the core elements we discussed to concretely define their semantics.

5.4   UML EXTENSIONS FOR REAL TIME

In this section, we discuss various extensions of UML and then show details from the recent profile for MARTE. The extension mechanism in UML (see Section 5.3.3) allows the definition of families of languages targeted to specific domains and levels of abstractions.

5.4.1 Overview of UML Profiles

The profile mechanism has been specifically defined for providing a lightweight extension mechanism to the UML standard via stereotypes, tagged values, and constraints as described in Section 5.3.3. For example, other work [34] presents a UML profile for a platform-based approach to embedded software development using stereotypes to represent platform services and resources that can be assembled together. There is an increasing number of profiles defined in various domains, resulting from either OMG standardization efforts or research outcomes. Different profiles may be overlapping and also inconsistent as each profile tailors the UML for a particular domain or platform. Some of the profiles emerged in previous versions of UML and then new profiles were defined to keep up with the latest changes in UML and to fill in the gaps identified in practice. In the following, we discuss some of the most commonly used profiles for real-time systems, with the historical background and the relationships between them.

The profile mechanism has been significantly refined starting with UML 2.0. Initially, UML 1.0/1.1 provided stereotypes and tagged values, but did not define the concept of a profile. Subsequent revisions of UML introduced the concept of a profile to provide structure to the extension elements. Moreover, to complete the previous versions of UML, the UML 2.0 Infrastructure and Superstructure specifications have defined the profile mechanism as a specific metamodeling technique. In addition, profile diagrams have been introduced in UML 2.0.

UML/Realtime (UML-RT) [35,36] extended UML 1.1–1.4 to support Real-Time Object-Oriented Modeling [37] concepts. It was also the UML dialect of the CASE tool Rational Rose®/RT. The extension used the standard UML mechanisms of stereotypes, tagged values, and constraints. Thus, UML-RT is a profile, although it was not called a profile initially because there was no profile concept in UML 1.1. For modeling architectural concepts, UML-RT introduced capsules (to model components), ports (to model the interaction of a capsule with its environment), connectors (communication channels between ports), and protocols (to model the behavior that can occur over a connector). A protocol comprises a set of participants (protocol roles), each specified by a set of signals received/sent by it. The corresponding communication sequence can be specified by a state machine and sequence diagrams. While UML-RT was a substantial improvement over the first generation of UML to model real-time systems, it still did not support all notations needed in modeling real-time systems. For example, Krüger et al. [38] presents an extension to UML-RT to support broadcasting. Because UML-RT was not based on an extensible framework such as the one supported by UML 2.0, extending it was an ad hoc process. UML-RT concepts were finally included in UML 2.0, which grants not only the ability to use its constructs in standard UML, but also the ability to systematically extend such constructs. For example, Krüger et al. [39] presents an approach to introducing broadcasting using UML 2.0 facilities similar to the UML-RT approach we mentioned earlier [38].

The UML profiles standardized by OMG include the UML Profile for Schedulability, Performance, and Time (SPT) [40], the UML profile for MARTE [18], and the UML profile for QoS and Fault Tolerance (QoS & FT) [41]. UML-RT focused on component-oriented development of communicating systems but left other aspects of embedded systems unaddressed. The SPT profile [40] was developed to define a resource model, time, and concurrency aspects in UML. MARTE is a new UML profile that updates the previous profile SPT for UML 2.x. MARTE is presented in more detail in Section 5.4.2. The QoS & FT profile [41] defines resource properties such as memory capacity and power consumption. The QoS & FT profile allows users to customize service characteristics (i.e., define new characteristics through specialization) and use tools to perform analyses such as performance and dependability. The QoS & FT profile allows defining a wide variety of QoS properties as compared to the SPT profile, which focused on schedulability and performance. In comparison, MARTE reuses concepts defined in both the SPT and the QoS & FT profile. Moreover, MARTE has the advantage that it allows modelers to attach directly to the design model additional information necessary for various analyses, rather than creating dedicated models for analysis.

OMG also provides the standard for the Systems Modeling Language™ (OMG SysML™) [42]. SysML reuses a subset of UML 2 (called UML4SysML) and provides additional extensions to address the concerns for systems engineering applications (called the SysML Profile). Therefore, SysML uses both UML extension mechanisms: the first-class extension via MOF is used to define the UML4SysML subset and then the profile extension mechanism is used not on UML but on UML4SysML. SysML does not use all of the UML diagram types and, thus, it is smaller and easier to learn than UML. In particular, SysML strictly reuses the UML use case, sequence, state machine, and package diagrams. SysML also modifies some of the UML diagrams. The SysML block definition diagram, internal block definition diagram, and activity diagrams extend the UML class diagram, composite structure diagram, and activity diagram, respectively. The SysML "block" is a significant extension in the direction of modeling complex systems. Blocks can be used to decompose the system into individual parts, with dedicated ports for accessing their internals. A block can represent almost any other type of structural entity. Furthermore, SysML allows the description of more general interactions than in software, for example, physical flows such as liquids, energy, or electricity. SysML activity diagrams add support for modeling continuous flows of material, energy, or information. By specifying a continuous rate, the increment of time between tokens approaches zero to simulate continuous flow. Nevertheless, SysML does not extend the time model of UML.

SysML adds two new diagram types: the requirements diagram and the parametric diagram. The requirements diagram captures text-based requirements and the relationships between them—requirements hierarchies and requirements derivation. A requirement can be related to a model element that satisfies or verifies the requirements. Thus, SysML requirements modeling not only supports the process of documenting requirements, but also provides traceability to requirements throughout the design flow. It furthermore provides tabular representations for requirements. Parametric diagrams allow the graphical specification of analytical relationships and constraints on system properties (such as performance and reliability) associated with blocks. Thus, parametric diagrams serve the integration of design models with analysis models.

SysML and MARTE are complementary and could be used together in a common modeling framework [43]. In general, there exist several profiles defined in various domains, and it is not clear how to combine multiple profiles when this is necessary for a particular interdisciplinary application. Therefore, Espinoza et al. [43] presents how SysML and MARTE can be combined and highlights the open issues in terms of convergence between the two profiles. For requirements engineering, SysML provides traceability relations, whereas MARTE provides ways to specify nonfunctional requirements. For system structure, modelers could start with a model specified with SysML blocks and then apply MARTE stereotypes to add additional semantics to these blocks. For system behavior, MARTE adopted the notion of port and flow from SysML, but they have different semantics. Therefore, when combining SysML and MARTE, it is required to define a common consistent semantics. Furthermore, significant differences exist in the specification of quantitative values for analysis (see Espinoza et al. [43] for a detailed analysis of the two profiles).

Even from this short exploration of UML extensions and profiles, it becomes clear that model consistency is a critical aspect of MBE—within *and* across profiles.

5.4.2 Modeling and Analysis of Real-Time and Embedded Systems

The UML profile for SPT [40] provides a framework for specifying time properties, schedulability analysis (rate monotonic analysis), and performance analysis (queuing theory). MARTE [18] updates SPT for UML 2.x. It allows for modeling of both software and hardware platforms along with their nonfunctional properties. It supports component-based architectures and different computational paradigms, and it allows more extensive performance and schedulability analysis.

In this section, we present an overview of MARTE's capabilities and we revisit the BART case study introduced in Section 5.3.2.3.

5.4.2.1 MARTE Basics

MARTE [18] is structured as a hierarchy of subprofiles (see the UML package diagram in Figure 5.12) with five foundation profiles and then further extensions used for design or analysis.

The foundation profiles are as follows:

- *Core Elements* define the basic elements used for structural and behavioral modeling. MARTE distinguishes between design-time Classifier elements and runtime instance elements created from the Classifiers. Behaviors are composed of actions and are triggered by events. Behaviors provide context for actions and determine when they execute and what inputs they have. In addition, MARTE supports modeling of operational modes (i.e., modal behavior), which are mutually exclusive at runtime. A mode defines a fragment in the system execution that is characterized by a given configuration when a set of system entities are active and have parameters defined for that mode.
- *Nonfunctional Properties Modeling (NFP)* supports the declaration of nonfunctional properties (such as memory usage and power consumption) as UML data types. The Value Specification Language (VSL) is introduced in MARTE to specify the values of those data types using a textual language

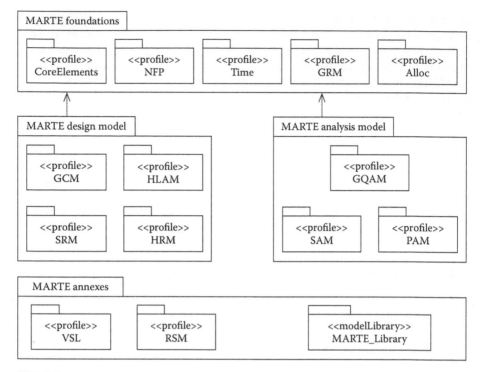

FIGURE 5.12 The architecture of the Modeling and Analysis of Real-Time and Embedded systems (MARTE) profile. (From Object Management Group, "UML Profile for MARTE: Modeling and Analysis of Real-Time Embedded Systems" Version 1.0, formal/2009-11-02, OMG, 2009. With permission.)

for specifying algebraic expressions. An annotated model (see Figure 5.13) contains annotated elements, which are model elements that have attached NFP value annotations for describing nonfunctional aspects (which can differ from one operational mode to another). The annotated model establishes the context for interpreting names used in the value specification. Examples of annotated elements (defined in other MARTE packages) used for performance analysis are step (a unit of execution), scenario (a sequence of steps), resource, and service (offered by a resource or component). A Modeling Concern establishes the ontology of relevant NFPs for a given domain used for the analysis. A domain model such as Figure 5.13 shows the concepts defined in MARTE. Then, a profile diagram defines the profile packages and how the elements of the domain model extend metaclasses of the UML metamodel. As explained in Section 5.3.3, the elements in the domain model are represented in the UML as stereotypes (e.g., the stereotypes «Nfp», «NfpConstraint», «Mode», etc.). However, not every element in the domain model results directly in a stereotype, because some of the domain concepts are abstract.

- The *Time* profile supports three models of time: chronometric, logical, and synchronous. It enriches the behavior specification from the core elements

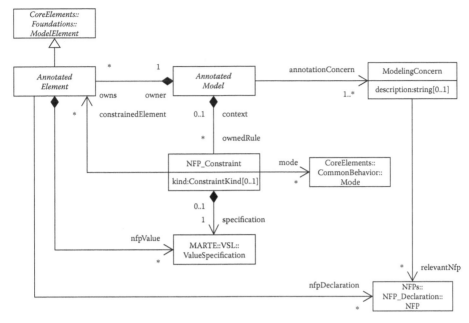

FIGURE 5.13 Nonfunctional properties annotations in MARTE. (From Object Management Group, OMG, 2009, "UML Profile for MARTE: Modeling and Analysis of Real-Time Embedded Systems" Version 1.0, formal/2009-11-02, 2009. With permission.)

with explicit references to time concepts. Time is represented as a partial ordering of instants. The occurrence of a time event refers to one instant. The basic model does not refer to physical time and, therefore, supports logical time, which is the basis also for synchronous languages. A time base is a set of instants where MARTE supports discrete and dense time bases (only countable sets). Physical time can be modeled as a dense time base. Clocks (logical or chronometric) use a discrete time base. For distributed systems, multiple time bases are supported.

Time can be used for triggering behaviors or observing event occurrences. MARTE defines the concepts for relating events, actions, and messages to time. A time constraint is specified as a predicate on timed observations (see Figure 5.14) and so the TimedObservation is a key concept introduced in MARTE. A constraint can be imposed on the occurrence of an event, the temporal distance between two events, or the duration of a behavior execution. A TimedObservation is a TimedElement, and, therefore, it has associated clocks used for observing time. A TimedObservation is the abstract superclass of TimedInstantObservation and TimedDurationObservation. For a behavior, observed events can be either its *start* or *finish* event. For a request, the possible events are its *send, receive,* or *consume* (the start of its processing by the receiver) events. Duration constraints can be defined on two events not necessarily occurring on the same clock.

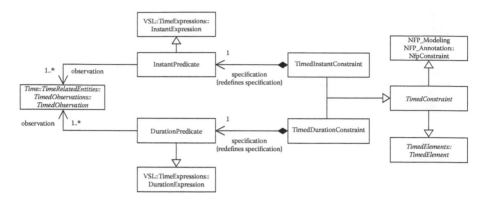

FIGURE 5.14   TimedConstraints as defined in MARTE. (From Object Management Group, "UML Profile for MARTE: Modeling and Analysis of Real-Time Embedded Systems" Version 1.0, formal/2009-11-02, OMG, 2009. With permission.)

- *Generic Resource Modeling (GRM)* provides an ontology of resources that allows the modeling of computing platforms, including computing resources, storage resources, communication media, and execution platforms.
- *Allocation Modeling (Alloc)* provides concepts for allocation of functionality to implementation entities. It includes space allocation and time allocation (i.e., scheduling). It also addresses the issue of refinement between models of different levels of abstraction. Nonfunctional properties (e.g., worst case execution time) can be attached to an allocation specification.

Based on the foundational profiles, MARTE provides two packages of extensions (see Figure 5.12): the MARTE design model supports model-based design of embedded applications and the MARTE analysis model supports model-based analyses and, thus, validation and verification.

For model-based design with MARTE, the *High-Level Application Modeling (HLAM)* subprofile provides extensions for real-time concerns such as Real-time Unit (using the stereotype «RtUnit») for concurrent computing units and Protected passive Unit («PpUnit») for shared information. An «RtUnit» owns one or several schedulable resources and can satisfy several requests from several real-time units at the same time, enabling intraunit parallelism if necessary. An «RtUnit» owns a single message queue for saving the messages it receives, and each message can be used to trigger the execution of a behavior owned by the unit. Real-time units and protected passive units may provide real-time services, which may specify real-time features such as deadlines and periods (with the ArrivalPattern data type). The stereotype for features («rtf») can be applied to multiple kinds of modeling elements (e.g., actions, messages, and signals). For example, the message *Status* from Figure 5.7 can be stereotyped as a real-time feature, indicating that is has a period of 500 ms (remember that the AATC system operates in half-second cycles). As a simple example, Figure 5.15 depicts the status message, omitting other messages exchanged in this scenario. We define *t*0

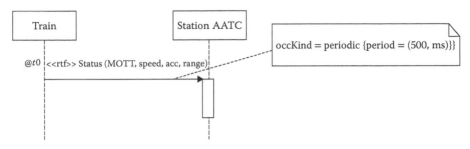

FIGURE 5.15 Bay Area Rapid Transit periodic feature.

as a TimedInstantObservation. Because the message is periodic, the period starts at time event *t*0[*i*]. Later in the chapter, Figure 5.17 shows an example for specifying timing constraints on the temporal distance between two events.

The *Generic Component Model (GCM)* subprofile supports component-based design, with both message and data communication between components. The MARTE component model adopted the concepts of ports and flows from SysML and added client/server ports. Furthermore, *Software Resource Modeling (SRM)* and *Hardware Resource Modeling (HRM)* allow designers to specify computing platforms. SRM allows modeling of elements such as tasks, semaphores, mailboxes, etc. The MARTE annexes feature supports modeling of OSEK, ARINC, and POSIX-compliant software computing platforms. The model of computation in MARTE is an asynchronous/event-based approach, but alternative models can be defined as extensions to the MARTE specification by using NFP, Time, and GRM packages.

Model-based analysis with MARTE is provided by the *Generic Quantitative Analysis Modeling (GQAM)* subprofile or by its two refinement subprofiles for schedulability and performance analysis. The analysis is based on the annotation mechanism in MARTE, which uses UML stereotypes. The model elements are mapped into analysis elements, which include the values for nonfunctional properties necessary for the analysis.

The Architecture Analysis and Design Language (AADL) [44] is an architecture description language defined ab initio (*not* as a UML profile) and standardized by the Society of Automotive Engineers. A system modeled in AADL consists of application software components (made of data, threads, and process components) bound to execution platform components (processors, memory, buses, and devices). Note that there is a MARTE rendering of AADL, formalized as a subset of MARTE.

5.4.2.2 MARTE Semantics

We discussed in Section 5.3.4 that UML 2.x defines a flexible semantics framework, which leaves several variation points open for profiles to specify. This approach leaves the developers of profiles free to adapt the general framework of UML behavior to the special needs of their target domains. In the case of MARTE, one of the most evident variations is the time model, which we discussed in Section 5.4.2.1. In fact, because the target domain includes real-time systems, the simplistic time model based on a global clock of the UML is not suitable.

While time is the most evident variation over standard UML semantics, MARTE defines semantics for many other variation points that the general UML specification left open. For example, we mentioned in Section 5.3.4 that the order in which messages are removed from the message pool and presented to the receiving instance is an open variation point. MARTE addresses this variation point by defining two default policies (first-in-first-out, FIFO and last-in-first-out, LIFO) and specifying that, by default, messages that arrive when the message pool is full will not be blocking and the message pool will silently drop the oldest message it contains to make room for the new one (see Section 12.3.2.6 in the MARTE profile specification [18]).

A complete discussion of all UML variation points fixed by MARTE is beyond the scope of this chapter. We recommend that the interested reader consult the MARTE profile specification [18] for a complete specification of the MARTE semantics.

5.4.2.3 MARTE Example

In this section, we revisit the BART example introduced in Section 5.3.2.3. We show different modeling perspectives using MARTE, give an example of timing constraints, and present an inconsistency that can arise when modeling behavior in the different diagrams.

Figure 5.16 depicts a component diagram with five components and the interface dependencies between them (we use the graphical notation of a ball-and-socket connection between a provided interface and a required interface). The Environmental Model component models the physical environment containing the trains. The Station AATC uses the Environmental Model to compute commands to send to trains. The Safety Control component checks all commands sent by the Station AATC for safety before forwarding them to each train. The safety computation is based on a simpler model than the one used to compute commands and only focuses on ensuring that all commands sent maintain the safety of each train. The last two components are deployed on the actual train. The Train Controller manages the train accelerations and decelerations, and the Emergency Brake is activated only in case of an emergency and stops the train as quickly as possible.

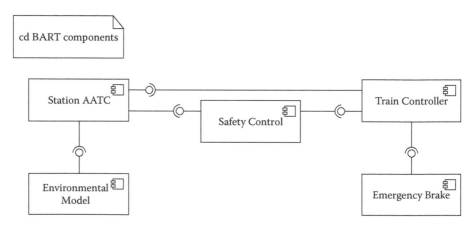

FIGURE 5.16   Component diagram for Bay Area Rapid Transit system.

Figure 5.17 depicts a sequence diagram capturing the common control scenario for the BART system. The figure represents the Issue New Commands diagram referenced earlier in Figure 5.6. This model is annotated with MARTE time constraints to specify the real-time requirements of the BART case study. The behavior specified in the diagram is the following:

- Station AATC sends a request to Environmental Model to compute the commands for the train.
- Environmental Model computes the commands, taking into account all parameters such as passenger comfort, schedule, etc.
- After receiving the commands from Environmental Model, Station AATC sends the commands to Safety Control to ensure the commands computed are safe.
- If the commands are safe, Safety Control forwards them to Train Controller.
- Train Controller informs Emergency Brake that the commands have been received.
- Emergency Brake acknowledges the commands received.
- Finally, Train Controller controls the train engine according to the commands received.

In Figure 5.17, we annotated two time instants $t0$ and $t1$ using Timed InstantObservations as defined in MARTE, which is indicated by the graphical representations @$t0$ and @$t1$. A TimedInstantObservation denotes an instant in time associated with an event occurrence (e.g., send or receive) and observed on a given clock. $t0$ is the instant when the message Compute Commands is *sent* by Station AATC, whereas $t1$ is the time instant when the message Commands Received is *received* by Emergency Brake.

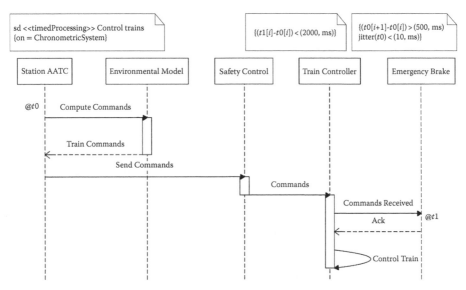

FIGURE 5.17   Sequence diagram for computing and delivering Train Commands.

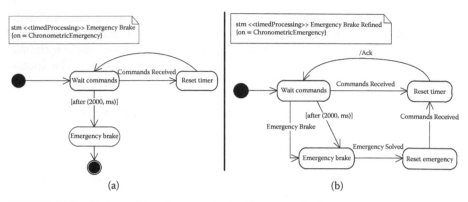

FIGURE 5.18   State-machine diagrams for the Emergency Brake.

Given those two instants, we leverage MARTE to define three time constraints in our system. With the time constraint $(t1[i] - t0[i]) < (2000, ms)$, we limit the duration of each iteration of this scenario to 2 s. The notations $t0[i]$ and $t1[i]$ represent the generic ith instantiation of the scenario (recall that the system operates in cycles). The second constraint, $(t0[i + 1] - t0[i]) > (500, ms)$, imposes that between each instantiation of the scenario, at least half a second passes. Finally, the last constraint, $jitter(t0) < (10, ms)$, limits the jitter of the $t0$ event, enforcing that between each iteration of the event at $t0$, there are between 500 and 510 ms.

Figures 5.18a and 5.18b present state-machine diagrams for the Emergency Brake system. These state machines are two different versions of the same perspective, where the one in Figure 5.18b is a refined version that enables restarting the system after an emergency brake. If we consider the three graphs from Figures 5.16, 5.17, and 5.18a together, we have an inconsistent model: the state machine diagram Figure 5.18a does not acknowledge the Commands Received call from Train Controller—contrary to what the sequence diagram from Figure 5.17 demands. Replacing the diagram from Figure 5.18a with Figure 5.18b, we obtain a consistent model.

5.5   DISCUSSION AND OPEN ISSUES

UML is a well-established language, with advanced tool support and an active community of developers that evolve the standard. The language has been developed with the goal of having a general purpose modeling language. It supports different types of systems, in different domains, and people with different backgrounds and expertise. This flexibility is the main strength of the UML. However, flexibility has a price. Many of the issues reported by researchers in working with the UML originate from the design decision of making it flexible and adaptable to multiple usage scenarios. In particular, problems arise in the field of UML model consistency, system synthesis, and model transformation.

In the following, we discuss to what degree UML meets the requirements for modeling languages identified in Section 5.2.

- *Consistency.* As discussed in Section 5.3.4, the semantics of UML is open for extensions. The UML standard from OMG does not provide a complete formal semantics and leaves many options in the language open for specialization. However, having formal semantics is a key enabler both to supporting model consistency and to leveraging UML in safety-critical systems. In fact, for safety-critical systems, it is often mandated to formally prove various properties. Moreover, in the domain of real-time systems, some level of semantics is required to support analysis tools, such as tools for scheduling. Because of its incomplete semantics, UML cannot be used directly for real-time systems simulation and verification. Profiles such as MARTE aim at adding sufficient semantics to the UML to support analysis tools.
- *Traceability.* UML use case diagrams have been traditionally used to document requirements, but they have a number of limitations including their lack of well-defined semantics. Other diagrams such as sequence diagrams can be used for requirements engineering to model the next level of detail from use cases. However, the relation between the requirements models (see Section 5.3.2.1) and the architectural ones (see Section 5.3.2.2) is typically difficult to trace in the UML. SysML adds requirements modeling as a key engineering activity and it provides traceability to requirements throughout the design flow; in SysML, a requirement can be related to a model element that satisfies or verifies the requirement. However, the requirements notion of SysML is again not a formal one, and thus not immediately amenable to a precise modeling approach in relation to the associated architectural models. Furthermore, MDA—also an OMG standard—supports traceability within the architectural models from PIM to PSM, both specified in UML. In fact, a PSM is obtained from a PIM through a series of transformations, although the platform model is often hard-coded in the transformation. Through its various subprofiles, MARTE provides support for making platform models explicit. Therefore, platform models can be an explicit input to the model transformation for the target platform.
- *Realizability.* UML models have several uses. In the real-time domain, they are often used to verify system properties (even before building the system) and to generate part of the implementation. A problem that can arise is that it is not possible to realize the model, given some implementation constraints. For example, constraints on the CPU could make it impossible to fulfill some timing constraint. Therefore, it is important to always revise the models during the development process to ensure that all assumptions taken in analyzing models are fulfilled by the implementation.
- *Distribution and integration.* A key issue in modeling large embedded systems is that the modeling activity is often distributed across different teams and even different organizations. In cases where the modeling activity is distributed, the UML must support independent modeling of different parts of the system, which will cause conflicts and inconsistences. Therefore, projects with distributed models must establish a model integration process and leverage tools to manage inconsistencies.

- *Interdisciplinary domains.* Through its built-in profile mechanism, UML can be tailored for various domains or different target platforms. Defining a custom domain-specific language ab initio has the advantage that the language can be optimized for the target domain, but complex systems often have various concerns that are modeled separately, and integrating models defined in different languages is not an easy task. The advantage of UML is that it provides an integrated framework, where profiles specialize UML concepts and, therefore, can be used with existing UML tools. Nevertheless, if the interdisciplinary application requires combining more than one UML profile, there are significant issues with overlapping and inconsistent language specification across the two profiles, especially when the two disciplines to be integrated are sufficiently close so that their domain representations in the respective profiles overlap. These issues necessitate careful, domain-specific integration of the profiles, because the UML itself provides no strong interface concept for profiles

- *Nonfunctional properties.* Several UML profiles including SPT, QoS & FT, and MARTE provide extensions to support various nonfunctional properties. MARTE allows specifying properties such as throughputs, bandwidths, delays, and memory usage, and these follow a well-defined textual syntax—MARTE introduces the VSL to formulate algebraic and time expressions. VSL is based on OCL but also provides time-related annotations. Section 5.4.2 provides more details on the specification capabilities of MARTE. Notably, UML models can be annotated with nonfunctional properties, given that annotation is a nonintrusive operation on an existing model to add additional information necessary for analysis. For example, MARTE can be used for requirements engineering to annotate nonfunctional requirements in use cases and the corresponding sequence diagrams.

- *Resource models.* MARTE provides a taxonomy for software and hardware resources, which support specifying computing platforms at different levels of abstractions and with their corresponding nonfunctional properties such as memory usage and power consumption. Support for modeling time, resources, allocation, and other qualitative and quantitative concerns is included in various subprofiles in MARTE. Other work [19] discusses the coverage provided by each subprofile. For example, resource modeling is addressed by the GRM, SRM, HRM, GQAM, SAM, PAM, and RSM subprofiles (see Section 5.4.2).

- *Timing.* The time model in UML is trivial, but MARTE provides significant extensions and supports three models of time: chronometric, logical, and synchronous. MARTE provides fundamental time notions such as time instant, duration, time bases, and clocks, while it supports multiple time bases for distributed systems. MARTE also introduces the key concept of observation, which enables a time constraint to be specified as a predicate on timed observations. Moreover, time concepts are not just annotations but they are defined as part of the system behavior.

- *Heterogeneous models of computation and communication.* Several embedded systems require more than one model of computation to reflect the nature of the application domain, whereas UML supports only event-based models. Therefore, several proposals have been made to extend UML with support for continuous time by using stereotypes to represent continuous variables, time, and derivatives [28]; a programming language for hybrid systems [29]; and a dataflow mechanism (distinguishing between signal ports and data ports) coupled with mathematical equations, called D-UML [30]. The model of computation in MARTE is an asynchronous/event-based approach but alternative models can be defined as extensions to the MARTE specification.

5.6 SUMMARY AND OUTLOOK

In this chapter, we have presented UML and some of its profiles specifically tailored for real-time systems. Using specific profiles, UML models can be leveraged for design and analysis during the development of real-time embedded systems. Models are a key asset in complex embedded systems, as they provide reusability across product generations, help in eliciting and specifying requirements, facilitate integration of components developed by different suppliers, support system verification before deployment, and support synthesis and deployment.

UML has evolved over the years to support more modeling techniques and adapt to novel tasks. In its latest incarnation 2.3, UML is a substantial improvement over the original version. As noticed in other works [10,45], UML has maintained many promises but has still some pitfalls that limit its utility.

The main promises it maintained are the support for multiple notations as part of a single language and the support for profiles to modify the language and its semantics. Furthermore, UML is supported by a plethora of tools to create and analyze models.

The main pitfalls attributed to UML are concerned with its complex and bloated metamodel, which makes it difficult to master the language in its details, and the lack of semantics associated to some elements (i.e., aggregation), which creates confusion and different interpretations from different tool manufacturers.

In this chapter, we presented the facilities of UML for dealing with time. UML 2.x can specify time properties by means of Time Duration and Observation (of time passing). The basic time facilities and time constraints are, however, too simple to address requirements of real-time systems. The UML standard itself suggests using a profile (see the UML Superstructure standard document [3] Chapter 13) to enhance the ability to specify time properties. Following this suggestion, MARTE adds a comprehensive ontology of time and a language, VSL, to write time constraints. In addition, SysML provides facilities to support continuous time in addition to discrete events.

More research is required to unlock the full potential of the UML and to make it more applicable to the real-time systems domain. In particular, the UML metamodel should be restructured and streamlined to simplify the definition of semantics variations and the creation of tools. Specifically, the current language specialization facilities (such as profiles) require a profound understanding of most of the complex UML metamodel. Furthermore, the specialization builds upon the full metamodel, adding further elements and relations (as we have shown for MARTE in this chapter). Therefore, the use of a particular profile implies a profound understanding of all of the UML intricacies plus the ones added by the profile. A substantial usability improvement would emanate from an option to eliminate parts of the metamodel that are irrelevant for a given application domain.

The other part of UML that requires substantial improvement is the semantics definition. The existence of alternative semantics for some elements and the absence of semantics for others confuse users of the standard. Profiles are supposed to lock down the semantics of some elements. However, the existence of different interpretations for the meaning of some modeling elements increases the chances of models being misinterpreted by users who are not experts in a particular profile.

Moreover, UML does not provide the necessary support for checking and enforcing model consistency. Modeling involves multiple diagram types (thus, multiple modeling languages) that focus on specific aspects of a system. Therefore, ensuring consistency between models is key for system synthesis and analysis. In Chapter 12, we will focus on the topic of model consistency for UML and cover the issues of consistency checking, assigning semantics to UML diagrams and integrating multiple models into a coherent system model.

ACKNOWLEDGMENTS

This work was partially supported by the NSF within projects CCF-0702791 and CNS-0963702 and by funds from the California Institute for Telecommunications and Information Technology (Calit2) at the University of California, San Diego. We are grateful to Barry Demchak and the anonymous reviewers for insightful comments.

REFERENCES

1. Giese, H., G. Karsai, E. Lee, B. Rumpe, and B. Schätz. 2011. *Model-Based Engineering of Embedded Real-Time Systems*, Vol. 6100. Berlin/Heidelberg: Springer.
2. Object Management Group. 2010. "Unified Modeling Language (OMG UML), Infrastructure, Version 2.3." formal/2010-05-03, OMG.
3. Object Management Group. 2010. "Unified Modeling Language (OMG UML), Superstructure, Version 2.3." formal/2010-05-05, OMG.
4. Sprinkle, J., B. Rumpe, H. Vangheluwe, and G. Karsai. 2011. "Metamodelling: State of the Art and Research Challenges." In *Model-Based Engineering of Embedded Real-Time Systems*, edited by H. Giese, G. Karsai, E. Lee, B. Rumpe, and B. Schätz, Lecture Notes in Computer Science, Vol. 6100, pp. 57–76. Berlin/Heidelberg: Springer.
5. MathWorks. 8 April 2011. MATLAB. 7.12. http://www.mathworks.com/products/matlab/.
6. MathWorks. 8 April 2011. Simulink. 7.7. http://www.mathworks.com/products/simulink/.
7. ETAS. April 2011. ASCET. 6.1.2. http://www.etas.com/sen/products/ascet_software_products.php.
8. Object Management Group. 2003. "Model Driven Architecture (MDA) v1.0.1." omg/03-06-01, OMG.

9. Nicolescu, G., and P. Mosterman. 2009. *Model-Based Design for Embedded Systems*, pp. 766. Boca Raton, FL: CRC Press.
10. France, R., and B. Rumpe. 2007. "Model-Driven Development of Complex Software: A. Research Roadmap." In *Proceeding of Future of Software Engineering (FOSE'07)*, L. Briand and A. Wolf (Eds.), pp. 37–54. Minneapolis, MN, USA, Los Alamitos, CA: IEEE Computer Society.
11. Sztipanovits, J., and G. Karsai. April 1997. "Model-Integrated Computing." *Computer* 30 (4): 110–11. Los Alamitos, CA: IEEE Computer Society.
12. Balasubramanian, K., A. Gokhale, G. Karsai, J. Sztipanovits, and S. Neema. 2006. "Developing Applications Using Model-Driven Design Environments." *Computer* 39 (2): 33–40 Los Alamitos, CA: IEEE Computer Society.
13. Fowler, M. 2004. *UML Distilled Third Edition: A Brief Guide to the Standard Object Modeling Language*. Boston, MA: Addison-Wesley.
14. Pretschner, A., M. Broy, I. H. Krüger, and T. Stauner. May. 2007. "Software Engineering for Automotive Systems: A Roadmap." In *Proceeding of Future of Software Engineering (FOSE'07)*, L. Briand and A. Wolf (Eds.), pp. 55–71. Minneapolis, MN, USA, Los Alamitos, CA: IEEE Computer Society.
15. Department of Defense. 2010. "The Department of Defense Architecture Framework (DoDAF) Version 2.02." DoD.
16. Boehm, B. W. 1988. "A Spiral Model of Software Development and Enhancement." *Computer* 21 (5): 61–72 Los Alamitos, CA: IEEE Computer Society.
17. Winter, V., F. Kordon, and M. Lemoine. 2004. "The BART Case Study." In *Formal Methods for Embedded Distributed Systems*, edited by F. Kordon, and M. Lemoine, pp. 3–22. New York: Springer.
18. Object Management Group. 2009. "UML Profile for MARTE: Modeling and Analysis of Real-Time Embedded Systems" Version 1.0, formal/2009-11-02, OMG.
19. Gérard, S., H. Espinoza, F. Terrier, and B. Selic. 2011. "Modeling Languages for Real-Time and Embedded Systems: Requirements and Standards-Based Solutions." In *Model-Based Engineering of Embedded Real-Time Systems*, edited by H. Giese, G. Karsai, E. Lee, B. Rumpe, and B. Schätz, Lecture Notes in Computer Science, Vol. 6100, pp. 129–54. Berlin/Heidelberg: Springer.
20. Horowitz, B. 2003. "Giotto: A Time-Triggered Language for Embedded Programming." Doctoral Dissertation, University of California, Berkeley.
21. Henzinger, T. A., C. M. Kirsch, M. A. A. Sanvido, and W. Pree. February 2003. "From Control Models to Real-Time Code Using Giotto." 23 (1): 50–64. New York, NY: IEEE Control Systems Magazine,
22. Henzinger, T. A., B. Horowitz, and C. M. Kirsch. 2001. "Embedded Control Systems Development with Giotto." In *Proceedings of the ACM SIGPLAN workshop on Languages, Compilers and Tools for Embedded Systems (LCTES'01)*, pp. 64–72, New York, NY, USA: ACM.
23. Templ, J. 2008. TDL: Timing Definition Language 1.5 Specification, preeTEC GmbH.
24. Farcas, E., C. Farcas, W. Pree, and J. Templ. 2005. "Transparent Distribution of Real-time Components Based on Logical Execution Time." In *Proceedings of ACM SIGPLAN/SIGBED Conference on Languages, Compilers, and Tools for Embedded Systems (LCTES)*, pp. 31–9. New York: ACM Press.
25. Object Management Group. 2006. "Meta Object Facility (MOF), Version 2.0." formal/2006-01-01, OMG.
26. Douglass, B. 2004. *Real-Time UML: Advances in the UML for Real-Time Systems*. Boston: Addison-Wesley.
27. Douglass, B. 2003. *Real-Time Design Patterns: Robust Scalable Architecture for Real-Time Systems*. Boston: Addison-Wesley.

28. Douglass, B. P. 1999. *Doing Hard Time: Developing Real-Time Systems with UML, Objects, Frameworks, and Patterns*, pp. 800, Boston: Addison-Wesley.
29. Warmer, J., and A. Kleppe. 1998. *The Object Constraint Language: Precise Modeling with UML*, pp. 144, Boston: Addison-Wesley.
30. ITU. 1996. "Message Sequence Charts (MSC)." ITU-TS Recommendation Z.120.
31. Krüger, I. H. 2000. "Distributed System Design with Message Sequence Charts." Doctoral Dissertation, Technical University of Munich, Germany.
32. Broy, M., M. Crane, J. Dingel, A. Hartman, B. Rumpe, and B. Selic. 2007. "2nd UML 2 Semantics Symposium: Formal Semantics for UML." In *Models in Software Engineering*, edited by T. Kühne, Lecture Notes in Computer Science, Vol. 4364, pp. 318–23. Berlin/Heidelberg: Springer.
33. Selic, B. 2004. "On the Semantic Foundations of Standard UML 2.0." In *Formal Methods for the Design of Real-Time Systems*, edited by M. Bernardo, and F. Corradini, Lecture Notes in Computer Science, Vol. 3185, pp. 75–6. Berlin/Heidelberg: Springer.
34. Chen, R., M. Sgroi, L. Lavagno, G. Martin, A. Sangiovanni-Vincentelli, and J. Rabaey. 2003. "UML and Platform-Based Design." In *UML for Real: Design of Embedded Real-time Systems*, pp. 107–26. Netherlands: Kluwer Academic Publishers.
35. Selic, B., and J. Rumbaugh. 1998. *Using UML for Modeling Complex Real-Time Systems*. Objectime Limited.
36. Selic, B. 1998. "Using UML for modeling complex real-time systems." In *Proceedings of the ACM SIGPLAN Workshop on Languages, Compilers, and Tools for Embedded Systems*, pp. 250–60. London, UK: Springer-Verlag.
37. Selic, B., G. Gullekson, J. McGee, and I. Engelberg. 1992. "ROOM: An Object-Oriented Methodology for Developing Real-Time Systems." In *Proceedings of 5th International Workshop on Computer-Aided Software Engineering*, pp. 230–40. IEEE Computer Society.
38. Krüger, I. H., W. Prenninger, R. Sandner, and M. Broy. 2002. "From Scenarios to Hierarchical Broadcasting Software Architectures using UML-RT." *International Journal of Software Engineering and Knowledge Engineering* 12 (2): 155–74 World Scientific Publishing.
39. Krüger, I., W. Prenninger, R. Sandner, and M. Broy. 2004. "Development of Hierarchical Broadcasting Software Architectures using UML 2.0." In *Integration of Software Specification Techniques for Applications in Engineering. Priority Program SoftSpez of the German Research Foundation (DFG). Final Report*, edited by H. Ehrig, W. Damm, J. Desel, et al., Lecture Notes in Computer Science, Vol. 3147, pp. 29–47. Berlin/Heidelberg: Springer.
40. Object Management Group. 2005. "UML Profile for Schedulability, Performance, and Time, Version 1.1." formal/05-01-02, OMG.
41. Object Management Group. 2008. "UML Profile for Modeling Quality of Service and Fault Tolerance Characteristics and Mechanism, Version 1.1." formal/2008-04-05, OMG.
42. Object Management Group. 2010. "Systems Modeling Language Version 1.2." formal/2010-06-01, OMG.
43. Espinoza, H., D. Cancila, B. Selic, and S. Gérard. 2009. "Challenges in Combining SysML and MARTE for Model-Based Design of Embedded Systems." In *Model Driven Architecture: Foundations and Applications*, edited by R. Paige, A. Hartman, and A. Rensink, Lecture Notes in Computer Science, Vol. 5562, pp. 98–113. Berlin/Heidelberg: Springer.
44. Feiler, P. H. 2009. *The SAE Architecture Analysis & Design Language (AADL)*, SAE International Document AS-5506A ed.
45. France, R. B., S. Ghosh, T. Dinh-Trong, and A. Solberg. 2006. "Model-Driven Development Using UML 2.0: Promises and Pitfalls." *Computer* 39 (2): 59–66 Los Alamitos, CA: IEEE Computer Society.

6 Modeling and Simulation of Timing Behavior with the Timing Definition Language

Josef Templ, Andreas Naderlinger, Patricia Derler, Peter Hintenaus, Wolfgang Pree, and Stefan Resmerita

CONTENTS

6.1 Introduction 160
6.2 Timing Definition Language 161
    6.2.1 TDL Properties 161
        6.2.1.1 Time and Value Determinism 161
        6.2.1.2 Portability 161
        6.2.1.3 Transparent Distribution 161
        6.2.1.4 Time Safety 161
        6.2.1.5 Compositionality 162
    6.2.2 TDL Language Constructs 162
        6.2.2.1 Modules 162
        6.2.2.2 Ports 162
        6.2.2.3 Tasks 162
        6.2.2.4 Modes 163
        6.2.2.5 Asynchronous (= Event-Triggered) Activities 164
    6.2.3 Example TDL Modules 164
    6.2.4 TDL Toolchain 166
        6.2.4.1 E-code 167
6.3 TDL Integration with MATLAB® and Simulink® 168
    6.3.1 Application Developer's Perspective 168
        6.3.1.1 Extension of the TDL Toolchain for MATLAB® and
                Simulink® 168
        6.3.1.2 Modeling 169
        6.3.1.3 Simulation 170
        6.3.1.4 Platform Mapping and Code Generation 170

        6.3.2   Implementation Perspective.. 171
                6.3.2.1   Resolving Data Dependencies 172
6.4   Tdl Integration With Ptolemy Ii... 172
        6.4.1   TDL Domain 173
6.5   Comparison between the Simulink® and the Ptolemy II Integration 174
6.6   Related Work 175
6.7   Conclusion 176
References.. 177

6.1   INTRODUCTION

Traditional development of software for embedded systems is highly platform specific. Exploiting a specific platform enables reducing cost of hardware to a minimum, whereas high development costs of software are considered acceptable in the case of large quantities of devices being sold. Nowadays, with ever more powerful processors in the low-cost range, we observe even more of a shift of functionality from hardware to software and a general tendency toward more ambitious requirements. Modern cars or airplanes, for example, contain dozens of the so-called electronic control units interconnected by multiple buses and are driven by several million lines of code. To cope with the increased complexity of the embedded software, a platform-independent "high-level" programming style becomes mandatory, as testing alone can never identify all the errors. In particular, in the case of safety-critical real-time software, this applies not only to functional aspects but to the temporal behavior of the software as well. Dealing with time, however, is not covered at all by any of the existing high-level imperative languages. Simulation environments that offer delay blocks allow at best the approximation of the simulated behavior to the behavior on the execution platform.

One reason is that execution and communication times related to computational tasks of an application can have a substantial influence on the application behavior that is unaccounted for in high-level models [1]. Consequently, the implementation of a model on a certain execution platform may violate requirements that are proved to be satisfied in the model. Explicitly considering execution times at higher levels of abstractions has been proposed as a way to achieve satisfaction of real-time properties [2]. One promising direction in this respect is the logical execution time (LET) [3].

This chapter presents the explicit specification of the timing behavior using the LET-based notation called Timing Definition Language (TDL). TDL is under active, commercially supported development [4]. As simulation is widely used in industry for testing and validation of complex systems (e.g., Ref. [5]), it is important to be able to simulate TDL-based systems. Thus, we describe how TDL has been integrated with two distinctly different simulation environments, that is, MATLAB® and Simulink® [6] from MathWorks® and Ptolemy [7], an open source environment developed at the University of California at Berkeley. Where MATLAB and Simulink is treated as yet another execution platform, Ptolemy is more closely aligned with TDL principles and so offers simulation capabilities that allow a more straightforward integration of TDL. We chose these two simulation environments to demonstrate two quite different approaches for simulating TDL-based systems.

6.2   TIMING DEFINITION LANGUAGE

The TDL [8] at its core follows the time-triggered programming model [3,9]. In a time-triggered system, all activities are triggered only by the ticks of a single global clock. To increase the range of applicability, TDL also supports a limited form of event-triggered programming, which allows, for example, responding to hardware interrupts.

6.2.1   TDL PROPERTIES

TDL programs that only rely on the time-triggered features exhibit the following properties by construction.

6.2.1.1   Time and Value Determinism

Value determinism means that a program provides the same outputs if it is provided with the same inputs. Time determinism means that a program provides the outputs at the same times if it is provided with the inputs at the same times, where all times are relative to the program start. TDL aims for both time determinism and value determinism. Thus, a TDL program provides the same outputs at the same times if it is provided with the same inputs at the same times. In other words, the chronologically ordered sequence of outputs (time plus values) of a TDL program, which is also referred to as the observable behavior of a program, is deterministic and platform-independent.

6.2.1.2   Portability

TDL programs represent a platform-independent description of the timing behavior of an application. Everything that is platform specific, for example, accessing sensors or actuators, is defined outside the TDL program. TDL programs behave exactly the same independent of the underlying CPU, network bandwidth, or operating system. Even when simulating a TDL application, for example, under MATLAB and Simulink, the application exhibits the same behavior.

6.2.1.3   Transparent Distribution

Since TDL abstracts from the execution platform, a TDL application shows the same observable behavior in the case of a distributed system as on a single-node system. Thus, the fact that a distributed system is used as an execution platform is transparent. It is the task of the TDL compiler to generate a suitable network communication schedule for maintaining the observable behavior of the application [10].

6.2.1.4   Time Safety

The TDL compiler provides a time safety check, which guarantees that a program behaves as expected for a particular target platform given that the worst case execution times for the tasks to be executed are known for that platform. In the case of a distributed platform, the compiler also guarantees that the network communication preserves the expected observable behavior of the application.

6.2.1.5   Compositionality

A TDL program consists of a set of so-called modules. All modules are executed in parallel, and the data flow between modules is handled by the TDL runtime system. Adding another module to the application does not change the observable behavior of the previously existing modules.

6.2.2   TDL LANGUAGE CONSTRUCTS

In the sequel, we introduce the individual TDL language constructs informally. For more details including a formal grammar, please refer to the TDL Language Specification [8].

6.2.2.1   Modules

At the outermost level, a TDL application consists of a set of modules[35]. Two modules can either be independent, that is, they share no data, or cooperating. Cooperating modules exchange data through ports. Statically, a module provides a namespace. Dynamically, modules are executed in parallel—possibly on different nodes in a distributed system. All modules share a common clock, which, in the case of a distributed execution platform, has to be distributed to the individual nodes of the platform. A module may encapsulate a finite state machine (FSM), where the states are denoted as modes. In a mode, the temporal aspects of all activities are defined.

6.2.2.2   Ports

Data flow within a single TDL module, between multiple modules, and between a TDL module and the physical environment is exclusively based on ports. A port is a typed variable that is accessed (read or written) at specific time instances only. Sensor and actuator ports (sensors and actuators for short) are the only means for a TDL module to communicate with the environment. A sensor declaration defines a typed read-only variable to represent a particular value in the physical environment and provides input to the TDL application. An actuator declaration is an initialized and typed write-only variable that influences a particular value in the physical environment and provides output from the TDL application to the environment. The access to the hardware is performed by user-provided setter and getter functions that are external to TDL.

6.2.2.3   Tasks

A task is the computational unit in TDL. It defines a namespace for input, output, and state ports. Each task is associated with a task function, that is, a stateless piece of code without any synchronization points. A single invocation of a task at runtime creates a *task activation*. A task activation lasts for a strictly positive amount of time that starts at the release time (the time when the task activation is released) and ends at the termination time (the time when the task activation is terminated). The time between these two instants is called the *logical execution time* (LET) of the task activation (Figure 6.1). At the release time, the input ports of the corresponding task are updated with the values read from the output ports of other tasks and sensors that have been passed as parameters by the task invocation. The actual execution of the

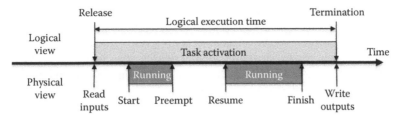

FIGURE 6.1 Logical execution time.

task function can be scheduled at will, as long as its execution starts after the release time and finishes before the termination time. The task activation locally buffers the output of the task function. At the termination time, the output ports of the task are updated with the values stored in the local buffers of the task activation. State ports hold data that must persist among multiple activations of a task.

6.2.2.4 Modes

Modules that encapsulate a state machine have a dedicated start mode each and can switch between modes independently of others. A mode m specifies a mode period P_m (in microseconds) and a set of activities. As long as a module remains in mode m, the activities associated with m are repeated with period P_m. A mode activity either is a task invocation, an actuator update, or a mode switch. Mode activities may be guarded. A *guard* is a function that returns either true or false. A guarded mode activity is executed only if its guard evaluates to true.

When defining a mode activity within a mode, a frequency f for this activity is specified. This frequency divides the mode period P_m into *slots* of duration P_m/f each. These slots define the times during the mode period at which the guard of the activity is evaluated and the activity is executed. Actuator updates and mode switches are executed at the end of a slot. Task invocations result in the release of a task activation at the beginning and the termination of the particular activation at the end. Thus, the LET of an activation spans the entire slot, and, by default, the task invocation is executed in each slot.

For control systems, the fixed relationship between the rate of task invocations and the LET of each task activation poses problems as the delay between reading the sensors and updating the actuators consumes phase reserve in the control loop (e.g., Ref. [11]). Besides increasing the frequency of a task invocation and thus the sampling rate, which may be prohibitive in terms of CPU load, TDL offers two mechanisms for dealing with this situation: slot selection and task splitting.

Slot selection is the explicit selection of the slots in which a mode activity should be executed. For task invocations, slot selection maintains the basic pattern of freezing the input ports of a task at the release of each of the task activations and updating the output ports (and actuators) at termination. It allows a separation between the specification of the LET of a task activation and the repetition rate of the associated task. Slots that are not selected go unused.

Task splitting means to split the single task function into two functions, one called fast step and the other called slow step. TDL assumes that the fast step does not consume any time. At release time, first the input ports of the task are updated and then the fast step is executed. As a modification of the basic actuator update pattern, there is the possibility to update an actuator with a value calculated in the fast step immediately after it has finished execution. The slow step is executed afterwards, during the LET of the activation. At the termination time, the output ports of the task are updated and further actuator updates may be performed. In a control system application, for example, the actual controller may be moved into the fast step. Thus, the delay between reading the output of the plant and updating the actuator that delivers the input to the plant is minimized. A state estimator (e.g., [11]), representing a higher computational load, will then be moved into the slow step.

6.2.2.5 Asynchronous (= Event-Triggered) Activities

In addition to time-triggered (alias synchronous) activities, it is often necessary to execute event-triggered (alias asynchronous) activities as well [12]. TDL supports asynchronous task invocations and actuator updates. Such an asynchronous activity is triggered either by an update of an output port, by the occurrence of a hardware interrupt, or by the tick of a timer that may potentially introduce its own time base.

By integrating asynchronous activities into TDL, the TDL runtime system is able to provide the synchronization of the data flow between synchronous and asynchronous activities. It has been shown in Ref. [13] that a lock-free synchronization approach with a negligible impact on the timing of the time-triggered activities is possible with the semantics outlined below.

Events may be associated with a priority and are registered in a priority queue when they arrive. Processing the events is delayed and supposed to be performed sequentially by a single background thread that runs whenever there are no time-triggered activities to perform. Input ports are read as part of the asynchronous execution, not at the time of registering an event. Output ports are updated immediately after an asynchronous task invocation has finished. If an activity is triggered again before it has started to execute, it will not be executed a second time but remains registered once. In the case of a distributed system, the communication of asynchronous output values to remote nodes is supposed to rely on asynchronous network operations. Since any network operation introduces a delay, the transparent distribution property (see Section 6.2.1.3) does not hold in the case of asynchronous activities [14].

6.2.3 Example **TDL** Modules

The following example of two TDL modules exemplifies the textual syntax of TDL (Figure 6.2). As an alternative to the textual representation, the TDL toolchain also provides a syntax-driven editor that supports a visual and interactive modeling of TDL modules (see Figure 6.6 later in the chapter).

```
module Sender {
  sensor int s1 uses getS1;
  actuator int a1 uses setA1;

  public task inc {
    input int i;
```

```
      output int o := 10;
      uses incImpl(o);
    }
  start mode main [period=5ms] {
      task
      [freq=1] inc(s1);//LET = 5ms/1 = 5ms
      actuator
      [freq=1] a1 := inc.o;
      mode
      [freq=1] if exitMain(s1) then freeze;
    }
  mode freeze [period=1000ms] {}
}
module Receiver {
  import Sender;
  ...
  task clientTask {
    input int i1;
    ...
  }
  start mode main [period=10ms] {
      task
      [freq=1] clientTask(Sender.inc.o); //LET = 10ms
    ...
  }
  ...
}
```

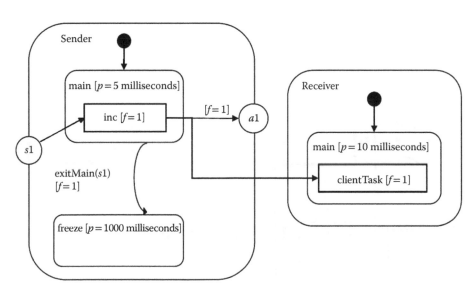

FIGURE 6.2   Sample TDL application.

Module *Sender* contains a sensor variable *s1* and an actuator variable *a1*. The value of *s1* is updated by executing the (platform-specific) function *getS1* and the value of *a1* is sent to the physical actuator by using the platform-specific function *setA1*. The declaration of task *inc* contains an input port *i* and an output port *o* with an initial value of 10. This task is invoked in the mode *main*, where it reads input from the sensor *s1*. In the same mode, actuator *a1* is updated with the value of the task's output port. The timing behavior of the mode activities is specified by means of individual frequencies within their common mode period. For example, with a frequency of 1, the activation of task *inc* is defined to have a LET of 5 milliseconds. The second module called *Receiver* imports the *Sender* module to connect the output of the task *inc* with the input of the task *clientTask*.

6.2.4 TDL TOOLCHAIN

TDL introduces appropriate abstractions to separate timing from functionality and platform-independent from platform-specific aspects. To obtain executable software, the textual TDL description must be compiled and combined with external functions that implement the required functionality. Figure 6.3 outlines the TDL toolchain. It shows as a central component the TDL compiler that compiles a textual TDL program to platform-independent embedded code (so-called E-code). The TDL compiler also offers a plug-in-architecture for generating target platform-specific output. For example, on an automotive platform with OSEK as the operating system, the platform-specific output could include the so-called OIL files [15]. The E-code together with the platform-specific output and the functionality code corresponding to task function implementations is used by the TDL runtime system, the so-called E-machine [16], to execute TDL applications.

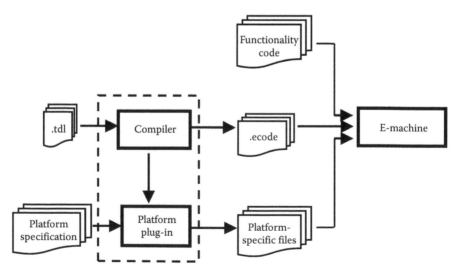

FIGURE 6.3   TDL toolchain.

6.2.4.1   E-code

The TDL compiler generates E-code for each mode of a module. The E-code covers a single mode period and is repeated by means of a jump instruction to the beginning of the mode. For every logical time instant at which E-code must be executed, one E-code block is generated. An E-code block is a list of E-code instructions. It specifies for one logical time instant the actions that must be taken by the E-machine to comply with the timing specifications and LET semantics. The following generic sequence of actions comprises one E-code block for a logical time instant t:

1. Update the output ports of all tasks that are defined to be terminated at t.
2. Update all actuators that are defined to be updated at t.
3. Switch mode if a mode switch is defined at t.
4. Update the input ports of all tasks that are defined to be released at t.
5. Release all task activations that are defined to be released at t.
6. Sleep until the next logical time instant that must execute E-code.

Sensors are read whenever their value is required. However, at one particular logical time, a sensor is read at most once.

In the following, we illustrate these actions for the module *Sender* from the previously described example. At time 0, actions in the start mode are processed. Output ports are initialized and connected actuators are updated. Then, the sensor $s1$ is read and an activation for task *inc* is released. There are no further actions to be processed at time 0. At time 5, which is the end of the LET of the first activation of task *inc*, the task's output port is updated. Following this, the actuator $a1$ is updated. Next, the mode switch condition in the guard function *exitMain* is evaluated. This causes sensor $s1$ to be read, and the value is provided as input to *exitMain*. If the guard evaluates to true, a mode switch to the empty mode *freeze* is performed, and no further actions are processed. Otherwise, the module remains in the mode, and the next activation for task *inc* is released. Figure 6.4 shows the periodic execution pattern of the task *inc* in mode *main* of module *Sender* and of the task *clientTask* of module *Receiver*.

In Section 6.2.4, we describe and compare the integration of TDL with the two simulation environments MATLAB and Simulink and Ptolemy. As these two environments are quite different, this requires two entirely different integration strategies. This applies to both modeling and simulation. Whereas Ptolemy is open source, targeted to support different models of computation, and highly adaptable even for fundamental elements, MATLAB and Simulink is proprietary and more restrictive in its adaptability.

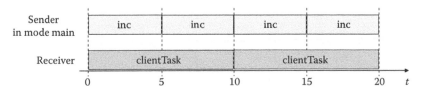

FIGURE 6.4   Periodic execution of the tasks *inc* and *clientTask*.

6.3 TDL INTEGRATION WITH MATLAB® AND SIMULINK®

Simulink builds on MATLAB (both products by MathWorks) [6] and has become the de facto standard for the modeling and simulation of real-time systems in various domains such as automotive, avionics, and aerospace. First attempts to integrate TDL with MATLAB and Simulink started as early as the initial development of TDL in 2003 [17]. In Simulink, systems are modeled in a visual and interactive environment using (mostly time-based) block diagrams. Code generators may then automatically translate the block diagrams into software, for example, into C code.

According to the results described in Ref. [18], manually implementing LET semantics in Simulink is strongly discouraged. Even simple models of single-mode systems are cluttered with additional blocks to ensure that the timing behavior in the simulation conforms to LET semantics. It turned out that it is practically infeasible to model LET-based applications with multimodal behavior by hand, even when using the Simulink extension Stateflow® [6]. In the following, we describe an approach that is based on an explicit timing specification with TDL. For the simulation, the TDL specification is automatically translated into a Simulink model with an E-machine implementation at its core. In this sense, MATLAB and Simulink represents yet another execution platform for TDL modules.

6.3.1 APPLICATION DEVELOPER'S PERSPECTIVE

We will start with a developer's perspective of the TDL integration with MATLAB and Simulink that covers the modeling, the simulation, and finally the platform mapping and the code generation.

6.3.1.1 Extension of the TDL Toolchain for MATLAB® and Simulink®

Figure 6.5 outlines the TDL toolchain when used together with MathWorks tools. Real-Time Workshop® Embedded Coder™ [20] (RTW-EC in Figure 6.5) can be used to generate the C source code for the TDL task implementations. The so-called

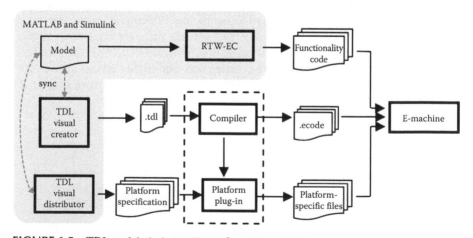

FIGURE 6.5 TDL toolchain in MATLAB® and Simulink®.

TDL:VisualCreator tool allows the visual and interactive modeling of TDL applications. For mapping TDL modules to specific platforms, that is, for organizing the various build steps, we provide the so-called TDL:VisualDistributor tool. It allows a developer to

- Define a hardware topology. This can be a single node or a cluster consisting of potentially heterogeneous nodes that are connected, for example, through a time-triggered Ethernet.
- Assign the individual TDL modules to their target nodes.

A built-in code and schedule generation framework generates platform-specific code, a communication schedule in the case of a distributed platform, makefiles, and any other output required for a particular platform [4].

6.3.1.2 Modeling

The modeling typically comprises two principal components: the controller and the plant. The plant is modeled as usual in MATLAB and Simulink. The controller is modeled with one or multiple *TDL module block(s)* (Figure 6.6) available in the TDL library.

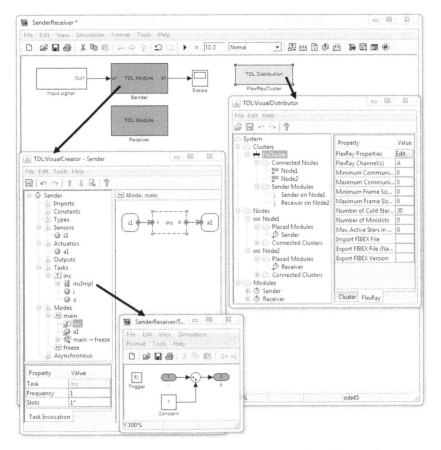

FIGURE 6.6 Modeling with the TDL toolchain in MATLAB® and Simulink®.

Each TDL module block represents one TDL module. Instead of using the textual representation of TDL, the module is edited using the TDL:VisualCreator tool that opens when double-clicking a TDL module block. The TDL module is two-way synchronized with Simulink. This means that a change in the TDL:VisualCreator, such as adding a sensor, is immediately reflected in the Simulink representation of the particular TDL module and vice versa. A sensor of the TDL module is represented as *Inport*, and an actuator is represented as *Outport* of the TDL module block. A task is represented as a *(Function-Call) Subsystem* that resides within the TDL module block. Within this subsystem, the functionality of the task may be modeled with appropriate library blocks (excluding those that comprise continuous-time behavior) with inherited sample time. Figure 6.6 shows the previously described TDL example within a MATLAB and Simulink model and the two tools TDL:VisualCreator and TDL:VisualDistributor. The TDL:VisualCreator tool lists the individual elements such as the sensor *s1* and the task *inc* of module *Sender* in a tree representation. The activities of mode *main*, for example, are shown in the right-hand half of the frame. Their timing is specified through properties in the table below the tree representation.

6.3.1.3 Simulation

The overall system can be simulated once the application developer has finished the modeling phase, that is, timing behavior has been specified using the TDL:VisualCreator, whereas controller functionality and plant behavior have been modeled with Simulink blocks. From the developer's point of view, there is no observable difference to starting a simulation if there were no TDL blocks present. This is achieved by an internal model translation (as sketched in the toolchain) to ensure that the simulation corresponds with the TDL specification. In fact, during the simulation, the compiled TDL program is executed within an E-machine encapsulated in a Simulink S-function. Details of how this is accomplished are described below.

As the TDL code and the schedule generators ensure that the timing behavior of TDL modules is equivalent when executed on a single node and when distributed among multiple nodes, the simulation can assume execution is on a single node and it is not necessary to account for any communication behavior.

6.3.1.4 Platform Mapping and Code Generation

To generate code for the application, the target platform must be specified. This is performed with the TDL:VisualDistributor tool that is integrated with Simulink as a *TDL Distribution block* (see the shaded box in the top-right corner of the Simulink model in Figure 6.6). The platform involves the specification of the (potentially heterogeneous) node platforms, their interconnections within the cluster, and the communication protocol. For this purpose, the developer may choose from a set of available node (such as a dSpace MicroAutoBox for prototyping) and cluster (such as FlexRay) plug-ins. After every TDL module of the Simulink model has been assigned to a node, the developer may start the generation of platform-specific code, a communication schedule in the case of distributed platforms, makefiles, and any other required output. If desired, the TDL:VisualDistributor tool can also trigger the RTW-EC to generate C code for all the tasks of the TDL modules.

6.3.2 IMPLEMENTATION PERSPECTIVE

On an embedded hardware platform, the E-machine represents the core piece for a LET-based execution. E-machine implementations exist for several different platforms [21]. For the TDL integration with MATLAB and Simulink, we implemented an E-machine that is based on a Simulink S-function [18]. An S-function is a Simulink block that references a user-defined functionality implemented in a programming language such as C and compiled by the *MATLAB EXecutable (MEX)* compiler. In this way, the built-in Simulink blockset can be extended. S-functions are composed of *callback methods* that the Simulink engine executes at particular points during the simulation.

As the S-function implements the E-code interpreter, it must be invoked whenever the simulation time matches the logical time of a TDL activity, as defined in the E-code. According to the E-code instructions, the S-function triggers the execution of Simulink Function-Call Subsystems. Each task and each guard is represented as a Function-Call Subsystem that is provided by the developer. Additional Function-Call Subsystems are generated automatically as part of the model translation when the simulation is started. They implement the port assignment operations, for example, to update an actuator port with the value of a task output port.

Figure 6.7 exemplifies this E-machine approach for a simplified application. The placement of the individual blocks conforms to the data flow, which is basically from left to right along the arrows from a source to a sink. The source value is read by a sensor, which provides the value to a guard and a task. The actuator block uses the output port of a task to write to a sink. The E-machine triggers the individual blocks according to the E-code resulting in the indicated order (1–6). The input port of such a generated subsystem is directly connected to the output port, which corresponds to an assignment in the imperative programming paradigm as soon as the system is triggered. This ensures the correct LET behavior of a task activation, for example, when triggering its release and termination at the correct time instants. Both fixed and variable sample time approaches for the E-machine are possible [22]. The suggested value for a fixed sample time is the GCD (greatest common divisor) of all activity periods.

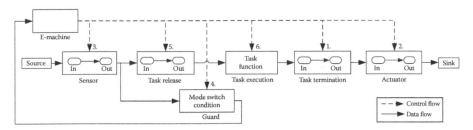

FIGURE 6.7 Basic principle of an E-machine as a Simulink® S-function and Function-Call Subsystems.

6.3.2.1 Resolving Data Dependencies

The S-function implementation of the E-machine for a simulation environment is analogous to E-machine implementations for hardware platforms. Compared to other simulation approaches, such a Simulink E-machine results in an efficient simulation model [18]. However, because of data dependency problems that can occur in simulation environments, the practical applicability of this basic mechanism turned out to be limited. We identified the following application scenarios that in general cannot be handled by this integration concept:

- Cyclic import relationships between LET-based controllers (TDL modules)
- Control loops involving plants without delay
- Control loops with mixed LET-based and conventionally modeled controllers

These cases are discussed in detail in Ref. [22]. They are all related to cyclic data flow dependencies and the ability of the simulation environment to find a valid strategy for executing each individual block. Like many other simulation environments, Simulink does not support cycles without a delay except for special cases [23]. Delays are introduced by explicit delay blocks or by other blocks whose output is not directly controlled by the input (although possibly dependent on the block state). Those blocks are said to have indirect (or nondirect) feedthrough. When Function-Call Subsystems are involved, Simulink reports a *data dependency violation*, which is similar to an *algebraic loop error* [24], when attempting to simulate a model with a direct data dependency cycle. From the control engineer's point of view, this appears to be counterintuitive, since the LET of a task is always greater than zero and thus should introduce the required delay. The problem is that the simulation environment is not aware of this LET characteristic.

In Ref. [25], we propose an E-machine implementation that consists of two interacting S-functions. Without violating the TDL specification, that is, without changing the timing behavior of the simulation, this approach introduces additional delay blocks to resolve the cyclic dependencies. This *two-step E-machine architecture* is capable of simulating these three scenarios with cyclic data flow dependencies and also supports TDL applications with mixed time- and event-triggered (asynchronous) activities [22]. In the case of simulating event-triggered activities, the simulation of events and the corresponding reaction cannot be guaranteed to match the behavior on a specific target platform. This is because asynchronously activated tasks do not have a LET and also because the simulation is not aware of any scheduling strategy, distribution topology, or CPU speed of the target platform.

6.4 TDL INTEGRATION WITH PTOLEMY II

Ptolemy II is the software infrastructure of the Ptolemy project [7], which studies modeling, simulation, and design of concurrent, real-time, and embedded systems. It is an open source tool written in Java that allows modeling and simulation of systems adhering to various models of computation (MoC). Conceptually, a MoC represents a set of rules, which govern the execution and interaction of model components.

The implementation of a MoC is called a *domain* in Ptolemy. Some examples of existing domains are Discrete Event (DE), Continuous Time (CT), Finite State Machines (FSM), and Synchronous Data Flow (SDF).

Ptolemy is extensible in that it allows the implementation of new MoCs. Most MoCs in Ptolemy support actor-oriented modeling and design, where models are built from actors that can be executed and that can communicate with other actors through ports. The nature of communication between actors is defined by the enclosing domain, which is itself represented by a special actor, called the domain director. Simulating a model means executing actors as defined by the top-level model director.

6.4.1 TDL DOMAIN

TDL is implemented as an experimental domain in Ptolemy. The TDL domain consists of three specialized actors: *TDLModule*, *TDLMode*, and *TDLTask*. The TDLModule actor (with the associated *TDLModuleDirector*) restricts the basic modal model behavior according to the TDL semantics. In modal models, mode switches are made whenever a mode switch guard evaluates to true, whereas in TDL modules, mode switches are only allowed at predefined points in time. Similar restrictions apply to port updates. To ensure the LET of a task activation, input ports are only allowed to be read once, at the beginning of the LET, and output ports are only allowed to be written at the end of the LET and not when a task finished its computation. TDL requires a deterministic choice of one of all simultaneously enabled transitions, which is not provided by the FSM domain. We resolve this by employing a convention similar to the one supported by Stateflow [19], where the outgoing transitions of the active mode are tested based on the graphical layout, in clockwise order starting from the upper left corner of the graphical representation of the mode. TDL timing information such as the mode period is associated with TDL actors in the model.

TDL activities are conceptually regarded as DEs that are processed in increasing time stamp order. Thus, a TDL module can be seen as a restricted DE actor. This enables the usage of TDL modules inside every domain that is amenable to DE actors. A TDL task is implemented as an SDF actor, which executes in logically zero time. The top-level director is a DE director. The DE director uses a global event queue to schedule the execution of actors in the model. The TDL module places events in this queue for every time stamp where at least one TDL action is scheduled.

The module *Sender* from the example above modeled in Ptolemy is shown in Figure 6.8. The model shown in the top-left box of the figure contains a TDL Module and two actors to provide sensor values and display actuator values. The TDL Module contains two modes *Main* and *Freeze* (see Figure 6.8, the bottom-left box). Both modes have their period indicated by an associated parameter. The *Main* mode contains the task and the association of sensor and actuator values to input and output ports of the task. The frequency of the task invocation, which determines the LET, is defined as a parameter.

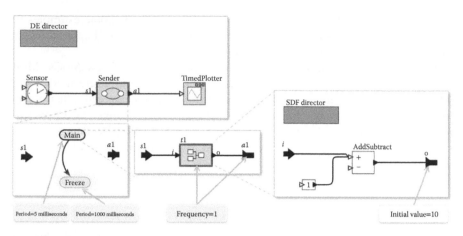

FIGURE 6.8 TDL module *Sender* in Ptolemy II.

Because of the iterative execution strategy of Ptolemy, problems with irresolvable cyclic dependencies as described for the Simulink integration do not arise. As the LET is always greater than zero, there is an actor without a direct dependency between inputs and outputs. Consequently, the DE director is able to resolve the loop and to process the TDL actions represented as events in the correct order and potentially interleaved with the plant.

6.5 COMPARISON BETWEEN THE SIMULINK® AND THE PTOLEMY II INTEGRATION

In the Simulink integration, the developer models the functionality (task implementations and the plant) as usual with Simulink blocks. Timing, mode switching logic, and the overall application data flow are expressed in TDL with the TDL:VisualCreator tool. The TDL toolchain automatically creates a simulation model that contains data flow and timing information. On the other hand, the open and extensible architecture of Ptolemy facilitates the expression of all TDL semantics directly in the model. In Ptolemy, the timing specifications of TDL are expressed as properties of the respective actors (i.e., within the simulation model), whereas in Simulink, we abstract from both the simulation and the execution platform.

Both approaches extend the existing simulation framework with new blocks (actors) that ensure the TDL semantics during the simulation. In Simulink, the timing requirements are encoded in a static E-code representation and enforced by an E-machine, resulting in a low computational overhead. In Ptolemy, TDL actions, such as releasing a task activation or performing a mode switch, are generated dynamically and are represented as DEs. They are enqueued in the event queue of the DE director, which then schedules the appropriate TDL actions at their corresponding time instants. This independence of E-code may make experiments with future TDL extensions more straightforward, because testing new features with Simulink potentially requires changing the compiler, the E-code instruction set, and the E-machine.

6.6 RELATED WORK

There is a Simulink integration of another LET-based language called HTL [26]. The HTL compiler is capable of compiling an HTL description into a Simulink model, which is then equipped with functionality for the control laws. The Simulink integration supports the hierarchical structure of HTL descriptions but restricts communication with the environment (plant model) to a single module, which limits the support for simulating distributed applications. Although HTL, similar to TDL, is based on an E-code variant, the HTL Simulink integration does not follow an E-machine approach but uses standard built-in MATLAB and Simulink blocks. This results in a simulation that does not exhibit the same behavior as the execution of the generated code, since the timing is distorted in several situations (e.g., by Unit-Delay blocks).

Simulink is closely related to synchronous languages. The simulation engine of Simulink executes subsystems implementing controller functionality in logical zero time. However, the semantics of Simulink are not defined formally [27], whereas synchronous languages are based on strict formal definitions and aim at formal verifiability. In several approaches, synchronous languages are combined with Simulink:

Argos, a synchronous language with a rigorously defined graphical notation, is prototypically embedded in Simulink as a less powerful but also less complicated alternative to Stateflow [28]. The implementation assumes that outputs calculated in response to an event are provided immediately. An extension [29] accounts for computational delays and corresponds to the implementation of the synchronous languages Argos and Esterel, where the response must be provided before a next event occurs. In both approaches, a synchronous program is embedded in Simulink.

The work by Caspi, Adrian, Maignan, Sofronis, and Tripakis [30] focuses on the conversion of a Simulink model into a synchronous program. Simulink systems consisting of a predefined subset of discrete-time blocks are translated automatically into Lustre, such that the original Simulink behavior is preserved. This enables the usage of tools for formal validation, simulation, synthesis, and so on. Another work [31] also includes Stateflow blocks. Caspi, Curic, Maignan, Sofronis, Tripakis, and Niebert [32] apply these results in the context of distributed systems. They present a layered end-to-end approach by translating Simulink models to the Lustre-based modeling environment SCADE in a first step. In a second step, the Lustre program is annotated to define timing assumptions and requirements and to specify the mapping to individual nodes. Finally, the application is mapped to a Time-Triggered Architecture (TTA) cluster [9].

The MathWorks product SimEvents® [6] extends Simulink with a mechanism for discrete-event simulation. This allows for simulating models that comprise continuous-time, discrete-time, and discrete-event components. Similar to the TDL integration with Ptolemy, an event-based scheduling can ensure the correct LET semantics of tasks in Simulink. However, based on the experience gained in the previous work, we doubt that a manual modeling of LET semantics scales to complex applications with multimodal behavior. A future research project may attempt to combine SimEvents with the TDL-based approach. Our Simulink integration could especially benefit from the event-based approach to simulate asynchronous activities of TDL. It is also conceivable that the E-machine is not sample-based but is itself triggered asynchronously by events.

TrueTime [33] is a Simulink toolbox for simulating networked control systems. It facilitates the cosimulation of control tasks, network transmissions, and continuous plant dynamics. TrueTime addresses the fact that traditional control design in MATLAB and Simulink often disregards temporal behavior and introduces a real-time kernel to be simulated in parallel with the plant. The *kernel block* is implemented as an S-function that simulates a flexible real-time kernel and provides support for A/D and D/A converters, network connections, and interrupt channels. It is event-driven and supports both periodic and aperiodic tasks. Different scheduling mechanisms, such as rate-monotonic or earliest-deadline-first, may be used. Control tasks are implemented in MATLAB functions, C, or Simulink blocks. For simulating distributed applications, network blocks support different network types. TrueTime allows a developer to experiment with various scheduling mechanisms and to investigate the true timing behavior of control applications in Simulink, taking into account latencies, execution times, jitter, and network effects. It requires execution times, distribution, and scheduling mechanism to be known in advance to approximate the behavior on the real execution platform. It is a platform-centered approach, whereas simulating TDL modules abstract from the platform and the network topology.

The TDL domain in Ptolemy II is related to the experimental Giotto domain in Ptolemy II [34]. The Giotto domain is designed based on basic Ptolemy II software components, whereas the TDL domain leverages the existing DE domain. The implementation of the TDL domain reflects the distinction between the fundamental concepts (LET, modes) and the manner in which these concepts are used (the operational semantics). The implementation is two-layered: the basic layer pertains to scheduling LET-based tasks grouped in modes, and the operational layer corresponds to a specific time-triggered programming model. The latter extends the basic layer by specifying additional operations as well as the order of data transfer and mode-change operations according to the programming model semantics. In principle, this forms the basis of domain controllers for any other time-triggered programming models (including Giotto) by extending the basic layer.

6.7 CONCLUSION

The LET allows the explicit specification of timing behavior independent of a specific platform, that is, LET abstracts from the physical execution time and thus from the execution platform and the communication topology. Thus, LET-based systems exhibit behavior equivalent or close to the platform without the necessity to specify platform details. Another significant advantage compared to the state-of-the-art is that the resulting embedded real-time systems generated from a LET-based language such as TDL are deterministic, composable, and portable. In this chapter, we presented the core concepts of TDL and its integration in two quite different modeling and simulation environments: MATLAB and Simulink and Ptolemy II. In both the cases, the integration concept and its implementation ensure that the simulation of time-triggered tasks exhibits the same behavior as the execution of the generated code on any potentially distributed hardware platform.

REFERENCES

1. Edmondson, James, and Doug Schmidt. 2012. "Towards Accurate Simulation of Large-Scale Systems via Time Dilation." In *Real-Time Simulation Technologies: Principles, Methodologies, and Applications*, edited by Katalin Popovici, Pieter J. Mosterman. Boca Raton, FL: CRC Press.
2. Stankovic, John A. 1988. "Misconceptions about Real-Time Computing: A Serious Problem for Next-Generation Systems." *Computer* 21 (10): 10–9.
3. Henzinger, Tom, Ben Horowitz, and Christoph Kirsch. 2003. "Giotto: A Time-Triggered Language for Embedded Programming." *Proceedings of the IEEE* 91: 84–99.
4. Chrona. 2011. http://www.chrona.com.
5. Zander, Justyna, Ina Schieferdecker, and Pieter J. Mosterman, eds. 2011. *Model-Based Testing for Embedded Systems*. Boca Raton, FL: CRC Press.
6. MathWorks. 2011. www.mathworks.com.
7. Eker, Johan, Jorn Janneck, Edward A. Lee, Jie Liu, Xiaojun Liu, Jozsef Ludvig, Stephen Neuendorffer, Sonia R. Sachs, and Yuhong Xiong. 2003. "Taming Heterogeneity: The Ptolemy Approach." *Proceedings of the IEEE, Special Issue on Modeling and Design of Embedded Software* 91 (1): 127–44.
8. TDL language report, white papers, tool demo, and tool download. 2011. http://www.chrona.com/en/resources/white-papers.
9. Kopetz, Hermann. 1997. *Real-Time Systems: Design Principles for Distributed Embedded Applications*. Norwell, MA: Kluwer Academic Publishers.
10. Farcas, Emilia, Claudiu Farcas, Wolfgang Pree, and Josef Templ. 2005. "Transparent Distribution of Real-Time Components Based on Logical Execution Time." *SIGPLAN Not* 40 (7): 31–9.
11. Franklin, Gene, Michael Workman, and David Powell. 1997. *Digital Control of Dynamic Systems*. 3rd ed. Boston, MA: Addison-Wesley Longman Publishing Co.
12. Scaife, Norman, and Paul Caspi. 2004. "Integrating Model-Based Design and Preemptive Scheduling in Mixed Time- and Event-Triggered Systems." In *Proceedings of the 16th Euromicro Conference on Real-Time Systems (ECRTS '04)*, Catania, Italy, pp. 119–126, Washington, DC: IEEE Computer Society.
13. Templ, Josef, Johannes Pletzer, and Wolfgang Pree. 2009. "Lock-Free Synchronization of Data Flow between Time-Triggered and Event-Triggered Activities in a Dependable Real-Time System." In *Proceedings of the Second International Conference on Dependability*, DEPEND, pp. 87–92. Washington, DC: IEEE Computer Society.
14. Templ, Josef, Johannes Pletzer, and Andreas Naderlinger. 2008. *Extending TDL with Asynchronous Activities*, Technical Report, University of Salzburg, http://www.softwareresearch.net/fileadmin/src/docs/publications/T022.pdf.
15. OSEK/VDX. 2004. *System Generation, OIL: OSEK Implementation Language, Specification Version 2.5*. www.osek-vdx.org.
16. Henzinger, Thomas A., and Christoph M. Kirsch. 2002. "The Embedded Machine: Predictable, Portable Real-Time Code." ACM Trans. Program. Lang. Syst., 29(6):33–61, October 2007.
17. Stieglbauer, Gerald, and Wolfgang Pree. 2004. *Visual and Interactive Development of Hard Real Time Code*. Automotive Software Workshop (ASWSD), San Diego.
18. Stieglbauer, Gerald. 2007. *Model-Based Development of Embedded Control Software with TDL and Simulink*. PhD thesis, University of Salzburg.
19. MathWorks. 2010. *Real-Time Workshop Embedded Coder 5, User's Guide*.
20. Farcas, Claudiu. 2006. *Towards Portable Real-Time Software Components*. PhD thesis, University of Salzburg.
21. Naderlinger, Andreas. 2009. *Modeling of Real-Time Software Systems Based on Logical Execution Time*. PhD thesis, University of Salzburg.

22. Denckla, Ben. 2006. "Many Cyclic Block Diagrams Do Not Need Parallel Semantics." *SIGPLAN Not* 41 (8): 16–20.

23. Mosterman, Pieter J., and John E. Ciolfi. 2004. "Interleaved Execution to Resolve Cyclic Dependencies in Time-Based Block Diagrams." In *CDC '04: Proceedings of the 43rd IEEE Conference on Decision and Control*, pp. 4057–62.

24. Naderlinger, Andreas, Josef Templ, and Wolfgang Pree. 2009. "Simulating Real-Time Software Components Based on Logical Execution Time." In SCSC '09: *Proceedings of the Summer Computer Simulation Conference*, pp. 148–155.

25. Iercan, Daniel, and Elza Circiu. 2008. "Modeling in Simulink Temporal Behavior of a Real-Time Control Application Specified in HTL." *Journal of Control Engineering and Applied Informatics (CEAI)* 10 (4): 55–62.

26. Carloni, Luca, Maria D. Di Benedetto, Roberto Passerone, Alessandro Pinto, and Alberto Sangiovanni-Vincentelli. 2004. *Modeling Techniques, Programming Languages and Design Toolsets for Hybrid Systems*. Technical Report IST-2001-38314 WPHS, Columbus Project.

27. Bourke, Timothy, and Arcot Sowmya. 2005. Formal Models in Industry Standard Tools: An Argos Block within Simulink. *International Journal of Software Engineering and Knowledge Engineering* 15 (2): 389–96.

28. Bourke, Timothy, and Arcot Sowmya. 2006. "A Timing Model for Synchronous Language Implementations in Simulink." In *EMSOFT '06: Proceedings of the 6th ACM & IEEE International Conference on Embedded Software*, pp. 93–101. New York: ACM.

29. Caspi, Paul, Adrian Curic, Aude Maignan, Christos Sofronis, and Stavros Tripakis. 2003. "Translating Discrete-Time Simulink to Lustre." In Rajeev Alur and Insup Lee, editors, EMSOFT, volume 2855 of Lecture Notes in Computer Science, pp. 84–99. Springer.

30. Scaife, Norman, Christos Sofronis, Paul Caspi, Stavros Tripakis, and Florence Maraninchi. 2004. "Defining and Translating a "Safe" Subset of Simulink/Stateflow into Lustre." In *EMSOFT '04: Proceedings of the 4th ACM International Conference On Embedded Software*, pp. 259–68. New York: ACM.

31. Caspi, Paul, Adrian Curic, Aude Maignan, Christos Sofronis, Stavros Tripakis, and Peter Niebert. 2003. "From Simulink to SCADE/Lustre to TTA: A Layered Approach for Distributed Embedded Applications." *SIGPLAN Not* 38 (7): 153–62.

32. Cervin, Anton, and Karl-Erik Arzén. 2009. *Model-Based Design for Embedded Systems,* chapter *TrueTime: Simulation Tool for Performance Analysis of Real-Time Embedded Systems*. Boca Raton, FL: CRC Press.

33. Liu, Xiaojun, Stephen Neuendorfer, Yang Zhao, Haiyang Zheng, Christopher Brooks, and Edward A. Lee. eds. 2007. *Heterogeneous Concurrent Modeling and Design in Java (Volume 3: Ptolemy II Domains)*. Berkeley, CA: EECS Department, University of California, UCB/EECS-2007-9.

34. Pree, Wolfgang, and Josef Templ. 2004. "Towards a Component Architecture for Hard Real Time Control Applications." In Manfred Broy, Ingolf Krüger, and Michael Meisinger, editors, Automotive Software, Connected Services in Mobile Networks, volume 4147 of Lecture Notes in Computer Science, pp. 74–85. Berlin, Heidelberg: Springer.

Section II

Real-Time Simulation
for System Design

7 Progressive Simulation-Based Design for Networked Real-Time Embedded Systems

Xiaolin Hu and Ehsan Azarnasab

CONTENTS

7.1 Introduction 181
7.2 PSBD for Networked Real-Time Embedded Systems 183
    7.2.1 PSBD Overview 183
    7.2.2 Bifurcated Design Process for Networked Real-Time Embedded
        Systems 185
7.3 Background on CR Design 187
7.4 PSBD of the CR Network 188
    7.4.1 Design Procedure and Implementation Environment 188
    7.4.2 Design of a Single Cognitive Modem 190
    7.4.3 Design of a CR Network 192
    7.4.4 Experiment Results 194
7.5 Conclusions 196
References 197

7.1 INTRODUCTION

Simulation has long been used to support design and analysis of complex engineering systems. Fast simulations allow designers to flexibly experiment with and analyze different design solutions without implementing the systems in hardware. Real-time simulations support designers to test the real-time features of a system that interacts with the physical world and/or other hardware components. The latter is especially useful for designing real-time embedded systems, such as mobile devices, manufacturing automation sensors/actuators, and the networked software-defined radio (SDR) system presented in this chapter. Design and implementation of these systems have been influenced by the increasing demand of new products and the recent advances in technologies. The complexity and multidisciplinary nature of these systems make analytical modeling and analysis infeasible. However, system engineers must assess a design before proceeding with implementation of an expensive solution. Although traditional modeling and simulation can help in this goal, its applicability has been limited because of the gap between simulation models and implementation in hardware.

One type of real-time simulation is hardware-in-the-loop (HIL) simulation, which is an advanced technique frequently used in embedded systems' development [1,2]. A HIL simulation refers to a system in which parts of a pure simulation have been replaced with actual hardware. This is based on the observation that once hardware is added to the loop, unmodeled characteristics can be investigated and controls can be further refined. HIL is typically aimed at developing a single module in a larger system. Although it is a useful technique that can greatly support an engineering design, it does not offer a general methodology that can scale to more complex systems. Systematic design processes and methodologies are necessary for designing complex systems that are characterized as large scale and networked with tight couplings between software and hardware. Motivated by this need, we developed a progressive simulation-based design (PSBD) methodology that goes beyond HIL simulation by gradually adding more hardware to the simulated system in a progressive manner and improving the co-simulated model in each step before continuing [3]. The design process of PSBD starts from all models being simulated on computers, proceeds by bringing real system components into the simulation to replace their virtual counterparts (models), and ends when all components are in their deployed form and the final system is tested in a physical environment. Throughout this process, model continuity is emphasized and the simulation model is continually updated whenever new design details are revealed. Several distributed robotic systems have been developed following the principles of the PSBD methodology [4–6].

This chapter presents the PSBD for networked real-time embedded systems. We give an overview of the PSBD methodology and show a bifurcated design process that implements PSBD for individual embedded devices and the networked embedded system. We then apply the PSBD methodology to the design of a networked SDR system. SDR technology refers to a radio communication system capable of transmitting and receiving different modulated signals across a large frequency spectrum using software programmable hardware [7]. Major components of a SDR board include a digital signal processor (DSP), field-programmable-gate-array (FPGA), and radio frequency (RF) front end that facilitates wireless communication among the nodes. A DSP or a conventional central processor unit computes baseband signal processing and implements the MAC layer for networking. An FPGA is used for fast parallel processing of incoming data and down-sampling the result to a lower sample rate suitable for baseband processing. SDR provides modem designers a great opportunity to build complex modems by programming in software the previously hardware components of the radio. This replacement of hardware by software is in line with the model continuity approach of PSBD, as the individual system components are code modules to be developed and tested along the design process. The advantage of PSBD becomes more explicit when multiple SDR nodes should collaborate to form a network. Hence, the interaction of many complex subsystems should be engineered for best performance of the entire system. We show how PSBD is applied to a single cognitive modem and a cognitive radio (CR) network. This chapter extends the previous work on the PSBD methodology [3] and the case study example on SDR [8].

We aim to show that PSBD is an effective methodology that can be applied to a wide range of networked real-time embedded systems and engineering applications.

The modeling and simulation environment that supports this work is based on Discrete Event System Specification (DEVS) [9,10]. DEVS is a formalism derived from generic dynamic systems theory and has well-defined concepts of coupling of components, hierarchical, modular model construction, and object-oriented substrate supporting repository reuse. DEVS models time explicitly. This makes it convenient to study timeliness, which is an essential property of real-time systems. The DEVSJAVA environment [11] is used in this work to support fast and real-time simulations in the design process. More information about the DEVS formalism can be found in Zeigler et al. [9]. Applying DEVS-based modeling and simulation to system design has been researched in previous work. For example, a DEVS-based model-driven approach for developing embedded real-time systems was presented in Wainer and coworkers [12,13] to improve the quality of a design and to reduce the need for expensive testing cycle. A methodology for developing hybrid hardware/ software systems was presented in Glinsky and Wainer [14], where techniques were developed to enable transition from the simulated models to the actual hardware counterparts and to allow developing models with different levels of abstraction. A stepwise model refinement approach was proposed in Schulz and Rozenblit [15] to support real-time embedded system development. In Kim et al. [16], a DEVS-based unified framework was presented for developing real-time software systems, in which logical analysis, performance evaluation, and implementation can be performed. Our work on PSBD emphasizes a systematic process that progressively transitions a design from models to system realization. We note that although the DEVS simulation environment is used in this work, the presented PSBD methodology is general and can be realized using other modeling and simulation environments.

7.2 PSBD FOR NETWORKED REAL-TIME EMBEDDED SYSTEMS

This section provides an overview of the PSBD methodology [3] and develops a bifurcated design process for networked real-time embedded systems.

7.2.1 PSBD OVERVIEW

PSBD views simulation as the driving force for designing and testing engineering systems. It provides a design process that explicitly focuses on systematic transitions from simulation models to system realization. As shown in Figure 7.1, the design process consists of three stages, each of which is characterized by the types of entities (virtual or physical) that are involved. The first stage is *conventional simulation* (in fast simulation mode), where simulation is carried out using all models. A major task of this stage is to develop the system model based on available knowledge and assumptions about the hardware and operating environment of the physical system. Often discrepancies between simulation models and physical system components exist. These discrepancies cause the models to comprise behavior different from the system behavior. To reveal such design discrepancies, the next stage of the design process is *virtual environment simulation* (in real-time simulation mode), where

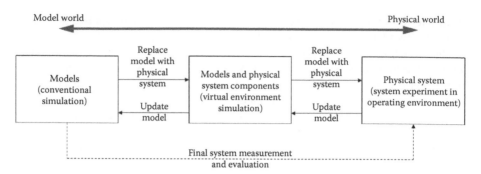

FIGURE 7.1 Progressive simulation-based design methodology.

simulation-based study is carried out in a virtual testing environment using combined models and physical system components. This stage brings simulation-based study one step closer to a realization by including physical system components into the simulation. The goal is to use physical system components to reveal overlooked design details, and thus help designers improve the control models/algorithms under development. The final stage is *physical system experiment*, where the physical system is tested in the target operating environment.

Along this design process, the PSBD methodology emphasizes two parallel activities in a progressive manner: *replace models with physical system components* and *update models*. As the design moves forward, physical system components are incrementally brought into the simulation to replace models. Simulations with these physical system components allow designers to validate their design assumptions and to reveal new design details. Such information is fed back to the previous stages to update the models if necessary. The updated model will then be used for follow-on design and test. This activity of model update allows designers to maintain a coherent model of the system under development. Thus, at the end of the design, not only the system is realized and tested, but also a system model that faithfully represents the system is developed. This system model can support final system measurement and evaluation (shown by the dashed line in Figure 7.1) and serve other purposes such as system maintenance and future development. It is important to note that each design stage is a dynamically evolving process by itself. For example, during the conventional simulation stage, it is common for designers to start from high level models and then refine them to more detailed models. Similarly, the virtual environment simulation stage that involves combined models and physical system components typically includes multiple phases too, for example, to start with replacing one physical system component first and then gradually add more.

Two important features of the PSBD methodology are *model continuity* and *virtual environment simulation* that supports simulation-based test with combined virtual and physical system components. Model continuity refers to the ability to transition as much as possible of a model specification through the stages of a development process [17]. For real-time embedded systems, we restrict model continuity to the models (software components) that implement the real-time control of the system. This means the control models of a real-time embedded system are designed, analyzed, and tested by simulation methods and then smoothly transitioned from simulation to hardware execution in the physical environment (see Hu et al. [4] and Azarnasab [6] for more details). To support model continuity, it is necessary to develop system models and run simulation-based tests in a systematic way. A modular design and well-defined interfaces are necessary to ensure the control models work with physical and simulated hardware in the same way at different stages of the design process [4,6]. The virtual environment simulation provides a virtual testing environment by using combined physical and virtual system components. It bridges the gap between conventional simulations that use all models and physical system experiments that use all physical system components. To support the virtual environment, simulation techniques must be developed that synchronize the physical and virtual system components. This includes allowing the physical and virtual components to "sense" each others' existence (see Hu and Zeigler [18] for an example of how physical and virtual robots are synchronized with each other). Meanwhile, time synchronization is also important. Since hardware components are included in the simulation-based study, real-time simulations are necessary to support the virtual testing environment.

7.2.2 BIFURCATED DESIGN PROCESS FOR NETWORKED REAL-TIME EMBEDDED SYSTEMS

Networked real-time embedded systems are characterized by a network of embedded devices interacting with each other and the tight couplings between software and hardware of those devices. Each of these devices is referred to as a *node* in this chapter. When commercial-off-the-shelf (COTS) nodes are not used, design of networked real-time embedded systems must design both the individual nodes and the networked system as a whole. Within this context, the PSBD methodology described above is elaborated to include a bifurcated design process as shown in Figure 7.2. The bifurcated design process explicitly differentiates the design of a single node (the bottom route in Figure 7.2) and the design of the networked system (the top route in Figure 7.2). The former mainly concerns designing the different functional modules, such as sensing, modulation/demodulation, and channel coding, of the embedded device. The latter focuses on how the multiple nodes work together as a whole, including designing and improving the communication protocols and the cooperative strategies among the nodes. Despite the different design focuses, both designs follow the PSBD process that starts from models and gradually adds more physical system components.

In Figure 7.2, the models are shown as white boxes; the physical system components are shown as grey boxes. The design starts and bifurcates into two routes: the design of a single node and the design of the networked system. For designing a single node, the first step is to model the functional modules of the embedded device and simulate how they work together to fulfill the functionality of the device. Then hardware components are brought into the design and HIL simulations are conducted to test how well the designed modules and algorithmic code work with the hardware. This proceeds in a stepwise fashion as more and more hardware components are included. Consider the CR (described later) as an example, first the channel sensing mechanism is implemented on the FPGA and DSP hardware, then the data transmission

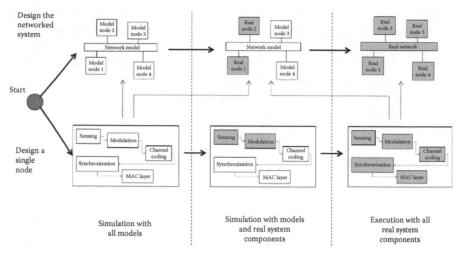

FIGURE 7.2   A bifurcated design process for networked real-time embedded system.

and data reception components are included, and finally the reporting mechanism is implemented. We use a signal generator and signal analyzer to help to test the developed components before a working node is complete. Eventually, all the code modules are implemented in the hardware and the embedded node is tested with all hardware components.

When designing the networked system, the first step is to develop individual node models and a model of the communicating network. The individual node model can reuse the model from designing the single node (as indicated by the dotted arrow in Figure 7.2). Alternatively, a different node model at a higher abstraction level (e.g., without including all the details of the devices' functional modules) can be used. Simulations with these models allow designers to test the networked system in the model world. The next step is to gradually include physical device nodes into the simulation-based study to conduct virtual environment simulations. The physical device nodes are either the ones from the single node design or COTS nodes. This continues until the entire system is realized and all designed nodes are tested in a physical network. It is important to note that the bifurcated design process provides a systematic view for designing general networked embedded systems. For a specific application, the design process can be tailored (e.g., some design stages are elaborated while others are omitted) to fit the specific design needs of the application. In Section 7.4, we provide an example of designing a CR system by starting from the design of a single radio modem and then proceeding to the design of a CR network. The single radio modem is also designed in progressive manner, where different functional modules are gradually implemented/tested in the hardware while other modules are provided by simulation models.

PSBD brings several advantages when designing complex networked engineering systems. Some of them are shared by the traditional HIL simulation. For example, it brings simulation-based study one step closer to the reality to provide useful information for designers. It also increases confidence by the designer about how the final system will operate. However, PSBD goes beyond that by emphasizing a systematic design process that gradually adds more physical system components to replace simulation models. The virtual environment simulation provides the flexibility for experimenting with a design in a virtual testing environment. It allows designers to use several, instead of all, physical nodes to carry out a system-wide test of a networked system. This is especially useful for large-scale networked real-time embedded systems whose complexity and scale severely limit experimentations in a physical environment using all physical nodes. As the scale of these systems increases, so does their design and test complexity. It is the intent of PSBD to systematically handle such design complexity in a progressive manner.

7.3 BACKGROUND ON CR DESIGN

The scarcity of unallocated frequency spectrum and the necessity for different radios to coexist in the shared unlicensed bands highlight the importance of the next generation of radios with dynamic spectrum access. Traditionally, spectrum is assigned to legacy devices, also called primary users (PUs). However, licensed frequency bands are rarely used everywhere [19,20] and over time lead to spectrum holes in time and space. To address this underutilization Federal Communications Commission has loosened the regulation to allow secondary users (SUs) to share some previously dedicated bands subject to minimal interference with legacy devices of the band [21]. Based on this definition of coexistence, SUs attempt to dynamically fill the spectrum holes over time and space [20], thus forming a CR [22]. For example, when a TV station (which is a PU of some frequency band) is not broadcasting or is in a location far from any TV broadcasting, SUs can instead use the spectrum in an opportunistic fashion. Spectrum access for first responders in a disaster scenario, where many wireless devices are active, is another application of CR [23].

A typical CR network consists of multiple SUs that coexist with PUs of a shared spectrum. PUs have priority access to the spectrum over SUs, that is, SUs should relinquish the spectrum when PUs begin transmission. The SU network should be designed to utilize more of the available bandwidth subject to minimum interference with the PUs. The hidden terminal problem (when a node is visible from a wireless access point (AP) but not from other nodes communicating with the AP) should be solved to minimize interference. For this purpose, SUs should form a CR network to sense the presence of active PUs and dynamically adapt to a suitable frequency, resulting in little or no interference with PUs. They collaboratively sense the spectrum and decide which part of the spectrum is available to them. Collaborative sensing involves signaling/reporting through a (possibly narrow band) control/reporting channel. To maximize the bandwidth efficiency of the SU network while minimizing interference with PUs, this signaling should be done in a reliable manner and in a minimum span of time. In addition to this requirement, it is desirable to have a system operating among radios with different modulations and protocols and supplied by various vendors. Coexistence and interoperability are two major design goals for a CR network. SDR technology best satisfies the required flexibility of CR, and thus is often used to implement CR network [24,25]. A SU that is realized on a SDR hardware board is called a cognitive modem. Design of a CR network must design the individual cognitive modems and the CR network as a whole.

The complexity of designing an individual cognitive modem is due to the many functional components and their complicated mutual interactions. These components can be implemented in parallel to expedite the manufacturing process. Parallelism can be achieved by simulating the entire system for a functional component that is being developed. Here the simulation closes the loop of system integration and provides a regression testing environment. In addition, as is common in system engineering design, not all hardware is available at the beginning of the project. Therefore, simulation is necessary to start the design with the partial hardware that is available. The rest of the functional components are provided by simulation. In our project, at first we received only one baseband module of SDR, capable of processing only the DSP algorithms such as polyphase implementation of a filterbank for channel sensing. We tested our sensing algorithm using a simulated fading channel and optimized it to some extent. The high dynamic range of filterbank sensing better detects low power PU; thus, it is less likely to interfere with legacy users. By the time we developed our fixed-point sensing method, we received one RF front end and applied sensing on a physical wireless channel to replace the previously simulated channel. The complexity of designing the CR network lies in the potentially large number of SUs and PUs that influence each other. By simulating the rest of the network before more hardware was available, we were able to test a CR network of many SUs when only one SDR board was available. Below we present how the progressive-based simulation design was applied to the design of both individual cognitive modems and the CR network. In this project, the code running on the hardware board was compiled by the Code Composer Studio (optimization level). The compiled code was then uploaded to the Small Form Factor (SFF) SDR [26] hardware platform provided by Lyrtech and Texas Instruments.

7.4   PSBD OF THE CR NETWORK

The PSBD methodology described above has been applied to the design of a CR network. We describe the overall design procedure, present the design of a single cognitive modem, followed by the design of a CR network, and show some experiment results.

7.4.1   DESIGN PROCEDURE AND IMPLEMENTATION ENVIRONMENT

The design started from a single cognitive modem and then proceeded to the CR network. As part of the simulation-based testing environment, simulated PUs generated data traffic based on a model of the application layer for realistic PUs on particular frequency channels. Depending on the channel, the traffic model can be a constant bit rate (CBR), a burst data, or a Poisson process. For example, Internet traffic is often modeled as bursts of data transfer, CBR is a good model for TDMA networks such as voice traffic, and Poisson process is a generic model of data arrival. Simulated SUs not only generate data traffic, but also are responsible for handling signaling packets over a control channel of the network.

In the first stage of PSBD, we implemented a single transceiver (transmitter and receiver radio). First, MATLAB® was used for conventional modeling and simulation of a generic transceiver modem and a generic channel. The goal was to find out the parameters to achieve the required bit error rate (BER) of the radio system. Second, one SDR board and simulation models formed a co-simulation carried out in DEVSJAVA [11]. In this stage, the signal processing and data acquisition components of the transceiver were substituted by physical hardware components (such as DSP and FPGA components) on the board. The remaining components were DEVSJAVA models. To support reuse, the DEVSJAVA models reused the corresponding MATLAB functions in the first stage. The MATLAB Builder™ for Java can build Java libraries from MATLAB functions and have them ready to use in DEVSJAVA. In our project, the MATLAB m-files developed for MATLAB simulation of the channel, some MATLAB visualization methods, and also the MEX files (that interfaced the SDR board) were compiled to Java methods and used inside the DEVSJAVA simulation. The embedded MATLAB code inside DEVSJAVA helped in generating precise communication-specific models such as traffic and random fading of multipath channel. In this way, we were able to avoid reimplementing and testing mathematical methods and could focus on higher level network modeling in DEVSJAVA. The integration of DEVS and MATLAB models decoupled the design of CR network model from the underlying complicated mathematical functions.

While MATLAB is good at mathematical modeling, it is not straightforward to model the event-driven nature of interconnected components. On the other hand, DEVSJAVA can naturally support network modeling and, thus, was used to model the network of CR nodes. After an initial cognitive modem was implemented (as one SU), the next step was to design the CR network. At first, a complete simulation of a CR network with SUs and PUs was developed in DEVSJAVA. The conventional simulation simulated the activities of channel assignments for SUs and PUs. An SU acting as a Secondary Base (SB) station model received the sensing information from all SUs and compiled channel state information. In our system, we had distributed sensing to avoid the hidden node problem as much as possible. A vector signal generator emulated the traffic of simulated PUs and simulated SUs on the physical channel. In modeling the network, CR nodes were modeled by DEVS atomic models and the network was modeled as a DEVS coupled model. The scheduled messages passing between the DEVS atomic models carried packets with different lengths. Based on the propagation rules, the channel model rescheduled the messages to the receiver. Note that DEVS modeling is homomorphic. This means the CR node could be modeled as a coupled model itself including its functional component models. The hierarchical model construction of DEVS helped in modeling the internal interactions of subsystems of a single cognitive modem and the interactions of multiple cognitive modems in a network. During the course of PSBD, we replaced two of the simulated SUs with physical CRs to carry out a virtual environment test. In the setup of the virtual environment simulation, the physical nodes used filterbanks to detect the presence of the PUs and sent the sensing information to the SB, which was simulated on a PC. In this way, we were able to develop the network and to carry out system-wide test without having to wait for all hardware to become available. Below we describe the two design routes in detail.

7.4.2 DESIGN OF A SINGLE COGNITIVE MODEM

Based on the PSBD methodology, a model of a modem was first developed at an abstraction level according to the design goal. This model included all the functional modules of the cognitive modem. First, we started from a conventional simulation of the cognitive modem in a simulated environment. In this example, the environment was defined by the frequencies in use and a channel model. It dynamically changed as a result of PU or SU transmission or channel fading. The model was simulated in Simulink® to test for the required BER [24].

Figure 7.3 shows the model of the cognitive modem, which includes physical layer and MAC layer modules. The MAC layer module is the core of this model. The cognition algorithm inside the MAC layer module is responsible for compiling the spectral information sensed by the filterbank sensing module to provide a utility function. This utility function is used to determine the channels that should be assigned and the duration for which those channels are valid. The filterbank sensing [7,23] module is applied directly on the channel to detect the power spectral density (PSD) and thus the active PUs. The synchronization module detects the incoming packets while correcting the residual carrier offset before passing data to an equalizer to compensate for the effect of channel on data. The communication channel in the simulation is considered as an additive white Gaussian (AWGN) model. Conventional simulations based on this model allow us to test the cognitive algorithm inside the MAC layer and the functionality of the physical layer.

The conventional simulation, however, is insufficient to design and test the features that are influenced by the implementing hardware. For example, tuning algorithm/protocol parameters to achieve optimal synchronization results depends on the real-time properties of the hardware that cannot be predetermined without employing the actual hardware. In our design, the MAC layer protocol is decentralized and consisted of three phases. In the *sensing* phase, all the nodes sense the channel. In the *reporting* phase, all the nodes communicate their cognition results over a control channel. The cognition algorithm uses these reports to assign channels. In the

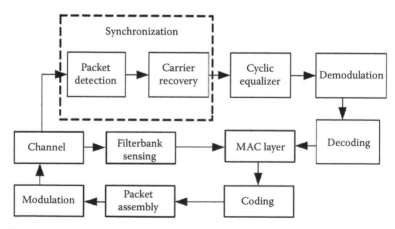

FIGURE 7.3 Simulation model of a single cognitive modem. It includes sensing, transmission, and reception.

channel usage phase, all the nodes start using the assigned channels. The purpose of the reporting period is to maximize cognition accuracy by sharing the cognition knowledge of individual SUs. Note that only the transmission time contributes to the effective bandwidth, and therefore, the time spent in reporting and sensing should be minimized. Since sensing takes a constant duration, an important task in designing the cognitive modem is to optimize the MAC layer so that reporting time is mini- mized [27]. Also we should note that the transmission time cannot last for a very long time, because a PU may return and thus SUs must release the occupied channels to the PUs. The time that SUs are permitted to transmit continuously without sensing is thus determined by a statistical model of PU arrivals in the bands of interest. To better design these features, the next stage is to include physical hardware compo- nents into the simulation for virtual environment simulation.

It is challenging to decide to what extent simulation-assisted design should be involved and to what level the system should be decomposed. A single cognitive modem comprises many individual radio modules, each of which is a candidate for virtual environment simulation along with other simulated objects and the environ- ment. Figure 7.4 shows how virtual environment simulation is carried out in design- ing a single modem. In this figure, one SDR board is used. The simulation runs on a single PC, while the SDR board (running implemented components) is connected to that PC via an Ethernet cable. Since sensing the channel is the most critical part of the CR, we implement this component in the earliest stage as shown in the rightmost section of Figure 7.4. We implement filterbanks sensing in DSP of the board. To test the sensing module, we emulate PU traffic on different frequency bands using a wide- band vector signal generator. The transmitted traffic of PU by the signal generator is a multiband waveform generated using a MATLAB script. In this stage, we adjust some design parameters such as analog-to-digital converter gains, frequency axis

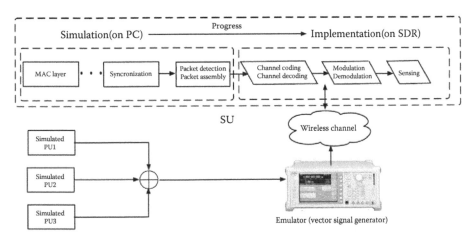

FIGURE 7.4 Progressive simulation-based design of a single cognitive modem. The implementation starts from the sensing module and progressively more of the simulated models (left dotted box) are implemented (right dotted box). The rectangles are Discrete Event System Specification models simulated on PC, and parallelograms are implemented modules on the software-defined radio board.

margins, and power threshold of PU detection. After sensing is implemented, other components such as modulation/demodulation and channel coding are implemented in the order depicted in the figure.

7.4.3 DESIGN OF A CR NETWORK

After the cognitive modem is implemented, we design the CR network using the developed cognitive modem. Similar to the design of a single modem, we start from conventional simulation of the cognitive network using the model of cognitive modem developed in the previous stage. In our design, first we consider a centralized network where some SUs were assigned to act as SB stations. This approach leads to a global common plane architecture for signaling and control to avoid the hidden node problem for PUs. This is essential in any CR network to avoid interfering with existing PUs. SB is in charge of synchronization and channel assignment. It compiles the channel state information based on sensing results of SUs and uses this knowledge to rank different parts of the spectrum and blacklist some active channels. After the role of each CR node is assigned, we model the cognitive network, which includes multiple PU and SU models, a SB model, and a Channel model that simulates the features of the wireless channel. Conventional simulations using this network model allows us to test if the SUs can successfully detect nonused bands and to dynamically use them without interfering with the PUs. A sample simulation result is provided in Section 7.4.4

The next stage is to introduce physical cognitive modems (SUs) in the simulation. The virtual environment simulation includes physical SUs, physical PUs, simulated SUs, simulated PUs, and simulated channels (used only between simulated SUs). Figure 7.5 shows the setup of a virtual environment simulation and the interconnections among simulated, emulated, and physical components. In Figure 7.5, two physical SUs (Lyrtech SFF SDR boards, denoted as *Real SU1* and *Real SU2*) and

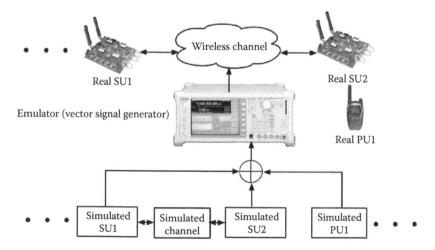

FIGURE 7.5 Virtual environment simulation of a cognitive radio network. The rectangles are Discrete Event System Specification models.

one physical PU (two-way radio, denoted as *Real PU1*) are used. The simulated SUs and PUs are emulated using a vector signal generator to generate spectral energy on the bands of the simulated agents. Therefore, the physical SUs would see the channel as if the simulated users exist. To test the cognitive modem against the generic PU channel usage pattern, the traffic of the PU (being simulated) is known and programmed in the PU model. The emulation of the PUs and SUs is necessary to test the sensing mechanism of physical SUs. The simulated SUs use the simulated channel and simulated traffic directly from the simulated PUs. The simulated channel is also used between two simulated SUs when transmitting a packet. A fading channel and different exponential PU traffic (with various mean for each channel) is implemented in MATLAB and compiled to be invoked by DEVSJAVA. Note that the emulator is considered as part of the testing environment and is not included in the system model.

To carry out virtual environment simulation, it is important to set up the environment so the physical and simulated nodes can "sense" each other's existence. For example, when a physical SU uses a band of the channel, the simulated SUs must know the band has been occupied (by the physical SU). The reverse is true too. Thus, a two-way communication is necessary between the physical SUs and the PC that hosts the simulated SUs. In our system, as shown in Figure 7.5, the emulator conveys the simulated environment and broadcasts the information sensed by the physical SUs. To support the communication from a physical SU to the PC, an Ethernet cable (not shown in Figure 7.5) is used to connect the physical SU with the PC. During the "report" stage of the MAC protocol, a physical SU sends its information to the PC through the Ethernet cable. Note that the Ethernet cable is not necessary in testing the physical system, in which case wireless communication is used. To synchronize the physical and simulated SUs (meaning to allow the physical and simulated SUs to know each other's existence) in a systematic way, each physical SU has a "shadow model" on the PC. This is similar to the robot-in-the-loop simulation [4–6], where each physical robot has a counterpart robot model in the simulation environment on the PC. The shadow model is responsible for receiving report information from the real SU and then passes that information to the Channel model similar as how other simulated SUs do.

Figure 7.6 shows a DEVS model of a CR network including two PUs, two SUs, and one SB in the virtual environment simulation. One of the SUs (SU2), the two PUs (PU1 and PU2), and also the SB station are simulated on the host PC. The other SU (SU1) has a hardware implementation and thus is executed using the co-simulation engine. The SU1 shown in Figure 7.6 is the shadow model of the real SU1. Using the message passing mechanism of DEVS between the models, in which messages are external events, and also exploiting the time-triggered message generation inside the modeled nodes, which are internal events, the simulation of the network is implemented. We use immediate messages for passing parameters between the models, while time scheduled messages are used to pass the data-carrying binary signals. For the simulated nodes, after a transmitter sends a packet of data, the Channel model passes the binary signal, along with the carrier frequency and other required parameters to MATLAB code that simulates a complete transmitter, channel, and receiver. The MATLAB code for the transmitter includes source

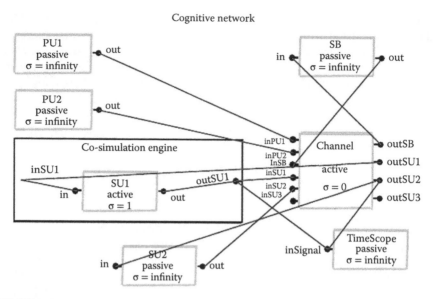

FIGURE 7.6 A cognitive network Discrete Event System Specification model with two primary users (PUs), two secondary users (SUs), and one Secondary Base (SB) station. SU1 is implemented on a small form factor software-defined radio board and emulated along with the other nodes.

coding, baseband modulation, upsampling and RF modulation, downsampling, etc. For the receiver, the necessary functions are also developed in MATLAB and the data (that may include error) are passed to the node in DEVS. The physical cognitive modem SU1 relies on its embedded software for data transmission, and the virtual environment simulation engine handles the interface between the shadow model SU1 and the physical SU1.

7.4.4 Experiment Results

Based on the PSBD methodology, we designed and implemented three cognitive SUs as follows. After initial simulation of one SU in MATLAB, we agreed on certain technical parameters. Then inside DEVSJAVA along with a generic simulation of one PU, we improved our SDR implementation of one SU along with its model. In the next step, we added one more SDR-based physical SU and more PU nodes along with one simulated SB to our network [24,28]. The simulated SB was in charge of transmitting channel assignment over the reporting channel. It combined the individual sensing result of the SU and assigned channels to them upon their request.

Figure 7.7 shows the result of a conventional simulation with all-simulated SUs and PUs. In this simulation, 2 SUs and 16 PUs were used. A flat fading wireless channel was used for simulation with white noise, and filterbanks were used for sensing. The simulation demonstrates how a SU can dynamically detect and change its band to

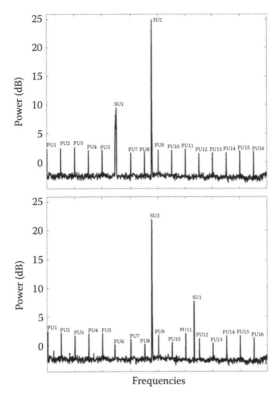

FIGURE 7.7   Simulation results using all models.

avoid interfering with a PU. As shown in the top part of Figure 7.7, initially PU6 was not present (its power was below the noise temperature) and SU1 was using its band. Then as shown in the bottom part of Figure 7.7, as soon as PU6 returned, SU1 dynamically found another unused band that was less likely to have a PU any time soon as described in Amini et al. [24] and changed to the new band. In this simulation, SU2 used a band that no other PU was using. Thus, SU2 did not have to change its band.

Figure 7.8 shows the result of a virtual environment simulation that included one physical SU and 12 simulated PUs. In this experiment, we implemented three sensing methods, including filterbank, Fast Fourier Transform (FFT), and FFT with Hanning, and compared them. The collected data was the PSD sensed from a densely populated spectrum for 256 subcarriers in an experiment. This figure shows that SB reassigned a new frequency band to the only SU when PU3 was detected in the adjacent frequency of the previous carrier (the dotted vertical line in the top figure). As a result, the SU changed to a new band (shown by the vertical dotted line in the bottom figure). The top and bottom figures show before and after the PU3 detection, respectively. The continuous transmission of voice during this action also proved the success of the frequency hopping. This experiment shows that the cognition modem was capable of locating a less active band in the spectrum where more than 10 active PUs were transmitting. In addition, as shown in this figure, the FFT

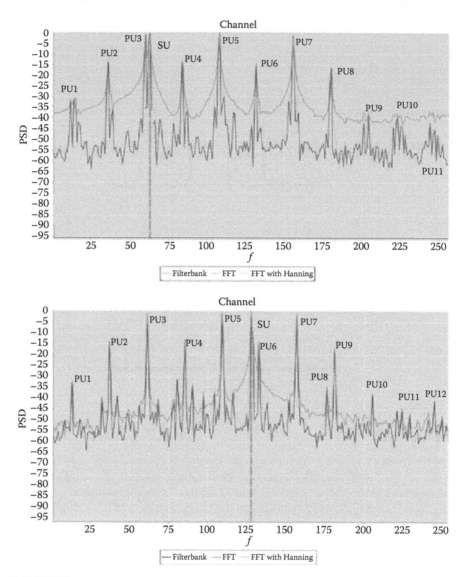

FIGURE 7.8 Simulation results using one physical secondary user (SU) among many primary users (PUs).

sensing could not detect two PUs on the right-hand side of the frequency spectrum, while the developed filterbank sensing could easily find them.

7.5 CONCLUSIONS

We present a PSBD methodology for designing networked real-time embedded systems. The methodology includes a bifurcated design process that implements PSBD for designing both individual embedded devices and the networked embedded system as a whole. We apply this methodology to the development of a CR network.

A single cognitive modem was first developed in a progressive manner and then the CR network was built in the same fashion. During this process, the MAC layer model was fine-tuned to increase the data rate of CR, while minimizing interference with PUs. Experimental results show that the designed CR system was able to respond dynamically to the changing environment to avoid active PUs on a real-time basis, while continuing the functionality of a wireless radio. The system was presented successfully at the 2007 Smart Radio Challenge [28] held by the SDR forum. This case study example shows the effectiveness of the PSBD methodology for developing complex networked real-time embedded systems.

Gradually including physical system components into a design to replace the simulation models should consider factors such as hardware availability, cost, speed, and required accuracy for a certain application. Strategies can be developed in the future to provide guidelines for including hardware in a stepwise fashion in the PSBD. Characterizing the types of systems that can take advantage of the PSBD methodology is another interesting topic asking for further research.

REFERENCES

1. Li, L., T. Pearce, and G. Wainer. 2003. "Interfacing Real-Time DEVS Models with a DSP Platform." In *Proceedings of the Industrial Simulation Symposium*, Valencia, Spain.
2. Upton, J. 1998. *Boeing 777 (AirlinerTech Series)*. 2nd ed. Voyageur Press (MN), vol. 2.
3. Hu, X. 2010. "From Virtual to Real: A Progressive Simulation-Based Design Framework," *Discrete-Event Modeling and Simulation: Theory and Applications*, edited by G. A. Wainer, P. J. Mosterman, Boca Raton, FL: CRC Press.
4. Hu, X., N. Ganapathy, and B. P. Zeigler. 2005. "Robots in the Loop: Supporting an Incremental Simulation-Based Design Process". In *IEEE International Conference on Systems, Man, and Cybernetics*, Hawaii.
5. Azarnasab, E. and X. Hu. 2007. "An Integrated Multi-Robot Test Bed to Support Incremental Simulation-Based Design." In *Proceedings of the IEEE International Conference on System of Systems Engineering*, San Antonio, TX.
6. Azarnasab, E. 2007. "Robot-in-the-Loop Simulation to Support Multi-Robot System Development: A Dynamic Team Formation Example." Master's thesis, Georgia State University, Department of Computer Science, Atlanta, GA.
7. Farhang-Boroujeny, B. 2008. *Signal Processing Techniques for Software Radios*. Morrisville, North Carolina: Lulu Publishing House.
8. Azarnasab, E., X. Hu, P. Amini, and B. Farhang-Boroujeny. 2008. "Progressive Simulation-Based Design: A Case Study Example on Software Defined Radio". In *Proceedings of the 2008 IEEE Conference on Automation Science and Engineering (IEEE-CASE 2008)*, Washington DC.
9. Zeigler, B., H. Praehofer, and T. Kim. 2000. *Theory of Modeling and Simulation*. 2nd ed. New York: Academic Press.
10. Zhang, Ming, Bernard Zeigler, Xiaolin Hu. 2012."A Formalized Approach for Design of Real-Time Distributed Computer System." In *Real-Time Simulation Technologies: Principles, Methodologies, and Applications*, edited by K. Popovici, P. J. Mosterman, CRC Press, ISBN 9781439846650.
11. Zeigler B. P. and H. S. Sarjoughian. 2005. *Introduction to DEVS Modeling and Simulation with JAVA: Developing Component-Based Simulation Models, Resources* http://www.acims.arizona.edu/SOFTWARE/devsjava_licensed/CBMSManuscript.zip (last accessed April, 2012) .

12. Wainer, G. A. and E. Glinsky. 2004. "Model-Based Development of Embedded Systems with RT-CD++." In *Proceedings of the WIP Session, IEEE Real-Time and Embedded Technology and Applications Symposium*, Toronto, ON, Canada.
13. Wainer, G. A., E. Glinsky, and P. MacSween. 2005. *A Model-Driven Technique for Development of Embedded Systems Based on the DEVS Formalism*. Springer-Verlag.
14. Glinsky, E. and G. A. Wainer. 2004. "Modeling and Simulation of Hardware/Software Systems with CD++." In *Proceedings of the 36th Conference on Winter Simulation*, Washington DC.
15. Schulz, S. and J. W. Rozenblit. 2002. "Refinement of Model Specifications in Embedded Systems Design." In *Proceedings of the 2002 IEEE Conference on Engineering of Computer-Based Systems*, 159–66. Sweden: Lund.
16. Kim, T. G., S. M. Cho, and W. B. Lee. 2001. "DEVS Framework for Systems Development: Unified Specification for Logical Analysis, Performance Evaluation and Implementation." In *Discrete Event Modeling and Simulation Technologies: A Tapestry of Systems and AI-Based Theories and Methodologies*. New York, NY: Springer-Verlag New York, Inc.
17. Hu, X. and B. P. Zeigler. 2005. "Model Continuity in the Design of Dynamic Distributed Real-Time Systems." *IEEE Transactions On Systems, Man And Cybernetics—Part A: Systems And Humans*, 35 (6), 867–78.
18. Hu, X. and B. P. Zeigler. 2005. "A Simulation-Based Virtual Environment to Study Cooperative Robotic Systems." *Integrated Computer-Aided Engineering (ICAE)* 12 (4), 353–67.
19. Brodersen, R., A. Wolisz, D. Cabric, S. Mishra, and D. Willkomm. 2004. *CORVUS: A Cognitive Radio Approach for Usage of Virtual Unlicensed Spectrum*, White paper, Berkeley, Technical Report. http://bwrc.eecs.berkeley.edu/Research/MCMA/CR_White_paper_final1.pdf (last accessed April, 2012).
20. Shankar, N., C. Cordeiro, and K. Challapali. 2005. "Spectrum Agile Radios: Utilization and Sensing Architectures." In *First IEEE International Symposium on New Frontiers in Dynamic Spectrum Access Networks (DySPAN)*, 160–9. Baltimore, Maryland.
21. FCC 2002. *Spectrum Policy Task Force*. ET Docket 02-135.
22. Haykin, S. 2005. "Cognitive Radio: Brain-Empowered Wireless Communications." *IEEE Journal on Selected Areas in Communications*.
23. Amini, P., R. Kempter, and B. Farhang-Boroujeny. 2006. "A Comparison of Alternative Filterbank Multicarrier Methods in Cognitive Radio Systems." In *Software Defined Radio Technical Conference*, Orlando, FL.
24. Amini, P., E. Azarnasab, S. Akoum, X. Mao, H. I. Rao, and B. Farhang-Boroujeny. 2007. "Implementation of a Cognitive Radio Modem." In *Software Defined Radio Technical Conference*, Denver, CO.
25. Su, H. and X. Zhang. 2008. "Cross-Layer Based Opportunistic MAC Protocols for QoS Provisionings over Cognitive Radio Wireless Networks." In *IEEE Journal on Selected Areas in Communications*, 26 (1), 118–29.
26. "Lyrtech SFF SDR Development Platform Technical Specs." 2007. Lyrtech Inc., Technical Report. http://www.lyrtech.com/ publications/sff_sdr_dev_platform_en.pdf, (last accessed February, 2007).
27. Azarnasab, E., R.-R. Chen, K. H. Teo, Z. Tao, and B. Farhang-Boroujeny. 2009. "Medium Access Control Signaling for Reliable Spectrum Agile Radios." In *IEEE Global Telecommunications Conference (GLOBECOM), ISSN: 1930-529X*. 1–5, Honolulu, Hawaii.
28. Azarnasab, E., P. Amini, and B. Farhang-Boroujeny. 2007. "Hardware in the Loop: A Development Strategy for Software Radio." In *Software Defined Radio Technical Conference*, Denver, CO.

8 Validator Tool Suite

Filling the Gap between Conventional Software-in-the-Loop and Hardware-in-the-Loop Simulation Environments

*Stefan Resmerita, Patricia Derler,
Wolfgang Pree, and Kenneth Butts*

CONTENTS

8.1 Solid System Verification and Validation Needs Improved
Simulation Support 200
    8.1.1 Real-Time Behavior in the Validator.. 201
8.2 Architecture of a Simulation with the Validator 203
    8.2.1 Basic Features of the Validator..205
8.3 Setup of a Simulation with the Validator...206
    8.3.1 Target Platform Specification 206
    8.3.2 Task Source Code Annotation 207
8.4 Embedded System Validation and Verification with the Validator..............208
    8.4.1 Simulation with the Validator as the Basis for
         Advanced Debugging 208
    8.4.2 Simulation with the Validator to Reengineer Legacy Systems 211
         8.4.2.1 Sample Analysis... 212
8.5 Related Work 214
    8.5.1 Co-Simulation.. 214
    8.5.2 Modeling and Simulating Legacy Code 214
8.6 Conclusions.. 215
Acknowledgments.. 216
References... 216

8.1   SOLID SYSTEM VERIFICATION AND VALIDATION NEEDS IMPROVED SIMULATION SUPPORT

An embedded system operates in a physical environment with which it interacts through sensors and actuators. An important class of real-time embedded systems is represented by control systems. In this case, the embedded software consists of a set of (controller) tasks and the physical system under control is referred to as the plant. Figure 8.1 sketches the typical architecture of an embedded control system.

Simulation is an approach for testing embedded systems before they are deployed in real-world operation. In simulation, the plant is represented by a software model executed on a host computer, typically a personal computer. In a hardware-in-the-loop (HIL) simulation, the entire embedded system (embedded software consisting of the controller tasks executing on the target platform) is operated in closed loop with the plant model, which is executed in real time on a dedicated computer [1]. Since the embedded software is executed in real time on the target platform, HIL simulations can be used to verify the real-time properties of the embedded system. Thus, any difference between the behavior of the embedded system in a HIL simulation and the corresponding behavior in the real world is due to the abstractions made in plant modeling.

In a software-in-the-loop (SIL) simulation, the embedded software consisting of the controller tasks is executed on a host computer other than the target platform, in closed loop with the plant model. Both simulations (of the controller and the plant model) are typically executed on the same host computer. The SIL model of an embedded system contains the embedded software and an abstraction of the target platform. This abstraction determines how close the software execution in the SIL simulation is to the HIL simulation, provided that the same plant model is used. It ranges from a minimal representation of the target platform that enables only testing of functional (transformational or processing) properties of the software, to full-fledged hardware simulators (called instruction set simulators, ISS), which lead to system behavior close to a HIL simulation, while offering better observability of software executions. Pure functional simulations are fast, but do not allow the testing of timing properties of the embedded system. ISS can be used for timing analysis [2], but they are extremely slow and expensive. Figure 8.2* summarizes the characteristics of conventional SIL, ISS-based SIL, and HIL simulations.

Figure 8.3 summarizes the features and the advantages and disadvantages of state-of-the-art SIL and HIL simulations in comparison with a Validator simulation. Note that the Validator replaces the ISS-based simulation.

Figure 8.4 refines the comparison between the Validator and SIL/HIL approaches. The Validator unifies characteristics of both a SIL and a HIL simulation. The Validator has the flavor of a SIL simulation as it does not require a target platform for executing the embedded software. On the other hand, the Validator separates the simulation of plant and controller tasks as in a HIL simulation.

---

* Parts of the picture are taken from http://www.mathworks.com/products/xpctarget/ and are courtesy of MathWorks.

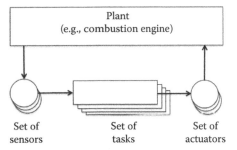

FIGURE 8.1 Typical architecture of an embedded system.

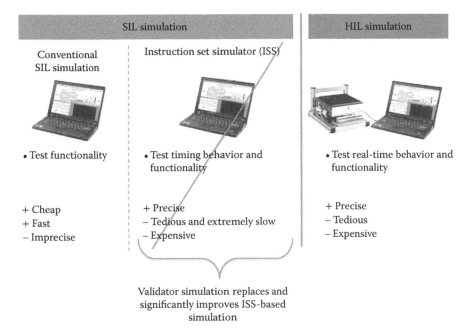

FIGURE 8.2 Conventional software-in-the-loop, instruction set simulators-based software-in-the-loop, and hardware-in-the-loop simulations.

8.1.1 REAL-TIME BEHAVIOR IN THE VALIDATOR

An important aspect is the simulation of real-time behavior. A simple example that illustrates in which respect the Validator is better than typical SIL simulation tools is shown in Figure 8.5. Consider three concurrent tasks, called DynamicsController (DC), MotorController (MC), and ParkingController (PC), that communicate through a shared (global) variable called *angle*. The code of the tasks is sketched in Figure 8.5a. The tasks DC and MC are periodic with periods equal to 5 and 1 ms, respectively. The task PC is event-triggered. Assume that they are deployed on a real-time operating system with fixed priority preemptive scheduling, where the priorities of the periodic tasks are assigned by a rate monotonic policy. Thus, the MC

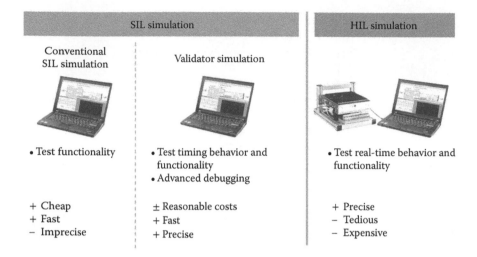

SIL simulation		HIL simulation
Conventional SIL simulation	Validator simulation	

• Test functionality

• Test timing behavior and functionality
• Advanced debugging

• Test real-time behavior and functionality

+ Cheap
+ Fast
− Imprecise

± Reasonable costs
+ Fast
+ Precise

+ Precise
− Tedious
− Expensive

FIGURE 8.3   Validator: advanced software-in-the-loop simulation in between conventional software-in-the-loop and hardware-in-the-loop simulations.

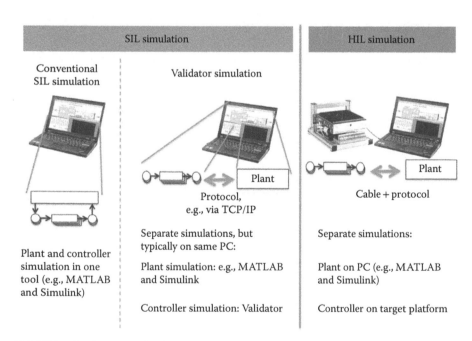

SIL simulation		HIL simulation
Conventional SIL simulation	Validator simulation	

Protocol, e.g., via TCP/IP

Cable + protocol

Plant and controller simulation in one tool (e.g., MATLAB and Simulink)

Separate simulations, but typically on same PC:

Plant simulation: e.g., MATLAB and Simulink

Controller simulation: Validator

Separate simulations:

Plant on PC (e.g., MATLAB and Simulink)

Controller on target platform

FIGURE 8.4   Structure of a simulation with the Validator.

task has a higher priority than the DC task. Moreover, consider that the PC task has highest priority.

A snapshot of real-time behavior of the application is depicted in Figure 8.5b, which indicates the sequence of accesses to the variable *angle* by the three tasks. Note that task DC is triggered first, at 5 ms, and then it is preempted by MC at 6 ms, before writing

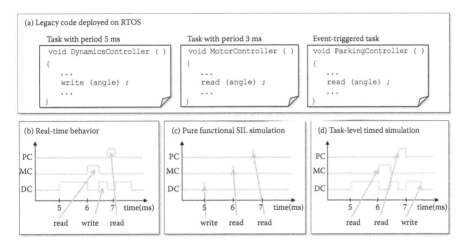

FIGURE 8.5 Examples of mismatch between simulated and real-time behaviors.

into the variable *angle*. Thus, MC reads from *angle* first. Thereafter, DC resumes and writes the angle, then it is preempted by PC, which reads the variable. Figure 8.5c shows the behavior of the application in a pure functional simulation, where each function is completely executed at the triggering time. In other words, the code is executed in logically zero time. Such a simulation can be obtained in Simulink® [3], for example, by importing the C code as so-called S-function(s) in Simulink. Figure 8.5d presents a timed-functional simulation, where each task has a specified execution time and sharing of processor time among tasks in the system is also simulated. A timed-functional model includes a scheduling component to decide when a triggered task obtains access to the processor. When the task is started, its code is still executed in zero time, thus using the inputs available at that moment; however, the outputs of the task are made available after the specified execution time has elapsed. Examples of SIL simulation environments that offer task-level timed simulation are the Timed Multitasking Ptolemy domain [4] and TrueTime [5]. Notice that the order in which the three tasks access the variable *angle* is different in the two simulations compared to the real-time case.

On the other hand, a simulation with the Validator would reflect the same order of accesses as in the real-time behavior shown in Figure 8.5b. This requires a detailed execution time analysis and a corresponding instrumentation of the embedded software by the Validator support tools as described in Section 8.3.2.

8.2 ARCHITECTURE OF A SIMULATION WITH THE VALIDATOR

Remember that a simulation with the Validator is a closed-loop co-simulation of the plant under control and the controller tasks. Currently, the Validator supports continuous-time plant models in MATLAB® [6] and Simulink. For that purpose, a communication interface was implemented as a MATLAB and Simulink S-function (see Figure 8.6). The underlying protocol for communication between the plant and the simulation with the Validator is TCP/IP. So both simulations can execute in parallel on the same computer or on different cores or, for example, for efficiency reasons,

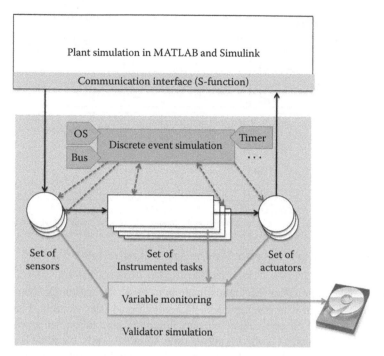

FIGURE 8.6   Architecture of a closed-loop simulation with the Validator.

on different computers. We will extend the Validator to support co-simulation with other simulation environments in the future. For that purpose, communication interfaces must be implemented for the particular simulation environment.

The Validator also offers a file reader for processing time-stamped values of input data from recorded signals. This is useful for regression testing as discussed in Section 8.4.2.

The Validator simulation engine is a discrete event simulation that takes the platform specifications into account. The platform specifications comprising the operating system (OS), the communication bus, hardware timers, etc. are plug-ins of the Validator simulation engine. The lower half of Figure 8.6 sketches this aspect of the architecture of a Validator simulation. The discrete event simulation controls which of the tasks are executed once the control flow gets back to the discrete event simulation from a task execution. For example, based on the scheduling strategy used by the OS, a higher priority task must interrupt one with a lower priority. In such a situation, the discrete event simulation will switch the execution to the appropriate task. Dashed arrows express this control flow between the instrumented tasks and the discrete event simulation in Figure 8.6. In an analogous way, the Validator takes care of the appropriate reading of sensor values and writing of actuator values.

Figure 8.7 illustrates the discrete event simulation in the Validator in more detail. As the discrete event simulation of the controller tasks proceeds from what we call a *spot* (S) within a task to the next spot, the control flow between tasks and the discrete event simulation constantly switches back and forth. In the sample scenario

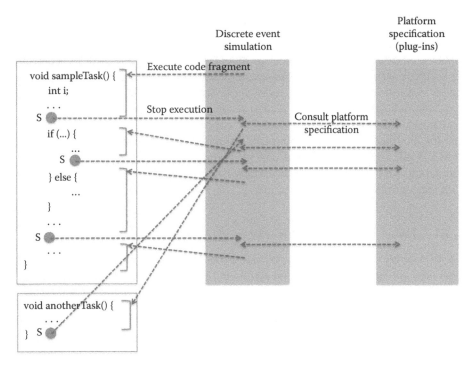

FIGURE 8.7 Sample switching between tasks at *spots* (S).

in Figure 8.7, the discrete event simulation starts executing sampleTask(). This task executes till it reaches the first spot and returns control to the discrete event simulation. Based on the platform specification, the discrete event simulation decides to interrupt sampleTask() and to give control to anotherTask(), for example, because that one was triggered for execution and has higher priority than sampleTask(). As anotherTask() has only one spot at the end, it executes completely and then returns the control back to the discrete event simulation. Now the discrete event simulation gives back control to sampleTask(). The discrete event simulation continues with the execution of sampleTask() also at the other two spots.

8.2.1 BASIC FEATURES OF THE VALIDATOR

Let us conclude this overview of the architecture and some core simulation concepts by summarizing the features that result from a bare-bones setup of the Validator:

Variable monitoring. The Validator allows the logging of the time-stamped values of selected variables (global variables and variables local to tasks) to a file. The variable monitor in Figure 8.6 corresponds to that functionality.

Stop and restart simulation runs. The Validator allows stopping a simulation and saving the state of the overall simulation, that is, the simulations of both the plant and the controller tasks. A simulation can later be restarted from a saved state.

8.3 SETUP OF A SIMULATION WITH THE VALIDATOR

The current version of the Validator basically supports the co-simulation of a plant represented as a variable-step model in MATLAB and Simulink with the controller software written in C. As an advanced feature for reengineering existing controller tasks or adding controller tasks, these tasks can be modeled in MATLAB and Simulink and the Validator then simulates the behavior of both the existing unchanged tasks and the modified or new controller tasks. The only constraint is that the modified or new controller tasks are modeled with discrete time semantics in MATLAB and Simulink. Let us now focus on the typical use case, that is, the co-simulation of a plant represented as variable-step model in MATLAB and Simulink with the controller software written in C.

8.3.1 TARGET PLATFORM SPECIFICATION

For an accurate simulation of the controller tasks on a virtual platform, the Validator must have configuration information about the target platform, which is provided by setting properties of the corresponding model components. To specify this information, we use Ptolemy's front end, as the original research prototype of the Validator was implemented harnessing Ptolemy's discrete event simulation. The screenshot in Figure 8.8 exemplifies the specification of the behavior of an interrupt service routine (ISR).

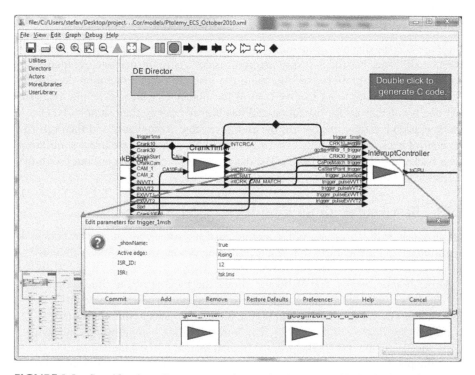

FIGURE 8.8 Specification of interrupt service routine behavior with the Ptolemy front end.

An ISR is represented as a so-called actor in Ptolemy. The actor-oriented programming model is in essence a dataflow-based programming model in which data flows from actor to actor. When activated, each actor performs its specific data processing. An actor-oriented environment such as Ptolemy must account for the execution order of actors. The Validator library, which is used to specify the target platform, comprises various kinds of actors:

- Hardware actors model functionality and timing of common hardware parts such as interrupt controllers, timers, bus controllers, hardware sensors, and hardware actuators.
- Operating system actors, which implement the functionality of the operating system on the target platform, including scheduling, resource management, and communication between tasks. Currently, the Validator provides actors for the OSEK operating system.

Note that actors are best understood as plug-ins to the discrete event simulation of the Validator, providing the various platform details.

8.3.2 TASK SOURCE CODE ANNOTATION

In addition to specifying the target platform, the source code of the controller tasks must be instrumented with callbacks to the simulation in the Validator. Details on which aspects require a callback is available in other work [7]. All the spots in the source code are instrumented with callbacks. An example of a type of spot is access to global variables. Between each pair of spots, the execution time must be determined. This is another crucial aspect of target platform information that the Validator must have to achieve its accuracy. From the Validator user point of view, it is only relevant that both the instrumentation and execution time estimation can be automated. Overall, a preparation of the controller tasks for a simulation with the Validator involves the following steps:

1. Execution time analysis of the application code. This is performed with existing program analysis tools such as AbsInt's Advanced Analyzer (a^3) tool [8]. To increase the accuracy of the estimates, generally details about the architecture of the execution platform must be made available to such tools.
2. Instrumentation of the code with execution time information.
3. Instrumentation with callbacks to pass control to the Validator simulation engine for the execution of the tasks.
4. Generation of what we call the Validator interface code between the Validator simulation engine and the tasks.

In the Validator, these steps are mostly automated by a tool set that achieves a straightforward preparation process. Nevertheless, this automation requires information about the hardware/software architecture, such as the list of lines of code where global variables are accessed.

8.4   EMBEDDED SYSTEM VALIDATION AND VERIFICATION WITH THE VALIDATOR

This section describes two principal usage scenarios where the Validator excels compared to the state-of-the-art SIL and HIL simulations: advanced debugging of embedded systems and the incremental reengineering of existing embedded systems, including regression testing. Case studies illustrate each particular usage scenario of the Validator.

8.4.1   SIMULATION WITH THE VALIDATOR AS THE BASIS FOR ADVANCED DEBUGGING

The key feature of the Validator that allows advanced debugging is that at every source code line in the controller tasks, the overall simulation, that is, of both the controller tasks and the plant, can be stopped. Then variables can be inspected and modified, external code can be executed, etc. Any C debugger can be attached to the Validator to perform the common debugging activities on the controller tasks. A state-of-the-art HIL simulation environment does not offer debugging capabilities. On the other hand, the impreciseness of the state-of-the-art SIL environments makes debugging unattractive or at least less helpful. The accuracy of a Validator simulation makes debugging a valuable means for the validation and verification of embedded systems. Figure 8.9 shows the schematic attachment of a debugger to the Validator. The jet fighter picture represents the plant model.

The screenshot in Figure 8.10 shows a sample set of controller tasks in Eclipse with the gnu debugger (gdb) plugin. The screenshot is discussed in more detail below. We have used gdb as it also supports reverse (or historical) debugging. This allows the following advanced debugging: Once you have turned on reverse debugging, the debugger records all state changes. So if the debugger stops execution at a breakpoint, you cannot only step forward as usual, but also step backward from that point in the code. For example, you want to find the cause of why an actuator value exceeds a certain limit. In this case you would set a conditional breakpoint where the actuator value is set. When the condition holds, the execution stops there and you can step back step by step.

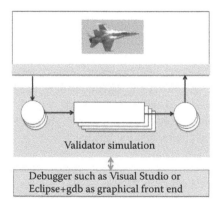

FIGURE 8.9   Attaching a debugger to the Validator.

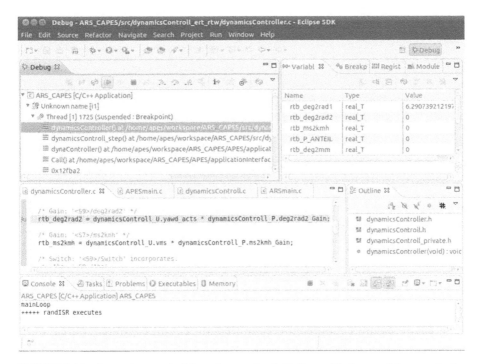

FIGURE 8.10   Start of a sample debugging session in Eclipse with the gnu debugger plugin.

FIGURE 8.11   gnu debugger controls before activating reverse debugging.

The following sequence of screenshots illustrates reverse debugging from a developer's point of view, with Eclipse and gdb as the graphical front end of the Validator.

In the state shown in Figure 8.10, we just entered the debugging mode by pressing the debug button (the bug in the menu bar on the top left side of the overall window). The execution stopped at an unconditional breakpoint in file dynamicsController.c. The statement at the breakpoint is an assignment statement in which the value of a variable called rtb_deg2rad2 is set. The line is highlighted in the tab labeled dynamicsController.c. According to the subwindow in the top right part of the window, the value of rtb_deg2rad2 is zero.

As a next step, we turn reverse debugging on by pressing the corresponding icon-button (see Figure 8.11). When stepping forward, the debug control panel changes to reflect the feature of reverse debugging, that is, being able to step forward and backward (see Figure 8.12).

FIGURE 8.12 gnu debugger controls for stepping forward *and* backward.

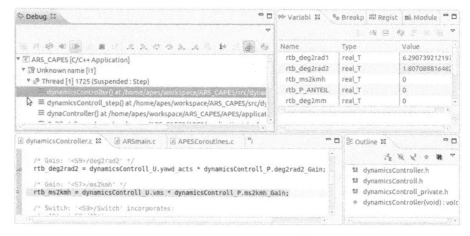

FIGURE 8.13 A step forward.

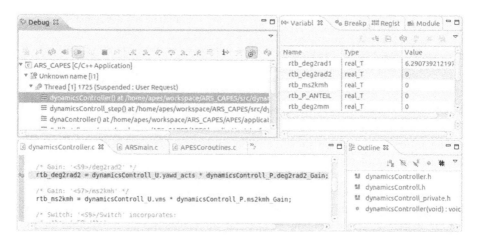

FIGURE 8.14 A step back.

Let us assume we just stepped forward one statement (see Figure 8.13). The assignment statement where we had originally stopped at the breakpoint has apparently changed the value of variable rtb_deg2rad2 from zero to approximately 1.807.

We can now press the button to go back one step in the debugging process. Figure 8.14 shows the result, that is, as expected, the value of variable rtb_deg2rad2 is again zero. Note that the discrete event simulation of the Validator was implemented such that reverse debugging also functions across multiple tasks. For example, if you have turned on reverse debugging, step forward, and the task is interrupted by another task, that is, the simulation switches to another task, you can still step back up to the point where you started reverse debugging.

As a future extension to the Validator, we will add the feature to also be able to set breakpoints in the plant simulation that halt the overall simulation.

8.4.2 SIMULATION WITH THE VALIDATOR TO REENGINEER LEGACY SYSTEMS

The initial motivation for developing the Validator was to provide solid support for the incremental migration of the Engine Controller System (ECS) of a large automotive manufacturer to a version in which the timing behavior is explicitly modeled with the Timing Definition Language (TDL) [9,10]. For that purpose, the behavior of the legacy ECS and the TDL-based ECS should be compared in detail in a SIL simulation, that is, as close to the behavior on the actual platform as possible. Figure 8.15 shows the generic setup for this kind of regression testing by means of the ECS example. The Validator can simulate both versions in parallel.

Note that the ECS comprises millions of lines of code, mostly written in C, and runs on top of an OSEK operating system. This required an efficient implementation of the discrete event simulation of the Validator and an efficient co-simulation with the automobile engine (=plant) model. The engine model is represented in MATLAB and Simulink. To accurately capture the times of the crank angle events, the engine model is simulated with a variable step solver.

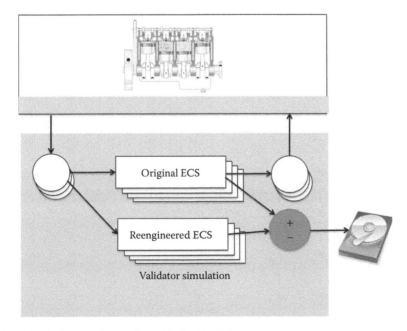

FIGURE 8.15   Regression testing with the Validator.

Implementation of the TDL semantics in the reengineered ECS [9] required a dedicated TDL component called *TDL-Machine* to be executed every 0.5 ms from the task with highest priority in the system. The TDL-Machine used additional global variables to store and restore values of original global variables at certain points in time.

The highest priority task in the original ECS had a period of 1 ms. To avoid introducing a new task, it was decided to change the original task to have a period of 0.5 ms, to call the TDL-Machine at every task invocation, and to execute the original task code every second invocation. Thus, the execution period of the original code was unaffected.

8.4.2.1 Sample Analysis

Let us take a look at one of the results of the regression tests: Figure 8.16 shows a selection of three signals monitored during a simulation of the two ECS versions with the Validator. Signals S1 and S2 are similar in the two versions. One can notice some delays introduced in the modified version by the execution times of the TDL-Machine. Signal S3 differs significantly towards the end of the simulated time frame.

The cause of the difference could be found by an investigation of task execution profiles. Task execution profile plots are shown in Figures 8.17a through c. In each plot, every task is assigned an identification number (ID). The execution state of a task with ID i is represented by a signal with y-coordinates between i and $i + 0.6$. A level of $i + 0.6$ indicates the execution mode E, a level of $i + 0.4$ means preempted mode P, a level of $i + 0.2$ indicates the waiting mode W, and a level of i means the task is in the suspended mode S. The plot in Figure 8.17a is obtained from the simulation of the original software, the plot in Figure 8.17b is obtained from the simulation of the software with the TDL-Machine and its execution time, and the plot in Figure 8.17c corresponds to the simulation of the software with the TDL-Machine but without its execution time. In case of Figure 8.17c, the TDL-Machine was executed in simulation, but its execution time was set to zero, for the purpose of debugging.

The ID of the highest priority task is 8. In the simulation represented by the plot in Figure 8.17a, the execution time of this task at time instant 5 is approximately 0.05 ms.

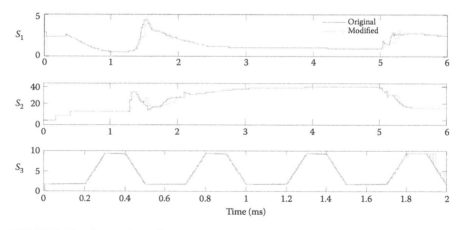

FIGURE 8.16 Comparison of three signals.

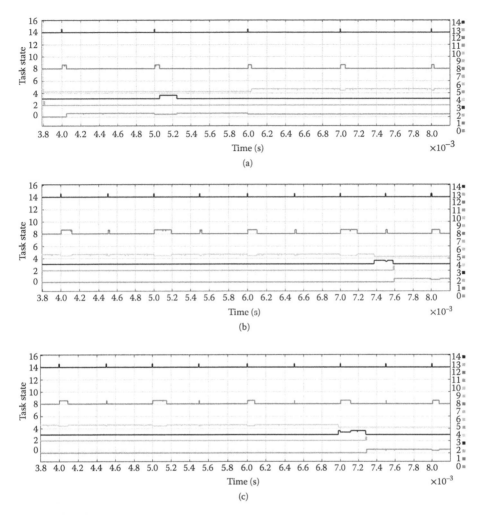

FIGURE 8.17 (a) Original ECS. (b) ECS with additional functionality. (c) ECS with additional functionality whose execution time is set to zero.

In the simulation of the plot in Figure 8.17b, the same execution requires about 0.20 ms. In Figure 8.17c, the same task execution takes about 0.17 ms. It follows that the main difference between the executions in Figures 8.17a and 8.17b is not given by the execution time of the TDL-Machine, but it is because the different state of the hardware platform resulted from the execution of the TDL function. Since the TDL-Machine performs many accesses to new memory locations, the main difference occurs most probably in the cache state. If the delays in the signals in the new version are not acceptable, then one should focus on minimizing the effect of the TDL-Machine on the platform state rather than on minimizing the execution time of the function code. For example, the TDL-Machine could be changed to operate only on local variables, or additional variables used by this function could be stored in the processor's internal memory space.

8.5   RELATED WORK

This section groups existing work in the area of simulation of embedded applications into two categories, related to the main features of the Validator: the ability to synchronize with an external plant model for closed-loop simulation (co-simulation) and the special focus on legacy software and its execution time.

8.5.1   CO-SIMULATION

The main purpose of co-simulation is validating the functionality of hardware (HW) and software (SW) components by simulating two or more system parts that are described on different levels of abstraction. The challenge is the interface between the different abstraction levels. A simulation of the system should be possible throughout the entire design process where the model of the same component is refined iteratively [11]. Co-simulation as a basis for co-design and system verification can be performed in various manners where typically a trade-off between accuracy and performance must be made [12]. Various commercial and academic co-simulation frameworks have been proposed in literature; surveys can be found in Edwards et al. [12], Huebert [13], and Bosman [14].

In HW/SW co-simulation, the processor model is responsible for connecting hardware and software models. The processor can be modeled at gate level, which is the most accurate but also the slowest solution, with clock cycle accuracy or on an instruction-set level. Faster co-simulation tools do not model the processor but implement a synchronization handshake [15]. Some co-simulation environments also provide a virtual operating system to emulate or simulate the hardware [16].

Many approaches use ISS to obtain correct timing information. However, ISS are slow because of the fine granularity of the simulation. Performance issues are addressed for instance with caching [17] and distributed simulation by applying distributed event-driven simulation techniques.

The co-simulation framework used by the Validator does not provide a model of the central processor unit and does not employ an ISS because of performance reasons. We work with execution time at the source code level. Hardware components are modeled at a higher level of abstraction. The simulation tool was in an original prototype implemented based on Ptolemy. Another Ptolemy-based co-simulation approach can be found in Liu et al. [17].

8.5.2   MODELING AND SIMULATING LEGACY CODE

There are various approaches that generate models from legacy code, but only a few of them include the timing aspect in the modeling. Some software reverse engineering efforts take the software and find equivalent modeling constructs in a modeling language to reconstruct the same behavior as exhibited by the software. An example is provided in Sangiovanni-Vincentelli [18], where C programs are reverse-engineered to Simulink models. This, however, usually leads to complex models that are not understandable and thus do not aid in gaining new insights in the embedded software system.

Code instrumentation and delaying of task execution to obtain a certain behavior is used by Wang et al. [19]. In this approach, the authors employ code instrumentation to generate deadlock-free code for multicore architectures. Timed Petri nets are generated from (legacy) code by instrumenting the code at points where locks to shared resources are accessed to model blocking behavior of software. A controller is synthesized from the code and used at runtime to ensure deadlock-free behavior of the software on multicore platforms by delaying task executions that would lead to deadlocks. The objective of the Validator is different: to replicate the real-time behavior of a given application (within certain accuracy limits). Thus, the Validator does not alter the functional behavior of the application.

In Sifakis et al. [20], a formal framework is described for building timed models of real-time systems to verify functional and timing correctness. Software and environment models are considered to operate in different timing domains that are carefully related at input and output operations. A timed automaton of the software is created by annotating code with execution time information. The tool presented in that work restricts the control part of task implementations and the plant model to Esterel programs. The authors state that for tasks written in general purpose languages such as C, an analysis must reveal observable states and computation steps. The Validator provides such an analysis and can be used for application code written in C and environment models in Matlab and Simulink.

The benefits of modeling all aspects of an embedded system at a suitable level of abstraction are well known and have been addressed in the platform-based design approach. Tools such as Metropolis [21] offer a framework for platform-based design where functionality and platform concerns are specified independently. A mapping between functionality and a given platform must be provided. Representation of components and the mapping between functional and architectural networks is possible at different levels of refinement. The purpose of Metropolis is to support the top-down design process. All system components are modeled in an abstract specification language, which is parsed to an abstract syntax tree and provided to back-end tools for analysis and simulation. Although Metropolis allows the inclusion of legacy components, its main goal is not a bottom-up analysis of legacy systems but a top-down specification of the required behavior and a specification of the platform. In the approach presented here, the behavior of the components such as the data flow or temporal constraints is retrieved from simulation. As opposed to Metropolis, we do not require a specification of the functionality in a metamodel language, which is further translated into SystemC [22] for simulation. Our approach directly includes the software as simulation components. Simulation tools related to the Validator are TrueTime [5] from academia and the commercial tool ChronSim from INCHRON [23]. Compared to these tools, the Validator offers a series of advanced features, as described in the chapter.

8.6 CONCLUSIONS

The Validator is tool suite for significantly improved SIL verification and validation of real-time embedded applications. The first prototype was developed with Ptolemy and was originally called the Access Point Event Simulator (APES). Chrona [24] developed the Validator as a product out of APES. Chrona's Validator has evolved to a product version that scales to the simulation of real-world embedded software consisting of millions of lines of code. The Validator achieves time-functional simulation of application software and execution platform in closed loop with a plant model. The Validator works independently of a specific domain as long as a simulation model for a particular plant is available.

A simulation with the Validator is based on a systematic manner to instrument the application code with execution time information and execution control statements, which enables capturing real-time behavior at a finer and more appropriate time granularity than most of the currently available similar tools. Chrona's Validator is able to simulate preemption at the highest level of abstraction that still allows for capturing the effect of preemption on data values, avoiding at the same time the slow, detailed simulation achieved by instruction set simulators. Moreover, the Validator can operate in closed loop with plant models simulated by a different tool. Also, Chrona's Validator enables traversing preemption points during forward and reverse debugging of the application. In the Validator, one can start a simulation from a previously saved state. Being implemented entirely in C, the Validator can be easily interfaced or even integrated with existing simulation tools such as MATLAB and Simulink.

ACKNOWLEDGMENTS

We would like to thank Edward Lee (University of California, Berkeley) for valuable insights during numerous discussions of the Validator concepts and his support in implementing an earlier prototype of the Validator with Ptolemy II.

REFERENCES

1. Mosterman, Pieter J., Sameer Prabhu, and Tom Erkkinen. 2004. "An Industrial Embedded Control System Design Process." In *Proceedings of the Inaugural CDEN Design Conference (CDEN'04)*, pp. 02B6-1 through 02B6-11, July 29–30. Montreal, Quebec, Canada.
2. Krause, Matthias, Dominik Englert, Oliver Bringmann, and Wolfgang Rosenstiel. 2008. "Combination of Instruction Set Simulation and Abstract RTOS Model Execution for Fast and Accurate Target Software Evaluation." In *Proceedings of the 6th IEEE/ACM/ IFIP International Conference on Hardware/Software Codesign and System Synthesis (CODES+ISSS '08)*, 143–48. New York: ACM.
3. MathWorks®. Simulink® 7.5 (R2010a). http://www.mathworks.com/products/simulink/. Accessed on September 10, 2010.
4. Liu, Jie, and Edward Lee. 2002. "Timed Multitasking for Real-Time Embedded Software." *IEEE Control Systems Magazine* 23: 65–75.
5. Cervin, Anton, Dan Henriksson, Bo Lincoln, Johan Eker, and Karl-Erik Årzén. June 2003. "How Does Control Timing Affect Performance? Analysis and Simulation of Timing Using Jitterbug and TrueTime." *IEEE Control Systems Magazine* 23 (3): 16–30.
6. MathWorks®. Matlab® 7.10 (R2010a). http://www.mathworks.com/products/matlab/. Accessed on September 10, 2012.
7. Resmerita, Stefan, and Patricia Derler. May 2011. "Wolfgang Pree: Validator—Concepts and Their Implementation." Technical Report, University of Salzburg.

8. AbsInt. 2011. aiT Worst-Case Execution Time Analyzer. http://www.absint.com/ait/. Accessed on July 12, 2010.
9. Resmerita, Stefan, Kenneth Butts, Patricia Derler, Andreas Naderlinger, and Wolfgang Pree. 2011. "Migration of Legacy Software Towards Correct-by-Construction Timing Behavior." In *Monterey Workshops 2010*, edited by R. Calinescu and E. Jackson, LNCS 6662, pp. 55–76. Springer Verlag, Heildelberg.
10. Templ, Josef, Andreas Naderlinger, Patricia Derler, Peter Hintenaus, Wolfgang Pree, and Stefan Resmerita. 2012. "Modeling and Simulation of Timing Behavior with the Timing Definition Language (TDL)." In *Real-time Simulation Technologies: Principles, Methodologies, and Applications,* edited by K. Popovici and P. Mosterman (this volume), pp. 157–176, Boca Raton, FL: CRC Press.
11. Kalavade, Asawaree P. 1995. "System-Level Codesign of Mixed Hardware-Software Systems." PhD thesis, University of California, Berkeley, Chair-Lee, Edward A.
12. Edwards, Stephen, Luciano Lavagno, Edward A. Lee, and Alberto Sangiovanni-Vincentelli. 1999. "Design of Embedded Systems: Formal Models, Validation, and Synthesis." In *Proceedings of the IEEE*, 366–90. IEEE, Washington, DC.
13. Huebert, Heiko. June 1998. "A Survey of HW/SW Cosimulation Techniques and Tools." Master's thesis, Royal Institute of Technology, Stockholm, Sweden.
14. Bosman, G. 2003. "A Survey of Co-design Ideas and Methodologies." Master's thesis, Vrije Universiteit, Amsterdam.
15. Mentor Graphics Corporation. 1996–1998. "Seamless Co-verification Environment User's and Reference Manual, V 2.2." Wilsonville, Oregon.
16. Saha, Indranil Saha, Kuntal Chakraborty, Suman Roy, B. VishnuVardhan Reddy, Venkatappaiah Kurapati, and Vishesh Sharma. 2009. "An Approach to Reverse Engineering of C Programs to Simulink Models with Conformance Testing." In *ISEC '09: Proceedings of the 2nd India Software Engineering Conference*, pp. 137–8. New York: ACM.
17. Liu, Jie, Marcello Lajolo, and Alberto Sangiovanni-Vincentelli. 1998. "Software Timing Analysis Using HW/SW Cosimulation and Instruction Set Simulator." In *CODES/CASHE '98: Proceedings of the 6th International Workshop on Hardware/Software Codesign*, pp. 65–9. Washington, DC: IEEE Computer Society.
18. Sangiovanni-Vincentelli, Alberto. February 2002. "Defining Platform-Based Design." *EEDesign of EE Times.*
19. Wang, Yin, Stephane Lafortune, Terence Kelly, Manjunath Kudlur, and Scott Mahlke. 2009. "The Theory of Deadlock Avoidance via Discrete Control." In *POPL '09: Proceedings of the 36th Annual ACM SIGPLAN-SIGACT Symposium on Principles of Programming Languages*, pp. 252–63. New York: ACM.
20. Sifakis, Joseph, Stavros Tripakis, and Sergio Yovine. January 2003. "Building Models of Real-Time Systems from Application Software." *Proceedings of the IEEE* 91 (1): 100–11.
21. Balarin, Felice, Massimiliano D'Angelo, Abhijit Davare, Douglas Densmore, Trevor Meyerowitz, Roberto Passerone, and Alessandro Pinto, et al. January 2009. "Platform-Based Design and Frameworks: Metropolis and Metro II." In *Model-Based Design of Heterogeneous Embedded Systems*, edited by G. Nicolescu and P. Mosterman. Boca Raton, FL: CRC Press.
22. Ghenassia, Frank. 2006. *Transaction-Level Modeling with Systemc: Tlm Concepts and Applications for Embedded Systems.* Secaucus, NJ: Springer-Verlag New York, Inc.
23. Inchron. 2011. The Chronsim Simulator. http://www.inchron.com/chronsim.html. Accessed on September 10, 2010.
24. Chrona's Validator tool suite. 2011. http://www.CHRONA.com. Accessed on September 10, 2010.

9 Modern Methodology of Electric System Design Using Rapid-Control Prototyping and Hardware-in-the-Loop

Jean Bélanger and Christian Dufour

CONTENTS

9.1 Introduction to Real-Time Simulation 219
    9.1.1 Timing and Constraints 221
    9.1.2 Analysis of Simulator Bandwidth Requirements 223
    9.1.3 Rapid Control Prototyping.. 224
    9.1.4 Hardware-in-the-Loop... 225
    9.1.5 Software-in-the-Loop 226
9.2 Real-Time Simulator Technology 226
9.3 Model-Based Design Using Real-Time Simulation... 227
    9.3.1 Model-Based Design.. 227
    9.3.2 Interaction with the Model 229
9.4 Application Examples... 230
    9.4.1 Power Generation Applications 230
    9.4.2 Automotive Applications 232
    9.4.3 All-Electric Ships and Electric Train Networks............................. 233
    9.4.4 Aerospace 235
    9.4.5 Electric Drive and Motor Development and Testing 235
    9.4.6 Mechatronics: Robotics and Industrial Automation....................... 236
    9.4.7 Education: University Research and Development........................... 236
    9.4.8 Emerging Applications 237
References... 239

9.1 INTRODUCTION TO REAL-TIME SIMULATION

A simulation is a representation of the operation or features of a system through the use or operation of another system [1]. For the types of digital simulation discussed in this work, it is assumed that a simulation with discrete-time and constant-step duration is performed. During discrete-time simulation, time moves forward with constant-step duration in set slices or steps of equal duration. This is commonly known as fixed time-step simulation [2]. However, it is important to note that other solving techniques that are unsuitable for real-time simulation make use of variable time steps, which helps in solving high-frequency dynamics and nonlinear systems [3]. This subject is not covered in this chapter, but it is covered in Crosbie [4].

To solve mathematical functions and equations at a given time step, each variable or system state is solved successively as a function of variables and states at the end of the preceding time slice. For the sake of clarity, we do not consider iterative solver methods that are not time-bounded. In real-time simulation, explicit or semiexplicit solvers such as State-Space Nodal [5] are preferred for the fact that their computational time at each iteration is predictable. During a discrete-time simulation, the amount of physical time, or real time, it takes to compute all equations and functions representing a system during a given time slice can be shorter or longer than the duration of the simulation time-step. Figures 9.1a and b represent these two possibilities. In Figure 9.1a, the computing time is shorter than a fixed time step (also referred to as accelerated simulation), while in Figure 9.1b, the computing time is longer. These two situations are referred to as offline simulation. In both cases, the moment at which a result becomes available is irrelevant. In fact, when performing offline simulation, typically the objective is to obtain results as fast as possible. The time that a system takes to produce a simulation depends on the computation power available and the mathematical complexity of the system model.

Conversely, during real-time simulation, the accuracy of the computations depends not only on the precise dynamic representation of the system, but also on

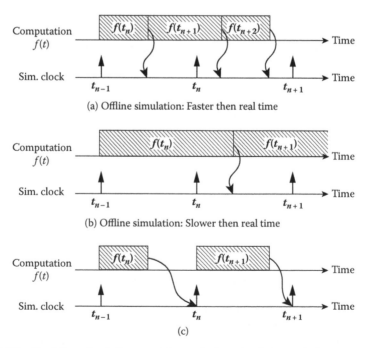

(a) Offline simulation: Faster then real time

(b) Offline simulation: Slower then real time

(c)

FIGURE 9.1   Real-time simulation requisites and other simulation techniques.

the time duration necessary to produce results [6]. Figure 9.1c illustrates the chronological principle of a real-time simulation. For a real-time simulation to be valid, the real-time simulator that is used must accurately produce the internal variables and outputs of the simulation and do so within the same length of time that its physical counterpart would. In fact, the time required to compute the solution at a given time step must be shorter than the physical time duration of the time slice. This permits the real-time simulator to perform all operations necessary to make a real-time simulation relevant, including driving inputs and outputs (I/O) to and from externally connected devices. For a given time slice, any idle time following (or preceding) simulator operation is literally lost, as opposed to accelerated simulation, where this idle time is used to compute the equations at the next time step. In case of idle time, the real-time simulator waits until the clock ticks to the next time-step. However, if all simulator operations are not achieved within the required fixed time step, the real-time simulation is considered erroneous. Such an occurrence is commonly defined as an "overrun."

Based on these basic definitions, it can be concluded that a real-time simulator is performing as expected if the equations and states of the simulated system are solved sufficiently accurately, that is, within a user specification, depending on the type of phenomena that the user wants to observe, with an acceptable resemblance to its physical counterpart and without the occurrence of overruns.

9.1.1 TIMING AND CONSTRAINTS

As previously discussed, real-time digital simulation is based on discrete time steps, where the simulator solves model equations successively. Proper step size duration must be determined to accurately represent system frequency response up to the fastest transient of interest. Simulation results can be validated when the simulator achieves real-time performance without overruns.

For each time step, the simulator executes the same series of tasks: (1) read inputs and generate outputs, (2) solve model equations, (3) exchange results with other simulation nodes, and (4) wait for the start of the next step. A simplified explanation of this routine suggests that the state(s) of any externally connected device is sampled once at the beginning of each time step of the simulation. Consequently, the state(s) of the simulated system is communicated to external devices only once per time step. If not all the timing conditions of real-time simulation are met, overruns occur and discrepancies between the simulator results and the responsive of its physical counterpart are observed.

The inherent constraint of today's real-time simulator is the inescapable use of a discrete-time-step solver, which can become a major limitation when simulating nonlinear systems, such as a High Voltage Direct Current (HVDC) electric power system [7], FACTS [8], active filters, or drives. When nonlinear events such as transistor switching are present in a real-time simulation, because of the nature of discrete-time-step solvers, numerical instability can occur. Different solving methods have been proposed to prevent this problem [6,9], but they cannot be used during real-time simulation.

Achieving real time is one thing, but achieving it synchronously is another. With nonlinear systems, there is no guarantee that switching events will occur (or should be simulated) at a discrete-time instance. Furthermore, multiple events can occur during a single time step, and without proper handling, the simulator may only be aware of the last one. Recently, real-time simulator manufacturers have proposed solutions to timing and stability problems. Proposed solutions generally known as discrete-time compensation techniques usually involve time-stamping and interpolation algorithms [8]. State-of-the-art real-time simulators take advantage of advanced I/O cards running at a sampling rate considerably faster than fixed-step simulation [10,11]. As the I/O card acquires data faster than the simulation, it can read state changes in between simulation steps. Then, at the beginning of the next time step, the I/O card passes on to the simulator not only state information, but also timing information regarding the moment at which the state change occurred. The simulator can then compensate for the timing error.

Figure 9.2 illustrates a classic case of simulation error caused by the late firing of a thyristor in a converter circuit. In this example, a thyristor is triggered at a

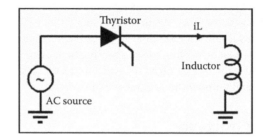

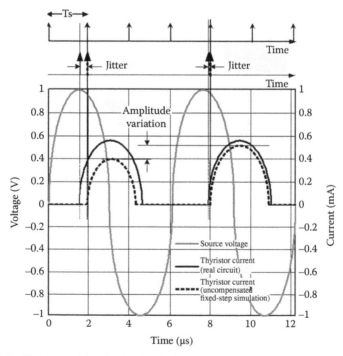

FIGURE 9.2 Timing problem in the simulation of a thyristor converter.

90-degree angle with respect to the AC voltage source positive zero-crossing. As soon as the thyristor is triggered, current starts flowing through it. The resulting load current obtained through uncompensated real-time simulation (dotted line) is represented with some degree of error in comparison to the current flowing through the physical circuit (plain black line). This is because the event at 90 electrical degrees does not occur synchronously to the simulator fixed time step. Thus, the thyristor gate signal is only taken into account at the beginning of the next time step. This phenomenon typically induces low-frequency spurious oscillations in the simulation, for example, causing "jitter" in the motor current [12]. When jitter occurs in a discrete-time simulation, subsynchronous or uncharacteristic harmonics (amplitude variations) may be visible in the resulting waveforms. In this case, variations are evident in the thyristor current.

Finally, the use of multiple simulation tools and different step sizes during real-time simulation can cause problems. When multiple tools are integrated in the same simulation environment, a method called co-simulation, data transfer between tools can present challenges since synchronization and data validity must be maintained [13]. Furthermore, in multirate simulations, where parts of a model are simulated at different rates (with different time-step durations), result accuracy and simulation stability are also issues [14]. For example, multirate simulation may be used to simulate a thermal system with slow dynamics along with an electrical system with fast dynamics [15]. The field of multirate simulation and co-simulation environments, where multiple tools are used side by side, is still an active topic of research.

9.1.2 Analysis of Simulator Bandwidth Requirements

The criteria that will dictate the capability, size, and consequently the cost of the simulator are (1) the frequency of the highest transients to be simulated, which in turn dictates minimum step size and (2) the size of the system to simulate (i.e., the number of differential/algebraic equation to compute), which along with the step size dictates the computing power required. The number of I/O channels required to interface the simulator with physical controllers or other hardware is also critically important, affecting the total performance and cost of the simulator.

Figure 9.3 outlines the typical step size and computing power required for a variety of applications. On the left-hand side of the chart, it is observed that mechanical systems with slow dynamics will generally require a simulation time step between 1 and 10 ms, according to the rule of thumb that the simulation step size should be smaller than 5% to 10% of the smallest time constant of the system. A shorter step size may be required to maintain numerical stability in stiff systems. In addition, when friction phenomena are present, simulation step sizes as low as 100–500 μs may be required.

It is a common practice with electromagnetic transient (EMT) simulators to use a simulation time step of 30–50 μs to provide acceptable results for transients up to 2 kHz [16]. Because greater precision can be achieved with smaller step sizes, simulation of transient phenomena with frequency content up to 10 kHz typically require a simulation time step of approximately 10 μs.

Accurately simulating fast-switching power electronic devices requires the use of very small time steps to solve system equations [12]. Offline simulation is widely used in

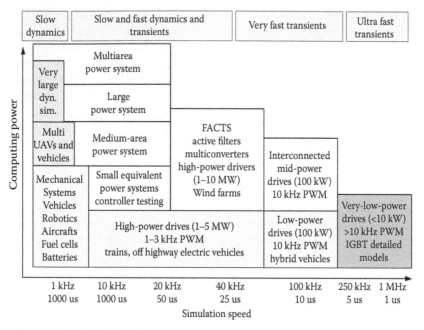

Slow dynamics	Slow and fast dynamics and transients	Very fast transients	Ultra fast transients

Computing power

Very large dyn. sim.	Multiarea power system		
	Large power system		
Multi UAVs and vehicles	Medium-area power system	FACTS active filters multiconverters high-power drivers (1–10 MW) Wind farms	Interconnected mid-power drives (100 kW) 10 kHz PWM
Mechanical Systems Vehicles	Small equivalent power systems controller testing		
Robotics Aircrafts Fuel cells Batteries	High-power drives (1–5 MW) 1–3 kHz PWM trains, off highway electric vehicles	Low-power drives (100 kW) 10 kHz PWM hybrid vehicles	Very-low-power drives (<10 kW) >10 kHz PWM IGBT detailed models

1 kHz	10 kHz	20 kHz	40 kHz	100 kHz	250 kHz	1 MHz
1000 us	1000 us	50 us	25 us	10 us	5 us	1 us

Simulation speed

FIGURE 9.3 Simulation time step by application.

the field, but it is time consuming if no precision compromise has been made on models (i.e., by the use of average models) [17]. Power electronic converters with a higher pulse width modulation (PWM) carrier frequency in the range of 10 kHz, such as those used in low-power converters, require step sizes of less than 250 ns in duration without inter-polation or 10 μs with an interpolation technique. AC circuits with higher resonance frequency and very short lines, as expected in low-voltage distribution circuits and electric rail power feeding systems, may require step sizes below 20 μs in duration. Tests that use practical system configurations and parameters are necessary to determine minimum step size and computing power required to achieve the time step chosen.

Regardless of the simulator used, both numerical solver performance and the bandwidth of interest are considerations when selecting the appropriate step size. The standard approach for selecting a suitable fixed step size for models with increasing complexity is a time-domain comparison of waveforms for repeated runs with different step sizes.

9.1.3 RAPID CONTROL PROTOTYPING

Real-time simulators are typically used in three different application categories [18–21], as illustrated in Figure 9.4. First, in rapid control prototyping (RCP) applications (Figure 9.4a), a plant controller is implemented using a real-time simulator and is connected to a real physical plant. RCP offers many advantages over implementing an actual controller prototype. A controller prototype developed using a real-time simulator is more flexible, less time consuming to implement, and easier to debug. The controller prototype can be tuned on the fly or completely modified with just a

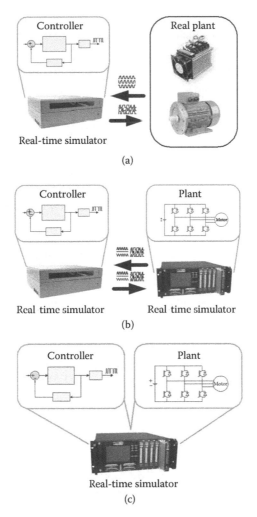

FIGURE 9.4 Application categories.

few mouse clicks. Also, since every internal controller state is available, an RCP can be debugged faster without having to take its cover off.

9.1.4 HARDWARE-IN-THE-LOOP

For hardware-in-the-loop (HIL), the second category of applications, a physical controller is connected to a virtual plant executed on a real-time simulator, instead of to a physical plant. Figure 9.4b illustrates a small variation to HIL. In this case, an implementation of a controller using RCP is connected to a virtual plant via HIL. In addition to the advantages of RCP, HIL allows for early testing of controllers when physical test benches are not available. Moreover, virtual plants usually cost less and have parameters that have less standard deviation of the parameters due to manufacturing process or caused by environmental variations. This allows for more repeatable results and provides for testing conditions that are unavailable on real hardware, such as extreme events testing in which a real device would be damaged for example.

9.1.5 Software-in-the-Loop

Software-in-the-loop (SIL) represents the third logical step beyond the combination of RCP and HIL. With a sufficiently powerful simulator, both controller and plant can be simulated in real time on the same simulator. SIL has the advantage over RCP and HIL that no I/O are used, thereby preserving signal integrity. Also, since both the controller and plant models run on the same simulator, timing with the outside world is no longer critical. The execution time can now be slower or faster than real time with no impact on the validity of the results. SIL can, therefore, be used for a class of simulation called accelerated simulation. In accelerated mode, a simulation runs faster than real time, allowing for a large number of tests to be performed in a short period of time. For this reason, SIL is well suited for statistical testing such as Monte Carlo simulations. SIL does not have to be done in real time because no physical device is connected to the process but because Monte Carlo simulations are very time consuming, involving typically several thousand simulation runs, a real-time simulator will result in shorter completion time than offline simulation; the simulation runs slower than real time.

9.2 REAL-TIME SIMULATOR TECHNOLOGY

Simulator technology has evolved from physical/analog simulators (e.g., HVDC simulators and transient network analyzer (TNA)) for EMT and protection and control studies, to hybrid TNA/analog/digital simulators with the capability of studying electromechanical transient behavior [22], to fully digital real-time simulators, as illustrated in Figure 9.5.

Physical simulators have served their purpose well. However, they were very large, expensive, and require highly skilled technical teams to handle the tedious jobs of setting up networks and maintaining extensive inventories of complex equipment. With the development of microprocessor and floating-point digital signal processor (DSP)

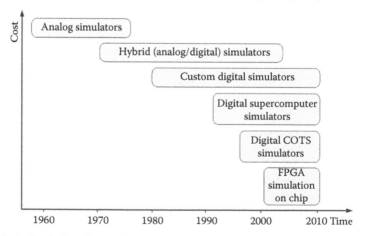

FIGURE 9.5 Evolution of real-time simulation technologies.

technologies, physical simulators have been gradually replaced with fully digital real-time simulators.

DSP-based real-time simulators developed using proprietary technology, and used primarily for HIL studies, were the first of the new breed of digital simulator to become commercially available [23]. However, the limitations of using proprietary hardware were quickly recognized, leading to the development of commercial supercomputer-based simulators, such as HYPERSIM™ from Hydro-Quebec [24] or RTDS™ real-time simulator [25]. Attempts have been made by a number of universities and research organizations to develop fully digital real-time simulators using low-cost standard PC technology in an effort to eliminate the high costs associated with the use of high-end supercomputers [16]. Such development was very difficult due to the lack of fast, low-cost intercomputer communication links. However, the advent of low-cost, easily obtainable multicore processors [26] (from INTEL and AMD) and related commercial-off-the-shelf (COTS) computer components has directly addressed this issue, clearing the way for the development of much lower cost and easily scalable real-time simulators. The availability of this low-cost, high-performance processor technology has also reduced the need to cluster multiple PCs to conduct complex parallel simulation. This reduces dependence on sometimes costly fast intercomputer communication technology.

COTS-based high-end real-time simulators equipped with multicore processors have been used in aerospace, robotics, automotive, and power electronic system design and testing for a number of years [27,28]. Recent advancements in multicore processor technology means that such simulators are now available for the simulation of EMT expected in large-scale power grids, microgrids, wind farms, and power systems installed in large electrical ships and aircraft. These simulators, operating under Windows, LINUX, and standard real-time operating systems, have the potential to be compatible with multi-domain tools, including simulation tools. This capability enables the analysis of interactions between electrical, power electronic, mechanical, and fluid dynamic systems.

The latest trend in real-time simulation consists of exporting simulation models to field programmable gate arrays (FPGA) [29]. This approach has many advantages. First, computation time within each time step is almost independent of the system size because of the parallel nature of FPGAs. Second, overruns cannot occur once the model is running and timing constraints are met. Last, but most importantly, the simulation step size can be very small, in the order of 250 ns. There are still limitations on model size since the number of gates is limited in FPGAs; nevertheless, this technique is promising for the future.

9.3 MODEL-BASED DESIGN USING REAL-TIME SIMULATION

This section explains how the model-based design methodology and real-time simulation technology are used together in the industrial development process.

9.3.1 MODEL-BASED DESIGN

Model-based design (MBD) is a mathematical and graphical method of addressing problems associated with the design of complex systems [30]. MBD is a methodology based on a workflow known as the "V" diagram, as illustrated in Figure 9.6 [21].

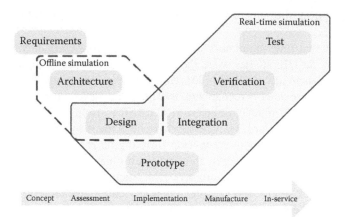

FIGURE 9.6   Model-based design workflow.

It allows multiple engineers involved in a design and modeling project to use models to communicate knowledge of the system under development in an efficient and organized manner [31]. Four basic steps are necessary in the process: (1) build the plant model, (2) analyze the plant model and synthesize a controller for it, (3) simulate the combination of plant and controller, and (4) deploy the controller.

MBD offers many advantages. By using models, it provides a common design environment available to every engineer involved in creating a system from beginning to end. Indeed, the use of a common set of tools facilitates communication and data exchange. Reusing older designs is also easier since the design environment can remain homogeneous through different projects. In addition to MBD, graphical modeling tools, such as the SimPowerSystems™ for Simulink™ from MathWorks™ [32], simplify the design task by reducing the complexity of models through the use of a hierarchical approach. Modeling techniques have also been used to embed independent coded models inside the power systems simulation tool PSCAD/ EMTDC [33].

Most commercial simulation tools provide an automatic code generator that facilitates the transition from controller model to controller implementation. The added value of real-time simulation in MBD emerges from the use of an automatic code generator [34,35]. By using an automatic code generator with a real-time simulator, an RCP can be implemented from a model with minimal effort. The prototype can then be used to accelerate integration and verification testing, something that cannot be done using offline simulation. HIL testing also offers a number of interesting possibilities. The HIL process, which is the reverse of the RCP methodology, involves implementing a plant model in a real-time simulator and connecting it to a physical controller or controller prototype. By using an HIL test bench, test engineers become part of the design workflow earlier in the process, sometimes before an actual plant becomes available. For example, by using the HIL methodology, automotive test engineers can start early testing of an automobile controller before a physical test bench is available.

Combining RCP and HIL while using the MBD approach has many advantages:

- Design issues can be discovered earlier in the process, enabling required trade-offs to be determined and applied, thereby reducing development costs.
- Development cycle duration is reduced because of parallelization in the workflow.
- Testing cost can be reduced in the medium- to long-term, since HIL test setups often cost less than the physical setups and the real-time simulator used can be typically used for multiple applications and projects.
- Testing results are more repeatable since real-time simulators evidence less variability because the dynamics do not change through time the way physical systems do.
- Tests that are too risky or expensive to perform using physical test benches become possible.

9.3.2 Interaction with the Model

Figure 9.7 illustrates the key advantage of real-time simulation: model interaction. These interactions can be (1) with a system user, (2) with physical equipment, or (3) with both at the same time.

It is important to note that a model executed on a real-time simulator can be modified online, which is not possible with a physical plant. Any parameter of the model can be read and updated continuously. For example, in a power plant simulation, the shaft inertia of a turbine can be modified during the simulation to determine its effect on stability, something impossible on a physical power plant. Furthermore, with a real-time simulator, all model variables are accessible during execution. As an example, in a wind turbine application, the torque imposed on the generator from the gearbox is available, since it is a modeled quantity. In a physical wind turbine, obtaining a precise torque value in real time may not be possible since the cost of a torque meter may well be prohibitive.

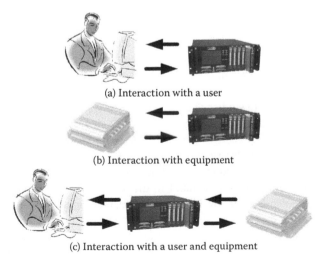

(a) Interaction with a user

(b) Interaction with equipment

(c) Interaction with a user and equipment

FIGURE 9.7 Types of simulator interaction.

Online model configuration and full data availability make previously unthinkable applications possible. An example of such an application is testing the ability of a controller to adapt to changes in a plant. It is, therefore, possible to verify if a controller can compensate for changes in the dynamics of the plant caused by component aging.

9.4 APPLICATION EXAMPLES

This section explains various applications of real-time simulation technology.

9.4.1 POWER GENERATION APPLICATIONS

Testing of complex electric systems shown in Figure 9.8 such as HVDC networks, static VAR compensators, static compensators, FACTS device control systems, and the integration of renewable energy sources in the grid, under steady state and transient operating conditions, is a mandatory practice during the controller development phase and before final system commissioning [30,36,37]. Testing is performed to reduce risk associated with conducting tests on physical networks. HIL testing must first be performed successfully with a prototype controller before the actual production controller is installed in the field. Thousands of systematic and random tests are typically required to test performance under normal and abnormal operating conditions [38]. This testing can also detect instabilities caused by unwanted interactions between control functions and the power system, which may include other FACTS devices that may interact with the system under test.

Protection and insulation coordination techniques for large power systems make use of statistical studies to deal with inherent random events, such as the electrical angle at which a breaker closes or the point-on-wave at which a fault appears [39]. By testing multiple fault occurrences, measured quantities can be identified, recorded, and stored in databases for later retrieval, analysis, and study. While traditional offline simulation software (e.g., ATP, EMTP) [40] can be used to conduct statistical studies during the development of protection algorithms, once a hardware relay is built, further evaluation and development may require using a real-time simulator. Typical studies include digital relay behavior evaluation in different power system operating conditions. Furthermore, relay action may influence the power system, increase distortions, and thus affect other relays. Closed-loop testing in real time is necessary for many system studies and for protection system development.

The integration of distributed generation (DG) devices, including microgrid applications and renewable energy sources (RES), such as wind farms, is one of the primary challenges facing electrical engineers in the power industry today [41,42]. It requires in-depth analysis and the contributions of many engineers from different specialized fields. With growing demand in the area, there is a need for engineering studies to be conducted on the impact that the interconnection of DG and RES will have on specific grids. The fact that RES and DG are usually connected to the grid using power electronic converters is a challenge in itself. Accurately simulating fast-switching power electronic devices requires the use of very short step sizes to solve system equations. Moreover, synchronous generators, which are typically the

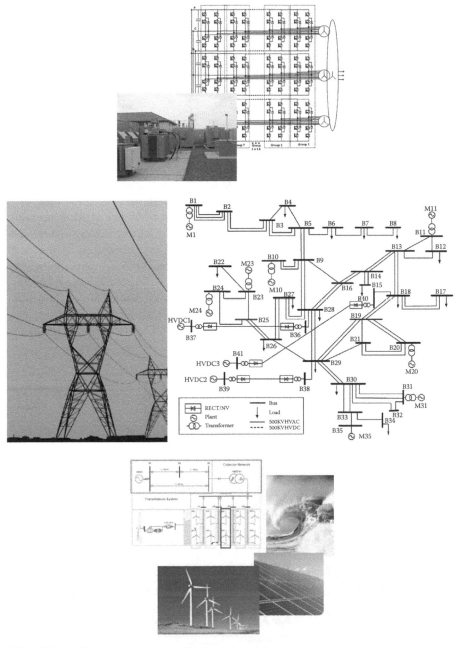

FIGURE 9.8 Power generation applications.

main generation sources on grids, have a slow response to EMT. The simulation of fast-switching power electronic devices in combination with slow electromechanical components in an electrical network is challenging for large grid benchmark studies, more so if proper computation resources are not available. Offline simulation is widely used in the field but is time consuming, particularly if no precision compromise is made on the models (e.g., use of average models for PWM inverters). By using real-time simulation, the overall stability and transient responses of the power system, before and after the integration of RES and DG, can be investigated in a timely matter. Statistical studies can be performed to determine worst-case scenarios, optimize power system planning, and mitigate the effect of the integration of these new energy sources.

9.4.2 AUTOMOTIVE APPLICATIONS

Internal combustion engine hybrid electric cars built by companies such as Toyota and Honda have become economically viable and widely available in recent years. At the same time, considerable research is underway toward the development of fuel cell hybrid electric vehicles, where the main energy source is hydrogen-based. Successful research and development of fuel cell hybrid electric vehicles requires state-of-the-art technology for design and testing. Lack of prior experience, expensive equipment, and shorter developmental cycles are forcing researchers to use MBD techniques for development of control systems [43]. For this reason, thorough testing of traction subsystems such as fuel cell hybrid vehicle is performed using HIL simulation [44], as illustrated by Figure 9.9. In this example, a real-time simulation of a realistic fuel cell hybrid electric vehicle circuit, consisting of a fuel cell, battery, DC–DC converter, and permanent

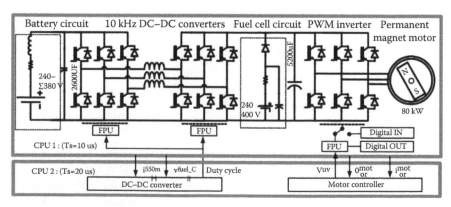

FIGURE 9.9 Automotive applications.

magnet motor drives, with a sufficient number of I/O for real controllers in HIL mode, can now (2010) be performed with a step size duration below 10 μs [45].

9.4.3 ALL-ELECTRIC SHIPS AND ELECTRIC TRAIN NETWORKS

Today, the development and integration of controllers for electric train and all-electric ship (AES) applications is a more difficult task than ever before. Emergence of high-power switching devices has enabled the development of new solutions with improved controllability and efficiency. It has also increased the necessity for more stringent test and integration capabilities since these new topologies come with less design experience on the part of system designers. To address this issue, real-time simulation can be a very useful tool to test, validate, and integrate various subsystems of modern rail (Figure 9.10a) and marine (Figure 9.10b) vehicle devices [46]. The requirements for rail/marine vehicle test and integration reaches several levels on the control hierarchy, from low-level power electronic converters used for propulsion and auxiliary systems to high-level supervisory controls.

The modular design and redundancies built into the power system of an AES designed for combat are critical in ensuring the ship's reliability and survivability during battle. For instance, auxiliary propulsion systems will dynamically replace the primary system in case of failure. This implies that such a power system be

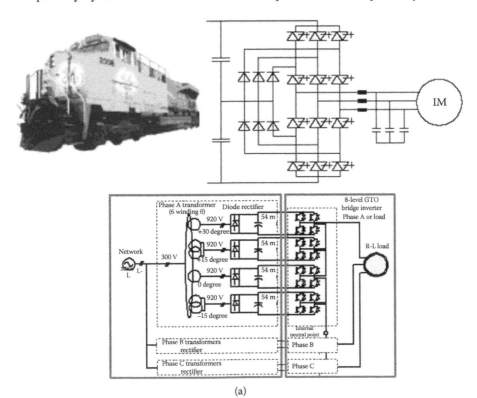

(a)

FIGURE 9.10   Train (a) and ship (b) applications.

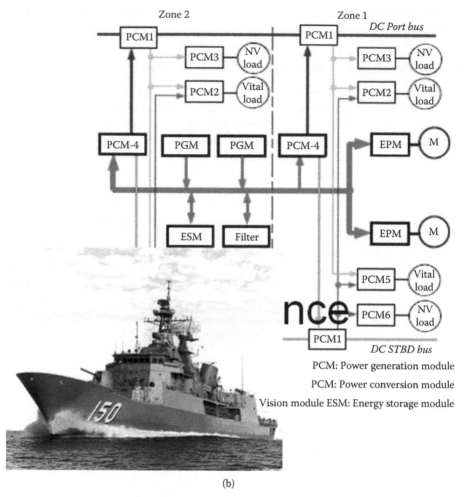

(b)

FIGURE 9.10   (*Continued*)

dynamically reconfigured, such as in zonal electric distribution systems (ZEDS) that have been recently designed for the U.S. Navy AES [47]. Therefore, power management operations must be highly efficient. Power quality issues must be kept to a minimum, and operational integrity must be as high as possible during transients caused by system reconfigurations or loss of modules.

The design and integration of an AES's ZEDS is challenging in many ways. Such a project requires testing of the interactions between hundreds of interconnected power electronic subsystems, built by different manufacturers. Large analog test benches or the use of actual equipment during system commissioning is, therefore, required at different stages of the project. A real-time simulator can be effectively used to perform HIL integration tests to evaluate the performance of some parts of these very complex systems, thereby reducing the cost, duration, and risks related to the use of actual equipment to conduct integration tests [40].

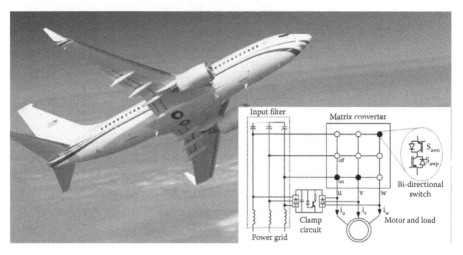

FIGURE 9.11   Aerospace applications.

9.4.4   AEROSPACE

Although most aerospace applications do not necessitate the extremely short step sizes required in power generation or automotive applications, repeatability and accuracy of simulation results is crucial. Safety is a critical factor in the design of aerospace systems (Figure 9.11). Accordingly, aircraft manufacturers must conform to stringent industry standards. Developed by the U.S.-based Radio Technical Commission for Aeronautics, the DO-178B standard establishes guidelines for avionics software quality and testing in real-world conditions [48]. DO-254 is a formal standard governing design of airborne electronic hardware [49].

The complex control systems found onboard today's aircraft are also developed and tested according to these standards. As a result, aerospace engineers must rely on high-precision testing and simulation technologies that will ensure this compliance. Of course, at the same time they must also meet the market's demands for innovative new products, built on time, according to the specifications, and within budget.

9.4.5   ELECTRIC DRIVE AND MOTOR DEVELOPMENT AND TESTING

A critical aspect in the deployment of motor drives lies in the detection of design defects early in the design process. Simply put, the later in the process that a problem is discovered, the higher the cost to fix it. Rapid prototyping of motor controllers is a methodology that enables the control engineer to quickly deploy control algorithms and find eventual problems. This is typically performed using RCP connected in closed loop with a physical prototype of the drive to be controlled as illustrated in Figure 9.12. This methodology implies that the physical motor drive is available at the RCP stage of the design process. Furthermore, this setup requires a second drive (such as a DC motor drive) to be connected to the motor drive under test to emulate the mechanical load. While this is a complex setup, it has proven very effective in detecting problems earlier in the design process.

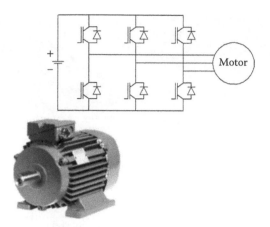

FIGURE 9.12   Smaller scale industrial and commercial applications.

In cases where a physical drive is not available or where only costly prototypes are available, an HIL-simulated motor drive can be used during the RCP development stage. In such cases, the dynamometer, physical insulated-gate bipolar transistor converter, and motor are replaced by a real-time virtual motor drive model. This approach has a number of advantages. For example, the simulated motor drive can be tested with abnormal conditions that would otherwise damage a physical motor. In addition, setup of the controlled-speed test bench is simplified since the virtual shaft speed is set by a single model signal, as opposed to using a physical bench, where a second drive would be necessary to control the shaft speed [50,51].

9.4.6   MECHATRONICS: ROBOTICS AND INDUSTRIAL AUTOMATION

Mechatronic systems that integrate both mechanical and electronic capabilities are at the core of robotic and industrial automation applications (Figure 9.13). Such systems often integrate high-frequency drive technology and complex electrical and power electronic systems. Using real-time simulation to design and test such systems helps ensure greater efficiency of systems deployed in large-scale manufacturing and for unique, but growing applications of robotics [17].

9.4.7   EDUCATION: UNIVERSITY RESEARCH AND DEVELOPMENT

To keep pace with the technological revolution that is underway, universities must change [52]. This transformation is necessary to create new ways to teach future engineers using a multidisciplinary approach, leveraging the possibilities offered by new tools that talented engineers are seeking, while providing them with practical experience that cultivates their creativity [53]. In this context, electronic circuit simulation programs have been in use as teaching aids for many years in electronics and control system classes. The workflow in these classes is quite straightforward: build the circuit with the circuit editor tool, run the simulation, and analyze the results. However, when it is necessary to study the effect of the variation of many parameters (e.g., oscillator frequency, duty cycle, discrete component values), this process can take a substantial

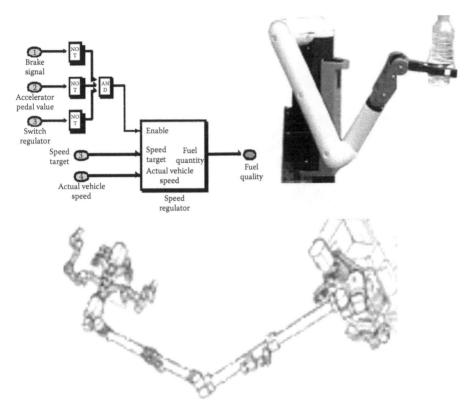

FIGURE 9.13 Mechatronic applications.

amount of time [54]. In such situations, an interactive simulation based on a real-time simulator that offers the capability to change model parameters at runtime becomes a valuable teaching tool. With such a tool, changes to the model are instantly visible, providing students with the live feedback that they are looking for to acquire a feel for how a system reacts to the changes that they make to it, as illustrated in Figure 9.14.

9.4.8 EMERGING APPLICATIONS

Real-time simulation is being used in two additional emerging applications. Since a real-time simulator can provide outputs and read inputs, it is an ideal tool for equipment commissioning and testing, as depicted in Figure 9.15a. Not only can it mimic a physical plant, it can also emulate other devices, play a recorded sequence of events, and record a device under test response. Modern simulators can also provide simulated network connections such as CAN, GPIB, and Ethernet. The application of real-time simulators to equipment commissioning and tests is common in the manufacturing of electronic control modules. For this application, the use of real-time simulators saves test bench costs and reduces testing time.

Operator and technician training is another way to put real-time simulation to good use, as illustrated in Figure 9.15b. While this application category is in an early growth

FIGURE 9.14   Education applications.

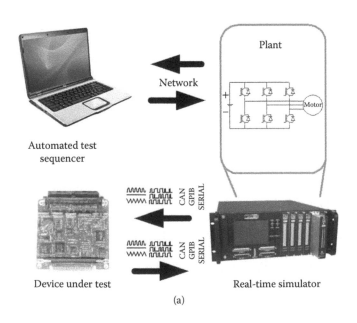

(a)

FIGURE 9.15   Emerging applications.

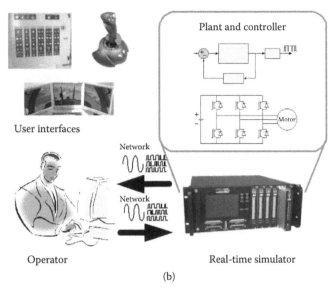

User interfaces

Plant and controller

Network

Network

Operator

Real-time simulator

(b)

FIGURE 9.15 *(Continued)*

stage, it offers terrific potential. For this category of application, both controller and plant are modeled in the same simulator using an SIL-like approach. The difference is that user interfaces are added to allow the operator to interact with the simulation in a user-friendly way. Interfaces such as control panels and joysticks manage user inputs, but must also provide feedback to the user about the simulation state. The advantage of using a real-time simulator for training is that the user can acquire a feeling for the controller and plant that correctly represents the real system, without the delays and limitations commonly found in training environments that are based on prerecorded scenarios.

REFERENCES

1. *The American Heritage Dictionary of the English Language.* 2000. Boston: Houghton Mifflin Company.
2. Dommel, H. W. 1969. "Digital Computer Solution of Electromagnetic Transients in Single- and Multiphase Networks." *IEEE Transactions on Power Apparatus and Systems*, vol. PAS-88.388–99.
3. Sanchez-Gasca, J. J., R. D. Aquila, W. W. Price, and J. J. Paserba. 1995. "Variable Time-Step, Implicit Integration for Extended-Term Power System Dynamic Simulation." In *Proceedings of the Power Industry Computer Application Conference 1995*, 7–12 May, pp. 183–89. Salt Lake City, UT.
4. Crosbie, R. 2012. "Real-Time Simulation Using Hybrid Models." In *Real-Time Simulation Technologies: Principles, Methodologies, and Applications*, edited by Katalin Popovici, Pieter J. Mosterman. CRC Press. ISBN 9781439846650.
5. Dufour, C., J. Mahseredjian, and J. Bélanger. 2011. "A Combined State-Space Nodal Method for the Simulation of Power System Transients." *IEEE Transactions on Power Delivery* 26 (2): 928–35.
6. Marti J. R. and J. Lin. 1989. "Suppression of Numerical Oscillations in the EMTP Power Systems." *IEEE Transactions on Power Systems* 4: 739–47.

7. Asplung G., et al. 1997. "DC Transmission Based on Voltage Source Converters." In *CIGRE SC-14 Colloquium*. Johannesburg.

8. Terwiesch, P., T. Keller, and E. Scheiben. 1999. "Rail Vehicle Control System Integration Testing Using Digital Hardware-in-the-Loop Simulation." *IEEE Transactions on Control Systems Technology* 7 (3), May 1999.

9. Noda, T., K. Takenaka, and T. Inoue. 2009. "Numerical Integration by the 2-Stage Diagonally Implicit Runge–Kutta Method for Electromagnetic Transient Simulations." *IEEE Transactions on Power Delivery* 24: 390–99.

10. Shu, Y., H. Li, and Q. Wu. 2008. "Expansion Application of dSPACE for HILS." In *IEEE International Symposium on Industrial Electronics, ISIE 2008*, 2231–35.

11. Rabbath, C. A., M. Abdoune, and J. Belanger. 2000. "Effective Real-Time Simulations of Event-Based Systems." In *2000: Winter Simulation Conference Proceedings*, Vol. 1, 232–38. Orlando.

12. Kaddouri, A., B. Khodabakhchian, L.-A. Dessaint, R. Champagne, and L. Snider. July 1999. "A New Generation of Simulation Tools for Electric Drives and Power Electronics." In *Proceedings of the IEEE International Conference on Power Electronics and Drive Systems (PEDS '99)*, vol. 1, 348–54. Hong Kong.

13. Faruque, M. O., V. Dinavahi, M. Sloderbeck, and M. Steurer. April 2009. "Geographically Distributed Thermo-Electric Co-simulation of All-Electric Ship." In *IEEE Electric Ship Technologies Symposium (ESTS 2009)*, 36–43. Baltimore, MD.

14. Bednar, R. and R. E. Crosbie. 2007. "Stability of Multi-Rate Simulation Algorithms." In *Proceedings of the 2007 Summer Computer Simulation Conference*, 189–94. San Diego.

15. Fang, R., W. Jiang, J. Khan, and R. Dougal. 2009. "System-Level Thermal Modeling and Co-simulation with Hybrid Power System for Future All Electric Ship." In *IEEE Electric Ship Technologies Symposium (ESTS 2009)*, 547–53. Baltimore, MD.

16. Hollman, J. A. and J. R. Marti. 2008. "Real Time Network Simulation with PC-Cluster." *IEEE Transactions on Power Systems* 18 (2): 563–69.

17. Jin, H. 1997. "Behavior-Mode Simulation of Power Electronic Circuits." *IEEE Transactions on Power Delivery* 12 (3).

18. Rath, G. 2012. "Simulation for Operator Training in Production Machinery." In *Real-Time Simulation Technologies: Principles, Methodologies, and Applications*, edited by Katalin Popovici, Pieter J. Mosterman. CRC Press. ISBN 9781439846650.

19. Şahin, S., Y. İşler, and C. Guzelis. 2012. "A Real Time Simulation Platform for Controller Design, Test and Redesign." In *Real-Time Simulation Technologies: Principles, Methodologies, and Applications*, edited by Katalin Popovici, Pieter J. Mosterman. CRC Press. ISBN 9781439846650.

20. Scharpf, J., R. Hoepler, and J. Hillyard. 2012. "Real Time Simulation in the Automotive Industry." In *Real-Time Simulation Technologies: Principles, Methodologies, and Applications*, edited by Katalin Popovici, Pieter J. Mosterman. CRC Press. ISBN 9781439846650.

21. Batteh, J., M. M. Tiller, and D. Winkler. 2012. "Modelica as a Platform for Real-Time Simulations." In *Real-Time Simulation Technologies: Principles, Methodologies, and Applications*, edited by Katalin Popovici, Pieter J. Mosterman. CRC Press. ISBN 9781439846650.

22. Su, H. T., K. W. Chan, and L. A. Snider. 2008. "Hybrid Simulation of Large Electrical Networks with Asymmetrical Fault Modeling." *International Journal of Modeling and Simulation* 28 (2): 124–31.

23. Kuffel, R., J. Giesbrecht, T. Maguire, R. P. Wierckx, and P. McLaren. 1995. "RTDS: A Fully Digital Power System Simulator Operating in Real Time." In *First International Conference on Digital Power System Simulators (ICDS '95)*, April 5–7, 1995, pp. 19–24. College Station, Texas, USA.

24. Do, V. Q., J.-C. Soumagne, G. Sybille, G. Turmel, P. Giroux, G. Cloutier, and S. Poulin. 1999. "Hypersim, an Integrated Real-Time Simulator for Power Networks and Control Systems." ICDS'99, 1–6. Vasteras, Sweden.

25. Forsyth, P. and R. Kuffel. "Utility Applications of a RTDS® Simulator." In *2007 International Power Engineering Conference, (IPEC2007)*, December 3–6, 2007, 112–17. Singapore.

26. Ramanathan, R. M. Pogo Linux. http://www.pcper.com/reviews/Processors/Intel-Shows-48-core-x86-Processor-Single-chip-Cloud-Computer?aid=825.

27. Bélanger, J., V. Lapointe, C. Dufour, and L. Schoen. 2007. "eMEGAsim: An Open High-Performance Distributed Real-Time Power Grid Simulator. Architecture and Specification." *Presented at the International Conference on Power Systems (ICPS'07)*, December 12–4, 2007. Bangalore, India.

28. Hugh H. T. Liu. 2012. "Interactive Flight Control System Development and Validation with Real-Time Simulation." In *Real-Time Simulation Technologies: Principles, Methodologies, and Applications*, edited by Katalin Popovici, Pieter J. Mosterman. CRC Press. ISBN 9781439846650.

29. Matar, M. and R. Iravani. 2010. "FPGA Implementation of the Power Electronic Converter Model for Real-Time Simulation of Electromagnetic Transients." *IEEE Transactions on Power Delivery* 25 (2): 852–60.

30. Dufour, C., S. Abourida, and J. Bélanger. 2006. "InfiniBand-Based Real-Time Simulation of HVDC, STATCOM and SVC Devices with Custom-off-the-Shelf PCs and FPGAs." In *Proceedings of 2006 IEEE International Symposium on Industrial Electronics*, 2025–29. Montreal, Canada.

31. Auger, D. 2008. "Programmable Hardware Systems Using Model-Based Design." In *2008 IET and Electronics Weekly Conference on Programmable Hardware Systems*, 1–12. London.

32. Sybille, G. and H. Le-Huy. 2000. "Digital Simulation of Power Systems and Power Electronics Using the MATLAB/Simulink Power System Blockset." In *IEEE Power Engineering Society Winter Meeting, 2000*, Vol. 4, 2973–81. Singapore.

33. Perez, S. G. A., M. S. Sachdev, and T. S. Sidhu. 2005. "Modeling Relays for Use in Power System Protection Studies." In *the 2005 Canadian Conference on Electrical and Computer Engineering*, 1–4 May 2005, 566–69. Saskatoon, SK, Canada.

34. Hanselmann, H., U. Kiffmeier, L. Koster, M. Meyer, and A. Rukgauer. 1999. "Production Quality Code Generation from Simulink Block Diagrams." In *Proceedings of the 1999 IEEE International Symposium on Computer Aided Control System Design, 1999*, 213–18. Kohala Coast, Hawaii, USA.

35. Sadasiva, I., F. Flinders, and W. Oghanna. 1997. "A Graphical Based Automatic Real Time Code Generator for Power Electronic Control Applications." In *Proceedings of the IEEE International Symposium on Industrial Electronics (ISIE '97)*, Vol. 3, pp. 942–47. Guimaraes, Portugal.

36. Liu, Y., et al. 2009. "Controller Hardware-in-the-Loop Validation for a 10 MVA ETO-Based STATCOM for Wind Farm Application." In *IEEE Energy Conversion Congress and Exposition (ECCE'09)*, 1398–1403. San-José, CA.

37. Etxeberria-Otadui, I., V. Manzo, S. Bacha, and F. Baltes. 2002. "Generalized Average Modelling of FACTS for Real Time Simulation in ARENE." In *IEEE 28th Annual Conference of the Industrial Electronics Society (IECON 02)*, Vol. 2. 864–69.

38. Zander, J., I. Schieferdecker, and P. J. Mosterman, eds.2011. *Model-Based Testing for Embedded Systems*. Boca Raton, FL: CRC Press. ISBN 9781439818459.

39. Paquin, J. N., J. Bélanger, L. A. Snider, C. Pirolli, and W. Li. 2009. "Monte-Carlo Study on a Large-Scale Power System Model in Real-Time Using eMEGAsim." In *IEEE Energy Conversion Congress and Exposition (ECCE'09)*, pp. 3194–3202. San-José, CA.

40. Paquin, J. N., W. Li, J. Belanger, L. Schoen, I. Peres, C. Olariu, and H. Kohmann. 2009. "A Modern and Open Real-Time Digital Simulator of All-Electric Ships with a Multi-Platform Co-simulation Approach." In *Proceedings of the 2009 Electric Ship Technologies Symposium (ESTS 2009)*, April 20–22, 2009, 28–35. Baltimore, MD.

41. Paquin, J. N., J. Moyen, G. Dumur, and V. Lapointe. 2007. "Real-Time and Off-Line Simulation of a Detailed Wind Farm Model Connected to a Multi-Bus Network." In *IEEE Canada Electrical Power Conference 2007*, October 25–26, 2007. Montreal, QC.

42. Paquin, J. N., C. Dufour, and J. Bélanger. 2008. "A Hardware-In-the-Loop Simulation Platform for Prototyping and Testing of Wind Generator Controllers." In *2008 CIGRE Conference on Power Systems*, October 19–21, 2008. Winnipeg, Canada.

43. Dufour, C., T. K. Das, and S. Akella. 2003. "Real Time Simulation of Proton Exchange Membrane Fuel Cell Hybrid Vehicle." In *Proceedings of 2003 Global Powertrain Congress*, Ann Arbor, MI.

44. Dufour, C., T. Ishikawa, S. Abourida, and J. Belanger. 2007. "Modern Hardware-in-the-Loop Simulation Technology for Fuel Cell Hybrid Electric Vehicles." In *IEEE Vehicle Power and Propulsion Conference (VPPC 2007)*, September 9–12, 2007, 432–39. Arlington, TX.

45. Dufour, C., J. Bélanger, T. Ishikawa, and K. Uemura. 2005. "Advances in Real-Time Simulation of Fuel Cell Hybrid Electric Vehicles." In *Proceedings of the 21st Electric Vehicle Symposium (EVS-21)*, April 2–6, 2005. Monte Carlo, Monaco.

46. Dufour, C., G. Dumur, J.-N. Paquin, and J. Belanger. 2008. "A Multi-Core PC-Based Simulator for the Hardware-in-the-Loop Testing of Modern Train and Ship Traction Systems." In *the 13th Power Electronics and Motion Control Conference (EPE-PEMC 2008)*, September 1–3, 2008, 1475–80. Poznan, Poland.

47. Sudhoff, S.D., S. Pekarek, B. Kuhn, S. Glover, J. Sauer, and D. Delisle. 2002. "Naval Combat Survivability Testbeds for Investigation of Issues in Shipboard Power Electronics Based Power and Propulsion System." In *IEEE Power Engineering Society Summer Meeting*, July 21–25, 2002, Vol. 1, 347–50. ISBN 0-7803-7518-1.

48. Radio Technical Commission for Aeronautics. 1992. "Software Considerations in Airborne Systems and Equipment Certification." Radio Technical Commission for Aeronautics Standard DO-178B.

49. Radio Technical Commission for Aeronautics. 2000. "Design Assurance Guidance for Airborne Electronic Hardware." Radio Technical Commission for Aeronautics Standard DO-254.

50. Bélanger, J., H. Blanchette, and C. Dufour. 2008. "Very-High Speed Control of an FPGA-Based Finite-Element-Analysis Permanent Magnet Synchronous Virtual Motor Drive System." In *IEEE 34th Annual Conference of Industrial Electronics (IECON 2008)*, November 10–13, 2008, 2411–16. Orlando, FL.

51. Harakawa, M., H. Yamasaki, T. Nagano, S. Abourida, C. Dufour, and J. Belanger. 2005. "Real-Time Simulation of a Complete PMSM Drive at 10 μs Time Step." *Presented at the International Power Electronics Conference, 2005 (IPEC 2005)*. Niigata, Japan.

52. Dufour, C., C. Andrade, and J. Bélanger. 2010. "Real-Time Simulation Technologies in Education: a Link to Modern Engineering Methods and Practices." In *Proceedings of the 11th International Conference on Engineering and Technology Education, (INTERTECH-2010)*, March 7–10, 2010. Ilhéus, Bahia, Brazil.

53. Min, W., S. Jin-Hua, Z. Gui-Xiu, and Y. Ohyama. 2008. "Internet-Based Teaching and Experiment System for Control Engineering Course." *IEEE Transactions on Industrial Electronics* 55 (6).

54. Huselstein, J.-J., P. Enrici, and T. Martire. 2006. "Interactive Simulations of Power Electronics Converters." In *12th International Power Electronics and Motion Control Conference (EPE-PEMC 2006)*, August 30–September 1, 2006, 1721–26. Portoroz, Slovenia.

10 Modeling Multiprocessor Real-Time Systems at Transaction Level

Giovanni Beltrame, Gabriela Nicolescu, and Luca Fossati

CONTENTS

10.1 Introduction 243
10.2 Previous Work... 244
10.3 Proposed Methodology... 246
    10.3.1 System Call Emulation 246
    10.3.2 Pthreads as a Real-Time Concurrency Model................................ 248
    10.3.3 Real-Time Concurrency Manager 250
    10.3.4 Interrupt Management 251
10.4 Experimental Results.. 252
10.5 Concluding Remarks 256
References.. 256

This chapter presents a transaction-level technique for the modeling, simulation, and analysis of real-time applications on multiprocessor systems-on-chip (MPSoCs). This technique is based on an application-transparent emulation of operating system (OS) primitives, including support for or real-time OS (RTOS) elements. The proposed methodology enables a quick evaluation of the real-time performance of an application with different design choices, including the study of system behavior as task deadlines become stricter or looser. The approach has been verified on a large set of multithreaded, mixed-workload (real-time and non-real-time) applications and benchmarks. Results show that the presented methodology (1) enables accurate real-time and responsiveness analysis of parallel applications running on MPSoCs, (2) allows the designer to devise an optimal interrupt distribution mechanism for the given application, and (3) helps sizing the system to meet performance for both real-time and non-real-time parts.

10.1 INTRODUCTION

Increasingly, large portions of electronic systems are being implemented in software, with the consequence that the software development effort is becoming a dominant factor in the development flow. The problem is aggravated by the strong trend from the "classical" embedded systems toward *real-time* (RT) systems with explicitly concurrent hardware that is more difficult to analyze and to model. In such systems, beyond the correctness of algorithms, early verification of the real-time behavior and timing constraints is essential. Guaranteeing the required properties with explicitly concurrent software and hardware adds a degree of complexity, which combines with the fact that software deployment and testing on target hardware is difficult and time consuming. To gather timing details and to validate the functionality of the overall system as soon as possible in the development process, high level models of the interactions among application, operating system, and hardware platform are necessary. Unfortunately, generally used methodologies suffer from the *code equivalence* problem (as presented in Yoo et al. [1]): the code executed by the virtual system is different from the code executed by the final deployed hardware, especially for code that concerns OS primitives. This may change the overall system behavior, leading to less-than-optimal or incorrect design choices. RT systems further complicate the situation as the correctness of the computation is highly dependent on its timing behavior, which implies a necessity for accurate modeling of scheduling choices, task interactions, and interrupt response times. Moreover, the implementation of critical RTOS parts can be carried out either in software or using hardware accelerators, with major consequences on system cost, behavior, and timing features.

In this work, we provide a codesign environment suitable for the development of multiprocessor systems with real-time requirements. The core idea consists in the *transparent emulation* of RTOS primitives on top of a virtual platform described at the transaction level; the implementation guarantees full compatibility with any POSIX-compliant application. Overall, our approach provides fast and accurate simulation results, allowing effective high-level design space exploration (DSE) for multicore RT systems. Our methodology can be applied to various tasks, such as analysis of system responsiveness in the face of different load and of varying frequency of external events, and as exploration of different scheduling policies.

This article is organized as follows: Section 10.2 describes previous research on the subject and Section 10.3 presents how the proposed methodology addresses the identified issues. Finally, Section 10.4 shows the experimental results and Section 10.5 draws some concluding remarks.

10.2   PREVIOUS WORK

A hardware/software (HW/SW) codesign flow [2] usually starts at system level, when the boundaries between the hardware and software parts of the final system have not yet been established. After functional verification, the HW/SW partitioning takes place and co-simulation is used to validate and refine the system. The tight time-to-market constraints, the high complexity of current designs, and the low simulation speed of instruction set simulators (ISSs) push for the addition of (RT)OS models in system-level hardware description languages (HDL) (such as SystemC or SpecC). This allows native execution of both the hardware and software models of the system, consistently accelerating simulation. In addition, as both hardware and software partitions are described using the same HDL, it is easy to move functionalities between them. Such ideas have been presented in work that models the application, the hardware, and the services of the RTOS using the same HDL [3–6]. Because of limitations in the typical HDL processing model, true concurrency is not achieved, and a trade-off has to be determined between simulation speed and accuracy of the intertask interaction models. When the design is refined, the RTOS model can be translated automatically into software services. Unfortunately, however, in practice, the use of a widely adopted RTOS is preferred, meaning that results taken during the modeling phase are no longer accurate.

An extension to these works has been implemented by Schirner and Domer [7] addressing the problem of modeling preemption, interrupts, and intertask interactions in abstract RTOS models. Their work mainly concentrates on simulating the system timing behavior, and the code equivalence problem is not taken into account.

He et al. [8] present a configurable RTOS model implemented on top of SystemC: as opposed to other approaches, only the software part of the system is modeled, while the hardware portion is taken into account only by means of timing annotations inside the RTOS model.

A different technique is used in Yoo et al. [1] for automatic generation of timed OS simulation models. These models partially reuse unmodified OS primitives, thus mitigating the code equivalence problem. High emulation speed is obtained, thanks to native execution on the host machine, but the timing of the target architecture is not accurately replicated and it does not allow precise modeling of multiprocessor systems. In contrast, we use ISSs in our approach, which implies lower simulation speed but also, as the assembly code of the final application is used, minimization of the code equivalence problem. Our approach is also OS-independent, enabling broader DSE.

An untimed abstract model of an RTOS is presented in Honda et al. [9]. The model supports all the services of the *μITRON* standard; therefore, it can be used with a wide range of applications, but this work is applied only to uniprocessor systems.

Similar goals are pursued in Posadas et al. [6] by mapping OS thread management primitives to SystemC. However, because of limitations in the SystemC process model, true concurrency cannot be achieved. Girodias et al. [10] also address the lack of support for embedded software development by working in the .NET environment and by mapping hardware tasks to eSys.net processes and software tasks to an abstract OS interface mapped onto .NET threads. A further refinement step allows the mapping of software tasks on a Win32 port of the chosen target OS.

All these approaches help the designer to perform and refine the HW/SW partition, but they do not help in the validation of the high-level design (for the code equivalence problem) and they are limited in the assessment of the system's timing properties. However, the execution on an ISS of the exact same software that will be deployed on the embedded system is seldom possible because the RTOS has to be already chosen and ported to the target hardware, meaning that it may be difficult or even impossible to refine the HW/SW partitioning (since the OS should be updated accordingly) and to explore alternative system configurations.

This chapter proposes a way to emulate RTOS primitives to minimize *code equivalence* issues while still maintaining both *independence from a specific OS* and *high timing accuracy*.

10.3 PROPOSED METHODOLOGY

For the implementation and evaluation of the design methodology presented in this study, we use the open source simulation platform ReSP [11]. ReSP is based on the SystemC library and it targets the modeling of multiprocessor systems. Its most peculiar feature consists in the integration of C++ and Python programming languages; this augments the platform with the concept of reflection [12], allowing full observability and control of every C++ or SystemC element (variable, method, etc.) specified in any component model. In this work, we exploit and extend ReSP's system call emulation subsystem to support the analysis of real-time systems and applications. The presented functionalities are used for preliminary exploration of the applications' behavior for guiding the designer in the choice of the target RTOS and as a support for early HW/SW codesign.

10.3.1 SYSTEM CALL EMULATION

System call emulation is a technique enabling the execution of application programs on an ISS without the necessity to simulate a complete OS. The low-level calls made by the application to the OS routines (*system calls*, SC) are identified and intercepted by the ISS and then redirected to the host environment, which takes care of their actual execution. Suppose, for example, that the simulated application program contains a call to the open routine to open file *"filename."* Such a call is identified by the ISS and routed to the host OS, which actually opens *"filename"* on the filesystem of the host. The file handle is then passed back to the simulated environment. A simulation framework with system call emulation capabilities allows the application developers to start working as early as possible, even before a definitive choice about the target OS is performed. This can also help in the selection, customization, configuration, and validation of the OS itself. Figure 10.1 shows an overview of ReSP's system call emulation mechanism. Here, each ISS communicates with one centralized trap emulator (TE), the component responsible for forwarding SCs from the simulated environment to the host. To ensure independence between the ISS and

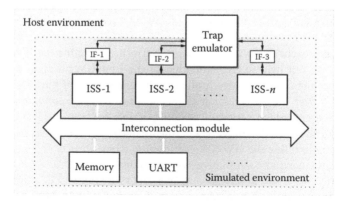

FIGURE 10.1 Organization of the simulated environment including system call emulation.

the TE, interfaces (IF-1,IF-2, etc.) are created and communication between the TE and the ISS exclusively takes place through them.

Figure 10.2 shows the system call emulation mechanism where each ISS communicates with the centralized TE, the component responsible for forwarding the system calls from the simulated to the host environment.

Instead of identifying the SCs through particular assembly instructions or special addresses (as in most simulator environments), we use the name (i.e., the symbol) of the corresponding routine. When the application program is loaded, the names of the low-level SCs (e.g., sbrk, _open, etc.) are associated with their addresses in the binary file and registered with the TE. At runtime, the ISS then checks for those addresses, and when one is found, the corresponding SC is emulated on the host environment.

The TE provides the emulation of concurrency management routines with an additional unit, called a *concurrency manager* (CM), where the TE intercepts calls for thread creation, destruction, synchronization etc. For this purpose, we created a placeholder library containing all the symbols (i.e., the function identifiers) of the POSIX-Thread standard, but without a corresponding implementation. This ensures that programs using the pthread library can correctly compile. During execution, all calls to pthread routines are trapped and forwarded to the CM. If the application software is compiled with a recent GNU GCC compiler (at least version 4.2), it is also possible to successfully emulate OpenMP directives. The CM is able to manage shared memory platforms with an arbitrary number of symmetric processors.

In addition to system call emulation, these functionalities can be used, through the CM, for the emulation of concurrency management (thread creation, destruction,

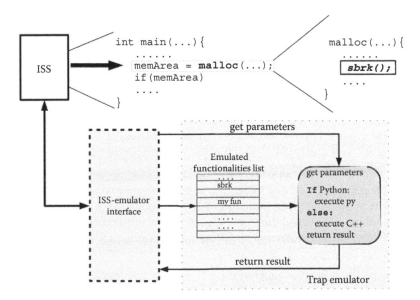

FIGURE 10.2 Internal structure and working mechanisms of the function trap emulator.

mutex lock, unlock, etc.) routines. With respect to previous work, these mechanisms demonstrate the following advantages:

1. *Independence from the cross-compiler toolchain*: since the names of the system call routines are used, it is not necessary to adhere to the conventions with which the software is built or to create fictitious jumps in the code.
2. *High interoperability with different ISS types*: the IF is the only component that requires customization to allow a new ISS to be integrated with the TE.
3. *Extensibility*: the presented mechanism can also be used for preliminary HW/SW partitioning. Moreover, by emulating the POSIX-Threads routines, multithreaded applications can be easily simulated.

Since only the low-level SCs (e.g., `sbrk`) are emulated and the rest of the OS code (e.g., `malloc`) is executed unmodified in the ISS, our method maintains high code equivalence with the final software, even at the assembly level.

Communication between the emulator and the ISS is a critical point in the overall design. On the one hand, it must be designed to be flexible and portable so that ISSs can be easily plugged into the system. On the other hand, it must be as fast as possible to guarantee high simulation speed. These are conflicting requirements and a proper trade-off must be determined. Two solutions were identified: the first is purely based on compiled C++, while the second, more flexible, one uses Python to unintrusively access the ISS internal variables.

To guarantee the timing accuracy of each input/output (I/O)-related SC (such as the `write` operation), which would generate traffic on the communication medium, we assume the SC is executed inside the processor, modeling only the data transfer from processor to memory and vice-versa. While this is only an approximation of an actual system, accuracy is not severely affected as shown by our experiments.

10.3.2 PTHREADS AS A REAL-TIME CONCURRENCY MODEL

Pthreads are a well known concurrent application programming interface (API) and, as part of the POSIX standard, are available for most operating systems (either natively or as a compatibility layer). The Pthread API provides extensions for managing real-time threads, in the form of two scheduling classes:

FIFO: threads of equal priority are scheduled following a first-come first-served policy; if a thread of high priority is created while one of lower priority is running, the running thread is preempted.
Round-robin: same scheduling policy of *FIFO*, with the difference that the processor is shared, in a round robin fashion, among threads of equal priority.

To manage these functionalities, Pthreads provide routines for setting/reading/changing thread priorities and scheduling policies. However, even when using the POSIX-Threads RT extension, the standard does not fully allow the management of

RT systems. Important features, such as task scheduling based on deadlines, are not present and this prevents an effective modeling and analysis of a wide range of RT systems. For this reason, our emulation layers extend the POSIX-Thread standard with the introduction of the *earliest deadline first* (EDF) [13] scheduling policy and with the possibility of declaring a task as unpreemptible. Theoretical results [14] expose that EDF scheduling brings better performance with respect to standard priority-based scheduling. In our implementation, the emulated RT features are compatible with the popular OS RTEMS task management policies.

This work enables the exploration, tuning, and analysis of RT systems. To effectively and efficiently perform such activities, we must be able to explore the task scheduling policies, their priorities, and in general, task attributes. As modifying the source code is not suitable for fast coexploration, thread attributes and scheduling policies can be specified also *outside* the simulation space, using ReSP capabilities. Figure 10.3 shows how ReSP can be used to tune a given system using an optimization loop: attributes are set, the system is simulated, the simulation results can be used to change the system parameters, and so on. The designer has two possible alternatives: (1) specifying the desired RT behavior directly in the application source code or (2) via Python scripting. The first mechanism is simply obtained by emulation of all threading-related primitives. In particular, the calls made by the application software to the functions for managing thread attributes are redirected to the CM, which takes care of managing and scheduling the tasks according to

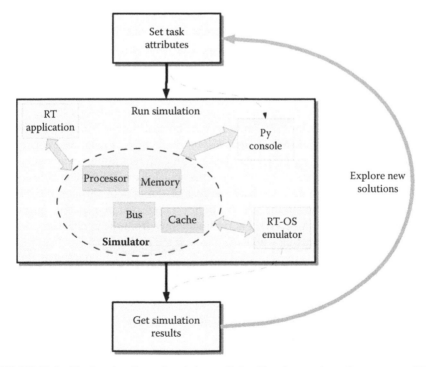

FIGURE 10.3 Exploration flow of real-time policies. Note how task attributes are modified at runtime through the Python console.

such attributes. The second method consists of using Python to directly export the internal structure of the CM to ReSP. As such, it is possible, either before or during the simulation, to modify the CM status and change the thread management policies without modifications to the application source code. In both cases, the system load and RT behavior can be modified during simulation, enabling an effective exploration of the system's real-time behavior.

10.3.3 REAL-TIME CONCURRENCY MANAGER

As mentioned in Section 10.3.1, the TE provides the emulation of concurrency management routines with an additional unit, the CM. The overall mechanism is analogous to the one depicted in Figure 10.2, but instead of trapping I/O or memory management, the TE traps routines for thread creation, destruction, synchronization, etc. During execution, all calls to `pthread` routines are trapped and forwarded to the CM in the simulator environment. If the application software is compiled with a recent GNU GCC compiler (at least version 4.2), it is also possible to successfully emulate OpenMP directives.

This CM was augmented to deal with real-time extensions and to correctly keep statistics about issues such as missed deadlines, serviced interrupts, etc. In particular, the following features were added:

- **Context Switch Capabilities**. To execute different threads on the same processor, *context switch* capabilities are necessary because a processor can switch between two threads either when the current thread is blocked (e.g., for synchronization) or when the time quantum associated with the current thread expires. Switching context consists of saving all the ISS registers and restoring the registers for the next thread, much like what would happen when using a nonemulated OS, with the only difference that registers are not saved on the stack in memory, but in the simulator's space.
- **Real-Time Scheduler**. We implemented the real-time scheduler in three different versions: *FIFO, Round-Robin*, and *EDF*. Each task can be assigned a scheduling policy and tasks with different policies can coexist in the system. Figure 10.4 shows how the scheduler is implemented inside the CM and communicates with the rest of the system through the TE. Each task, according to the selected policy, is inserted in a specific queue. Policies of tasks can be varied at runtime either from the application code or by directly interacting with ReSP through the Python console. The latter mechanism has been implemented to enable flexible task management, thus allowing an effective and efficient exploration of the different scheduling policies and priorities and the different RTOS configurations. Tasks with the EDF policy are assigned the highest priority. The scheduler is able to manage shared memory platforms with an arbitrary number of symmetric processors. Since scheduling and, in general, task management operations are performed in the host environment, it is possible to add features such as deadlock and race-condition detection without altering the system behavior. Because of this, our system can also be successfully employed for

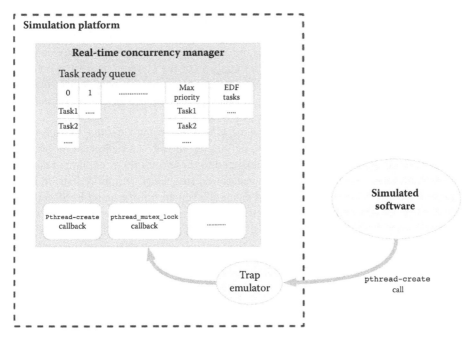

FIGURE 10.4 Detailed structure of the real-time concurrency manager (CM). The CM communicates with the simulated application through the trap emulator. In the CM, RT tasks are organized in queues of different priorities.

verification of system correctness. In contrast, if such features as deadlock and race-condition detection were implemented in the simulated software, system behavior would be affected and disallow the verification.

- **Interrupt Management**. The interrupt management is composed of an emulated interrupt generator and an interrupt service routine (ISR) manager. While the former is present only to emulate external events and to force execution of ISRs (to enable the analysis of the system behavior under different practical environmental conditions), the latter feature is used to deal with ISRs, no matter how they are triggered. No major modifications were necessary to the system to control ISRs since, after creation, they are treated as standard real-time tasks.

- **Python Integration**. This feature enables task control from outside of the simulated application. This means that from ReSP's interactive shell it is possible to manage task priorities, deadlines, etc. As such, it is not necessary to modify the simulated software to perform an effective exploration and to analyze the effects of different scheduling policies and/or priorities.

10.3.4 Interrupt Management

Reactivity to external events is a fundamental feature in embedded systems, especially for what concerns real-time applications; most of the time such systems have to react in a timely and predictable manner to inputs coming from the outside world.

The interrupt management system implemented in ReSP's CM has two operating modes: interrrupts can either be triggered by simulated peripheral components described in SystemC (thus mimicking the actual system behavior) or they can be artificially raised by the CM, to ease the analysis of the systems' real-time behavior under particular stress conditions. The latter mechanism is particularly useful to quickly emulate, explore, and analyze the behavior in different environmental conditions.

In both ways ISRs are managed by the CM and they are treated like normal tasks, meaning that any function can be defined as an ISR. As such, custom priorities, scheduling policies, etc. can be associated to these routines as explained above for standard tasks. Parameters such as generation frequency, temporal distribution, interrupt type, etc. can be easily set and changed even at runtime using ReSP reflective capabilities, thus allowing an effective exploration of the configuration alternatives and a simple emulation of the possible environmental conditions.

10.4   EXPERIMENTAL RESULTS

The methodology has been tested on a large set of OpenMP-based benchmarks (namely the OMPScr suite) and a large parallel application, namely ffmpeg (video encoding/decoding).

The basic assumption of this work is that the system is subject to a mixed application workload: a computationally intensive element with soft real-time constraints and a set of elements with very strict hard real-time characteristics, here called computational and real-time parts, respectively. The number and parameters of both the computational and real-time parts varies and strictly depends on the system being considered. This model well represents applications such as observation spacecraft payload, where massive data processing is required with high availability, while response to external stimuli within a given time is paramount (e.g., for the spacecraft's navigation system).

The purpose of our methodology is to answer a set of key questions during the development of real-time applications running on an MPSoC:

1. What is the performance of the real-time applications? Is the system missing any deadlines with the current hardware and scheduling setup?
2. What is the performance of the computational part? Is it performing within requirements?
3. How much performance can the current hardware and software setup deliver? Is it possible to add additional computational or real-time tasks without affecting global performance? Can we reduce the number of hardware resources? What is the benefit of moving parts of the application or OS to hardware?

All tests have been executed using ReSP on a multi-ARM architecture consisting of a variable number of cores with caches, and a shared memory, all interconnected by a shared bus. Simulations where timing was recorded were run on a Core 2 Duo 2.66GHz Linux machine.

To evaluate the performance and accuracy of OS emulation with respect to a real OS, twelve OmpSCR benchmarks were run with the real-time operating system eCos [15], using a 4-core platform. A large set of eCos system calls were measured running six of these benchmarks as a training or calibration set, and the average latency of each class of system calls was determined. The Lilliefors/Van Soest test of normality [16] applied to the residuals of each class shows evidence of nonnormality ($L = 0.30$ and $L_{critical} = 0.28$ with $\alpha = .01$), but given that the population variability remains limited (with a within-group mean square $MS_{S(A)} = 7602$ clock cycles), it can be assumed that each average latency is representative of its class.

The derived latencies were introduced for each system call in our OS emulation system, and the remaining six benchmarks (used as a validation set) were executed. Since profiling did not include all functions used by the OS and for which the latency was considered zero, the overall results were uniformly biased for underestimation. This bias can be easily corrected considering the average error, leading to an average error of 6.6 ± 5.5%, as shown in Figure 10.5. Even with this simple scheme, the methodology can very well emulate the behavior of a specific OS with minimal error, especially considering that full code equivalence is present for the application and library functions, but threading, multiprocessor management, and low-level OS functions are emulated.

In addition, the use of the OS emulation layer introduces a noticeable speedup (13.12 ± 6.7 times) when compared to running the OS on each ISS. This is because of several factors, including the absence of some hardware components such as debugging UARTs and timers (the TE implements a terminal in the host OS and the configuration manager uses SystemC and its events to keep track of time), and the fact that, in our mechanism, idle processors do not execute busy loops but they are, instead, suspended. The latter is implemented by trapping the busy loop wait function in the TE and redirecting it to a SystemC wait() call.

Using the proposed methodology, a designer can verify the real-time performance of a multiprocessor system under load and explore the use of different interrupt distribution and handling schemes. As proof-of-concept, we ran the benchmarks as computationally intensive applications, while the real-time tasks are implemented

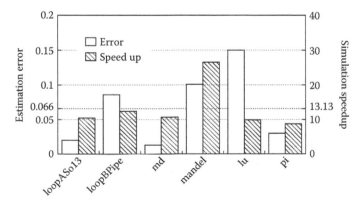

FIGURE 10.5 Simulation speedup and estimation error using the emulation layer instead of eCos.

by synthetic functions, with varying deadlines. These functions can be categorized as (1) housekeeping (scheduled regularly, perform sanity checks, repetitive tasks, etc.) and (2) response to external events (when an alarm is fired, its response is usually required within a given deadline).

A first analysis that is performed with the current methodology is to run the real-time part separately from the computational part, reducing all OS-related latencies (such as the latency of the mutex lock operation) to zero. The obtained *concurrency profile* shows the number of active PEs in time, that is, the effective utilization of the system resources. A similar graph is derived for the computational part, allowing the designer to determine if sufficient resources are available to run the application within its performance constraints. Finally, the computational and real-time parts are combined together and the concurrency profile is drawn as shown in Figure 10.6. This graph helps the designer tweak the hardware and software to match the desired requirements. As an example, the combination diagram (Figure 10.6b) can show a

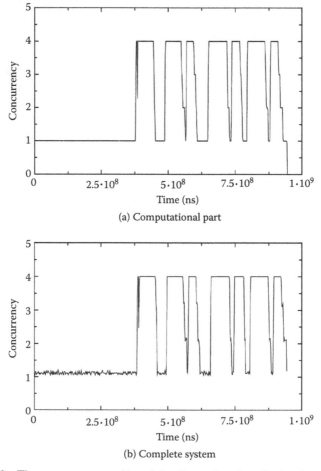

(a) Computational part

(b) Complete system

FIGURE 10.6 The concurrency profiles of the ffmpeg benchmark, showing the computational parts (a) and the combination of the computational and real-time parts (b).

lower-than-expected utilization in case access to a shared bus represents a bottleneck in the system. If, instead, utilization is already at a maximum, the designer can conclude that more processing elements are necessary to reach the performance requirements. Simulating the system with realistic OS-related latencies (that can be targeted to any possible OS choice) leads to determining the best OS choice for the current application.

Figure 10.7 shows how the methodology is used to determine the best scheduler for the system where the performance is graphed for two schedulers (Priority and EDF).

Running the application with and without RT tasks shows the different computational performance of the system, as depicted in Figure 10.8. As in our methodology, the RT or non-RT status of a task can be changed without modifications to the code,

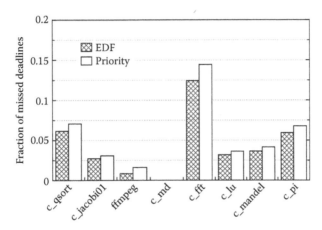

FIGURE 10.7 Fraction of missed deadlines with different schedulers and high real-time workload (1 kHz).

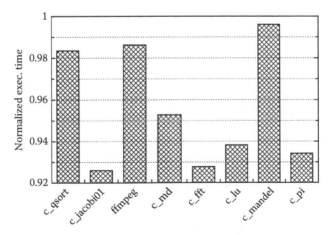

FIGURE 10.8 The performance impact of the real-time (RT) part on the computational part, that is, their relative execution time when compared to execution without the RT part.

and this evaluation is simply made with two runs of the simulator. The designer can see how changes in the OS scheduling affect the performance of the system.

10.5 CONCLUDING REMARKS

In this chapter we presented an innovative mechanism for RTOS emulation inside ISS. In addition to being nonintrusive in the ISS source code, the described techniques are extended for the emulation of real-time tasks. High code equivalence is maintained, enabling fast and accurate simulation of real-time applications. This powerful RTOS emulation mechanism allows the early DSE of the RTOS and its configuration. In particular, it helps the designer determine the best OS parameters (such as scheduling mechanism, tick frequency, etc.) and allows the early evaluation of important performace metrics (e.g., missed deadline count and interrupt response time). Our methodology has been applied to many mixed-workload (real-time and non-real-time) applications and benchmarks, showing how it can be used to analyze their real-time behavior and how the system can be parametrized to obtain the desired performance for both the real-time and non-real-time parts.

REFERENCES

1. Yoo, S., G. Nicolescu, L. Gauthier, and A. A. Jerraya. 2002. "Automatic Generation of Fast Timed Simulation Models for Operating Systems in SoC design." In *Proceedings of the Conference on Design, Automation and Test in Europe* (DATE '02), pp. 620–627. IEEE Computer Society, Washington, DC.
2. Wolf. W., 2003. "A Decade of Hardware/Software Codesign." *Computer* 36(4): 38–43.
3. Moigne, R. L., O. Pasquier, and J. Calvez. 2004. "A Generic RTOS Model for Real-Time Systems Simulation with SystemC." In *Proceedings of the Conference on Design, Automation and Test in Europe*, Volume 3 (DATE '04), Vol. 3, pp. 82–87. IEEE Computer Society, Washington, DC.
4. Huck, E., B. Miramond, and F. Verdier. 2007. "A Modular SystemC RTOS Model for Embedded Services Exploration." In *First European Workshop on Design and Architectures for Signal and Image* Processing *(DASIP)*, ECSI, Grenoble, France.
5 Gerstlauer, A., H. Yu, and D. Gajski. 2003. "RTOS Modeling for System Level Design." In *Proceedings of the Conference on Design, Automation and Test in Europe* (DATE '04), pp. 130–135. IEEE Computer Society, Washington, DC.
6. Posadas, H., J. Adamez, P. Sanchez, E. Villar, and F. Blasco. 2006. "POSIX Modeling in SystemC." In *ASP-DAC '06: Proceedings of the 2006 Asia and South Pacific Design Automation Conference*, pp. 485–90. Yokohama, Japan: IEEE Press.
7. Schirner, G., and R. Domer. 2008. "Introducing Preemptive Scheduling in Abstract RTOS Models Using Result Oriented Modeling." In *Proceedings of the Conference on Design, Automation and Test in Europe* (DATE '08). ACM, New York.
8. He, Z., A. Mok, and C. Peng. 2005. "Timed RTOS Modeling for Embedded System Design." In *Proceedings of the Real Time and Embedded Technology and Applications Symposium* 448–457. IEEE Computer Society, Washington, DC.
9. Honda, S., T. Wakabayashi, H. Tomiyama, and H. Takada. 2004. "RTOS-Centric Hardware/Software Cosimulator for Embedded System Design." In *CODES+ISSS '04: Proceedings of the 2nd IEEE/ACM/IFIP International Conference on Hardware/Software Codesign and System Synthesis*, pp 158–163. IEEE Computer Society, Washington, DC.

10. Girodias, B., E. M. Aboulhamid, and G. Nicolescu. 2006. "A Platform for Refinement of OS Services for Embedded Systems." In *DELTA '06: Proceedings of the Third IEEE International Workshop on Electronic Design, Test and Applications*, pp. 227–236. IEEE Computer Society, Washington, DC.
11. Beltrame, G., L. Fossati, and D. Sciuto. 2009. "ReSP: A Nonintrusive Transaction-Level Reflective MPSoC Simulation Platform for Design Space Exploration." *Computer-Aided Design of Integrated Circuits and Systems, IEEE Transactions on* 28(12): 1857–69.
12. Foote, B. and R. E. Johnson. 1989. "Reflective Facilities in Smalltalk-80." *Proceedings OOPSLA '89, ACM SIGPLAN Notices* 24: 327–35.
13. Kargahi, M. and A. Movaghar. 2005. "Non-Preemptive Earliest-Deadline-First Scheduling Policy: A Performance Study." In *MASCOTS '05: Proceedings of the 13th IEEE International Symposium on Modeling, Analysis, and Simulation of Computer and Telecommunication Systems*, pp. 201–10. IEEE Computer Society, Washington, DC.
14. Buttazzo, G. C. 2005. "Rate Monotonic vs. EDF: Judgment Day." *Real-Time Systems* 29(1): 5–26.
15. eCos operating system. http://ecos.sourceware.org/.
16. Lilliefors, H. 1967. "On the Kolmogorov-Smirnov Test for Normality with Mean and Variance Unknown." *Journal of the American Statistical Association.* Vol. 62, No. 318 (Jun., 1967), pp. 399–402.

11 Service-Based Simulation Framework for Performance Estimation of Embedded Systems

Anders Sejer Tranberg-Hansen and Jan Madsen

CONTENTS

11.1 Introduction 260
    11.1.1 System-Level Performance Estimation 260
    11.1.2 Overview of the Framework 261
    11.1.3 Organization of the Chapter 262
11.2 Service Models 263
11.3 Service Model Interfaces 264
11.4 Service Requests 265
11.5 Service Model Implementations 267
    11.5.1 Model-of-Computation 268
    11.5.2 Composition 270
11.6 Application Modeling 271
11.7 Architecture Modeling 272
11.8 System Modeling 273
11.9 Simulation Engine 274
    11.9.1 Discrete Event Simulation 274
    11.9.2 Representation of Time 275
    11.9.3 Simulation 276
    11.9.4 Service-Based Model-of-Computation for
         Architecture Modeling 277
11.10 Producer–Consumer Example 277
11.11 Industrial Case Study 284
    11.11.1 Mobile Audio Processing Platform 284
    11.11.2 Accuracy 286
    11.11.3 Simulation Speed 286

11.12 Conclusion 287
Acknowledgments 287
References 287

11.1 INTRODUCTION

The advances of the semiconductor industry seen in the last decades have brought the possibility of integrating evermore functionality onto a single chip. These integration possibilities also imply that the design complexity increases and so does the design time and effort. This challenge is widely acknowledged throughout academia and the industry, and to address this, novel frameworks and methods that will both automate design steps and raise the level of abstraction used to design systems are being called upon.

11.1.1 SYSTEM-LEVEL PERFORMANCE ESTIMATION

To allow efficient system-level design, a flexible framework for performance estimation providing fast and accurate estimates is required. Several methods have been presented in recent years allowing performance estimation through formal analysis or simulations of architectures at high levels of abstraction [1–6].

Recently, approaches that rely, at least partly, on formal methods of analysis to allow performance estimation have been presented [4]. In theory, these approaches eliminate the need for simulations to predict performance. However, in most cases, the accuracy of these approaches only justifies their use in the very early stages of the system design phase, where they can be used to reduce the number of potential candidate architectures as is done in Kunzli et al. [4], and the detailed performance estimates are obtainable only through simulation in the later design stages.

The majority of the approaches based on fast simulations, for example [1,3], are using high speed instruction set simulators with high-level modeling of data memories, caches, interconnect structures, etc. They are performing a number of abstractions and thereby trading accuracy for simulation speed. These approaches have their merit, especially in the early design stages. Often, they even allow software developers to start the target-specific software development in parallel with the hardware developers long before low-level register transfer level descriptions of the platform exist or the actual hardware bringup.

The high-level models fulfill the needs for early software development and initial architectural exploration. However, in many cases, one must be able to generate accurate performance estimates to reason about the actual performance of the system so as to verify architectural design choices. To do so, cycle accurate models are required, implying that, currently, register transfer level descriptions of the architectural elements of the target platform are often the only viable solution. The simulation of large-scale systems described at the register transfer level, however, suffers from tremendous slowdown in the simulation speed compared to the high-level simulations. Even worse, the development of such detailed descriptions is long and costly, which implies that when these are finally available, often at a very late stage of the development phase, changes of the architecture are very difficult to incorporate, resulting in limited opportunities for design space exploration.

Thus, there exists a gap between the fast semiaccurate methods, which are highly useful in modern design flows, allowing the construction of high-level virtual platforms, in which rough estimates of the performance of the system can be generated, and the detailed and very accurate estimates that can be produced through register transfer level simulations.

This chapter introduces a compositional framework for system-level performance estimation, first presented in Tranberg-Hansen et al. [7], for use in the design space exploration of heterogeneous embedded systems. The framework is simulation-based and allows performance estimation to be carried out throughout all design phases, ranging from early functional to cycle accurate and bit true descriptions of the system. The key strengths of the framework are the flexibility and refinement possibilities and the possibility of having components described at different levels of abstraction to coexist and communicate within the same model instance. This is achieved by separating the specification of functionality, communication, cost, and implementation (which resembles the ideas advocated in Keutzer et al. [8]) and by using an interface-based approach combined with the use of abstract communication channels. The interface-based approach implies that component models can be seamlessly interchanged. This enables one to investigate different implementations, possibly described at different levels of abstraction, constrained only by the requirement that the same interface must be implemented. Additionally, the use of component models allows the construction of component libraries, with a high degree of reusability as a result.

11.1.2 OVERVIEW OF THE FRAMEWORK

The framework presented, illustrated in Figure 11.1, is related to what is known as the Y-chart approach [9,10]. However, in our case, the application model is refined in its own iteration branch as step one, verifying the functionality of the application model only. Once this step is completed, the application model is left unchanged and only the mapping and platform model are being refined in step two. A need to change the application model implies that the functionality of the application has changed. Hence, step one must be redone to verify the new functionality before repeating step two.

A key concept in the framework is the notion of service, which plays an important role to achieve a decoupling of the specification of functionality, communication, cost, and implementation. A service is defined as a logical abstraction that represents a specific functionality or a set of functionalities offered by a component. In this way, services are used to abstract away the implementation details of the functionality that is offered by the component. Thus, the service abstraction allows two different models to offer the same services, having the same functional behavior but with a different implementation, cost, and/or latency associated. Consequently, different implementations of a model can be investigated easily.

The functionality of the target application is captured by an *application model*. Application models are composed of a number of tasks, each represented by a service model. The tasks serve as a functional specification of the application only specifying a partial order of service requests necessary to preserve the functionality to capture the functionality of the application. No assumptions on who will provide the required services are made, thus separating the specification of functionality and implementation.

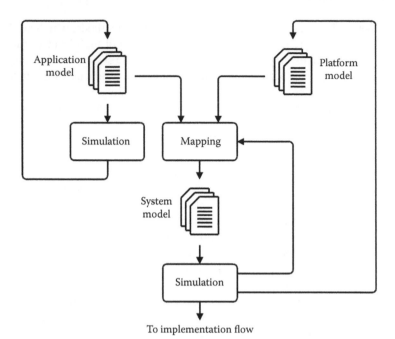

FIGURE 11.1 Overview of the presented system-level performance estimation framework.

The target architecture is modeled by a *platform model* that is composed of one or more service models. The service models of a platform model can be described at arbitrary levels of abstraction. In one extreme, they only associate a cost with the execution of a service, while on the other end of the spectrum, the service request is modeled in the platform model both cycle accurate and bit true. Costs can be associated with service requests irrespective of whether they are computed dynamically or precomputed. It is the cost of the execution of a service that differentiates various implementations of the particular service.

Quantitative performance estimation is performed at the system level through the simulation of a *system model*. A system model is constructed through an explicit mapping of the components of an application model onto the components of a platform model. The components of an application model, when executed, request the services offered by the component onto which they are mapped. This models the execution of the requested functionality, taking the implementation-specific details and required resources into consideration and associates a cost with each service requested. In this way, it becomes possible to associate a quantitative measure with a given system model and, hence, it becomes possible to compare systems and select the best-suited one from well-defined criteria.

11.1.3 ORGANIZATION OF THE CHAPTER

The remaining part of this chapter is organized as follows: First an introduction to the modeling framework and its components is given followed by a small example of its use. Then, an extract of an industrial case study performed is presented and finally conclusions are given.

11.2 SERVICE MODELS

A service model captures the functional behavior of a component through the notion of services. Services are used to represent the functionality offered by a given component without making any assumptions about the actual implementation. Concrete examples of services are functions of a software library, arithmetic operations offered by a hardware functional unit, or the instructions offered by a processor depending on the level of abstraction used to model the component.

The functionality offered by a service model, represented as services, can be requested by other service models through requests to the offered services. The detailed operation of a service model, that is, the implementation of the services offered, is hidden to other models. In that sense, a service model can be viewed as a black box component where the services offered are visible but not their implementation. Service models can be described at multiple levels of abstraction and need not be described at the same level to communicate. The clear separation of functionality and implementation through the concept of services implies that there is no distinction between hardware or software components seen from the point of view of the modeling framework. It is the cost associated with each service, which eventually dictates whether a component is modeling a hardware or software component.

A service model is composed of one or more *service model interfaces* and a *service model implementation* as illustrated in Figure 11.2. Interfaces are used to connect models, allowing models to communicate through the exchange of service requests. In this way, there is a clear separation of how behavior and communication of a component is described. The service model interfaces provide a uniform way of accessing the services offered and enabling service models to communicate and facilitate structural composition by specifying the sets of services that are offered and are required by the model. It is the active interfaces that specify required services and the passive interfaces that specify the offered services. In this way, structural composition is

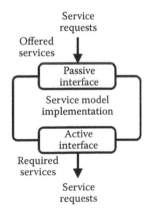

FIGURE 11.2   The service model basics.

dictated by the interfaces that each service model implements and subsequently by the services offered and required. Only models fulfilling the required–provided service constraints can be connected. Although the interfaces dictate the rules of composition, the service model implementation is responsible for specifying the actual behavior of the service model. The specification implements the functionality of each of the services that are offered while optionally associating a cost with them.

The use of services allows a decoupling of the functionality and of the implementation of a component. In this fashion, several service models can offer the same service, implemented differently, however, and thus having different associated cost. In this way different implementations can be compared and evaluated based on a specific, preferred cost metric.

Intermodel communication is handled through service requests that have the benefit of allowing the initiator model to request one of the services specified, as required by the model, without knowing any details about the model that provides the required service or how it is implemented. When the service is requested, the service model providing the required service will execute the requested service according to the specification of the model. When the service has been executed, the initiator model is notified that the execution of the service requested has completed.

11.3   SERVICE MODEL INTERFACES

To facilitate structural composition of models and to allow communication between models, possibly described at different levels of abstraction, interfaces are used. The interfaces directly impose the rules of composition that specify how models can be connected by allowing only interfaces fulfilling the required–provided service constraints to be connected.

The use of interfaces also implies that multiple implementations of a service model can be constructed and be seamlessly interchanged, allowing different implementations to be investigated and described at different levels of abstraction, constrained only by the requirement that the service model implementations considered must implement the same interfaces.

Two types of interfaces are defined:

- The *passive service model interface* specifies a set of services that the model implementing the interfaces offers to other models. The passive interface also includes structural elements allowing the interface to be connected to an active interface and provide means for requesting the services offered.
- The *active service model interface* specifies a set of services that are required to be available for the model implementing the interface. The set of required services becomes available for the model that implements the active interface, when the active interface is connected to the passive interface of a service model, which offers a set of services in which the required set of services is a subset.

Active service model interfaces can only be connected to passive service model interfaces in which the set of services required by the active service model interface is a subset of the services offered by the passive service model interface. In essence, the composition rules, which specify which active service model interfaces can be connected to which passive service model interfaces, is dictated by the services required and services offered by the two connecting interfaces.

11.4 SERVICE REQUESTS

As mentioned briefly before, all intermodel communication is modeled using service requests exchanged via active–passive interface connections.

A service request can be viewed as a communication transaction between two components, similar to the concept of transaction-level modeling [11]. However, it is the implementation of the two service models that will determine if the service request will be implemented as a function call from, for example, a sequential executing piece of code to a function library or if it will be a bus transfer. Communication refinement is supported in several ways. Service requests can include arbitrary data structures as preferred by the designer of the model. If a communication channel must be modeled, an extra service model can be inserted between the two primary communicating service models as illustrated in Figure 11.3. The extra service model

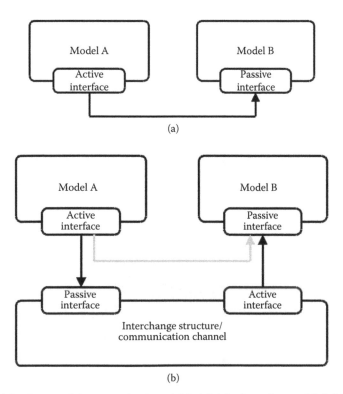

(a)

(b)

FIGURE 11.3 Intermodel communication. a) Model A is the active model, initiating communication with model B. b) The communication medium is now modeled explicitly using a separate service model.

inserted will be transparent to the two communicating service models. This allows, for example, simple properties such as reliability of the communication channel to be modeled. More elaborate communication interconnects such as buses and network-on-chips can also be modeled.

A service request specifies the requested service, a list of arguments (which can be empty), and a unique request number used to identify the service request, for example, to annotate it with a cost. The argument list can be used to provide input arguments to the implementation of a service, for example, to allow modeling of dynamic dependencies or arithmetic operations on actual data values. Depending on the implementation of the service model, an arbitrary number of service requests can be processed in parallel, for example, modeling operating system schedulers, pipelines, very long instruction word, single instruction multiple data, and super scalar architectures.

A service request can be requested as either blocking or nonblocking. It is the designer of a model who determines whether a service request is requested as a blocking or a nonblocking request. The request of a blocking service request implies that the process of the source model that requested the service request is put into its blocked state until it has been executed, indicated by the destination model, as illustrated in Figure 11.4. A nonblocking service request, on the other hand, will be requested and the process of the source model, which requested the service request, will proceed—not waiting for the execution of the request to finish.

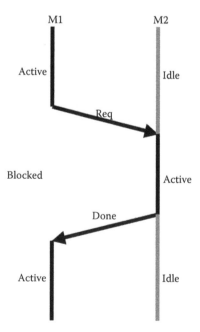

FIGURE 11.4 A process of model $M1$ requests a blocking service request from $M2$. $M1$ is blocked till the completion of the request.

A number of events are associated with a service request to notify the requester and receiver model of different phases of the lifetime of the service request. The lifetime and corresponding events of a service request are as follows:

1. The service request is being requested, indicated by a service request requested event.
2. The service request is being accepted for processing of the model by which it is requested, indicated by a service request accepted event.
3. The service request may be blocked, indicated by a service request blocked event.
4. The service request has been executed, indicated by a service request done event.

When a service request is being requested at a model interface, the receiver model has the possibility of receiving a notification to change its status to active. The requesting service model will similarly have the possibility of being notified when the request is accepted for processing in the receiver model (i.e., before the actual execution of the service request) and when it has finished executing the service request. During the evaluation of a service request, the request itself can become blocked because of one or more requirements not being fulfilled (e.g., because of mutually exclusive access to resources and missing availability of data operands). When a service request is being blocked during evaluation, the source model is notified to allow it to take appropriate actions, if any. However, the author of the requesting service model need not be interested in receiving these notifications and, hence, the model is allowed to ignore these. In this way, it is the designer who chooses event sensitivity for the individual processes of the service model implementation.

To handle multiple simultaneous service requests, the designer of a service model must incorporate a desired arbitration scheme. The arbitration scheme may be an integrated part of the model or it may be a separate service model itself. The latter is often advantageous if different arbitration schemes are to be investigated.

11.5   SERVICE MODEL IMPLEMENTATIONS

To capture the behavior of a service model, the service model implementation must be defined. It is the service model implementation that captures the actual behavior of a service, possibly taking implementation-specific details into consideration. A service model must provide an implementation of all services offered by the passive interfaces implemented and optionally specify the latency, resource requirements, and cost of each service. There are no restrictions on how a service should be implemented. The abstract representation of the functionality, using services, implies that there is no immediate distinction between the representations of the hardware or software components. Whether a service model represents a hardware or software component is determined solely by its implementation and, eventually, its cost and thus an elegant unified modeling approach can be achieved.

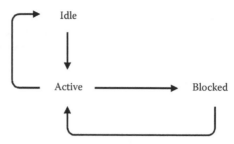

FIGURE 11.5   The possible states of a process of a service model.

11.5.1   MODEL-OF-COMPUTATION

To capture the implementation of a service model, concurrently executing processes are used as the general execution semantics. The service model implementation can contain one or more processes that each have the possibility of executing concurrently. A process executes sequentially and interprocess communication within a service model is done through events communicated via channels in which the order of events is preserved. All intermodel communication between processes residing in different service models is done using service requests via the service model interfaces defined by the model in which the processes reside.

A process can be in one of three states: idle, active, or blocked as shown in Figure 11.5, which also shows the valid transitions between the possible states. If a process is idle, it indicates that it is inactive but ready for execution upon activation. If a process is active, it is currently executing. If a process is blocked, it is currently waiting for a condition to become true and will not resume execution until this condition has been fulfilled.

To overcome the problem of finding a single golden model-of-computation for capturing all parts of an embedded system, the service model concept supports the existence of multiple different models-of-computation within the same model instance. Figure 11.6 shows an example of a system composed of service models, each described by a different model-of-computation. The service model concept allows these to coexist and communicate through well-defined communication semantics in the form of service requests being exchanged via active–passive interface connections.

Interesting work on supporting multiple different models-of-computation within a single model instance is presented in the theoretically well-founded tagged signal model [12] and the absent-event approach [13]. Both approaches show that it is possible to allow models-of-computation, defined within different domains, to be coupled together and allowed to coexist within the same model instance. In principle, the service model concept does not impose a particular model-of-computation. The individual models can be described using any preferred model-of-computation as long as communication between models is performed using service requests.

Currently, the focus of the framework is the modeling of the discrete elements of an embedded system only, that is, hardware and software parts, which can be represented by untimed, synchronous, or discrete event–based models-of-computation.

To synchronize models described using different models-of-computation and ensure a correct execution order, the underlying simulation engine is assumed to

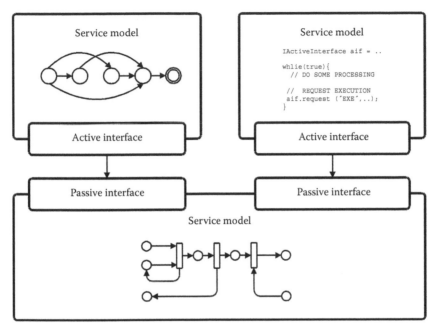

FIGURE 11.6 The service model concept provides support for heterogeneous models-of-computation to coexist.

be based on a global notion of time that is distributed to all processes, no matter which model-of-computation is used. This contrasts with both the tagged signal model and the absent event approach. The former distributes time to the different processes through events, while the latter relies on the special absent event. The drawback of using a global notion of time is that processes cannot execute independently, which impacts simulation performance—it is very hard to parallelize such a simulation engine—the advantage, however, is that great expressiveness can be obtained in such a simulation engine. The simulation engine also tags all events with a simulation time value so that the event can be related to a particular point of simulation time, no matter which model-of-computation was used to describe the process that generated the event. However, this does not mean the individual service models must use this time tag and it is merely a practical requirement to schedule the execution order of the processes of the individual service models.

Service models that have processes described using untimed models-of-computation obviously have no notion of time and perform computation and communication in zero time. This implies that a process of a service model that is described using an untimed model-of-computation is activated on the request of services offered by the model only (and not based on a specific point in time). The corresponding process then evaluates and produces possible outgoing service requests immediately.

Service models having processes described using synchronous models-of-computation do not use an explicit notion of time. Instead, a notion of time slots is used and each execution cycle lasts one time slot. In order for such models-of-computation to be used, the service model using a model-of-computation within this domain must specify the frequency of how often the processes of the model should be allowed to execute. The simulation engine will then ensure that the processes are evaluated at the specified frequency, in this way implicitly defining the actual time of the current time slot of the model.

Discrete-time models-of-computation are supported directly by the simulation engine, which provides a global notion of time that can be accessed from all processes. In this way, a process can describe behaviors that use timing information directly.

The generality of service models imposes few restrictions on the model-of-computation used to capture the behavior of the component being modeled. New models-of-computation can be added freely under the constraint that they *must* implement intermodel communication through the exchange of service requests and they must fit under the general execution semantics defined—that is, it must be possible to implement the preferred model-of-computation as one or more concurrently executing processes. It is the implementation of the service models that determines their actual behavior, and thus, it is the designer of the service model implementation who determines the model-of-computation used.

11.5.2 COMPOSITION

To tackle complexity, the service model concept supports both hierarchy and abstraction-level refinement. Here, the term abstraction-level refinement covers the process of going from a high level of abstraction to a lower level through gradual refinements of a given component, where a component may be replaced by a more detailed version (Figure 11.7).

This type of refinement is supported quite easily by the service model concept because of the fundamental property of the service model in which the functionality offered by a model is separated from the implementation. Two service models implementing the same set of interfaces, and thus offering the same set of services, can be freely interchanged, even though they differ in the level of detail used to model the functionality offered.

Furthermore, service models can be constructed hierarchically to investigate different implementations of a specific subpart of the model or to hide model complexity. One service model can be composed of several subservice models. However, it will then be only the interfaces implemented by the topmost model in the hierarchy that dictate which services are offered to other models. The hierarchical properties, combined with the use of interfaces, imply that designers who are using a model need only know the details of the interfaces implemented by the model and need not be concerned about the implementation details at lower levels in the model hierarchy.

To summarize, service models can be viewed as black-box components. The behavior of a service model is determined by the services requested via its active interfaces. The use of interfaces and service request, on the one hand, implies that there are no restrictions on the service model implementation, which is part of the service model that actually determines the behavior. This also implies that in principle there are no restrictions on which model-of-computation is used to describe the implementation of a service model.

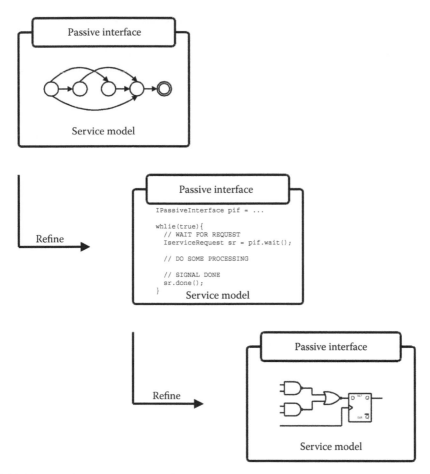

FIGURE 11.7 Abstraction-level refinement.

11.6 APPLICATION MODELING

Applications are represented by *application models* that are composed of an arbitrary number of components that can be executed in parallel. These executable components are referred to as tasks, each represented by a service model. Application models are used to capture the functional behavior and communication requirements of the application only. The tasks of an application model communicate through the exchange of service requests making communication explicit and implementation independent.

In the general case, a strict separation of the functional behavior, communication requirements, and implementation of an application must be applied in order for the application to be platform-independent. Thus, no assumptions on how the application is implemented should be made. However, in some cases, it may be desirable to include platform-specific information in the application model (e.g., in case an existing platform is modified) and, thus, support for including implementation-specific details in the application model is provided. However, including such detail will reduce the number of platforms onto which the application model can be mapped.

Application models can be executed and used for verifying the functional behavior of the model. However, at the application level of abstraction, there is no notion of time, resources, or other quantitative performance estimates. To obtain these, the service models of the application model must be mapped onto the service models of a platform model. When the tasks of an application model are mapped onto the processing elements of a platform model, the tasks, when executed, can request the services offered by the processing elements, modeling the execution of a particular functionality or set of functionalities.

11.7 ARCHITECTURE MODELING

In contrast to the application models, the goal of the platform model is to capture a specific implementation of the functionality offered by the target architecture. In order for a platform model to be valid, it must offer *all* the services required by the application models that are mapped onto the platform. Several platform models can offer the same set of services, representing the same functionality, through different implementations. The differentiating factor will then be the cost associated with the execution of the applications on each of the candidate platforms.

The platform model of a given target architecture is implemented as a service model having one or more passive service model interfaces, as illustrated in Figure 11.8. Platform models are composed of an arbitrary number of service models, each modeling a component of the architecture, thereby forming a hierarchical model. The compositional properties of service models even allow multiple

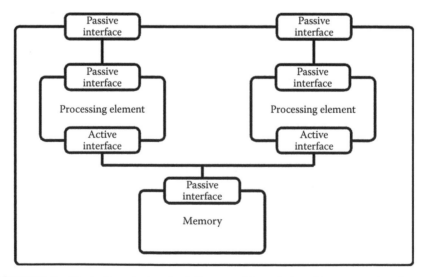

FIGURE 11.8 Illustration of a simple platform model consisting of two processing elements both connected to a block of shared memory.

platform models to be merged into a single platform model. In this way, a modular approach can be taken in which subblocks of the target architecture are modeled and explored individually, if preferred.

The resulting set of services offered by a platform model is dictated by the composition of internal service models. From the tasks of the application models, which are mapped to the platform, the services of the platform model are accessible through the passive service model iinterfaces. This allows the tasks to request the services offered during simulation and so models the execution of the tasks.

The platform model also specifies how the service models, of which it is composed, are interconnected, thereby specifying the communication possibilities of the models. There are restrictions neither on how intercomponent communication is modeled, nor on the level of abstraction that is used. This implies that, in principle, all types of intercomponent communication methods are supported. In this way, platform models can represent arbitrary target architectures.

11.8   SYSTEM MODELING

A system model is constructed by mapping the service models of one or more application models onto the service models of a platform model. If multiple service models from an application model are mapped onto the same service model of a platform model, a scheduler must be provided as part of the platform model. Schedulers are implemented as separate service models, for example, acting as abstract operating systems.

An application model is mapped onto a platform model by specifying how the active service model interfaces of an application model should be connected to the passive service model interfaces of the platform model. Mappings are specified in a separate XML file, which allows different mappings to be easily investigated without changing the actual model descriptions.

The service models of the platform model allow designers to associate quantitative cost with the execution of an application model. Without a platform model, the service models of an application model simply execute in zero time and have no cost associated. When the service models of an application model are mapped onto the service models of a platform model, it becomes possible to capture resource requirements, quantitative cost, and associate a time measure with the execution of the application model on the target platform. In this way, it is the platform model that introduces a notion of time into the simulations. Thus, the platform model orchestrates the execution of a given application model with respect to the timing and resource requirements defined by the individual parts of the application model that are executing on the platform model.

A given mapping is required to be valid in order for the resulting system model to be used for performance estimation through simulations. A mapping is said to be valid if and only if all the requested services of a given application model are offered by the processing elements of the platform model onto which it is mapped. However, it is *not* a requirement that all service models of an application model are mapped to the platform model. In the case where a service model of an application model is unmapped, only the functional behavior of the service model is modeled and no performance estimates are associated with that particular task. Currently, the validity of a given mapping is determined during runtime only, resulting in an error during simulation in the case of an invalid mapping. In the future, however, tools for performing such checks before runtime will be constructed. Also, future work will include the possibility of performing checks for the semantic validity of specific mappings to ensure, for example, that no deadlocks occur in a given mapping.

11.9 SIMULATION ENGINE

This section describes the simulation engine used for simulation of the models constructed in the presented system-level performance estimation framework with the objective to obtain quantitative performance estimates.

11.9.1 Discrete Event Simulation

To support multiple different models-of-computation, a discrete event simulation engine has been chosen for coordinating the execution of the individual service models and their internal processes. Discrete event modeling is used within a range of different application domains and described thoroughly in [14,15].

A custom simulation engine for prototype use supporting the modeling framework presented has been implemented during the course of this project. The reason for implementing a custom simulation was to obtain maximum control over the simulation kernel in order to be able to validate the ideas of the framework. The implemented discrete event simulation kernel keeps track of simulation time and schedules the execution of the individual processes of the service models according to their sensitivity to events.

As already described in Section 11.4, a number of events are associated with the lifetime of a service request, which by their occurrence can trigger processes sensitive to the event. The occurrence of a given event allows a waiting process to become activated at different phases of the simulation as described by the designer of the model and depending on the desired behavior.

The simulation engine provides basic semantics for expressing the behavior of a service model through a process-like behavior for modeling concurrency and a number of waitFor statements as listed in Table 11.1, which will cause the execution of the model to block.

waitFor(time) causes the process to block until the specified time has passed. The effect is that an event is being scheduled with a time tag of the current simulation time plus the specified time value. The simulation engine will fire the event when the specified simulation time has been reached and the process will resume its execution.

TABLE 11.1

Process Wait Types

waitFor(time)
waitFor(service request, event type)
waitFor(interface, service request type, event type)
waitFor(event)

waitFor(service request, event type) causes the process to block till the specified event type of the specific service request instance is being fired. When this occurs, the process will resume its execution and the time tag of the event will be decided by the simulation engine.

waitFor(interface, service request type, event type) causes the process to block till the specified service request type and event type is being fired at the specified service model interface. In this case, the simulation time will also be annotated by the simulation engine when the event is fired.

waitFor(event) causes the process to block till the specified event is fired. When the event is fired, the process will be notified and continue its execution.

11.9.2   REPRESENTATION OF TIME

The simulation engine is implemented as a discrete event simulation engine using a delta-delay based representation of time [13]. The delta-delay based representation divides time into a two-level structure: regular time and delta-time. Between every regular time interval, there is a potentially infinite number of delta-time points $t + \delta$, $t + 2\delta$, ... Each event is marked with a time tag that holds a simulation time value, indicating when the event is to occur. Every time an event is fired and new events are generated having the same regular time value as the current time value of the simulation (e.g., in case of a feedback loop), the new event will have a time stamp with the same regular time value but now with a delta value incremented by one. The use of delta-delays ensures that no computations can take place in zero-time, but will always experience at minimum a delta-delay. The delta-delay based representation is making simulations deterministic because the use of delta-time makes it possible to distinguish between two events generated at the same point in time, which is to be processed first by looking at their delta-value.

Application models, on the one hand, have no notion of time. It is only when the application model is mapped onto a platform model that it becomes possible to annotate the execution time of the application model by relating the execution of its tasks and the generation of service requests to discrete time instances.

In the platform model, on the other hand, time is represented explicitly using the delta-based representation of time. Each model of the platform can be modeled with arbitrary delays or specify a clock frequency at which they want to be evaluated. In this way, it is possible to model synchronous components in the platform model, which are activated only at regular discrete time instances. Thus, the tasks of the application model are blocking while the service requests are being processed in the platform model, which enables associating an execution time metric with tasks. Similarly, it is possible to annotate tasks with other types of metrics such as power costs, etc.

A special event *type* is used to represent hardware clocks used for example by models described using clocked synchronous models-of-computation. These models do not use time explicitly, instead they represent time as a cycle count. To use synchronous models of computation in the currently implemented simulation engine, these models must specify a clock frequency that determines how often they are to be evaluated. Regular events are removed from the event list, executed, and then disposed. Because events are bound to a unique time instance, they can thus occur only once. The special event type used to represent hardware clocks, however, is implemented as a reschedulable event in the simulation engine, which takes care of handling the uniqueness of each instance of this event type. Such a clock event is automatically rescheduled and reinserted into the event list. Each clock event object has a list of active processes that are to be evaluated when the clock ticks. This list is updated dynamically during simulation, and in case a clock has no active processes in the active list, the clock itself is removed from the pending list of events to increase simulation performance. When a clock event object is inserted into the event list, it will be sorted according to the simulation time of the next clock tick of the clock event object. It is also possible to have a clock object that contains no static period. In this case, the clock ticks can be specified as random time points or as a list of periods that can be used once or repeated.

Of course, the platform model can also contain service models that are activated on the arrival of service requests only triggered by the event associated with the request of the service. In these cases, the occurrence of such an event, at a specific simulation time, will activate the blocking process. This blocking allows modeling of, for example, propagation delays associated with combinational hardware blocks, etc. The modeling of such elements can also be handled in zero regular time by means of the delta-delay mechanism.

11.9.3 SIMULATION

The simulation engine uses two event lists for controlling the simulation: one for delta events and one for regular events. The delta event list, on the one hand, contains only events with a time tag equal to the current simulation time plus one delta cycle. The regular event list, on the other hand, contains pending events with a time tag in which the regular time is greater than the current simulation time. The regular event list is sorted according to the time stamp of the events in increasing order. In this way, the head of the regular event list always points to the event with the lowest time tag.

The simulation engine always checks the delta event list first. If the list is not empty, the delta cycle count is incremented and all events contained in the delta event list are fired one by one till the list is empty, implying that all events belonging to the same delta cycle have been fired.

If no delta events are pending (i.e., the delta event list is empty), the simulation engine removes the first event in the regular event list and advances the simulation time to the time specified by the time tag of the event. Also, the current delta count is cleared and the event is then fired.

Delta-delay based discrete simulation engines suffer the risk of entering infinite loops where regular time is not advanced and only the delta cycle is incremented. A naive approach to handle this is implemented, allowing the designer to specify a maximum number of delta cycles to be allowed before the simulation engine quits the simulation.

The firing of an event can, of course, cause new events to be generated and scheduled in the simulation engine. If a new event is generated having its time tag set to the same regular time value as the current simulation time, it is added to the delta event list. Otherwise, it is scheduled according to its time tag and inserted into the regular event list according to the time tag specified. The delta event list contains only events with a time tag equal to the current simulation time plus one delta cycle.

After an event has been fired, it is checked if the event type of the event is periodic. Events belonging to a periodic event type (e.g., the special clock event object described in Section 11.9.2) are then automatically rescheduled and inserted into the event list at the correct position.

11.9.4 SERVICE-BASED MODEL-OF-COMPUTATION FOR ARCHITECTURE MODELING

One example of a custom model-of-computation for describing service models is the model-of-computation developed for modeling synchronous hardware components presented in [16,17], which is based on Hierarchical Colored Petri Nets (HCPNs) [18]. HCPNs have been selected because of the great modeling capabilities with respect to concurrency and resource access and the compositional properties that match the requirements of service models well. Traditional HCPNs, however, suffer from a number of inadequacies with respect to model complexity and especially the obtainable simulation speed. These issues are addressed by, for example, defining special execution semantics. Details of the construction, simulation, and workings of service models based on HCPNs can be found in [16,17].

11.10 PRODUCER–CONSUMER EXAMPLE

To illustrate the use of the framework and elaborate on the different elements, a simple producer–consumer application is considered in the following.

The current implementation of the framework is in Java [19] and models are specified directly in Java as well through the use of a number of libraries developed for supporting the concept of service models, application models, platform models, and system models. In addition to the libraries, a simple graphical user interface has been implemented. The graphical user interface is built on the Eclipse [20] platform. It allows simulations to be controlled and inspected in a more convenient way than pure text-based simulations.

The first step to start using the presented framework is to construct an application model. The application model captures the functional behavior of the application in a number of tasks. It also specifies the communication requirements of the individual tasks explicitly, without any assumptions on the implementation, following the principle, on which the framework is founded, of separating the specification of functionality, communication, cost, and implementation. The application model serves as the functional reference in the refinement steps toward the final implementation. However, at this application level of abstraction, there is no notion of time or physical resources—hence only very rough performance estimates can be generated from a profiling of the application model.

Figure 11.9 shows an application model composed of a consumer task and a producer task, which communicate through an abstract buffer. Both the tasks and the abstract buffer are modeled by service models. The producer has an active interface that is connected to the passive write interface of the buffer offering a write service.

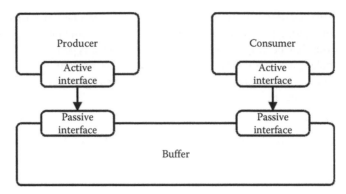

FIGURE 11.9   A simple application model consisting of a producer and a consumer communicating through an abstract buffer.

Similarly, the consumer has an active interface that is connected to the passive read interface of the buffer model offering a read service. The producer and consumer models are the only active models, that is, only the producer and consumer models can initiate the request of services.

The service models are specified as one or more concurrently executing processes. Processes execute till they are blocked, waiting for some condition to become true. When the condition becomes true, the process can continue. Such behavior can be implemented using threads. The thread-context switching required, each time a process is being blocked or activated, has a high impact on simulation performance. Because we currently employ a global notion of time in the discrete event simulation engine, most processes will execute in a lock-step fashion. As a consequence, blocking and unblocking require frequent context switches. At the same time, threads provide more functionality than actually necessary now that only one process can be active at a time in the *physical* simulation engine. Thus, the desired behavior can be implemented in the much simpler concept of coroutines, which allows execution to be stopped and continued from the point where it was stopped. The current implementation of the simulation kernel thus uses a concept very similar to coroutines implemented in Java [19]. This requires that the designer of a process must divide the code body of the process into blocks. Blocks are executing sequentially and the execution of a block cannot be stopped. When the code contained in a block completes execution, the block returns a reference to the next block to be executed. In this way it is possible to model the blocking of a process on some condition and then when the condition becomes true, the process will resume its execution by executing the block of code returned by the previous executing block. Listing 11.1 shows the description of the main body of the producer service model.

In this case a single process is used to capture the behavior of the producer. A similar description is made for the consumer which, however, is not shown. The main body of the producer is actually an infinite loop in which the producer first calculates some value, then instantiates a write service request with the calculated value as argument and then requests the write service via an active service model interface. It is possible to request a service and then continue the execution directly,

```
...
private IActiveServiceModelInterface fWriteInterface;
...
public final IBlock execute() {
  // Calculate data
  ...
  // Create write service request
  IServiceRequest sr = fWriteInterface.createServiceRequest (
    "WRITE", new Object[]{...});
  // (Request write i. e. production of data)
  fWriteInterface.request(sr);
  // (Wait for write request to be done)
  waitFor(sr, IService Request.EventType.DONE);
  ...
}
```

LISTING 11.1 Body of the producer service model main process.

however. In such a case, a waitFor statement is used to block the producer service model until the requested write service has been executed, signaled by the firing of a service-done event. When the service-done event is fired, the producer will be activated and rescheduled for execution in the simulation engine. The producer will then resume execution continuing from the point at which it was blocked.

As can be seen, the write service request is used for both signaling the request of a write service and for transporting the actual data values that are to be exchanged between the producer and buffer service model and that will be written into the buffer, eventually. In this way, arbitrary data and objects can be transferred.

The buffer model is activated only when a service request is requested through one of its passive interfaces. In such a case, the behavior of the read and write services depends on the implementation of the buffer. So, by simply interchanging the buffer model, different blocking or nonblocking read and write schemes can be conveniently investigated.

As an example, the buffer service model described in Listing 11.2 uses a blocking write policy.

Again, the main body is actually executing an infinite loop. The main body starts executing a waitFor statement, blocking the execution till the arrival of a write service request, indicated by the firing of a service request requested event of type write. In this case, the waitFor statement is not instance-sensitive as in the case of the producer, which was blocking for a service request done event on the instance of the requested write service request. Instead, it is blocking until a service request requested event of the specified type is fired, indicating a request through the active service model interface. This implies that whenever a service request requested event of type write is fired, the write process of the buffer service model will be scheduled for execution in the simulation engine and the write process will become active and continue its execution from the point at which it halted. It then starts the actual execution of the requested write service. If the buffer is already full, the write process will block once more until an empty slot in the buffer becomes available. Otherwise, the actual write

```
...
private IPassiveServiceModelInterface fWriteInterface;
private IserviceRequest fWriteServiceRequest;
public final IBlock execute() {
  // (Wait till a write service is requested)
  waitFor (fWriteInterface, "WRITE",
    IServiceRequest.EventType.REQUESTED);
  ...
  if (fFull) {
    // Wait till notified
    waitFor();
    ....
  }
  else{
    // Perform write
    ....
    // Signal done
    fWriteServiceRequest.done();
    ....
  }
}
```

LISTING 11.2 Main body of the write process of the buffer service model.

to the buffer will be executed and a service request done event will be fired to notify the requester, in this case the producer, that the service request has been executed. A read process that implements a blocking read policy can be described analogously.

To generate quantitative performance estimates, the task and buffer service models of the application model must be mapped to the service models of a platform model, which creates a system model. Performance estimates relevant for evaluating the different platform options can then be extracted from the simulation of the system model. When the service models of an application model are mapped to the service models of a platform model, the service models of the application model can, when executed, request the services offered by the platform service models. In this way, the functionality of, for example, a task, is represented by an arbitrary number of requests to services. When these requests are executed, they model the execution of a particular operation or set of operations. The execution of a service in the platform model can include the modeling of required resource accesses and latency only, or, depending on the level of abstraction used to describe the service model, even include the actual functionality including bit true operations.

The mapping of the service models of the application model to the service models of the platform model need not be complete, that is, it is allowed that only a subset of the service models of the application model are mapped.

In this case, the unmapped service models of the application model represent all functionality (i.e., both the control flow and data operations within the application model) and have no costs associated. In Figure 11.10, the consumer service model is an example of an unmapped service model.

On the other hand, the service model of the producer in Figure 11.10 is an example of a mapped task. As a first refinement step, the service model of the platform model,

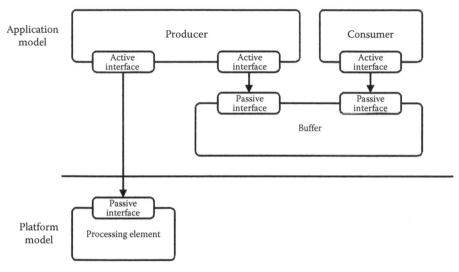

FIGURE 11.10   Only the producer task is mapped to the processing element of the platform model. The consumer task is unmapped and modeled functionally only.

```
IActiveServiceModelInterface fExecute;
...
IServiceRequest sr = fExecute.createServiceRequest
(
   "PRODUCE", new Object[] {...});
fExecute.request(sr);
..
```

LISTING 11.3   The refined production of data values, now depending on the platform model mapping.

onto which the producer service model is mapped, is only used for latency modeling and for associating a cost with the execution of the producer. This is achieved by modeling a processing element that offers the service produce through a passive service model interface.

The producer service model, as shown in Listing 11.3, then requests the produce service via its active service model interface during execution of the produce service, modeling a particular implementation in terms of cost and latency. In this example, the produce service is a very abstract service. In general, it is up to the designer to determine the level of abstraction used to associate with each service. One could also imagine the specification of required services for each arithmetic operation required to calculate the produced value. Currently, required services are specified manually, and to have a valid mapping of a service model from the application model onto the service model in a platform model, all required services must be provided by the service model of the platform model. The higher level of abstraction used to specify required services also implies that there are more options for mapping a model because the separation of functionality and implementation is retained.

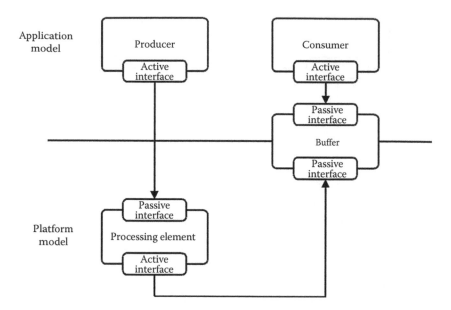

FIGURE 11.11 Only the producer task is mapped onto a processing element of the platform model. The consumer task is unmapped and modeled functionally only. Similarly, the FIFO buffer is only partially mapped, allowing it to be accessed from the active interface of the processing element in the platform model and at the same time be accessible from the active interface of the consumer task in the application model.

If preferred, it may also be possible to refine the processing element to include the calculation of actual data values. In this case, part of the behavior of the producer is modeled according to the functional specification in the application model, while another part is modeled according to a particular implementation in the platform model. As a result, different levels of abstractions are mixed seamlessly. This is particularly useful in the early stages of the design process where rough models of the platform may be constructed. In such a scenario, fast estimations of the effect of adding redundant hardware support for specific operations or even rough estimates of the effect of using multiprocessor systems, the effect of buffer sizes, etc. can be explored. This scenario may be refined to a level where cycle accurate and bit true models are described, but still leaving the control flow of the application to be handled in the application model and modeling only the cost of control operations in the platform model.

Such an example is seen in Figure 11.11, where the producer task could be implementing, for example, loop control in the application model directly. At the same time, it only implements part of the functionality of the loop body through requests to services offered by the service model of the processing element in the platform model onto which it is mapped. Figure 11.11 also shows how it is possible for the producer, mapped to and executing on the processing element, to communicate through a partially mapped buffer with the functional consumer service model, which is modeled only in the application model.

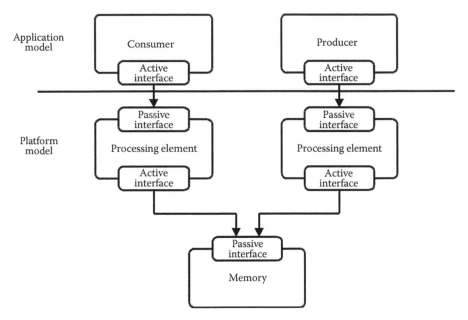

FIGURE 11.12   Both the producer and consumer tasks are now mapped onto processing elements of the platform model. The FIFO buffer is mapped to the service model, modeling a block of shared memory to which both processing elements have access.

Finally, as shown in Figure 11.12, it is possible to model a full mapping of an application model onto a platform model, including both the active service models, in this case the producer and consumer running on one or more processing elements, and the passive service models, in this case the buffer mapped onto a block of memory. Still, it is possible to mix partially functional behavior modeled in the application model with the actual behavior of the implementation modeled in the platform model. In the most extreme case, the producer and consumer is represented solely by a set of requests to the services offered by the processing element of the platform model onto which it is mapped. In this case, the complete functionality is modeled in the platform model. Consequently, the complete task is represented by a service request image, directly equivalent to a binary application image of a processor. Another advantage of the support for such compiled tasks is that a platform model described at this level of abstraction can also be used for performance estimation of compiler technologies.

It is important to notice that service models described at different levels of abstraction can be mixed freely, providing great expressiveness and flexibility. Thus, depending on the level of abstraction used to describe the service models of a platform model, the execution of a requested service can model the actual functionality of the target architecture. Furthermore, the required resources and cost in terms of, for example, latency or power of the service can be included. Thereby, the functionality of the application model mapped onto the platform model is modeled according to the actual implementation, including, for example, the correct bit widths and availability of resources, simply by refining the platform model without changing the application model. It then becomes possible to annotate a given task in the application model with the cost of execution. This then adds a quantitative cost measure for use in the assessment of the platform.

11.11 AN INDUSTRIAL CASE STUDY

As an illustration of the usage of the presented framework, an extract of the industrial case study presented in Tranberg-Hansen et al. [21] is given, in which a mobile audio processing platform developed by the Danish company Bang & Olufsen ICEpower was explored.

11.11.1 A MOBILE AUDIO PROCESSING PLATFORM

The mobile audio processing platform, illustrated in Figure 11.13, is comprised of a digital front end and a class D amplifier including the analog power stage on-chip. The platform offers stereo speaker and stereo headphone audio processing, resulting in a total of four audio channels being processed.

The application receives one common stereo audio stream, consisting of a left and right audio stream. After the processing of algorithm **A**, these are split into four separate audio streams. Two are for speakers and two are for headphones, to allow a separate processing of the speaker and the headphone streams. The resulting application model is shown in Figure 11.14.

The platforms considered here are based on the use of an application-specific instruction set processor (ASIP), which is optimized to execute the type of algorithms necessary in the case-study application. The ASIP has a shallow three-stage pipeline and offers 61 different instructions. Even though the pipeline is shallow, the instructions are relatively complex: the processor offers a dedicated second-order IIR filter instruction that is used heavily in the application considered. Two versions of the ASIP were described: (1) a high-level model that models only the latency of each

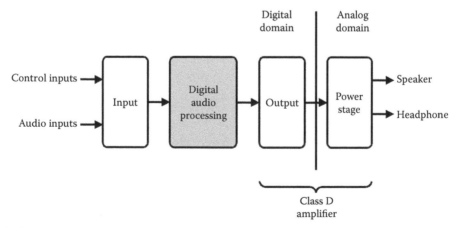

FIGURE 11.13 Overview of the case-study platform.

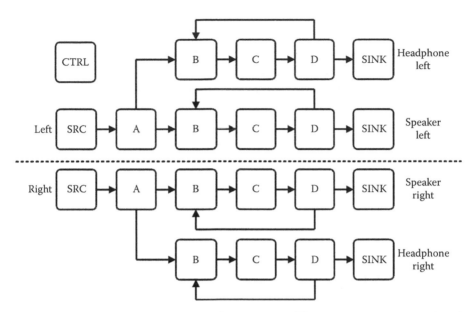

FIGURE 11.14 The application model of the case study. The application model contains 14 tasks ($2 \times A$, $4 \times B$, $4 \times C$, $4 \times D$) for modeling the processing of the audio streams, 6 tasks ($2 \times SRC$, $4 \times SINK$) for modeling the audio interfaces to and from the environment, as well as the environment itself, and finally an additional task (CTRL) for modeling the changes of the application state and/or control parameter changes that influence the individual audio processing parts.

instruction offered and (2) a detailed cycle accurate and bit true model. The latency models do not model the actual functionality but only the resource access and latency of each service without any modeling of data dependencies.

A number of platform models were constructed consisting of one to four ASIPs (ASIPx1-ASIPx4). Subsequently, the application model, capturing the functional requirements of the application, was mapped onto the platform models based on the latency ASIP service models. Because the latency-based service models do not model the actual functionality, the control flow of the application must be handled in the application model and the tasks mapped to the individual latency-based service models. This group of system models is hence named latency. The four platforms were then refined to use the bit true and cycle accurate service model of the ASIP using the compositional properties of the framework. The tasks of the application models were mapped to the ASIP service model processing elements. The tasks were represented as service request images generated using the existing compiler infrastructure associated with the ASIP. In this way, a one-to-one correspondence with the physical execution of the application can be obtained. These system models are referred to as compiled system models.

To relate the simulation speed and quality of the performance estimates produced by the models described using the modified HCPN model-of-computation, a register transfer level implementation of the ASIPx1 platform was created in the hardware description language Verilog, referred to as the RTL model in the following.

TABLE 11.2

Estimated Number of Cycles for the Processing of 3000 Samples in One Audio Channel at Three Levels of Abstraction. A, B, C, and D Are the Different Tasks of the Application Being Executed

	Latency	Compiled	RTL
A	84,034	102,034	102,034
B	123,041	129,384	129,384
C	33,011	33,011	33,011
D	153,051	168,056	168,056
Total	393,137	429,143	429,143

11.11.2 ACCURACY

The simulations performed, using the RTL model, were compared with the results obtained from the performance estimation framework.

Table 11.2 shows the estimated number of cycles used to process a stereo audio channel produced by the framework for the latency and the compiled ASIPx1-system model and estimates extracted from the RTL model simulations. The table shows that the cycle estimates obtained from the latency model in which only the latency-based service models are used is *not* cycle accurate. The cycle estimates produced by the latency model are, in general, too optimistic. This is caused by the fact that the latency-based model does not take data dependencies into account. In the other range of the scale, in terms of accuracy, the table also shows the cycle estimates of the refined compiled model in which a cycle accurate and bit true modeling of the components was used. In this case, the cycle estimates are identical with the estimates obtained from the RTL model as can be seen from the table.

The RTL model constructed was also used to make a comparison of the functional results produced by the framework using the compiled version of the ASIPx1 platform, which proved to be 100% identical to the results obtained from the RTL model. Furthermore, the time required to describe the model using the presented framework is much less than the time required to describe the equivalent RTL model because of the higher level of abstraction used. More importantly, it is significantly faster to modify a model of the framework in case of bug fixes or functionality extensions.

11.11.3 SIMULATION SPEED

Table 11.3 shows the measured simulation speeds expressed as cycles per second for the individual system models investigated. Simulations were performed on a 2.0 GHz Intel Core 2 Duo processor with 2 GB RAM. What is most interesting is the substantial speedup seen when comparing the simulation speed of a detailed bit true and cycle accurate version of the ASIPx1-system model as opposed to the equivalent register transfer level simulation. The ASIPx1-system model runs at approximately 20 million cycles per second with all algorithms enabled, including data

TABLE 11.3

Obtainable Simulation Speed (Simulation Cycles per Second) of the Investigated System Models

	Latency	Compiled	RTL
ASIPx1	21.7M	19.9M	15,324
ASIPx2	18.6M	15.8M	N/A
ASIPx3	17.2M	13.8M	N/A
ASIPx4	15.9M	12.4M	N/A

logging, and the functionally equivalent RTL description runs with approximately 15,000 cycles/second resulting in a speedup of more than 1000× compared to the RTL simulation.

11.12 CONCLUSION

This chapter has introduced a novel compositional modeling framework for system-level performance estimation of heterogeneous embedded systems. The framework is simulation-based and allows performance estimation to be carried out throughout all design phases ranging from early functional to cycle accurate models. Using a separation of the specification of functionality, communication, cost, and implementation, combined with an interface-based approach, provides a very flexible framework with great refinement possibilities. Also, the framework enables the possibility for components, described at different levels of abstraction, to coexist and communicate within the same model instance.

ACKNOWLEDGMENTS

The authors acknowledge the industrial case provided by the DaNES partner Bang & Olufsen ICEpower A/S.

The work presented in this chapter has been supported by DaNES (Danish National Advanced Technology Foundation), ASAM (ARTEMIS JU Grant No 100265), and ArtistDesign (FP7 NoE No 214373).

REFERENCES

1. Benini, Luca, Davide Bertozzi, Alessandro Bogliolo, Francesco Menichelli, and Mauro Olivieri. 2005. "Mparm: Exploring the Multi-Processor SoC Design Space with SystemC." *The Journal of VLSI Signal Processing-Systems for Signal, Image, and Video Technology* 41 (2): 169–82.
2. Cesario, W. O., D. Lyonnard, G. Nicolescu, Y. Paviot, Sungjoo Yoo, A.A. Jerraya, L. Gauthier, and M. Diaz-Nava. 2002. "Multiprocessor SoC Platforms: A Component-Based Design Approach." *IEEE Design and Test of Computers* 19 (6): 52–63.
3. Gao, Lei, K. Karuri, S. Kraemer, R. Leupers, G. Ascheid, and H. Meyr. 2008. "Multiprocessor Performance Estimation Using Hybrid Simulation." *2008 45th ACM/IEEE Design Automation Conference*, 325–330.

4. Kunzli, S., F. Poletti, L. Benini, and L. Thiele. 2006. "Combining Simulation and Formal Methods for System-Level Performance Analysis." *Proceedings of the Design Automation & Test in Europe Conference* 1: 1–6.

5. Mahadevan, Shankar, Kashif Virk, and Jan Madsen. "Arts: A SystemC-Based Framework for Multiprocessor Systems-on-Chip Modelling." *Design Automation for Embedded Systems* 11 (4): 285–311.

6. Pimentel, A. D., C. Erbas, and S. Polstra. "A Systematic Approach to Exploring Embedded System Architectures at Multiple Abstraction Levels." *IEEE Transactions on Computers* 55 (2): 99–112.

7. Tranberg-Hansen, Anders Sejer, and Jan Madsen. "A Compositional Modelling Framework for Exploring MPSoC Systems." In *CODES+ISSS '09: Proceedings of the 7th IEEE/ACM International Conference on Hardware/Software Codesign and System Synthesis*, 1–10. New York, NY, USA: ACM.

8. Keutzer, K., A. R. Newton, J. M. Rabaey, and A. Sangiovanni-Vincentelli. 2000. "System-Level Design: Orthogonalization of Concerns and Platform-Based Design." *IEEE Transactions on Computer-Aided Design of Integrated Circuits and Systems* 19 (12): 1523–43.

9. Balarin, Felice, Massimiliano Chiodo, Paolo Giusto, Harry Hsieh, Attila Jurecska, Luciano Lavagno, Claudio Passerone et al. 1997. *Hardware-Software Co-Design of Embedded Systems: The POLIS Approach*. Norwell, MA, USA: Kluwer Academic Publishers.

10. Kienhuis, B., E. Deprettere, K. Vissers, and P. Van Der Wolf. 1997. "An Approach for Quantitative Analysis of Application-Specific Dataflow Architectures." In *ASAP '97: Proceedings IEEE International Conference on Application-Specific Systems, Architectures and Processors*, pp. 338–349.

11. Cai, L. and D. Gajski. 2003. *Transaction Level Modeling: An Overview*. In *CODES+ISSS '03: Proceedings of the 1st IEEE/ACM/IFIP international conference on Hardware/software codesign and system synthesis*, pp. 19–24.

12. Lee, E. A. and A. Sangiovanni-Vincentelli. 1998. "A Framework for Comparing Models of Computation." *Computer-Aided Design of Integrated Circuits and Systems, IEEE Transactions on* 17 (12): 1217–29.

13. Jantsch, Axel. 2003. *Modeling Embedded Systems and SoCs—Concurrency and Time in Models of Computation*. Morgan Kaufmann Burlington, MA: Systems on Silicon.

14. Cassandras, Christos G. 1993. *Discrete Event Systems: Modeling and Performance Analysis*. Aksen Associates, Inc., Homewood, IL.

15. Wainer, G. A. and P. J. Mosterman. *Discrete Event Simulation and Modeling: Theory and Applications*. Computational Analysis, Synthesis, and Design Of Dynamic Systems Series. CRC Press, Taylor and Francis Group, 2010.

16. Tranberg-Hansen, Anders Sejer, and Jan Madsen. 2008. "A Service Based Component Model for Composing and Exploring MPSoC Platforms." In *ISABEL '08: Proceedings of the First IEEE International Symposium on Applied Sciences in Bio-Medical and Communication Technologies*, pp. 1–5.

17. Tranberg-Hansen, Anders Sejer, Jan Madsen, and Bjorn Sand Jensen. A Service Based Estimation Method for MPSoC Performance Modelling. In *Proceedings of the International Symposium on Industrial Embedded Systems*, SIES 2008, pp. 43–50.

18. Jensen, Kurt. 1992. *Coloured Petri Nets. Basic Concepts, Analysis Methods and Practical Use. Volume 1, Basic Concepts*. Springer-Verlag Germany.

19. Sun Mircrosystems. Java SE 1.6, 2008. http://www.java.sun.com.

20. Eclipse Foundation. Eclipse. http://www.eclipse.org/.

21. Tranberg-Hansen, Anders Sejer, and Jan Madsen. 2009. "Exploration of a Digital Audio Processing Platform Using a Compositional System Level Performance Estimation Framework." *2009 IEEE International Symposium on Industrial Embedded Systems*, *SIES '09*, pp. 54–7.

12 Consistency Management of UML Models

Emilia Farcas, Ingolf H. Krüger, and
Massimiliano Menarini

CONTENTS

12.1 Introduction 289
    12.1.1 Multiview Models and Consistency Challenges..............................290
    12.1.2 Inconsistency Example 292
    12.1.3 Outline 295
12.2 State of the Art in Consistency and Semantics...295
    12.2.1 UML Model Consistency Requirements 296
    12.2.2 Consistency Checking 297
    12.2.3 Semantics..301
12.3 Solving UML Consistency 304
    12.3.1 Queries and Constraints Semantics 305
        12.3.1.1 Notational Preliminaries and System Formalization.........308
        12.3.1.2 Abstract Specification Language 309
        12.3.1.3 Specification Language Semantics 310
    12.3.2 Notion of Consistency... 312
    12.3.3 Example of Consistency Management... 315
12.4 Discussion..323
12.5 Summary and Outlook 325
Acknowledgments..326
References..326

12.1 INTRODUCTION

Model-based engineering (MBE) [1] approaches support the use of models throughout the development process from requirements elicitation, to architecture specification, to system analysis, and deployment. Models can be leveraged in a variety of ways. During requirements elicitation, models help in defining the problem domain and communicating with stakeholders. During development, models can be used to define clear interfaces between components developed by separate suppliers. Modeling can also provide support for system synthesis by using model transformations and automatic code generation. For analysis, models can be used to verify system properties before the system is actually deployed. Moreover, models can also be used at runtime to monitor the system behavior and identify compliance to behavior interfaces and QoS properties.

Notably, modeling languages are most useful when they fulfill some key requirements. Two important requirements for modeling real-time embedded systems are a clear definition of the semantics of models together with the ability to guarantee that models are consistent. In fact, without a consistent set of models and a precise and complete semantics it is impossible to leverage models to synthesize code or to verify properties of the modeled system, which are two important activities in MBE of embedded systems. It is worth noting that consistency and semantics are two closely related aspects of a language. In fact, without a formal semantics it is impossible to prove the absence of inconsistencies.

The OMG's Unified Modeling Language™ (UML®) [2,3] is a general-purpose modeling language widely used across application domains. It is a family of graphical notations underpinned by a single metamodel [4]. It provides 14 types of diagrams, which support modeling structural (i.e., static) and behavioral (i.e., dynamic) views of a system. The UML provides a built-in extension mechanism through profiles, which allows tailoring the UML for a particular domain or target platform. The recent UML profile for Modeling and Analysis of Real-Time and Embedded Systems (MARTE) [5] allows specifying timing properties, supports component-based architectures and different computational paradigms, allows for modeling of both software and hardware platforms along with their nonfunctional properties, and supports schedulability and performance analysis.

In Chapter 5, we presented the modeling capabilities of UML and MARTE and discussed how ensuring consistency and defining a formal semantics is still an open issue in UML. In this chapter, we discuss various approaches to model consistency and we present an innovative solution. We picked the topic of model consistency to explore in detail because we believe it is paramount for a comprehensive modeling methodology.

In the following paragraphs, we discuss the consistency problem in UML and give an example of inconsistent models.

12.1.1 MULTIVIEW MODELS AND CONSISTENCY CHALLENGES

To tackle complexity, MBE approaches support multiple perspectives with associated modeling languages, each focusing on a particular subset of system properties. Each perspective can cover a separate aspect of the same part of the system or depict the same aspect with different notations to clarify or stress a modeling concept. For instance, we could use a sequence diagram to show the communication protocol between two class instances and two state-machine diagrams to describe the proper ordering of the method calls upon each class. These two perspectives clearly overlap. This overlapping requires that all models are consistent.

The UML standard from the Object Management Group® (OMG®) comprises many languages (14 types of diagrams), each emphasizing a different structural or behavioral modeling aspect. The most recent version is UML 2.3, whose specification consists of the UML Superstructure [3] defining the notation and semantics for diagrams and the UML Infrastructure [2] defining the language on which the Superstructure is based. Constraints can be expressed in the textual Object Constraint Language [6].

When using multiple modeling perspectives, the central question from an engineering point of view is this: Is the modeled system realizable? However, UML does not provide a complete formal semantics, which leaves issues such as model consistency unsolved.

In this chapter, we will focus on the issues related to consistency of UML models. We discuss the UML consistency problem in detail, explain how it originates or is worsened by the trade-offs in the language design, and propose an avenue to solve it. We chose this problem for its importance not only in the real-time systems domain but also in most areas where MBE is applied. One important example is the field of feature modeling. In this field, different software functionalities (features) are modeled separately and programs are generated by composing features. A recent article [7] argues that model consistency is still an open and important issue in this area.

Furthermore, Model-Driven Architecture® (MDA®) [8] is an MBE approach that distinguishes between a platform-independent model (PIM) and a platform-specific model (PSM). PIM captures the core system entities and their interactions without specifying how these are implemented. PIM can be mapped to multiple PSMs, each capturing all aspects of a particular deployment architecture. UML is the language choice of MDA, where both PIM and PSM are expressed as UML models. The distinction between PIM and PSM also introduces a requirement for model consistency.

The first issue to solve when discussing model consistency for UML is to clearly define what kind of consistency we are interested in and how to effectively determine whether or not a UML model is consistent. Of course, UML is a broad-spectrum language with an informally defined semantics, which serves the goal to be inclusive with respect to modeling styles and domains. However, this creates the first hurdle we have to overcome in our consistency definition. Any approach aiming at defining consistency needs to explicitly or implicitly define a more precise semantics for UML. A rich body of work exists in the literature on defining multiview or multiperspective consistency based on UML semantics definitions. In Section 12.2, we will examine this related work.

Although the consistency problem has been extensively studied in the literature, a solution has been elusive—especially in the context of the UML with its rich set of interrelated description techniques for system structure and behavior. Existing approaches to defining UML model consistency lead to complex definitions of the notion of consistency or address only a subset of the available modeling notations. Our goal is to create a consistency checking approach that is flexible enough to be able to target the full UML language. However, we do not want the engineer to be forced to fully define the semantics of all UML notations, only the semantics of a subset (profile) of the UML language used in the specification should be defined.

The main novelty of the consistency checking approach we present in this chapter is in the comprehensive, yet simple mechanism we introduce for specifying consistency rules. Instead of analyzing the semantics of the UML at the metamodel level and extracting consistency rules between different diagram types, we define a simple execution framework (similar to a "virtual machine"), based on a target ontology whose concepts map one-to-one to elements of the system class we are interested in modeling, i.e., distributed, reactive systems. All UML diagram types are then treated as model generators for this virtual machine; each diagram selects entities of the virtual machine and constrains their structure or behavior. Model consistency is then simply defined as the presence of virtual machine behaviors under the specified constraints.

12.1.2 INCONSISTENCY EXAMPLE

We revisit the example from the Bay Area Rapid Transit (BART) [9] system, introduced in Chapter 5, Sections 5.3.2.3 and 5.4.2.3. BART is the commuter rail train system in the San Francisco Bay area. The BART system automatically controls over 50 trains on a large track network with several different lines. We show three modeling perspectives of BART using UML 2.3 and the MARTE profile: component, sequence, and state-machine diagrams (see Figure 12.1). We present an inconsistency that can arise when modeling behavior in the different diagrams, namely sequence and state-machine diagrams.

We focus on the Advanced Automatic Train Control (AATC) system, which controls the train movement for BART. The AATC system consists of computers at train stations, a radio communications network that links the stations with the trains, and AATC controllers on board of each train. Most of the control computation is done at the stations. Each station is responsible for controlling all trains in its area. Trains receive acceleration and brake commands from the station via the radio communication network. The train controller is responsible for operating the brakes and motors of all cars in the train. Controlling the trains must occur efficiently with a high throughput of trains on the congested parts of the network, while ensuring train safety. The station's control algorithm takes the track information, train speed and acceleration, train position estimation, and information from the neighboring stations into account to compute new commands that never violate the safety conditions. To ensure this, each station computer is attached to an independent safety control computer that validates all computed commands for conformance with the safety conditions.

The component diagram for AATC is depicted in Figure 12.1a. It has three nodes: two for the train station and one for the train. The first node, Fast Computer, represents the station computer that computes the commands to be sent to all trains under the control of that station. It contains two components: one represents the Station AATC control system and the other called Environmental Model, which models the physical environment of a station. The Station AATC uses the Environmental Model to compute commands to send to trains. The second node, Slow Safety Computer, contains the Safety Control component, which checks all commands sent by the Station AATC for safety before forwarding them to each train. The safety computation is based on a simpler model than the one used to compute commands and, therefore, requires less computation resources. However, the Slow Safety Computer

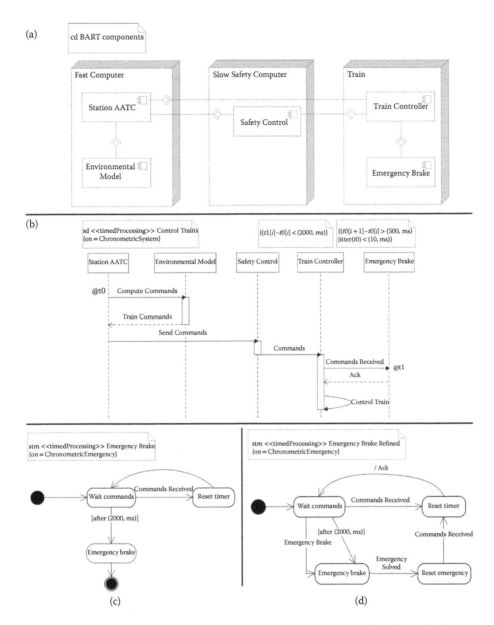

FIGURE 12.1 Three different perspectives of the Bay Area Rapid Transit (BART) case study: (a) component diagram defining the structure, (b) sequence diagram describing the train commands computation and delivery, and (c, d) state-machine diagrams describing the Emergency Brake system.

is required to have high reliability. The third node in the figure is the Train. It has two components: the Train Controller manages the train accelerations and decelerations, and the Emergency Brake is activated only in case of an emergency and stops the train as quickly as possible.

The AATC system operates in half a second cycles. In each cycle, the station receives train information, computes commands for all trains under its control, and forwards these commands to the train controllers. The Station AATC system obtains the status information regarding train speed, acceleration, and range by using the radio network, which allows the system to track train positions. The Station AATC system computes the train position from the status information and updates its Environmental Model. Then, the Station AATC interacts with the Environmental Model and the Safety Control components to compute and send the new commands, as depicted in the sequence diagram from Figure 12.1b. The behavior specified in the diagram is the following:

- Station AATC sends a request to Environmental Model to compute the commands for the train.
- Environmental Model computes the commands, taking into account all parameters such as passenger comfort (e.g., not too strong braking and acceleration changes), train schedule, engine wear, and most importantly safety.
- After receiving the commands from Environmental Model, Station AATC sends the commands to Safety Control to ensure the commands computed are safe.
- Safety Control checks that the commands do not exceed maximum bounds for safety. If the commands are safe, Safety Control forwards them to Train Controller.
- Train Controller informs Emergency Brake that the commands have been received.
- Emergency Brake acknowledges the commands received.
- Finally, Train Controller controls the train engine according to the commands received.

The model in Figure 12.1b is annotated with MARTE time constraints to specify the real-time requirements of the BART case study. We annotated two time instants $t0$ and $t1$ using TimedInstantObservations as defined in MARTE, which is indicated by the graphical representations @$t0$ and @$t1$. A TimedInstantObservation denotes an instant in time associated with an event occurrence (e.g., send or receive of a message) and observed on a given clock. $t0$ is the instant when the message Compute Commands is *sent* by Station AATC, whereas $t1$ is the time instant when the message Commands Received is *received* by Emergency Brake. Because the system operates in cycles, the notation $t0[i]$ and $t1[i]$ represents the generic ith instantiation of the interaction scenario.

Given those two instants, we leverage MARTE to define three time constraints in our system. Commands to trains become invalid after 2 seconds. If a train does not receive a valid command within 2 seconds, it goes into emergency braking. Therefore, with the time constraint $(t1[i] - t0[i]) < (2000, ms)$, we limit the duration of each iteration of this scenario to 2 seconds. The AATC control algorithm needs to take this timing constraint, track information, and train status into account to compute new commands that never violate the trains' safety.

The second constraint, $(t0[i + 1] - t0[i]) > (500, ms)$, imposes that between each instantiation of the scenario at least half a second passes. Finally, the last constraint, $jitter(t0) < (10, ms)$, limits the jitter of the $t0$ event enforcing that between each iteration of the event at $t0$ there are between 500 and 510 ms.

In normal operations, the AATC system computes the train commands in fixed time cycles. However, in case of a detected emergency condition, the system has to react immediately and take appropriate measures to ensure maximum safety of passengers and equipment. Figure 12.1c and Figure 12.1d present state-machine diagrams for the Emergency Brake component. A train will continue to exercise a command until a new one arrives or until that command expires, 2 seconds after the originating time. The state-machine diagram for the Emergency Brake has states for waiting for commands and entering emergency mode if the timer of 2 seconds expires. When commands are received, the timer is reset. These state machines are two different versions of the same perspective where the one in Figure 12.1d is a refined version that enables restarting the system after an emergency brake. If we consider the three graphs from Figure 12.1a, 12.1b, and 12.1c together, we have an inconsistent model: the state-machine diagram Figure 12.1c does not acknowledge the Commands Received call from Train Controller—contrary to what the sequence diagram from Figure 12.1b demands. Replacing the diagram from Figure 12.1c with Figure 12.1d, we obtain a consistent model.

12.1.3 OUTLINE

In the preceding paragraphs, we discussed the role of modeling, multiview models in UML, and the problem of consistency management. We also presented an example of inconsistent models, which we will use later in the chapter as an example for applying our consistency checking solution.

The remainder of the chapter is structured as follows. In Section 12.2, we identify a set of core requirements to address the consistency problem of UML adequately and we analyze related approaches in the literature. Because none of the approaches fully addresses all requirements, we propose an innovative approach to UML consistency in Section 12.3. Our approach is based on defining (1) an explicit ontology that captures the target domain we are modeling and (2) a simple execution framework based on this target ontology. Sections 12.4 and 12.5 contain discussions and outlook.

12.2 STATE OF THE ART IN CONSISTENCY AND SEMANTICS

In this section, we analyze the details of the UML consistency problem and its relation with the UML semantics. We discuss alternative solutions proposed in the research literature. The solutions presented address three specific problems related to UML consistency: (1) verifying the consistency of different views of the same system, (2) giving a complete and consistent semantics to UML diagrams, and (3) integrating models of different subsystems into a coherent, implementable model of the system.

In the following, we present a set of requirements we use to evaluate approaches to UML consistency. We identified these requirements by surveying the literature and analyzing strengths and weaknesses of each approach proposed. Although each requirement is fulfilled by at least a few approaches, none of the approaches we surveyed performs well in all the areas. Thus, we use these requirements as a tool for comparing the different approaches proposed by different researchers and to identify areas that can be improved. In Section 12.3, we propose an approach that can address all requirements by leveraging different ideas surveyed in our state of the art overview.

12.2.1 UML MODEL CONSISTENCY REQUIREMENTS

We have identified 12 important requirements (collected in Table 12.1) by analyzing the requirements discussed in the literature for current approaches to model consistency. Requirements R1 to R3 in Table 12.1 originate from the observation that any strategy to manage model consistency should not limit the freedom of developers. This entails that developers should be allowed to modify models even if they introduce some inconsistencies. This idea is introduced in Finkelstein et al. [10], where the authors observe that inconsistency is necessary and often desirable in some phase of the development cycle. For example, in the inception phase of a large project with different stakeholders involved, each stakeholder pursues different goals and, during the collection of requirements, this can lead to inconsistent views that must be identified and reconciled in subsequent iterations. Other arguments in support of Requirements R1 to R3 have been documented elsewhere [11–13]. The common denominator of all arguments is that effective modeling techniques must support decomposing the problem into independent subproblems. This is the case when in

TABLE 12.1
Requirements for UML Consistency Management

	Requirement Description
R1	*Inconsistent* models can be *introduced and kept in the system* specification for a certain amount of time
R2	*Inconsistencies* should be *discovered automatically and tracked* during the evolution of model
R3	Support should be provided to the developer to *resolve inconsistencies when convenient*
R4	Support *multiple* modeling *languages* (for example, different UML notations or even non-UML languages)
R5	Support *different levels* of abstraction
R6	Support the *extension or specialization* of languages
R7	Support *horizontal* consistency
R8	Support *vertical* consistency
R9	Support *static* consistency
R10	Support *dynamic* consistency
R11	*Tool* support (or translations to available tools)
R12	*Scalability* to large models

order to solve complex problems, engineers decompose various aspects of the system and reason about each aspect in isolation. Alternatively, this occurs when in order to solve complex problems efficiently, different teams work in parallel on different aspects of the system.

A second observation is that each model caters to different needs that arise during the development process. For example, informal models are used to gather requirements and exchange ideas between stakeholders and developers during requirements gathering [14]. Later in the development process, more formal models are used to describe the structure or the behavior of certain parts of the system. In this phase, formal models are used to verify properties of a system or to generate part of the implementation code. This second observation is the source of the additional requirements R4 to R6 in Table 12.1.

To evaluate consistency management techniques, the notion of consistency must be clearly defined. The scientific literature examines different notions of consistency. A distinction can be made between *horizontal* and *vertical* consistencies [15,16]. *Horizontal* consistency involves different perspectives on the same system model. For example, on the one hand, to describe the communication between a client and a server, it is possible to use a UML sequence diagram to capture the protocol and a state diagram to capture the server behavior. The two diagrams are different views on the same system and should be horizontally consistent. On the other hand, *vertical* consistency addresses views of the same aspect of one system, but at different levels of abstraction, often in relation to the evolution of one model during different phases of the development process. For example, an abstract model created during requirements gathering must agree with a more detailed model used for code generation in a later step of the development process. Another important distinction is between *static* and *dynamic* consistency [17]. *Static* consistency addresses syntactical and structural model dependencies, whereas *dynamic* consistency ensures the consistency of executable models. We introduce four requirements (R7 to R10 in Table 12.1) to capture these four notions of consistency.

The final two requirements address practical use of consistency management techniques. Requirement R11 recognizes that consistency checking must be supported by a tool chain. Requirement R12 recognizes that industrial systems are large scale and this implies they have large system models. Therefore, scalability of the chosen technique to large models is an important requirement.

12.2.2 CONSISTENCY CHECKING

In the literature, a number of promising approaches for model consistency checking exist. The existing approaches can be divided into two categories: rule-based approaches and translation-based approaches. The first category uses rule-based systems to define consistency rules directly on the modeling language. In general, rule-based approaches scale well to large models but are limited to addressing only *static* consistency (failing requirement R10). On the other hand, translation-based approaches leverage a target language with a formal semantics and translate models into this language. Tools that support automated reasoning or formal verification in the target language can be used to reason about the consistency of the original models. Translation-based approaches can address both *static* and *dynamic* consistencies; however, they have very limited scalability and tend to be restricted to a subset of the UML languages (failing requirement R4).

Rule-based approaches translate a modeling language into a logic framework on which the system of rules is defined. The goal of the translation is to be able to apply rules defined on a logic framework. This translation does not aim at defining a formal semantics for the modeling language and does usually limit the expressiveness of consistency rules to structural elements of the language. On the other hand, translation-based approaches leverage languages with formal semantics and the translation process implies the definition of a formal semantics for the UML. These approaches enable reasoning about consistency of behavioral models but tend to work only on subsets of UML and impose the resolution of variation points. Because translation-based approaches are closely related to the problem of defining a formal semantics for UML, we discuss them further in the Subsection 12.2.3.

Table 12.2 summarizes the approaches to consistency that we discuss in this section. Each requirement we identified in the previous section is a column of the table, whereas each of the approaches we discuss here is represented as a row. This section and Table 12.2 cover consistency management approaches that use rule-based systems; a second table in the next section covers approaches that use translation to formal languages. These approaches are important to know, especially if you want to understand the fundamental issues with consistency arising from the flexible, but complex, UML metamodel. Additional consistency problems, such as timing consistency, arise when using profiles for real-time systems, such as MARTE. These problems are not directly addressed by any of the methodologies surveyed.

A logic framework for capturing models and rules has been proposed by Van Der Straeten et al. [18]. In this work, the authors propose the use of description logic (DL), which is less powerful than first-order logic, but is decidable. The approach, which targets a subset of the UML language, proposes to encode UML models and metamodels as well as consistency rules in DL. The focus of this appproach is on both software evolution (vertical consistency) and consistency between different views of the same specification (horizontal consistency). To this end, the UML metamodel is enhanced with classes capturing horizontal and vertical relations. This approach is used, for example, to support model refactoring in [19], which documents how the RACER tool is used to enable inference from the DL knowledge bases. The translation between UML and DL is performed by the RACOoN tool, which enables refactoring and is integrated into the Poseidon UML modeling tool.

Sabetzadeh and Easterbrook [20] present a requirements merging approach based on category theory. The approach aims at merging inconsistent and incomplete views. Views are expressed as graphs, and relations between views are expressed by an interconnection diagram that relates common elements of different views. A colimit operation on the views combined with the interconnection diagram return the integrated view. To address inconsistencies, the methodology supports annotating elements of the graphs to identify where inconsistencies stem from. Annotations are captured via a knowledge lattice that identifies proposed, repudiated, affirmed, and disputed elements. The article, however, does not discuss how this technique can be applied to address the consistency of views expressed in different languages.

TABLE 12.2

Evaluation of Consistency Checking Approaches Based on Rule Systems

	Rules Specification Language	R1	R2	R3	R4	R5	R6	R7	R8	R9	R10	R11	R12
Van Der Straeten et al. [18], Van Der Straeten and D'Hondt [19]	Description logic	×	×	×	✓	×	×	✓	✓	✓	×	✓	?
Sabetzadeh and Easterbrook [20]	Interconnection diagram describes relations between views, a knowledge lattice captures additional information on each element	✓	×	×	×	×	✓	✓	✓	✓	?	✓	?
Nentwich et al. [21]	Rules expressed in a reduced first-order logic, XPATH expressions identify where to apply them	✓	×	×	✓	✓	×	✓	×	✓	×	✓	×
Engels et al. [22]	Dynamic metamodeling (rules defined on the UML metamodel)	✓	×	✓	✓	✓	×	✓	×	✓	✓	✓	×
Egyed [23,24]	Rules are stateless and deterministic	✓	✓	✓	✓	✓	✓	✓	×	✓	×	✓	✓

Note: The symbol ✓ indicates that the requirement is met, the symbol × indicates that the requirement is not met, and the symbol ? indicates that it is not clear from the literature if and to which degree the requirement is met.

Other approaches proposed in the literature define consistency rules on models represented by XML documents. These approaches can be adapted to work on XML representations of UML models serialized in the XML Metadata Interchange (XMI®) [25] interchange format. For example, a general XML-based consistency checking tool called xlinkit has been used to verify structural consistency; Nentwich et al. [21] present the tool, an incremental algorithm for checking, and a case study. To make rules decidable, they are based on a restricted version of first-order logic. The tool can check XML documents or other types of models by using a "fetcher" that reads the code and creates an in-memory Document Object Model (DOM) from it. Then, xlinkit uses XPATH (a query language for DOM) to specify the parts of the memory model where the rules must be applied. As a result of the rules evaluation, a set of links between conflicting elements is created. This enables the user to navigate the model and resolve inconsistencies without forcing the immediate resolution of every inconsistency. To improve scalability, an incremental algorithm is used to detect which rules have to be rechecked after a change. However, once the tool decides that a rule must be reevaluated, it is reevaluated on the entire model. As a result, when complex rules (such as the language constraints defined by the UML standard) have to be reevaluated, the benefits of this incremental algorithm are limited. The authors claim that the cause of this drawback is the complexity of the XMI representation. The benefits of this technique are the independence from a specific modeling tool (XMI is an interchange standard supported by many tools) and the possibility of distributing the documents over the Internet. Major limitations are the fact that only structural consistency is addressed and that different rule evaluation orders can lead to different results. Moreover, XMI is not interpreted consistently by different tools. Therefore, it is not functioning as an interchange language in practice and consistency checking results could also be influenced by the tool chosen for creating the XMI file.

Other approaches define rules at the UML metamodel level instead of choosing a specific UML representation, such as the XML-based approaches of the previous paragraph. For example, Engels et al. [22] present an approach based on dynamic metamodeling (DMM). Rules are defined on the graphical representation of the UML metamodel using an extended class diagram notation. The extensions include the ability to annotate elements of the metamodel with *new* and *delete* keywords. These keywords specify that the corresponding elements are respectively created or removed when the configuration captured by the metamodel is instantiated. The verification of consistency is then performed by testing. Overlapping models are verified in pairs where one is executed according to the DMM semantics, whereas the other is checked to identify violations. This approach has the great benefit of using a graphical model with which UML users are familiar. Drawbacks are that many rules must be defined for UML and that testing is not complete. In particular, the paper does not discuss any form of coverage metric to assess the confidence in the verification process.

Other approaches extend modeling tools. For example, Egyed [23,24] presents an extension to the Rational Rose tool. This extension, which is based on an algorithm presented in Ref. [26] by the same author, supports incremental consistency checking by using stateless deterministic rules. The incremental rule checking algorithm can analyze which part of the model is accessed during the evaluation of a rule and use the information to reevaluate rules only on the relevant portion of the model when changes happen. For each inconsistency identified, the tool creates an inconsistency annotation in a report. This tool is highly scalable and has been used on actual large models in industrial case studies.

12.2.3 SEMANTICS

The second type of consistency checking approaches is closely related to the problem of defining a formal semantics for UML languages. The research community has been very active in defining semantics for UML [27–31]. Broy et al. [29–31] give a formal model for UML based on stream semantics [32]. This approach bases the semantics on the elements of the UML metamodel and maps the UML to a formal model that closely matches each element of the metamodel itself. This results in a very complete semantic framework that formally captures every detail of the language, enabling the reasoning about every feature of the UML. However, for the goal of checking dynamic consistency of models, this framework is difficult to use. In fact, model checking tools are used to verify dynamic consistency. Because the translation to the formal model suggested by Broy et al. [29–31] contains all details of the UML, the state space of the models tends to be very large and properties are difficult to verify. Other work by Krüger and Menarini [28] proposes to map each diagram element to a very simple domain model specifically developed for the application domain the model is describing. This simplifies the automatic reasoning about the model properties but loses details about the language. In general, to use the UML and provide a formal semantics, the variation points are resolved and a profile is created to match the modeling needs.

Translation-based approaches for consistency management have been pursued by various authors. Instead of defining consistency rules based on the source models, they translate models into various formal languages and logics. The consistency is then implied by contradiction in the semantics of the target language. One of the main benefits of such approaches is that the translation to the target formal language defines the semantics of behavioral models. Therefore, on the one hand, all approaches following this avenue support reasoning about dynamic consistency. On the other hand, because in the translation the mapping to the original model can be lost, such approaches have greater difficulty in dealing with syntactic and structural rules required by static consistency.

Table 12.3 summarizes the consistency approaches discussed in this section. These approaches are all based on giving a formal semantics to UML by providing a translation to some formal language.

An example of an approach that involves translation is presented by Easterbrook and Chechik [33]. They use the Xbel framework to reason about state machines encoding inconsistent multiview requirements. The approach proceeds as follows. First, multiple state-machine views are merged into a single state machine. Because views can be inconsistent, the system is based on a multivalued logic called quasi-Boolean logic. Next, a multivalued model checker (called Xchek) is used to verify temporal properties expressed in XCTL (an extension of computation tree logic [38]). Translating state machines into their multivalued logic counterparts enables the

TABLE 12.3

Evaluation of Consistency Checking Approaches Based on Translation to a Different Semantics

Translation Target Language	R1	R2	R3	R4	R5	R6	R7	R8	R9	R10	R11	R12
Easterbrook and Chechik [33] — State machines in multivalued quasi-Boolean logic	✓	✓	×	×	×	×	✓	×	✓	✓	✓	×
Inverardi et al. [34] — State machines encoded in Promela, sequence diagrams encoded in linear temporal logic	×	×	×	×	×	×	✓	×	×	✓	✓	×
Ossami et al. [35] — Translate UML to B	×	×	×	×	×	×	✓	?	?	✓	✓	×
Engels et al. [36] — Communicating Sequential Processes	×	×	×	×	×	×	✓	✓	✓	✓	?	×
Paige et al. [37] — Prototype Verification System or Eiffel	×	×	×	×	×	×	✓	✓	✓	✓	?	×

Note: The symbol ✓ indicates that the requirement is met, the symbol × indicates that the requirement is not met, and the symbol ? indicates that it is not clear from the literature if and to which degree the requirement is met.

merger of inconsistent and incomplete models. The Xchek model checker enables the analysis of how inconsistencies affect the execution behavior of the state machines and helps in resolving them.

Inverardi et al. [34] use the SPIN [39] model checker to verify consistency of the behavior expressed by overlapping state machines and sequence diagrams. The approach uses Milner's [40] Calculus of Communicating Systems notation to describe state machines and a stereotype of the UML for sequence diagrams. The paper proposes to translate state machines into Promela (the input language of the SPIN model checker) and to translate sequence diagrams into linear temporal logic (LTL) formulae. SPIN is used to verify consistency by checking that the Promela model satisfies the properties expressed by the LTL formulae. A tool called Charmy supports the creation of sequence diagrams and state machines while it also generates the Promela code and the LTL formulae for verification.

A different solution to consistency checking is proposed by Ossami et al. [35]. The semantics of the UML is formalized by using the B [41] language. This enables developers to verify properties of their systems by using the verification tools supporting B. However, often a developer must modify the B specification to complete the proof. In this case, the two views of the system (UML and B) can become inconsistent. The solution proposed by the authors is to create development operators that map each requested transformation in one domain to equivalent transformations in the other. Therefore, once the correctness of the development operator is proven, it can be applied many times without introducing any inconsistency. This constructive approach has the benefit of reducing the amount of verification to perform. On the other hand, the developer is limited by the existing operators in how to modify the model because the solution does not allow consistency to be broken at any time. The approach of creating development operators is useful in the specific application domain; however, it is not applicable in development processes where inconsistencies are tolerated.

Engels et al. [36] present a translational approach to reasoning about the consistency of UML behavioral models. The approach is used for UML protocol statecharts. In particular, the approach is demonstrated in presence of the inheritance of classes whose protocol behavior is defined by statecharts. Inheritance imposes a refinement relation between statecharts. The paper focuses on vertical consistency, but the same technique addresses horizontal consistency as well. The target language for the translation is Communicating Sequential Processes (CSP) [42]. Rules are defined on the UML metamodel, which defines the UML graphical syntax used to translate the graphical models into the CSP textual notation. Then a tool called FDR (for Failures-Divergence Refinement) is used to verify the refinement relation in CSP.

Two other methodologies making use of translational approaches are presented by Paige et al. [37]. They share the same idea of translating the metamodel of the modeling language into a formal language supported by verification tools. Then each model—instead of being expressed in the original modeling language—is translated into the target language. The modeling language used is Business Oriented Notation (BON); however, UML could be supported as well. In fact, the semantics of the modeling language are completely defined by the metamodel expressed in the new formalism. The two approaches in Ref. [37] differ in the target language. The first approach translates models and metamodels into Prototype Verification System (PVS), a theorem prover language that supports semiautomatic proofs. The second uses Eiffel as a target language. Eiffel is object oriented and supports annotations to define method preconditions, postconditions, and class invariants. On the other hand, translation into PVS is the more flexible of the two approaches as it can encode all details of the metamodels and the verifications are complete. The limitation of PVS is the complexity of the translated models and the fact that it requires manual intervention in the verification process. On the other hand, Eiffel is a programming language and the "verifications" are performed by testing, which can be automated but the results are not complete. The general benefit of this type of approaches is that, because both syntax and semantics are captured by the translated metamodel, it is possible to check all consistency properties of the models. On the other hand, especially if applied to the large UML metamodel, scalability is a serious issue. Moreover, to perform a complete consistency check, developers are forced to use the PVS translation. This creates tremendous issues for the maintainability of large models, especially because the PVS translations cannot be kept from the developer as it is necessary to manually intervene in the verification effort.

From this comparison we see that addressing consistency for a large standard such as the UML is very difficult—even more so when fundamental modeling concepts such as time are not inherent language concepts but are bolted on using profiles and stereotypes. In the following section, we will present an alternative solution to the consistency management of the UML.

12.3 SOLVING UML CONSISTENCY

None of the approaches surveyed in the previous section fully address all requirements of Table 12.1. We believe that the common challenge of previous work is in losing track of the abstractions implemented in the models that are checked for consistency.

Previous work has taken two routes: either analyzing the semantics of the diagrams at the metamodel level (or defining consistency rules between different notation types from there) or translating the models into an existing formal language leveraged for verification. In contrast, we propose an approach that defines an explicit ontology that captures the target domain we are modeling and, further, we define a simple execution framework (similar to a "virtual machine") based on this target ontology. The ontology concepts map one-to-one onto elements of the system class we are interested in modeling.

The main novelty of the consistency checking approach presented here is in the comprehensive, yet simple mechanism we introduce for specifying consistency rules. By defining a simple "virtual machine" containing the abstraction used in our models, we can treat all UML diagram types as model generators for this virtual machine. Each diagram selects entities of the virtual machine and constrains their structure or behavior. Model consistency is then simply defined as the presence of virtual machine behaviors under the specified constraints.

We encode constraints as a set of logic propositions over elements of our target ontology and reduce the verification of virtual machine behaviors to a satisfiability (SAT) problem. Although the work presented here is specific to UML, the same approach can be leveraged to integrate other modeling languages with UML-like models.

For the proposed approach to work, we first tailor the UML to the target domain that we are interested in—we leverage the UML MARTE profile to target embedded real-time systems. For the purposes of this chapter, we limit ourselves to a subset of the MARTE notations, rich enough for us to show the value of our consistency notion. In particular, the subset we demonstrate in this chapter includes state diagrams, component diagrams, and interaction diagrams. In Section 12.4, we will analyze avenues to extending this approach to a richer subset of UML 2.0 and to other modeling languages.

12.3.1 Queries and Constraints Semantics

To provide the backdrop for our definition of model consistency, we provide a formal semantic framework based on an abstract model of distributed reactive systems, similar to a "virtual machine." We call this model of our target domain the "abstract semantic space." In this space, we show how each element of a model can be interpreted as a constraint on the system. The consistency property can then be trivially defined over the "abstract semantic space" as the existence of a system in that domain satisfies all constraints imposed by the models.

Our semantics is based on two elements: queries and constraints. Each model element of a UML specification is interpreted as a set of (query, constraint) tuples. Each query selects some elements in the "abstract semantic space" that we have defined where the corresponding constraint defines a restriction on the structure or behavior of these elements in a system satisfying the specification. The key benefits of our approach are (1) a mathematically simple, yet comprehensive definition of consistency, (2) the ability to tie the reasoning about consistency to entities of the target domain—resulting in a nongeneric model subclass to which the consistency notion applies, and (3) the interpretation of model elements as constraints over the target domain.

Our consistency checking approach contrasts with other translation-based approaches that we surveyed in the literature in the way we perform the translation. In fact, the target model of our translation abstracts the main components of the target implementation domain. The semantics is then specified by directly mapping each element of the UML model onto some configuration of the target model. Our first step is to define an ontology for real-time distributed systems. This ontology is used to assign precise semantics to the UML models we use and is formalized with Queries and Constraints. This step allows us to formally reason about the specification (using first-order logic). After the formalization, we present the grammar of a language to describe systems based on the target ontology formalism. This step enables the translation of UML models to the new domain. The final steps are the definition of the semantics for our abstract language and, based on such semantics, the definition of consistency.

Figure 12.2 captures the core elements of our ontology for distributed systems with real-time constraints. A real-time system in our ontology is described by five types of elements: two elements, Entities and Channels, form the structural configuration

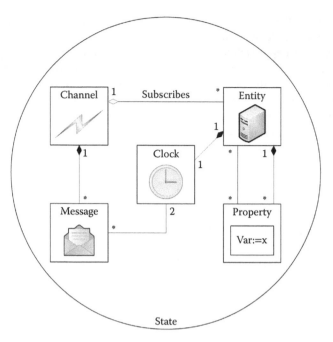

FIGURE 12.2   Core elements.

of the system; another two, Messages and Properties, define the behavior; and the Clock captures real-time constraints.

An Entity captures the concept of a process in a distributed system. An Entity has local variables, captures state information, has computational capabilities, and can communicate with other Entities by means of sending and receiving messages over channels. Channels are the communication infrastructure. Each entity that must send or receive messages subscribes to a channel. Channels transport Messages. When a message is sent on a channel, all entities that have subscribed to the channel eventually receive the message. Properties can be used to capture variables and their state. Each entity has a named set of properties that can be evaluated at runtime. Finally, the Clock captures the time relative to an entity. We could have used different notions of time, the choice depends on the type of system we intend to model and the profile of UML we use. MARTE supports not only the type of time we model here but also other time models, for example, modeling of synchronous reactions.

Figure 12.2 shows these five core elements forming the *abstract* state of the system. At each instant, the structural part of the system state is defined by the existing Entities and Channels and by the subscription of Entities to Channels. The behavioral part is defined by the Messages exchanged on each Channel and by the internal state of each Entity defined by the valuation of its Properties. Timing relations are expressed by the collections of all clocks associated to entities. Each Entity has its own reference of time given by the clock. At any given instant, when we capture the state of the system, different clocks can have different time values. It is interesting to note that, because the state comprises both a behavioral and a structural part, it is possible to represent a reconfiguration of the system as a change of state.

Based on the concept of state, we can now define a run as an infinite sequence of states (cf. Figure 12.3). In turn, we now define the semantics of a system based on runs. In Figure 12.4, we show the full ontology that we use to assign a semantics to UML. A system is defined by a set of runs. A specification defines a set of acceptable

Run

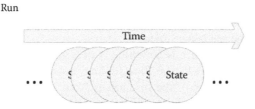

FIGURE 12.3 Definition of a run.

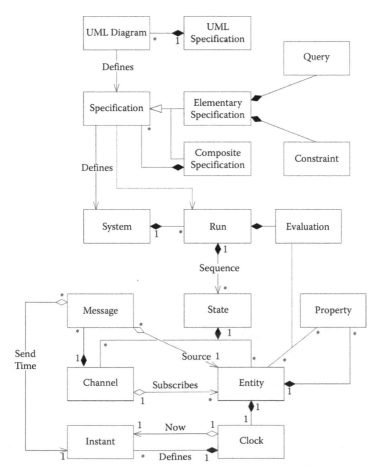

FIGURE 12.4 Ontology for distributed real-time system semantics.

runs. The specification can constrain the acceptable runs by specifying the initial states and the acceptable transitions.

Another interesting element of Figure 12.4 is the definition of Specification. A Specification can either be composite or elementary. Every Elementary Specification is made up of two elements: Query and Constraint. A Query selects states from all possible runs while the corresponding Constraint defines the characteristics for the run to be acceptable. We can think of the selector as an operator that is applied to all possible runs. All states selected by the Query are compared with the rules specified in the constraints. If they match, the run is accepted as part of the system whereas if they do not match, the run is discarded.

An important point to notice is how time is treated in the ontology. Each Entity has access to one private Clock. The Clock defines a series of Instants. At any given time, the Clock refers to one of the instants as Now. Each message has one Send Time (an instant on the clock of the entity that sends the message) and one Source Entity (the sender of the message). Therefore, it is possible to reason about when each Message was sent and by which Entity. Messages can be received by different Entities at different times. When an Entity that has subscribed to a Channel receives a Message, it can identify the local Entity time using its Clock and obtain the time of the sending Entity from the Message. Depending on the system and the requirements, it is possible to define synchronization strategies between the Clocks so as to be able to reason about times of events across different Clocks.

We can now give a formalization of the semantics informally described above. To this end, we first formalize the concepts of state and run as a foundation for the semantics of a distributed system specification. Then we present a simple grammar for a specification based on Queries and Constraints and use the formal definitions introduced before to provide a semantic for it.

12.3.1.1 Notational Preliminaries and System Formalization

We represent sets with capital Greek letters. For instance, the set of properties will be represented by Ψ. Each element of the set will be represented by the corresponding lowercase letter. For instance, a property in Ψ would be represented by ψ. A function from a domain A to a codomain B is expressed as $\varphi: A \to B$. A tuple is defined as $y = (y_1, y_2, \cdots) \in Y_1 \times Y_2 \times \cdots$ and $\pi_i \cdot y = y_i$ is the projection operator returning the ith element of the tuple. Given a set X, $\mathcal{P}(X)$ is the power set of X, where $|X|$ returns the cardinality of X. Furthermore, with $\mathbb{B}$ we indicate the set of Boolean values (true and false), $\mathbb{N}$ the set of natural numbers, $\mathbb{N}_+$ the set of natural numbers without 0, and with $\mathbb{N}_\infty$ the set of natural numbers with its supremum ∞.

A stream [32] is a finite or infinite sequence of Messages. Given a set of Messages M, we indicate with M^* the set of finite sequences over M, with M^∞ the set of infinite sequences, and with M^ω the union of those two sets. We can obtain the ith element of a stream x by using the infix dot operator $x.i$. The notation $x \downarrow i$ returns the prefix stream of length i, whereas $x \uparrow i$ returns the tail stream obtained by removing the first i elements from x. The concatenation of two streams x and x' is denoted as $x \frown x'$. We overload this notation to work with sets of streams $X \frown X'$ such that the resulting set contains all streams of the form $x \frown x$, where $x \in X \wedge x' \in X'$.

We can now give a formal definition of the elements of our ontology. For the two structural elements, Entities and Channels, we define two sets: the set E of Entities and the set X of Channels. For each Channel χ we have a set sigma Σ_χ keeping track of the Entities subscribed to χ. A Channel valuation relates the Channels (elements of the set X) to Messages exchanged over the Channels. Because a Channel can be used to send multiple Messages at any given moment, for every Channel χ we define a set M_χ of Messages currently sent over it. Furthermore, for each Entity ε we define a set Ψ_ε of Properties. A special Property v_ε encodes the current time of entity ε's Clock.

State is defined by (1) a structural configuration formed by Entities, Channels, and the subscriptions of Entities to Channels; (2) a behavioral configuration formed by Messages on each Channel, and valuation of Properties for each Entity; and (3) the current time value of the Clock property for each Entity.

Properties are intended to encode the state of an Entity. To abstract from the concrete data types used to define the variable space we define a set of functions Φ. Each $\varphi \in \Phi$ is a function defined from the values of a tuple of Properties to a Boolean: $\forall \varphi \in \Phi : (\{\varphi : \Psi \times \Psi \times \cdots \to \mathbb{B}\})$. This allows for easy translation of UML specifications. For instance, if we want to model a UML Deployment Diagram specifying that a node would run a particular program P, we can define a function run and have it evaluate to $true$ on the entity corresponding to the node ($run(P) = true$). The evaluation of the function set Φ over an entity ε is defined as $\Phi_\varepsilon \equiv \{(\varphi, \varphi_\varepsilon) : \varphi \in \Phi \wedge \varphi_\varepsilon \in \mathbb{B}\}$.

We can now define structural configuration as

$$Conf_{Structural} \equiv \left(E, X, \{\Sigma_\chi : \chi \in X\}\right)$$

We define behavioral configuration as

$$Conf_{Behavioral} \equiv \left(\{M_\chi : \chi \in X\}, \{\Phi_\varepsilon : \varepsilon \in E\}, \{v_\varepsilon : \varepsilon \in E\}\right)$$

We define state as

$$State \equiv (Conf_{Structural}, Conf_{Behavioral}) \in StateUniverse$$

where $State$ is an element of the $StateUniverse$ set containing all possible states.

We can now define the concept of a run using streams: $Run \in StateUniverse^\infty$. The semantics of a system specification in this framework emerges as the set of admissible runs:

$$System \in \mathcal{P}(StateUniverse^\infty)$$

12.3.1.2 Abstract Specification Language

We now define the abstract language we use to specify Queries and Constraints (and, therefore, systems). The benefits of defining this language are twofold. First, it provides an explicit context for mapping specifications (both composite and elementary) to systems in the semantic framework. Second, it provides a target language for the UML translation. The goal of the language is not to introduce a new textual syntax, and, therefore, we keep it simple by ignoring punctuation and other syntactic sugar necessary for a complete textual language definition.

We present the grammar of the language in a Backus–Naur Form using production rules of the following form:

$$\langle N \rangle ::= alt_1^{\langle N \rangle} \| alt_2^{\langle N \rangle} \| \cdots \| alt_n^{\langle N \rangle}$$

Nonterminals are enclosed in angular brackets, the symbol ‖ separates alternative productions, optional terms are enclosed in square brackets, and the notation {T}* represents the repetition of term {T} for 0 or more times.

$\langle \text{ELEM-SPEC} \rangle \quad ::= \langle \text{QUERY} \rangle \langle \text{CONSTRAINT} \rangle$

$\langle \text{SPEC} \rangle \quad\qquad ::= \langle \text{SPEC} \rangle \langle \text{SPEC} \rangle \| \langle \text{ELEM-SPEC} \rangle$

$\langle \text{QUERY} \rangle \quad\qquad ::= \{ \langle \text{MSG} \rangle \}^* \langle \text{ASSERTION} \rangle$

$\langle \text{CONSTRAINT} \rangle \ ::= [\exists \| \neg \exists] \{ [\neg] \langle \text{MSG} \rangle \}^* \langle \text{ASSERTION} \rangle$

$\langle \text{MSG} \rangle \quad\qquad\quad ::= \langle \text{MSGCONTENT} \rangle \langle \text{CHANNEL} \rangle$

$\langle \text{MSGCONTENT} \rangle ::= \langle \text{MSGNAME} \rangle (\langle \text{SENDER} \rangle \langle \text{TIME} \rangle \{ \langle \text{PARAM} \rangle \}^*)$

$\langle \textit{ASSERTION} \rangle \quad ::= \langle \text{FUNCTION} \rangle (\{ \langle \text{PROPERTY} \rangle \}^*) \|$

$\qquad\qquad\qquad\qquad \langle \text{UN-OPERATOR} \rangle \langle \text{ASSERTION} \rangle \|$

$\qquad\qquad\qquad\qquad \langle \text{ASSERTION} \rangle \langle \text{BIN-OPERATOR} \rangle \langle \text{ASSERTION} \rangle$

Operator definitions are not part of this grammar. Instead, they will be introduced when necessary in the translation of UML. In particular, we express all unary operators with the nonterminal $\langle \text{UN-OPERATOR} \rangle$ and binary operators with $\langle \text{BIN-OPERATOR} \rangle$. $\langle \text{FUNCTION} \rangle$ is a Boolean formula from property names to Boolean. Using this grammar, we can specify a system based on the ontology we have devised using Queries and Constraints. In the next section, we define the semantics of such specifications.

Using the $\langle \text{CONSTRAINT} \rangle$ optional operators $\exists$ and $\neg \exists$, it is possible to affect the structure of the system. We use $\exists$ to create new entities and channels, $\neg \exists$ to remove them.

Time is addressed in this language as a property of entities. In particular, we use the notation $next(t)$ to indicate the value of an entity clock in the first state where the value is greater than t. With $next$ we are able to reason about next states without constraining their occurrence to a particular time value. Moreover, the messages contain the $\langle \text{SENDER} \rangle$ entity and the sending $\langle \text{TIME} \rangle$ of the message in its parameter list.

12.3.1.3 Specification Language Semantics

An elementary specification $\langle \text{ELEM-SPEC} \rangle$ is captured in our abstract language by a tuple $\langle \text{QUERY} \rangle$, $\langle \text{CONSTRAINT} \rangle$. The goal of a specification is to define what runs are part of a system implementing such a specification. The $\langle \text{QUERY} \rangle$ identifies what parts of the run the specification is constraining, whereas the ⟨CONSTRAINT⟩ specifies how those parts are constrained. A run that fulfills a pair of query and constraint is such that in all states following a state where the query is true the constraint is true. Therefore, an ⟨ELEM-SPEC⟩ encodes a transition function between two states.

We define a ⟨QUERY⟩ as a communication context selecting the states that follow a particular message interaction, and a Boolean formula over properties, which identifies states to constrain. A query thus addresses both the contents of channels (the channel history) and predicates over the local data state of the relevant entities. We first define the channel configuration Xc as

$$Xc \equiv \left(X, \{ M_\chi : \chi \in X \}, \{ \Sigma_\chi : \chi \in X \} \right)$$

This definition captures the part of a state S that specifies the channel configuration and the messages being exchanged in the given state. The semantics $[\![q]\!]$ of a ⟨QUERY⟩ q is, therefore,

$$\forall q \in \langle \text{QUERY} \rangle, [\![q]\!] \equiv (h \in \mathcal{P}(Xc^*), a : \mathcal{P}(\Psi) \to \mathbb{B})$$

where Xc^* is a finite stream of channel configurations, the channel history $h \in \mathcal{P}(Xc^*)$ is a set of such streams, and the assertion a is a function from a set of properties to Boolean values.

We define a helper function

$$query : (\mathcal{P}(X^*), \mathcal{P}(\psi) \to \mathbb{B}) \times StateUniverse^\infty \to \{ \mathcal{P}(E) \times \mathbb{N}_\infty \}$$

that, given a ⟨QUERY⟩ semantics and a run, returns a set of tuples containing (1) the indexes of the states where one of the message histories is matched and (2) the corresponding set of entities for which the evaluation of the function is true. This helper function gives us all states in the run where we have to constrain the next state, as well as the corresponding entities to be constrained.

⟨CONSTRAINT⟩ is defined as a tuple of channel configurations, Boolean functions over properties, and one of the three quantifiers $\{ \exists, \neg \exists, - \}$. Similar to what we did for queries, we define the semantics of ⟨CONSTRAINT⟩ as

$$\forall c \in \langle \text{CONSTRAINT} \rangle, [\![c]\!] \equiv (Xc, a : \mathcal{P}(\Psi) \to \mathbb{B}, \{ \exists, \neg \exists, - \})$$

We can define a helper function

$$constr : (Xc, a : \mathcal{P}(\Psi) \to \mathbb{B}, \{ \exists, \neg \exists, - \}) \times \{ \mathcal{P}(E) \times \mathbb{N}_\infty \} \times StateUniverse^\infty \to \mathbb{B}$$

where $constr$ takes as arguments a run, the result of a query operation, and the semantics of a constraint. This function returns true if the constraint is satisfied. To be satisfied, the channel configuration of the selected states must match the Xc specified by the constraint. Moreover, how the rest of the constraints is satisfied depends on the choice among the three quantifiers $\{\exists, \neg\exists, -\}$. If the chosen quantifier is $-$, the assertion s must evaluate to true in all entities selected. If the quantifier is $\exists$, the assertion s must evaluate to true in some entities not part of the selected ones. Finally, if the quantifier is $\neg\exists$ the selected entities must not be present in the selected states.

Now we can define a $\langle \text{SPEC} \rangle$ in the semantic domain as a set of tuples of the form (query, constraint), and the system corresponding to the specification as the set of all possible runs that fulfill all such tuples (query, constraint) of the set.

Formally,

$$\langle \text{SPEC} \rangle \subseteq \{(\langle \text{QUERY} \rangle, \langle \text{CONSTRAINT} \rangle)) : \langle \text{QUERY} \rangle, \langle \text{CONSTRAINT} \rangle\}$$

$$[\![\langle \text{SPEC} \rangle]\!] \equiv \{Run : Run \in StateUniverse^{\infty} : \forall s \in \langle \text{SPEC} \rangle,$$
$$\forall q \in query(s.0, Run), constr(s.1, q, Run)\}$$

12.3.2 NOTION OF CONSISTENCY

We are interested in defining dynamic consistency for real-time distributed systems. This is the reason why we have tailored our semantic framework to this domain rather than staying within the generality of the UML language metamodels. Given the semantic framework presented in the previous section, it is now straightforward to define dynamic consistency for models in this system class. We will first define horizontal consistency and then vertical consistency.

We can define horizontal consistency as follows: a specification is horizontally consistent if the system it defines admits at least one run. This is consistent with the definition of horizontal consistency we gave before. In fact, a specification $\langle \text{SPEC} \rangle$ is formed of multiple views at the same level of abstraction (in our formalism this means multiple sets of query and constraint tuples).

Definition 12.1

A specification $\langle \text{SPEC} \rangle$ such that $[\![\langle \text{SPEC} \rangle]\!] \in \mathcal{P}(StateUniverse^{\infty})$ is horizontally consistent $iff [\![\langle \text{SPEC} \rangle]\!] \neq \emptyset$.

This definition captures the idea that the specification is implementable. There are two possibilities for a system to fulfill this property. Either there are no contradictions in the specification or the admissible runs do not match any query that defines inconsistent constraints. There is nothing wrong in using different perspectives to constrain the system behavior specified by other perspectives. However, if a perspective constrains the behavior of the system such that no run satisfying the specifications of that perspective is allowed in the final system, there can be a consistency problem. A stricter rule for horizontal consistency requires that the system has at least one run admissible for each perspective, meaning that there is at least one run satisfying some queries of each perspective specification.

Definition 12.2

A specification $\langle\text{SPEC}\rangle$ such that $[\![\langle\text{SPEC}\rangle]\!] \in \mathcal{P}(StateUniverse^\infty)$ and $\langle\text{SPEC}\rangle$ made of N specifications $\langle\text{PERSP}_i\rangle$ called perspectives such that $[\![\langle\text{SPEC}\rangle]\!] = \bigcap_{i\in N}[\![\langle\text{PERSP}_i\rangle]\!]$ is horizontally consistent *iff* $\forall \langle\text{PERSP}_i\rangle, \exists\, Run \in [\![\langle\text{SPEC}\rangle]\!] \wedge \exists s \in \langle\text{PERSP}_i\rangle$ such that $query(s.0, Run) \neq \varnothing$.

A possible problem with our first definition of horizontal consistency is that we could have a system specification with no runs satisfying any query of the general specification. The consistency specification for such a system is vacuously satisfied (i.e., runs are possible because selectors never match). The second definition solves this problem requiring that some runs matching the specification queries are present.

The two definitions of horizontal consistency we gave support two different usage scenarios. In fact, we can identify two main reasons to create a specification. First, we can be interested in constraining how the system works in a given scenario. The scenario we want to constraint must, therefore, be possible and the corresponding query must select some runs. For this type of usage we should use consistency Definition 2. A different use case is when we want to specify recovery from some failure of the system. For example, we may identify that a given interaction can happen as a result of a failure even if the specification would not allow for it. In this case, the goal is to describe the detection and recovery from a given failure. In this case, we can use consistency Definition 1.

Vertical consistency is defined between two specifications at different levels of abstraction. We can define this consistency notion by a containment relation between runs. If we have a more abstract specification SPEC_a and a more concrete specification SPEC_c, we define vertical consistency as follows: a concrete specification SPEC_c is consistent with an abstract specification SPEC_a if all runs allowed in the concrete system specification are also allowed in the abstract one. Moreover, the abstract system allows runs that the concrete system does not allow. This definition requires that the concrete systems admit a strict subset of the runs admitted by the abstract one.

Definition 12.3

Two specifications $\langle\text{SPEC}_a\rangle$ and $\langle\text{SPEC}_c\rangle$, where the first is the abstract and the second the concrete specification, are vertically consistent *iff* $[\![\langle\text{SPEC}_a\rangle]\!] \subseteq [\![\langle\text{SPEC}_c\rangle]\!]$.

Given the definitions of $\langle\text{SPEC}\rangle$ and $[\![\langle\text{SPEC}\rangle]\!]$ of the previous section, we can now define a modularity theorem. We first observe that each specification has a set of tuples containing one query and one constraint. Therefore, each of these tuples defines a set of runs. From the definition of $[\![\langle\text{SPEC}\rangle]\!]$, we can infer a lemma asserting that the semantics of a complex $\langle\text{SPEC}\rangle$ (i.e., formed by multiple tuples of query and constraint) is the intersection of the semantics of all the subspecifications formed by single query/constraint tuples. The modularity theorem states that for any complex specification $\langle\text{SPEC}\rangle$ we can always identify two subspecifications such that the intersection of the runs permitted by the two contains exactly the runs permitted by the original specification. Moreover, the theorem states that, to obtain such subspecifications we can simply take two subsets of the tuples of the original specification, provided that all tuples of the original specification are in at least one of the two subspecifications. Now we can formally define the lemma and the theorem as follows.

Lemma 12.1

Given a specification $\langle SPEC \rangle$

$$[\![\langle SPEC \rangle]\!] = \bigcap_{\forall t \in \langle SPEC \rangle} [\![\{t\}]\!]$$

Proof: Lemma 1 can be proved by observing that the definition of $[\![\langle SPEC \rangle]\!]$ is such that if a specification contains a single query/constraint tuple t, the $\forall$ quantification in $\forall s \in \langle SPEC \rangle$ return a single element. Therefore, we have

$$[\![\{t\}]\!] \equiv \{Run : Run \in StateUniverse^{\infty} : \forall q \in query(t.0, Run), constr(t.1, q, Run)\}$$

given the definition of intersection: $\bigcap_{\forall s \in S} s = \{e : \forall s \in S, e \in s\}$. Replacing the specification of the semantics of a query/constraint tuple into the definition of intersection we obtain

$$\bigcap_{\forall t \in \langle SPEC \rangle} [\![\{t\}]\!] = \{e : \forall t \in \langle SPEC \rangle,$$

$$e \in \{Run : Run \in StateUniverse^{\infty} :$$

$$\forall q \in query(t.0, Run), constr(t.1, q, Run)\}\}$$

From this, by replacing e with the definition of Run we obtain

$$\bigcap_{\forall t \in \langle SPEC \rangle} [\![\{t\}]\!] = \{Run : \forall t \in \langle SPEC \rangle, Run \in StateUniverse^{\infty} :$$

$$\forall q \in query(t.0, Run), constr(t.1, q, Run)\}$$

which is our definition of $[\![\langle SPEC \rangle]\!]$.   ∎

The Modularity theorem asserts that complex query/constraint specifications can be split into two simpler ones without losing information.

Theorem 12.1

Modularity. Given a specification $\langle SPEC \rangle$ such that $|\langle SPEC \rangle| > 1$ (i.e., the specification is complex), $\forall \langle SPEC_1 \rangle, \langle SPEC_2 \rangle$ such that

$$\langle SPEC_1 \rangle \subset \langle SPEC \rangle \wedge \langle SPEC_2 \rangle \subset \langle SPEC \rangle \wedge$$

$$|\langle SPEC_1 \rangle| > 0 \wedge |\langle SPEC_2 \rangle| > 0 \wedge \langle SPEC_1 \rangle \cup \langle SPEC_2 \rangle = \langle SPEC \rangle$$

$$[\![\langle SPEC \rangle]\!] = [\![\langle SPEC_1 \rangle]\!] \cap [\![\langle SPEC_2 \rangle]\!].$$

The proof of Theorem 12.1 derives easily form Lemma 12.1. In fact, because the semantics of a specification is equivalent to the intersection of the semantics of all its constituent query and constraint tuples, we can use the commutative and associative properties of intersection to prove Theorem 12.1.

12.3.3 EXAMPLE OF CONSISTENCY MANAGEMENT

To show how the methodology outlined in this chapter applies to consistency checking in the context of UML for real time, we are required to provide a translation from UML and from its MARTE profile to the abstract language we introduced. Translating the entire UML and MARTE metamodels is beyond the scope of this chapter. Instead, we chose a simple subset of UML and MARTE that uses three graphical notations: component diagrams, sequence diagrams, and state diagrams, which we used in the example of Figure 12.1. Furthermore, we translate MARTE timed constraints as we used them in our example.

The translation from UML models to our query and constraint language assigns a precise semantics to each model. Several options for assigning semantics to each notation exist. For the sequence diagram, for instance, we have many possibilities for interpreting them existentially (at least the specified behavior must be possible) or universally (precisely the specified behavior is required) [43]. Notice that the decision of interpreting the diagrams existentially or universally depends on what the goal of the specification is. For example, in a requirements document an interaction can exemplify one of many possible scenarios and the existential interpretation would be correct. For real-time systems modeling we interpret sequence diagrams universally. All messages exchanged in the system must be represented in diagrams. This interpretation of sequence diagrams is viable for our application domain. In fact, one of the key uses of models of communication in real-time systems is to analyze the network traffic and ensure that real-time constraints can be met. To this end a complete view of which messages are exchanged over the communication channels is necessary.

Our translation strategy interprets every element of a UML graph as a query and constraint tuple. We introduce an operator to compose those elementary specifications—this closes the loop with the introduction of the abstract query/constraint syntax. For demonstration purposes, we introduce the parallel operator. This operator is applied between any two specifications in our translation and returns the specification containing all query and constraint tuples of the operand specifications.

$$\alpha, \beta \in \langle \text{SPEC} \rangle$$

$$[\![\alpha \text{ PAR } \beta]\!] \equiv \{ s : s \in \alpha \vee s \in \beta \}$$

In Table 12.4, we provide translation rules for some of the interesting model elements used in our example. The entire set of rules is beyond the scope of this chapter. Each rule provides a set of query/constraint tuples that can be composed in a specification using the parallel operator. To support the translations, we define a small set of helper functions.

TABLE 12.4

Translation Rules for UML Metamodel Elements

Name	Metamodel Element	Translation
UML::BasicComponents:: Component	Figure 12.5	$Q : \{ \}$ *true* $C : \exists \{ \} EType = Component.name$
UML::BasicInteractions:: MessageOccurrence Specification	Figure 12.6	$MOS \equiv MessageOccurrenceSpecification$ $t \equiv$ time of last message received in history $Q : ExtractHistory (\{MOS\}) \ Clock > t$ $C : toMSG(\{MOS\}) \ true$
UML:: BehaviorStateMachines:: Transition	Figure 12.7	$TR \equiv Transition, B \equiv TR.effect$ $s_1 = TR.source, s_2 = TR.target$ $Q : ExtractHistory(\{TR.trigger\}) \ State = s1 \wedge Clock = t$ $C : toMSG(B) \ State = s2 \wedge Clock = next(t)$
TimedConstraints::Timed Constraint	Figure 12.8	$PR \equiv TimedInstantConstraint.specification$ $Q : MsgFromObservations(PR.observation)$ $PropFromObservations(PR.observation)$ $C : \{ \} evalVSL(PR) = true$

The function *toMSG*() is used to convert two elements of the UML metamodel, MessageOccurrenceSpecification and Triggers, into objects suitable for our abstract language. Informally, we can think of MessageOccurrenceSpecification as representations on sequence diagram lifelines of the events related to message sending and receiving (plus execution of actions and other details we do not consider in our simplified model). The function *toMSG*() expresses the translation from OccurrenceSpecification elements of the UML metamodel to messages in our abstract language specification.

Similarly, the *ExtractHistory*() function applied to a model element of type MessageOccurrenceSpecification returns the sequence of messages that maps to the Events in the lifeline before the one defined by the given MessageOccurrenceSpecification. Intuitively, this function returns the history necessary for a query to select the correct interactions before applying the constraint to match the message event defined by the MessageOccurrenceSpecification model element. We do not describe the details of how this translation is performed because it is beyond the scope of this chapter. In fact, the UML metamodel is very complex. Extracting relations between events and specification elements in different diagrams often requires the exploration of a deep class hierarchy. For example, by inspecting the metamodel of Figure 12.6, we can observe that to extract the history of events before a given message in a sequence diagram, we have to identify the Lifeline the OccurrenceSpecification is covered by. Then, leveraging the fact that the set of events of a lifeline is ordered, we could extract all the OccurrenceSpecifications that precedes the one for which we are creating the history. Once we have the ordered list of OccurrenceSpecifications in the history, we can navigate their event property to obtain the corresponding Events. By reflection, we can identify the events that are related to sending and receiving messages and use this information to generate the list of message specifications.

The four translations given in Table 12.4 map the elements of UML and MARTE metamodels depicted in Figures 12.5 through 12.8 to query and constraint tuples. The first line of the table gives a translation for Figure 12.5. This part of the metamodel defines UML components in a component diagram. The simple model in the figure captures the relation between Components and Interfaces, which can be required or provided by the Component. Our translation simply asserts that a specification of a component always imposes the existence of an entity with a property called *EType* and value equal to the component name in the UML diagram.

The translation of line 2 of Table 12.4 defines constraints imposed by a MessageOccurrenceSpecification in a UML sequence diagram. The query extracts the message history before the given MessageOccurrenceSpecification. As we already mentioned discussing ExtractHistory, this is not a trivial operation. Figure 12.6 presents the relevant subset of the UML model for sequence diagrams. Interactions are the type of behavior specified by this type of diagram. In particular, an Interaction is a type of InteractionFragment that can be composed of other such fragments. Special types of InteractionFragments are OccurrenceSpecifications which reference communication Events and Lifelines. An example of such specifications is MessageOccurrenceSpecifications, which represent messages exchanged according to the interaction modeled. The constraint in our translation is the existence of the message corresponding to the

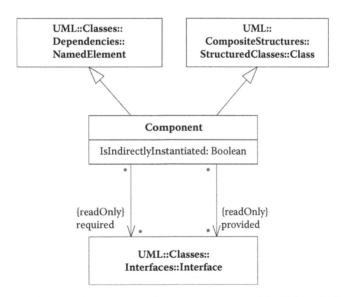

FIGURE 12.5 Subset of the UML Component metamodel. (Adapted from Object Management Group, "Unified Modeling Language (OMG UML), Superstructure, Version 2.3." 2010, formal/2010-05-05, OMG.)

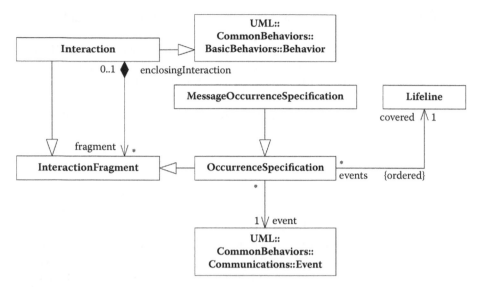

FIGURE 12.6 Subset of the UML Message metamodel. (Adapted from Object Management Group, "Unified Modeling Language (OMG UML), Superstructure, Version 2.3." 2010, formal/2010-05-05, OMG.)

MessageOccurrenceSpecification. This translation covers only events that are messages. Other types of events cause properties in some entity to be set and are not covered in our example.

The third line of Table 12.4 defines a Translation for state-machine transitions. To this end, we introduce an entity property named State. Figure 12.7 depicts the relevant subset of the UML metamodel for state-machine diagrams. According to the UML metamodel, a Transition has a source and target Vertex, and State is a type of Vertex. A Transition can be taken only if the guard constraint is true and, in this case, is taken when a given trigger occurs. Moreover, a Transition can have an effect. The effect is a Behavior. An example of Behavior is the Interaction (as depicted in Figure 12.6), which can contain message-related events (because MessageOccurrenceSpecifications are also InteractionFragments). For our case study, we simplify the translation to address just triggers and effects that are messages. The query part of the translation of Transition selects entities, where the State variable coincides with the source state of the model. Other propositions in the query can be used to restrict the selection to only specific entities. In fact, state diagrams define the behavior of particular model elements. For example, in our case study we want to apply the state diagram of Figure 12.1c only to the component Emergency Brake. In this case, the query should also limit the selection to states of the entity Emergency Brake. To this end, we can add to the query another clause that selects only entities of the correct type (i.e., *EType* = "Emergency Brake"). The other part of the query limits the selection to states where the trigger message is present. The constraint simply forces the next state of the selected entities to have the target state in the State property.

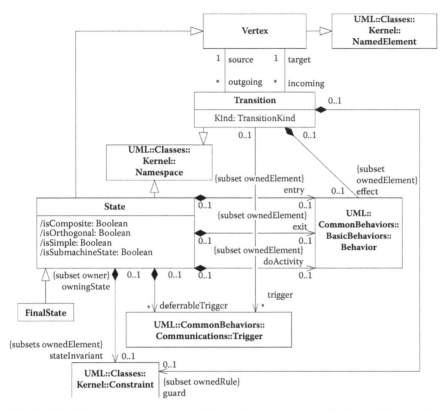

FIGURE 12.7 Subset of the UML Transition metamodel. (Adapted from Object Management Group, "Unified Modeling Language (OMG UML), Superstructure, Version 2.3." 2010, formal/2010-05-05, OMG.)

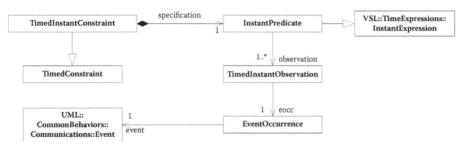

FIGURE 12.8 Subset of the MARTE TimedConstraints metamodel. (Adapted from Object Management Group, "UML Profile for MARTE: Modeling and Analysis of Real-Time Embedded Systems, Version 1.0." 2009, formal/2009-11-02, OMG.)

Finally the last line of Table 12.4 defines the translation for MARTE TimedInstantConstraint. Figure 12.8 shows the relevant MARTE metamodel. A TimedInstantConstraint has a specification that is a predicate over a set of observations (TimedInstantObservations). Each observation identifies an event occurrence. EventOccurrences relate MARTE observations to UML Event elements.

MARTE introduces the Value Specification Language (VSL) to formulate algebraic and time expressions. The translation of MARTE InstantPredicates has to interpret the VSL InstantExpression. To this end, in our translation, we use an *evalVSL* Boolean function that evaluates a VSL expression. Moreover, we use two functions, *MsgFromObservations* and *PropFromObservations*, to obtain the messages and properties that correspond to the events referred to by the observation of a predicate. From Figure 12.8, we can observe the complexity of obtaining an event from a TimedConstraint. Obtaining messages and properties from events requires a good understanding of the UML metamodel and the exploration of many nested relations. Although complex, those functions can be implemented in a program. The translation then selects the correct messages and entities in the query part of the specification and asserts that the specification in VSL evaluates to true in the constraint.

We can now show how to detect inconsistency with this query and constraint framework using the example of Figure 12.1. Thanks to the modularity theorem defined in the previous section, we can split each specification into simpler specifications. In particular, because the intersection of the specifications obtained with the modularity theorem is equivalent to the original specification, we can prove inconsistency by just translating a subset of the model and proving that such subset is inconsistent (no runs allowed).

For example, we can translate the model element of Figure 12.1b that represents the sending of an Ack message from the Emergency Brake to the Train Controller. This translation, according to Table 12.4, would look like

$$Q : \{\} * \{(\text{Commands Received (Train Controller, } t_1))\} \, EType$$
$$= \text{Emergency Brake} \wedge Clock > t_1$$

$$C : \{(\text{Ack(Emergency Brake, } t_{ack}))\} \, EType = \text{Emergency Brake}$$

The translation of the Figure 12.1c transition triggered by the Commands Received message is

$$Q : \{(\text{Commands Received (Train Controller, } t'))\} \, EType = \text{Emergency Brake} \wedge State$$
$$= \text{Wait Commands} \wedge Clock = t'' \wedge t'' > t'$$

$$C : \{\} \, EType = \text{Emergency Brake} \wedge State = \text{Reset Timer} \wedge Clock = next(t'')$$

and

$$Q : \{\} \, EType = \text{Emergency Brake} \wedge State = \text{Reset Timer} \wedge Clock = t'''$$

$$C : \{\} \, EType = \text{Emergency Brake} \wedge State = \text{Wait Commands} \wedge Clock = next(t''')$$

Let us analyze the type of runs that satisfy the translation of the sequence diagram. We can observe that, for a run to satisfy the specification, if in a state there is a Commands Received message received by the Emergency Brake component, it must send an Ack message. In the sequence diagram translation, we do not specify if there is some other action local to the Emergency Brake. In fact, our simplified translation for sequence diagrams deals only with messages sent and received, not local actions. So the message can be returned immediately (next state) or after some local transitions (that is the meaning of { }*, which represents a sequence of zero or more states where the channel is empty). The specification, however, is clear in identifying that no other messages are sent or received by Emergency Brake before returning a message.

The translation of the state diagram of Figure 12.1c triggers a transition from Wait Commands to Reset Timer when the Commands Received message is received by Emergency Brake. We can identify the inconsistency by observing that all runs that fulfill our translation for Figure 12.1c never send the Ack message. The intersection of sets of runs identified by the two specifications is, therefore, empty. Thus, the two specifications are inconsistent.

In our formalism, we can prove consistency by composing query and constraint tuples and identifying contradictions. In particular, we chose to encode queries and constraints using Propositional LTL formulae [44]. The encoding changes for each definition of consistency. We can then prove that a system is consistent according to the chosen definition by proving that the LTL formula that encodes such definition is satisfiable. This proof can be automated by means of a SAT solver for LTL formulas. Examples of algorithms for assessing SAT of propositional LTL formulas and tools implementing them can be found in Goranko et al. [45] and Rozier and Vardi [46].

In this chapter, we do not give a complete translation for all definitions. Instead, we use the example of inconsistent specification from Figure 12.1b and 12.1c and encode the query and constraint specification to prove inconsistency according to Definition 1. For each tuple of query (Q) and constraint (C), we create the implication $Q \Rightarrow \mathbf{X}C$, where $\mathbf{X}$ is the next operator in LTL. If we can find a set of variables that satisfies the disjunction of all these implications, the specification is consistent according to Definition 1.

We capture this in the following theorem.

Theorem 12.2

Consistency D1 Satisfiability. Given a specification $\langle\mathrm{SPEC}\rangle$, $\langle\mathrm{SPEC}\rangle$ is consistent according to consistency Definition 12.1 *iff* the expression $\bigwedge_{\forall(Q,C)\in\langle\mathrm{SPEC}\rangle} Q \Rightarrow \mathbf{X}C$ is satisfiable.

In this theorem, we assume that messages in the channel history are encoded using appropriate variables and nested temporal operators. The exact discussion of how to encode these messages is beyond the scope of this chapter. The proof of Theorem 12.2 follows from the definition of $[\![\langle\mathrm{SPEC}\rangle]\!]$. In fact, the semantics of $\langle\mathrm{SPEC}\rangle$ is defined as the set of runs that satisfy all query/constraint tuples. We encode each tuple as an implication in LTL that is true if a run satisfies it. The conjunction of all the LTL implications is true only if a run satisfies all of them. If the formula in Theorem 12.2 is not satisfiable, there exists no run that can satisfy all implications at the same time, thus $[\![\langle\mathrm{SPEC}\rangle]\!]$ is empty. On the other hand, if the expression is satisfiable, there exists at least one run that can satisfy all queries and constraints, thus $[\![\langle \mathrm{SPEC}\rangle]\!]$ is non-empty. This proves Theorem 12.2.

Now we can consider how Theorem 12.1 and Theorem 12.2 apply to our example. From Theorem 1 we know that to prove inconsistency we are not required to compose all the queries and constraints. Instead we can split the specification into two sub-specifications, and the original one will be equivalent to the intersection of the new specifications. Then, if we can prove that one of the two is empty we know that the full specification must be inconsistent. We chose to compose only the specifications of Figure 12.1b and 12.1c. We prove that this subspecification is inconsistent (i.e., has an empty set of runs) and from Theorem 1 we obtain that the full specification is also inconsistent.

Consider all runs satisfying the translation of the transition from Wait Commands to Reset Timer in Figure 12.1c. We identify all runs with a trigger message Commands Received and a transition in the entity Emergency Brake with State changing from "Wait Commands" to "Reset Timer."

Because the constraint of this specification is the query of the translation for the transition from Reset Timer to Wait Commands in Figure 12.1c, if we compose the two specifications, we obtain all runs where Emergency Brake reacts to a Commands Received by changing two states without sending any message.

We can now compose the current system into the translation of the Ack message specification in Figure 12.1b and discover that one of the next states of the runs selected must send an Ack message before any other messages are received by Emergency Brake. However, from state Wait Commands the system can exit only if the trigger message Commands Received is received. Therefore, by exploring specification tuples we can argue that because the Clock time greater than t_1, at which the Ack message must be sent by the sequence diagram constraint, is finite and the specification of the state diagram does not allow any transition that sends messages without receiving anything from the state that it enters after the trigger message, we have a contradiction, and, therefore, the specifications are inconsistent.

If we observe the specification of the state diagram in Figure 12.1d, we can give the following translation for the transition from Reset Timer to Wait Commands:

$$Q : \{\} \; EType = \text{Emergency Brake} \wedge State = \text{Reset Timer} \wedge Clock = t'''$$

$$C : \{(\text{Ack}(\text{Emergency Brake}, t_{ack}))\} \; EType = \text{Emergency Brake} \wedge State$$
$$= \text{Wait Commands} \wedge Clock = next(t''')$$

With this change, the composition of the specifications for the state machine identifies a sequence of states initiated by the trigger message Commands Received that ends with the sending of an Ack message. In the composition with the specification from Figure 12.1b, we have that the state where the Ack message is sent must happen at a time $next(next(t''))$ that is greater than t_1. Therefore, there is no contradiction between constraints and, thus, no inconsistency. To prove that the entire specification is consistent we would be required to add the translations of the remaining elements. Although this process is long and error-prone if performed by hand, the existence of automated tools for solving the SAT problem makes it a viable solution.

12.4 DISCUSSION

In this section, we discuss our consistency checking approach, including its strengths and weaknesses. We demonstrated our approach for consistency management based on queries and constraints on a reduced subset of UML and its MARTE profile. The goal of this work was to demonstrate the feasibility of the approach by providing a case study where we were able to identify inconsistencies in UML models. Thus, the translation we gave assigned a semantics only to a subset of the modeling elements defined in UML and MARTE. However, even using this reduced subset we were able to detect and formally verify the inconsistency between models of our BART case study including timing constraints. Because our translation binds query and constraint tuples to single entities in the UML metamodel, an extension to the full language definition of UML 2.0 and its different profiles is straightforward, albeit complex. Such an extension requires us to decide on a precise semantics for each diagram. Therefore, we must fix how each syntactic element of each diagram contributes to its semantics.

Tailoring our consistency notion to a particular target domain (real-time distributed systems in this case) may, at first, seem limiting. However, we argue that a completely general definition of consistency for a general purpose language such as UML ultimately limits the applicability of consistency checking to very abstract models or to purely structural notions of consistency (without taking the notion of behavior into account). This claim is supported by our analysis of the related work. We observe in Tables 12.2 and 12.3 that all approaches that dealt with behavior consistency translated UML to some other language supporting a specific target domain. Similarly, our approach precisely defines the semantic space for both structural and behavioral consistency—albeit at the expense of complete generality. If we decide to target a different domain, a new target model and updated translation rules are necessary.

Different decisions in how to interpret diagrams can lead to different translations. For example, we decided to interpret sequence diagrams universally regarding the messages exchanged. Each message represented in the diagram is exchanged and messages not represented are not. In contrast, state transitions are not part of our translation of sequence diagrams. This is why we set the Clock in the query of row 2 of Table 12.4 as greater than the time the previous message was sent without setting a specific interval. This is equivalent to a commitment to eventually have a state in which the constraint is true.

The definitions of horizontal and vertical consistencies we gave seem adequate for the domain of real-time systems. However, when a richer subset of UML will become amenable to translation and more experience is acquired in verifying it, we see potential for reevaluating the definitions. One possible area of concern with the current definition arises when we allow side effects between the queries and constraints of multiple diagrams, in other words, nonlocal constraints. In this case we could change the definition of horizontal consistency, for instance, to yield inconsistency if the majority of the queries do not match.

The benefit of moving from the abstract domain of UML metamodels to the query and constraint abstract language is that the translation rules define the semantics and implicitly also the consistency rules. We can then avoid enumerating a long list of consistency rules and obtaining a very simple definition of consistency.

With this approach, we have converted the problem of detecting the consistency of graphs based on the UML metamodel to verifying emptiness of sets. The sets are defined by logical formulae, each defining the effect of one model element on the system runs. The composition of specifications is defined by set intersection. Additionally we have presented a modularity theorem (Theorem 12.1) that enables reasoning on separate subsets of the query/constraint specifications. This setup is amenable to translation into propositional LTL and supports use of many automatic formal verification tools, such as SAT solvers. We have provided Theorem 12.2 that affirms the equivalence of proving that an LTL expression is satisfiable with horizontal consistency of the corresponding specification.

We evaluate the query and constraint approach we proposed by identifying how it addresses the 12 requirements we identified in Table 12.1.

R1. Support inconsistent models. We address this requirement by not forcing the user to remove inconsistencies. Models that are inconsistent can be identified by identifying the tuples that are in contradiction. More modeling elements can be added and more contradictions detected before the system is made consistent.

R2. Automatic inconsistency discovery. Inconsistencies are discovered by hand in our example. The goal was to show the complexity of the problem and a possible solution. It is possible, however, to automate translation (which leverages the UML metamodel used by all UML modeling tools) and detection to discover inconsistencies automatically. Furthermore, inconsistencies can be tracked by identifying the subset of specifications that are in contradiction.

R3. Support inconsistency resolution. The support to resolve inconsistencies is provided by the ability to identify a small subset of the specification that is sufficient to prove the inconsistency (this property stems from the Modularity Theorem).

R4. Support multiple modeling languages. The query and constraint approach supports multiple languages by creating different translation rules from the UML metamodel to the abstract target language. It could also support languages that are not UML as long as they are based on a metamodel and a translation is provided.

R5. Support different levels of abstraction. We have identified different consistency rules and translation rules to support different levels of abstraction.

R6. Support extensions. We demonstrated the support for extensions of UML providing a translation rule for the MARTE profile.

R7. Support horizontal consistency. We provided two horizontal consistency definitions.

R8. Support vertical consistency. We provided one definition for vertical consistency.

R9. Support static consistency. This approach supports static consistency by querying entity properties and channel messages and by constraining them.

TABLE 12.5

Evaluation of the Consistency Checking Presented in This Chapter

Translation Target Language	R1	R2	R3	R4	R5	R6	R7	R8	R9	R10	R11	R12
Queries and Constraints	✓	✓	✓	✓	✓	✓	✓	✓	✓	✓	✓	✓

Note: The symbol ✓ indicates that the requirement is met.

R10. Support dynamic consistency. The approach supports dynamic consistency by constraining the properties of different states in admissible runs. By leveraging LTL logic and the Clock, it is possible to set constraints on consecutive states or future states.

R11. Provide tool support. While we have not provided any tool support for this approach, we have demonstrated that a translation of the consistency problem to SAT of LTL formulae exists (Theorem 12.2). The translation from UML to another domain can be automated and because we can encode queries and constraints in LTL, we can use existing SAT solvers for this logic to automate the verification.

R12. Address scalability. Thanks to the modularity theorem our approach does not require reasoning about the entire model to identify inconsistencies. This makes it applicable to large models. However, depending on how the different specifications are interconnected, to ensure that no inconsistency exists it may be necessary to compose a large number of tuples, which could slow down the identification of inconsistencies on some models.

From this requirements analysis, summarized in Table 12.5, we conclude that the query and constraint approach proposed is a step toward a more comprehensive consistency management approach for UML models. However, more work is required to implement tools to automate the approach and experiment with the effective scalability of such tools by testing them on large industrial-scale system models.

12.5 SUMMARY AND OUTLOOK

UML is a widely accepted language that can be used at different stages of the development process. In fact, while the current state of the art provides profiles that support real-time systems modeling, some issues in ensuring consistency, defining semantics, understanding models, and integrating models of large-scale systems still hamper the applicability of UML to this domain. To address these shortcomings, we discussed the consistency problem and steps toward a comprehensive solution for it in more detail.

From our analysis of the various techniques to address consistency management in UML, we infer that, even if it is theoretically possible to manage consistency in UML, the problem has not been solved yet completely. Most of the problems in dealing with consistency, and in general in developing analysis tools and techniques for UML arise from the size and complexity of the UML metamodel and from the lack of a comprehensive semantics definition. Consistency of models and correctness of analysis and synthesis tools are key requirements for modeling real-time systems—hence, research into increased consistency provides significant leverage for improved quality for this systems class.

We proposed a solution for consistency checking of UML based on the definition of an abstract model of the application domain that we want to target and a translation of the UML metamodel to this abstract domain. We also defined two theorems that apply to our consistency checking solution: the first enables checking consistency on subsets of system models, the second enables using SAT solvers for LTL to automate consistency checking. Although we believe the solution we proposed to be a significant step in the right direction, to make its use practical, providing a tool that supports automatic translation and detection of inconsistencies is important future work.

ACKNOWLEDGMENTS

This work was partially supported by the National Science Foundation within projects CCF-0702791 and CNS-0963702, as well as by funds from the California Institute for Telecommunications and Information Technology (Calit2) at the University of California, San Diego, CA. We are grateful to Barry Demchak and the anonymous reviewers for insightful comments.

REFERENCES

1. Giese, H., G. Karsai, E. Lee, B. Rumpe, and B. Schätz. 2011. *Model-Based Engineering of Embedded Real-Time Systems*. Lecture Notes in Computer Science, Vol. 6100. Berlin/ Heidelberg: Springer.
2. Object Management Group. 2010. "Unified Modeling Language (OMG UML), Infrastructure, Version 2.3." formal/2010-05-03, OMG.
3. Object Management Group. 2010. "Unified Modeling Language (OMG UML), Superstructure, Version 2.3." formal/2010-05-05, OMG.
4. Sprinkle, J., B. Rumpe, H. Vangheluwe, and G. Karsai. 2011. "Metamodelling—State of the Art and Research Challenges." In *Model-Based Engineering of Embedded Real-Time Systems*, edited by H. Giese, G. Karsai, E. Lee, B. Rumpe, and B. Schätz, 57–76. Lecture Notes in Computer Science, Vol. 6100. Berlin/Heidelberg: Springer.
5. Object Management Group. 2009. "UML Profile for MARTE: Modeling and Analysis of Real-Time Embedded Systems, Version 1.0." formal/2009-11-02, OMG.
6. Warmer, J., and A. Kleppe. 1998. *The Object Constraint Language: Precise Modeling with UML*. Wokingham, U.K.: Addison-Wesley.
7. Batory, D., D. Benavides, and A. Ruiz-Cortes. 2006. "Automated Analysis of Feature Models: Challenges Ahead." *Commun ACM* 49 (12): 45–7.
8. Object Management Group. 2003. "Model Driven Architecture (MDA), Version 1.0.1." omg/03-06-01, OMG.
9. Winter, V., F. Kordon, and M. Lemoine. 2004. "The BART Case Study." In *Formal Methods for Embedded Distributed Systems*, edited by F. Kordon, and M. Lemoine, 3–22. New York, NY: Springer US.
10. Finkelstein, A. C. W., D. Gabbay, A. Hunter, J. Kramer, and B. Nuseibeh. 1994. "Inconsistency Handling in Multiperspective Specifications." *IEEE Transactions on Software Engineering* 20 (8): 569–78.

11. Easterbrook, S., and B. Nuseibeh. 1996. "Using ViewPoints for Inconsistency Management." *Software Engineering Journal* 11 (1): 31–43.
12. Easterbrook, S. 1996. "Learning from Inconsistency." In *Proceedings of the Eighth International Workshop on Software Specification and Design,* March 22–23, 1996, pp. 136–40. Schloss Velen, Germany. Washington, DC, USA: IEEE Computer Society.
13. Easterbrook, S., J. Callahan and V. Wiels. 1998. "V&V through Inconsistency Tracking and Analysis." In *Proceedings of the Ninth International Workshop on Software Specification and Design,* April 16–18, 1998, pp. 43–9. Mic, Japan, Washington, DC: IEEE Computer Society.
14. Evans, E. 2004. *Domain-Driven Design: Tackling Complexity in the Heart of Software,* 560. Wokingham, U.K.: Addison-Wesley.
15. Hongyuan, W., F. Tie, Z. Jiachen, and Z. Ke. 2005. "Consistency Check between Behaviour Models." In *IEEE International Symposium on Communications and Information Technology,* pp. 486–9.
16. Engels, G., J. M. Küster, R. Heckel, and L. Groenewegen. 2001. "A Methodology for Specifying and Analyzing Consistency of Object-Oriented Behavioral Models." *SIGSOFT Software Engineering Notes* 26 (5): 186–95.
17. Malgouyres, H., and Motet, G. 2006. "A UML Model Consistency Verification Approach Based on Meta-Modeling Formalization." In *Proceedings of the 2006 ACM Symposium on Applied Computing,* April 23–27, 2006, pp. 1804–9. Dijon, France, New York: ACM.
18. Van Der Straeten, R., T. Mens, J. Simmonds, and V. Jonckers. 2003. "Using Description Logic to Maintain Consistency between UML Models." In *"UML" 2003—The Unified Modeling Language, 2003,* pp. 326–40.
19. Van Der Straeten, R., and D'Hondt, M. 2006. "Model Refactorings through Rule-Based Inconsistency Resolution." In *Proceedings of the 2006 ACM Symposium on Applied Computing,* April 23–27, 2006, pp. 1210–7. Dijon, France, New York: ACM.
20. Sabetzadeh, M., and S. Easterbrook. 2006. "View Merging in the Presence of Incompleteness and Inconsistency." *Requirements Engineering* 11 (3): 174–93. New York: Springer-Verlag.
21. Nentwich, C., W. Emmerich, A. Finkelstein, and E. Ellmer. 2003. "Flexible Consistency Checking." *ACM Transaction Software Engineering Methodology* 12 (1): 28–63. New York: ACM.
22. Engels, G., J. H. Hausmann, R. Heckel, and S. Sauer. 2002. "Testing the Consistency of Dynamic UML Diagrams." In *Proceedings of the 6th International Conference on Integrated Design and Process Technology (IDPT),* Pasadena, California, June 23–28, 2002.
23. Egyed, A. 2007. "Fixing Inconsistencies in UML Design Models." In *Proceedings of the 29th International Conference on Software Engineering,* May 20–26, 2007, pp. 292–301. Los Alamitos, CA: Minneapolis, IEEE Computer Society.
24. Egyed, A. 2007. "UML/Analyzer: A Tool for the Instant Consistency Checking of UML Models." In *Proceedings of the 29th International Conference on Software Engineering,* pp. 793–6.
25. Object Management Group. 2007. "MOF 2.0/XMI Mapping, Version 2.1.1." formal/2007-12-01, OMG.
26. Egyed, A. 2006. "Instant Consistency Checking for the UML." In *Proceedings of the 28th International Conference on Software Engineering,* May 20–28, 2006, pp. 381–90. Shanghai, China: ACM Press.
27. "A Formal Semantics for UML." 2006. Workshop at MoDELS. May 12–19, 2001. Toronto, Ontario, Canada. Washington, DC: IEEE Computer Society.
28. Krüger, I. H., and M. Menarini. 2007. "Queries and Constraints: A Comprehensive Semantic Model for UML2." In *Models in Software Engineering,* edited by T. Kühne, 327–8. Lecture Notes in Computer Science, Vol. 4364. Berlin/Heidelberg: Springer.

29. Broy, M., M. Cengarle, and B. Rumpe. 2006. *Towards a System Model for UML. The Structural Data Model.* Technical Report TUM-I0612. Munich, Germany: Munich University of Technology.

30. Broy, M., M. Cengarle, and B. Rumpe. 2007. *Towards a System Model for UML. Part 2. The Control Model.* Technical Report TUM-I0710. Munich, Germany: Munich University of Technology.

31. Broy, M., M. Cengarle, and B. Rumpe. 2007. *Towards a System Model for UML. Part 3. The State Machine Model.* Technical Report TUM-I0711. Munich, Germany: Munich University of Technology.

32. Broy, M., and K. Stølen. 2001. *Specification and Development of Interactive Systems: Focus on Streams, Interfaces, and Refinement.* Berlin/Heidelberg: Springer.

33. Easterbrook, S., and M. Chechik. 2001. "A Framework for Multi-Valued Reasoning Over Inconsistent Viewpoints." In *Proceedings of the 23rd International Conference on Software Engineering*, May 12–19, 2001, pp. 411–20. Toronto, Ontario, Canada. Washington, DC: IEEE Computer Society.

34. Inverardi, P., H. Muccini, and P. Pelliccione. 2001. "Automated Check of Architectural Models Consistency Using SPIN." In *Proceedings of the 16th Annual International Conference on Automated Software Engineering*, November 26–29, 2001, pp. 346–9. San Diego, California, edited by H. Muccini. Washington, DC: IEEE Computer Society.

35. Ossami, D., J. Jacquot, and J. Souquières. 2005. "Consistency in UML and B Multi-View Specifications." *Integrated Formal Methods* 3771: 386–405.

36. Engels, G., R. Heckel, and J. M. Küster. 2001. "Rule-Based Specification of Behavioral Consistency Based on the UML Meta-Model." In *Proceedings of the 4th International Conference on UML 2001—The Unified Modeling Language. Modeling Languages, Concepts, and Tools.* October 1–5, Toronto, ON, Canada, pp. 272–86.

37. Paige, R. F., P. J. Brooke, and J. S. Ostroff. 2007. "Metamodel-Based Model Conformance and Multiview Consistency Checking." *ACM Transaction Software Engineering Methodology* 16 (3): 11.

38. Clarke, E., and E. Emerson. 1982. "Design and Synthesis of Synchronization Skeletons Using Branching Time Temporal Logic." In *Logics of Programs*, edited by D. Kozen, 52–71. Lecture Notes in Computer Science, Vol. 131. Berlin/Heidelberg: Springer.

39. Holzmann, G. J. 2004. *The SPIN Model Checker: Primer and Reference Manual*, p. 608. Wokingham, U.K.: Addison-Wesley Professional.

40. Milner, R. 1982. *A Calculus of Communicating Systems*, April 1, 1985, p. 256. Secaucus, NJ: Springer-Verlag New York, Inc.

41. Abrial, J. R. 1996. *The B-Book: Assigning Programs to Meanings.* Cambridge, MA: Cambridge University Press.

42. Hoare, C. A. R. 1985. *Communicating Sequential Processes.* Upper Saddle River, NJ: Prentice Hall.

43. Ingolf Heiko Krüger. 2000. "Distributed System Design with Message Sequence Charts." PhD dissertation. Fakultät für Informatik, Technischen Universität München.

44. Pnueli, A. 1977. "The Temporal Logic of Programs." In *Proceedings of the 18th Annual Symposium on Foundations of Computer Science (FOCS 1977)*, October 31–November 2, 1977. pp. 46–57. Long Beach, CA: IEEE Computer Society.

45. Goranko, V., A. Kyrilov, and D. Shkatov. 2010. "Tableau Tool for Testing Satisfiability in LTL: Implementation and Experimental Analysis." *Electronic Notes in Theoretical Computer Science* 262: 113–25. Elsevier Science Publishers B. V.

46. Rozier, K. Y., and M. Y. Vardi. 2007. "LTL Satisfiability Checking." In *Proceedings of the 14th International SPIN Conference on Model Checking Software*, pp. 149–67. Berlin/Heidelberg: Springer-Verlag.

Section III

Parallel and Distributed Real-Time Simulation

13 Interactive Flight Control System Development and Validation with Real-Time Simulation

Hugh H. T. Liu

CONTENTS

13.1 Flight Control System Development Process 331
13.2 Validation and Verification 332
13.3 Modeling and Simulation Platforms 333
13.4 Interactive Flight Control System Development Test Bed 334
13.5 Design Case Study 338
13.6 Challenges and Lessons Learned 341
13.7 Conclusions 346
References 347

This chapter describes the process of flight control system (FCS) development where simulation plays a critical role in interactive design, validation, and verification. Some simulation-based platforms are introduced. As a case study, one such interactive FCS development and real-time simulation test bed is presented.

13.1 FLIGHT CONTROL SYSTEM DEVELOPMENT PROCESS

Because of the complexity of the system, modern aircraft development often adopts a systematic engineering process. The systems engineering has evolved to become a specialized engineering discipline. There are several proposed standards such as IEEE standard (IEEE-Std-1220-1998) and SAE standard (SAE-ARP-4754). Different standards may adopt slightly different practice. However, the core features of a valuable systems engineering process (SEP) should remain unchanged. The purpose of the SEP is to provide a systematic approach for product development and management from initial concept to final delivery. It is a generic problem-solving process that provides the mechanisms for identifying and evolving the product and process definitions of a system. The SEP applies throughout the system life cycle to all activities associated with product development, verification/test, manufacturing, training, operation, support, distribution, disposal, and human systems engineering. A typical SEP is described by a "V"-shape diagram, where the top-down line (the left-hand stroke) represents the process from top-level requirement analysis all the way down to detailed design, and the bottom-up line (the right-hand stroke) represents the process of verification and validation starting from the component level at the bottom all the way up to the top system level performance evaluation.

The FCS development process is no exception. The aim is at finding subsystems (control channels) to represent a solution, given the inputs and desired outputs or tolerable errors, and to integrate them into a functional system that performs its assigned tasks associated with flight mission. The design process can be broken into several phases that are extensively interrelated and interconnected [1,2]. After the overall system requirements are established, the basic functional block diagrams are determined, when characteristics of component parts that become unalterable are fixed. Once a best system has been selected through competing system assessment and trade-off study, the detailed design must be carried out and be validated, often through a series of system simulations and eventually ground and flight testing.

On the basis of operational needs, the FCS offers different functions at different levels. On the one hand, an automatic control (autopilot) provides "pilot relief" in continuous slow mode control such as pitch attitude hold, altitude hold, bank angle hold, turn coordination, and heading hold. The flight control can also be used as a landing aid in terms of glide slope control, localization to align the aircraft in the lateral direction, and the flare control. On the other hand, the FCS includes both stability and control augmentation, to provide stability or ensure appropriate handling qualities, and to provide a desired response to a certain control input. Examples of stability augmentation include the pitch damper, roll damper, and yaw damper. Examples of control augmentation may involve the pitch or roll rate control, or normal acceleration [3].

As a result, both the system design process and the flight control development require interactions between design and validation, often through extensive simulations. The effectiveness and efficiency of the development therefore depends on the smoothness in the iterations and cycles.

13.2   VALIDATION AND VERIFICATION

Requirement validation and verification (RV&V) is an integrated part of the SEP. On the one hand, *validation* is the process of ensuring the requirements are correct and complete and also ensuring compliance with system and airplane-level requirements. On the other hand, *verification* is the process of ensuring that an item complies with all of its design requirements. Requirement validation is embedded in the program risk management. It generally consists of two types of activities: (1) evaluation to ensure correctness and (2) assessment to ensure completeness and necessity. The major validation methods include test, analysis, demonstration, and inspection. The requirement verification determines whether the requirements have been fully complied with the design. The major verification methods include test, analysis, examination, and demonstration.

RV&V methods very much depend on the development environment that the systems engineer chooses. For control systems development, we are particularly interested in the simulation environment that supports validation and verification activities. The concept of "systems simulation" abandons a single, monolithic system development strategy and instead prescribes multiple, incrementally delivered development. It requires that individual control channels and specific functions are tested at an overall system level such that its correctness, completeness (validation), and compliance (verification) are evaluated. It is the highest level of integration and testing before the ground and flight-testing phase is initiated. Several systems simulation models are valuable in the FCS development.

Before a design becomes available or finalized, a *rapid prototyping model* allows the simulation model that represents the component under development be placed into the system platform for testing and demonstration. The design parameters are able to be tuned under this framework, and it is convenient to finalize the component product prototype. In a different scenario, some components are difficult to model in software, or the level of details and complexity can only be represented by the actual physical product. A *hardware-in-the-loop* (HITL) model provides a solution that the hardware be inserted into the simulation environment, to replace the simplified component software. It is expected that the HITL simulation not only provides physical component (the actual prototype) representation but also provides an opportunity to verify the accuracy of the model describing the component product. At a later stage, the *pilot-in-the-loop* (PITL) flight simulation would enable the pilot(s) to operate the virtual aircraft and evaluate its handling qualities.

The challenges when dealing with a heterogeneous simulation environment involve several fronts. From the modeling perspective, the level of fidelity shall be identified. From the integration point of view, special attention shall be placed on the interface design to ensure consistency and compatibility. On the RV&V side, the quantification of requirements, the specification executability, and the traceability shall be represented properly. On the simulation itself, real-time simulation and automation (e.g., automated code generation) aspects shall be considered. In addition, the simulation platform being developed shall also be extensible, scalable, and interoperable. If possible, the commercial-off-the-shelf (COTS) products or modules shall be looked into to take advantage of their maturity, standard, and cost benefits.

13.3   MODELING AND SIMULATION PLATFORMS

The engineering systems development process typically has a waterfall view as if the different development stages were performed chronologically and independent of each other, when in practice they are extensively interrelated and interconnected. Moreover, a system often consists of multiple subsystems and components that are also interacting or even have conflicting characteristic features. Therefore, the entire development relies on iterative cycles between design modifications and integration and testing verifications, until the final design "converges." Unfortunately, this also makes the design process time-consuming and fragile: a slight change may require a completely new cycle of redesign. The necessity for an integrated development process arises from complexity of engineering intensive applications such as in aerospace.

Several research programs that aim at developing an integrated development framework have been proposed and developed over the years. Since 1994, NASA has been funded and tasked at a national priority level to develop and deploy advanced technologies in the general aviation industry. The Advanced General Aviation Transport Experiment (AGATE) seeks to improve utility, safety, performance, and environmental compatibility [4]. A joint development of a state-of-the-art computational facility for aircraft flight control design and evaluation is referred to as CONDUIT (Control Designer's Unified Interface) [5]; it provides an environment for design integration and data resource management. A Defense Advanced Research Projects Agency (DARPA) research program entitled Software-Enabled Control (SEC) [6] was initiated to integrate multimodal and coordinated operation of subsystems and enable large-scale distribution of control. Under this program, an open control platform (OCP) was developed to integrate control technologies and resources. The approach is to use a hierarchical control structure where mission planning situation awareness is at the highest level, the flight control is at the lowest level, and a midlevel controller coordinates the transitions between model selection and model implementation through mode switching and reconfigurable control. The emerging field of multiparadigm modeling addresses the directions of research in model abstraction, transformation, multiformalism modeling, and metamodeling [7]. They are concerned with the models of system behavior, the relationship between models at different levels of details, conversion of models expressed in different formalism, and the description of classes of models, respectively.

13.4 INTERACTIVE FLIGHT CONTROL SYSTEM DEVELOPMENT TEST BED

One of the key requirements for an integrated development process is the "plug-and-play" capability, which enables decoupled component (subsystem) design and integrated testing based on the same model. Therefore, design decisions are gradually finalized through iterations between component-level and system-level developments. For complex control systems development, strategies have been proposed or developed including the computer-automated multiparadigm modeling, actor-oriented control system design approach, cosimulation, code integration, model encapsulation, and model translation ([8] and the references therein). Generally speaking, not only are these efforts time-consuming, but they are also still facing technical difficulties to reach seamless integration. Furthermore, the auto-generated code is often not sufficiently optimized to be used in a production environment. As the complexity of the system is increasing rapidly, these challenges become more and more evident. In this section, we introduce an interactive design and simulation platform for flight vehicle systems development. The proposed platform adopts the cosimulation concept and avoids code generation challenges faced by other integration approaches. It enables the component design "plug-and-play" in a systems simulation environment, thereby bringing the "systems simulation" into the "control design" for integrated development. Moreover, the seamless interactive design and simulation is achieved by an adaptive "connect-and-play" capability. The following presentation is based on the work by Liu and Berndt [8].

A popular approach in FCS development follows the concept of *code encapsulation* in principle. The controller is designed and validated in isolation by desktop off-line simulation. The controller algorithm codes are generated (in C code as one example).

Then, encapsulated as a monolithic submodel, it is integrated into the model of an enclosing system for systems validation and verification. The challenges of the code-generation-based integration approaches include the level of automation, compatibility, and synchronization of the models and simulations. Since design codes are used as the media, one must ensure that the generated codes can work on heterogeneous simulation platforms, with proper interfaces. If so, it is expected that the code generation can be processed automatically, to avoid tedious manual labor and errors. These challenges are still open research topics. In this section, instead, we adopt a different integration strategy that is similar to cosimulation in principle. As shown in Figure 13.1a, this proposed interactive platform allows the component model to be simulated (plug-and-play) in a different, system-level environment. Moreover, the platform is adaptable such that the systems simulator can "connect" to the design model directly. We believe that this "connect-and-play" capability is one significant improvement over the "plug-and-play" capability. Since there is only one physical design model that takes residence at the component level, one can work with this model to make modifications and perform testing "online" without the intermediate code-generation process. Obviously, the "connect-and-play" property and the adaptability make the design and simulation platform truly interactive and integrated in development. Under the proposed platform (Figure 13.1b), the integrated FCS results in a multiparadigm control framework. It represents a standard FCS block diagram with some special features.

The blocks with a drop-down shadow represent "swapping" features. The guidance and command block represents the flight path generation (guidance) or command inputs (for controller design). The actuation and sensor blocks can be replaced by

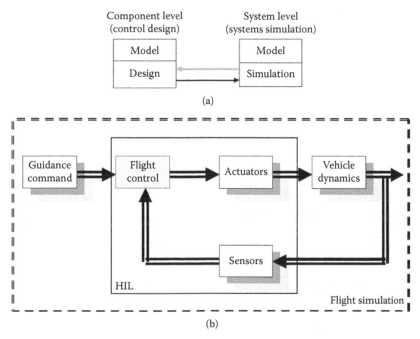

FIGURE 13.1 (a) An illustration of cosimulation concept; (b) Interactive flight control system design and simulation platform.

software modules with different levels of fidelity or even hardware equipment. The vehicle dynamics module can also be replaced by different software modules for different simulation purposes. A simplified linearized dynamics model is used for control system design, while full-scale nonlinear flight equations will be used for high-fidelity simulations such as flight simulations. To emulate the reality that different flight systems components are physically installed in different locations and their interactions are communicated through mechanical links or electrical bus, the proposed framework allows for a distributed modeling structure. Each block can be individually modeled, as one software module in different processors. Therefore, it is possible to distribute different parts of a computing task across individual processors operating at the same time, or "in parallel," and thus reduce the overall time to complete the task. Furthermore, the distributed modeling structure makes it feasible to "swap" different modules of the same block, including the HITL simulation. Because of the distributed modeling and "swapping" feature, it is possible to replace block modules developed under different platforms, and even to run simulations on machines from different manufacturers. Therefore, the proposed framework supports heterogeneous simulations.

To demonstrate the proposed FCS framework and the interactive design and simulation platform, an experimental test bed is set up. A real-time systems simulator and a flight training device (RTSS-FTD) are equipped to provide a suitable proof-of-concept facility, as shown in Figure 13.2.

The RTSS facility is a networked cluster of high-end COTS computers. Its core computing features include three host computers, each having dual-Pentium processors running Windows 2000 OS; four real-time computers, each having dual-Pentium processors running the QNX real-time operating system; and real-time nodes that are directly connected by 400 Mbit/s FireWire and communicate with hosts over a dedicated 100-Mbit/s Ethernet network. Furthermore, the system consists of a 108 multiple channel input/output (I/O) system for HITL simulation. The RTSS is also connected through a 1.25 Gb/s Giganet to a similar facility to share data and sources, and it is connected to a 56-alpha-processor high-power computer for off-line computing and

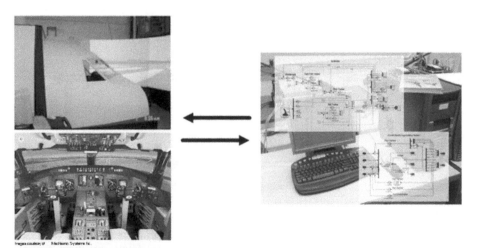

FIGURE 13.2 Integrated flight control system development test bed.

simulation, as well as data storage. This configuration provides the following key capabilities to support our proposed framework.

A separate FTD is set up for flight simulation. It simulates the operation of a generic jet aircraft within the tolerances and conditions set out by the Transport Canada Authority. The major aircraft subsystems include the automatic controls; the auxiliary power unit (APU); doors; the engine indication and crew alerting system (EICAS); the electrical systems; the environmental control systems; flight controls and flight instruments; the fuel, pneumatic, and hydraulic systems; the landing gear; the lighting; and the navigation and communications systems. The design of the FTD is such that all the simulated functionality is concentrated in the software model running on the host computer. This software model contains all the mathematical and logic modeling to make the FTD behave like the Generic Jet aircraft. All the other computers and hardware are I/O interfaces between the pilot/copilot and the model software running on the host computer. The control loading is handled by a PC on the network. It communicates with the host on the Ethernet switch. This computer has digital wiring running to the primary flight controls in the cockpit. The computer systems are networked through a 100 Mbaud Ethernet switch. All the simulated aircraft panels are intelligent; they each contain an embedded CPU that manages their local I/O and communicates with the host computer through a CAN bus network. The aircraft flight and subsystem models are developed using the C programming language. The visual database is developed using the MultiGen paradigm (www.presagis.com). The control system is developed under the MATLAB® (version 7.6) and Simulink® (version 7.1) platform. Both MATLAB and Simulink are software packages provided by MathWorks Inc. (www.mathworks.com). A fully functional FTD also includes an instructor operation station (IOS).

In summary, the RTSS is able to simulate the aircraft systems and flight maneuvers. The features of reconfigurability, modeling, and customization of cockpit displays are critical to our systems integration research. The FTD presents a more complete and realistic aircraft model, which includes factors not taken into account in the RTSS development. It offers a different perspective as the flight mission may be observed from a cockpit with out-the-window visual and instrument displays. The RTSS and FTD facilities are connected through Ethernet cables to form a networked RTSS-FTD test bed for integrated modeling and simulation activities.

To use the FTD as a test bed for interactive controller design and simulation, a network connection is established for "connect-and-play." As introduced before, the control development environment uses the MATLAB and Simulink platform that offers an application programming interface called "S-function" that can be used to integrate blocks with user-defined behavior into a Simulink block diagram. The idea is to let an S-function for MATLAB and Simulink work as a network I/O layer, which outputs the current state vector of the FTD and takes control commands as inputs, as it is commonly done with HITL approaches (Figure 13.3). Note that the control input can carry additional payload, if necessary. In particular, the S-function allows for three values of wind components, which, if given, will be used to overwrite the built-in wind model.

It was then decided that a transmission control protocol and Internet protocol (TCP/IP) connection offers the robustness needed for controller operation. The packet format used for the network connection is simple, yet extensible. A fixed-length

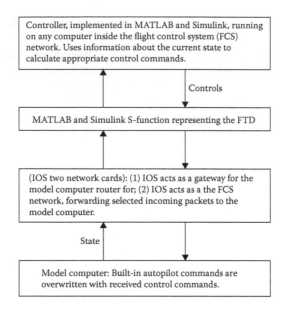

FIGURE 13.3   Network communication.

integer specifying the total packet length is followed by an arbitrary number of tri-ples specifying an identifier, a value, and a delimiter. The S-function is written in the programming language C and can be further customized at compile time using pre-processor macros. This way, among other preferences, maximum allowable packet length and floating point data formats can be adjusted.

13.5   DESIGN CASE STUDY

We provide a business jet aircraft model to illustrate the interactive FCS develop-ment. The 50-seat Canadair Regional Jet 200 (CRJ200) series was announced by Bombardier in 1995 and introduced into service in 1996. The design of the CRJ200, which evolved from the Challenger 604 business jet, is for regional airline operations. The aircraft is equipped with high-efficiency engines giving a range up to 3050 km. The wings are transonic and are fitted with winglets for efficient high-speed flight.

The general equations of motion of this flight vehicle are described by the general rigid-body dynamics when choosing the body reference frame:

$$X - mg \sin\theta = m\left(\dot{U} + QW - RV\right)$$
$$Y + mg \cos\theta \sin\phi = m\left(\dot{V} + RU - PW\right) \quad (13.1)$$
$$Z + mg \cos\theta \cos\phi = m\left(\dot{W} + PV - QU\right)$$

$$\begin{cases} L = I_{xx}\dot{P} + I_{xy}\left(\dot{Q} - PR\right) + I_{xx}\left(\dot{R} + PQ\right) + I_{yz}\left(Q^2 - R^2\right) - \left(I_{yy} - I_{zz}\right)QR \\ M = I_{yy}\dot{Q} + I_{yz}\left(\dot{R} - PQ\right) + I_{yz}\left(\dot{P} + QR\right) + I_{xx}\left(R^2 - P^2\right) - \left(I_{zz} - I_{xx}\right)PR \quad (13.2) \\ N = I_{zz}\dot{R} + I_{zx}\left(\dot{P} - QR\right) + I_{zy}\left(\dot{Q} + PR\right) + I_{xy}\left(P^2 - Q^2\right) - \left(I_{xx} - I_{yy}\right)PQ \end{cases}$$

where the aerodynamic forces and moments are expressed in the body frames $[X\ Y\ Z]^T$ and $[L\ M\ N]^T$, respectively. In most cases, a perturbed fluid-aerodynamic force (moment) is a function of perturbed linear and angular velocities and their rates are as follows:

$$\Delta F_{\text{aero}} = F\left(u, v, w, p, q, r, \dot{u}, \dot{v}, \dot{w}, \dot{p}, \dot{q}, \dot{r}\right) \tag{13.3}$$

Thus, the aerodynamic force at time t_0 is determined by its series expansion of the right-hand side of this equation:

$$\Delta F_{\text{aero}} = F_u u + F_v v + F_w w + F_{\dot{u}} \dot{u} + F_{\dot{v}} \dot{v} + F_{\dot{w}} \dot{w} + \ldots \tag{13.4}$$

$$F_u = \left(\frac{\partial F}{\partial u}\right)_{(t=t_0)} = \left(\frac{\partial F}{\partial u}\right)_0 \tag{13.5}$$

and so on are known as the *stability derivatives*, or more generally as *aerodynamic derivatives*. Because of the assumed symmetry of the vehicle, derivatives of X, Z, M with respect to motions out of the longitudinal plane are zero. This may be visualized by noting that X, Z, M must be symmetrical with respect to lateral perturbations. In other words, we neglect the symmetric derivatives with respect to the asymmetric motion variables, that is, for aerodynamic force X, $X_v = X_p = X_r = 0$, and so on:

$$\begin{bmatrix} \Delta X \\ \Delta Z \\ \Delta M \end{bmatrix} = \begin{bmatrix} X_u & X_w & X_q & X_{\dot{u}} & X_{\dot{w}} & X_{\dot{q}} \\ Z_u & Z_w & Z_q & Z_{\dot{u}} & Z_{\dot{w}} & Z_{\dot{q}} \\ M_u & M_w & M_q & M_{\dot{u}} & M_{\dot{w}} & M_{\dot{q}} \end{bmatrix} \cdot \begin{bmatrix} u \\ w \\ q \\ \dot{u} \\ \dot{w} \\ \dot{q} \end{bmatrix} + \begin{bmatrix} \Delta Xc \\ \Delta Zc \\ \Delta Mc \end{bmatrix}$$

$$\begin{bmatrix} \Delta Y \\ \Delta L \\ \Delta N \end{bmatrix} = \begin{bmatrix} Y_v & Y_p & Y_r & Y_{\dot{v}} & Y_{\dot{p}} & Y_{\dot{r}} \\ L_v & L_p & L_r & L_{\dot{v}} & L_{\dot{p}} & L_{\dot{r}} \\ N_v & N_p & N_r & N_{\dot{v}} & N_{\dot{p}} & N_{\dot{r}} \end{bmatrix} \cdot \begin{bmatrix} v \\ r \\ p \\ \dot{v} \\ \dot{r} \\ \dot{p} \end{bmatrix} + \begin{bmatrix} \Delta Yc \\ \Delta Lc \\ \Delta Nc \end{bmatrix}$$

The fully nonlinear dynamic model of the aircraft is implemented in MATLAB and Simulink, as shown in Figure 13.4, where the states also include the inertial position displacement x_E, y_E, altitude h, and control inputs are specified by the control surface deflection angles:

$$\underline{x} = \begin{bmatrix} x_E & h & u & w & q & \theta & y_E & \psi & v & p & r & \phi \end{bmatrix}^T$$
$$\underline{u} = \begin{bmatrix} \delta_e & \delta_p & \delta_a & \delta_r \end{bmatrix}^T$$

In this example, we design and test the altitude hold and steady-turn (heading hold) autopilot function (Figure 13.5). To carry out the control design, we start by

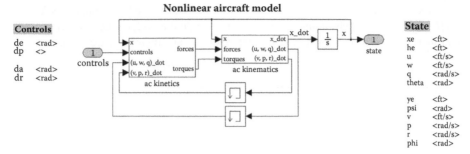

FIGURE 13.4   Nonlinear aircraft model.

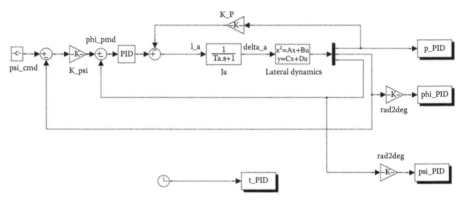

FIGURE 13.5   Autopilot in heading control.

developing a linearized lateral system design model about a reference equilibrium. The reference steady flight condition is cruising at 33,000 ft altitude, Mach 0.74 (or speed of 726.5928 fps). The linearized $G(s)$ is represented by

$$\dot{x}_{hdg} = A_{hdg}x_{hdg} + B_{hdg}u_{hdg}, \quad x_{hdg} = \begin{bmatrix} v \\ p \\ r \\ \phi \\ \psi \end{bmatrix}, \quad u_{hdg} = \delta_a$$

$$y_{hdg} = Cx_{hdg}, \quad y_{hdg} = \begin{bmatrix} p \\ \phi \\ \psi \end{bmatrix}$$

where ψ is the heading angle to be regulated.

We then use the linear system platform LinAC _ Hdg _ Hold _ Sample.mdl to design a heading-hold controller. During the steady flight at 33,000 ft, command the airplane for a steady 90-degree turn under the heading-holding autopilot, that is, the autopilot controls the heading step change of $\psi = 0 \Rightarrow 90$ (deg).

The design requirements are as follows:

1. The aircraft is stable during the motion.
2. The heading overshoot is less than 10%.
3. The steady heading error is less than 5%.

A sample proportional-integral-derivative (PID) control structure is provided. Once the design is complete and satisfactory, the validation is carried out through simulations on the nonlinear Simulink model nonLinAC _ Hdg _ Hold _ Sample.mdl (Figures 13.6 and 13.7). Two test cases should be performed:

1. Heading command of $\psi = 90$ (deg) only while maintaining altitude at 33,000 ft.
2. Heading command of $\psi = 90$ (deg) with altitude climb to 35,000 ft.

The linear simulation result is shown in Figure 13.8; the nonlinear simulation results are presented in Figures 13.9 and 13.10.

The next step is to incorporate the design into the interactive FCS platform as described before. To assure that there exists a stable solution and it is feasible to run successful flight simulation, we demonstrate a successful design, using a linear quadratic regulation (LQR) approach. In this case, both nonlinear off-line simulation results (Figure 13.11) and flight simulation results (Figure 13.12) are quite satisfactory. Note that the flight simulation model is similar to Figure 13.7 except the aircraft model inside the box will directly connect to the FTD according to the communication protocol. The snapshots of the FTD simulation results can either be read from the cockpit panel or recorded at the remote desktop computer, as shown in Figure 13.13.

13.6   CHALLENGES AND LESSONS LEARNED

To develop the interactive FCS and implement it to the flight simulation platform, a number of challenges were encountered and have been addressed in the research investigation.

- The design, simulation, and integration architecture. The I/O interface is required to be consistent, such that the model can be substituted seamlessly.
- The switch between the simulator built-in autopilot function and the designed one must be smooth. It is activated by a command to turn on the design (and automatically overwrites the default one). It is automatically turned off once the testing phase is complete.
- Version control. Even though it is a relatively trivial issue, we did experience on a couple of occasions that the off-line design and simulation model was not compatible with the nonlinear platform for flight simulation. It was solved by specifying the version and release of the software environment.

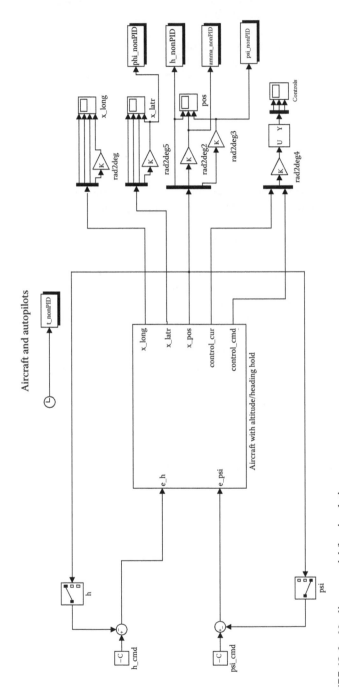

FIGURE 13.6   Nonlinear model for simulation.

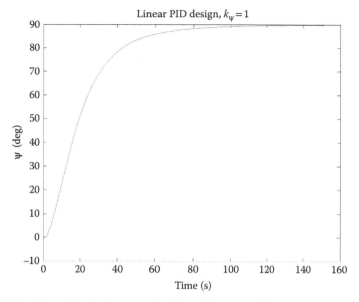

FIGURE 13.7 Control design implemented in nonlinear model.

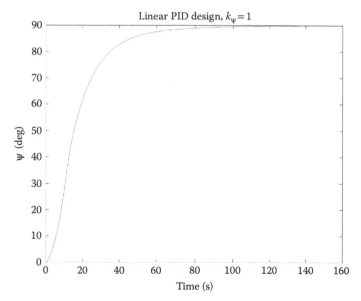

FIGURE 13.8 Linear simulation result: ψ.

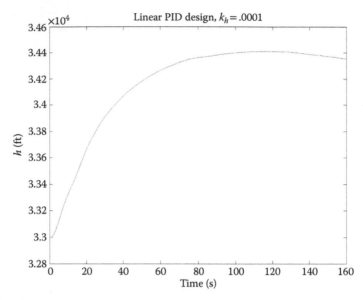

FIGURE 13.9   Nonlinear simulation result: ψ.

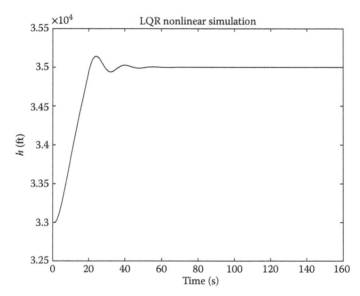

FIGURE 13.10   Nonlinear simulation result: h.

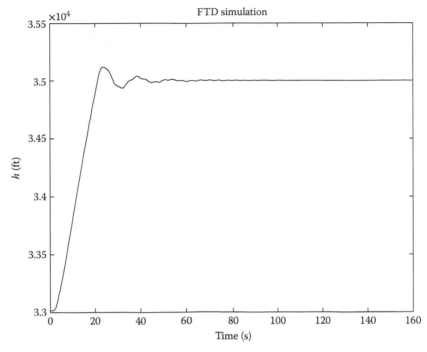

FIGURE 13.11 Nonlinear simulation of altitude from linear quadratic regulation design.

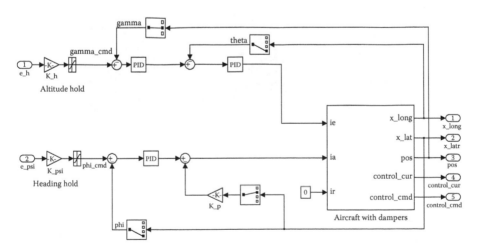

FIGURE 13.12 Flight simulation of altitude from linear quadratic regulation design.

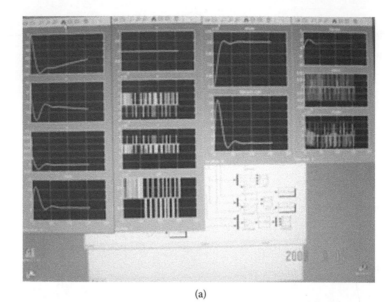

(a)

(b)

FIGURE 13.13 Interactive flight control system flight training device simulation (a) real-time simulation results display; (b) real-time cockpit display.

13.7 CONCLUSIONS

This chapter presents an FCS development using an innovative, interactive simulation platform. It integrates design, testing, and flight simulation to provide a close-to-reality engineering development practice. Valuable lessons are learned through "flight testing" experience. It also provides feedback for design fine-tuning and improvement.

REFERENCES

1. McRuer, Duane, Irving Ashkenas, and Dunstan Graham. 1973. *Aircraft Dynamics and Automatic Control*. Princeton, NJ: Princeton University Press.
2. Nelson, Robert C. 1998. *Flight Stability and Automatic Control*. 2nd ed. Boston, MA: WCB McGraw-Hill.
3. Pratt, Roger W. 2000. *Flight Control Systems*. Reston, VA: AIAA.
4. Thompson, J. Garth. 1996. "Aircraft/Control System Simulation." In *IEEE International Conference on Control Applications*, Dearborn, MI, September 15–18, 1996, pp. 119–24.
5. Tischler, Mark B. et al. 1999. "A Multidisciplinary Flight Control Development Environment and Its Application to a Helicopter." *IEEE Control Systems* 19: 22–33.
6. Schrage, Daniel P., and George Vachtsevanos. 1999. "Software-Enabled Control for Intelligent UAVs." In *International Symposium on Computer Aided Control System Design*, Hawaii.
7. Mosterman, Pieter J., Janos Sztipanotits, and Sebastian Engell. 2004. "Computer Automated Multiparadigm Moldeing in Control Systems Technology." *IEEE Transactions on Control Systems Technology* 12 (2): 223–34.
8. Liu, Hugh H.T., and Holger Berndt. 2006. "Interactive Design and Simulation Platform for Flight Vehicle Systems Development." *AIAA Journal of Aerospace Computing, Information, and Communication* 3 (1): 550–61.

REFERENCES

[References text illegible]

14 Test Bed for Evaluation of Power Grid Cyber-Infrastructure

David C. Bergman and David M. Nicol

CONTENTS

Definitions 350
14.1 Introduction 350
14.2 Simulating DNP3 352
    14.2.1 RINSE 352
    14.2.2 DNP3 Overview 354
        14.2.2.1 Data Link Layer 355
        14.2.2.2 Psuedo-Transport Layer 356
        14.2.2.3 Application Layer 356
    14.2.3 Attacking DNP3 356
    14.2.4 Modeling DNP3 357
        14.2.4.1 Approach 357
        14.2.4.2 Trade-offs 358
14.3 Virtualizing Hosts 358
    14.3.1 Virtual Relays 358
    14.3.2 Virtual Data Aggregators 359
    14.3.3 State Server 360
14.4 Development and Validation 361
    14.4.1 Lab Setup 361
    14.4.2 Interoperability 363
    14.4.3 Tuning Parameters 364
    14.4.4 Expert Analysis 364
14.5 Sample Workflow 364
    14.5.1 Create the Basic Network 364
    14.5.2 Integrate New Technology 367
    14.5.3 Evaluate Experiment 367
14.6 Concluding Remarks 368
    14.6.1 Conclusion 368
    14.6.2 Future Work 368
References 369

DEFINITIONS

- *DETER*–Cyber Defense Technology Experimental Research: A computer network, used for network security experiments, at the University of Utah, which contains physical compute nodes and a highly configurable network topology.
- *DNP3.0*–A protocol used to gather data and issue control commands between master and slave devices. Also goes by DNP v3.0, DNP3, or just DNP.
- *Power Grid*–The collection of buses, generators, lines, transformers, and so on that comprises the system of power delivery from power stations to power consumers.
- *Substation*–Also known as an outstation, the substation is an unmanned location in charge of controlling and collecting data from intelligent electronic devices (IEDs) in the immediate area.
- *Data Aggregator*–A SCADA device at the substation level, which exists to compile data from the RTUs and report the data to the control station.
- *RINSE*–Real-Time Immersive Network Simulation Environment.
- *Modbus*–A legacy protocol that operates point-to-point, generally over a serial connection such as RS-232.
- *RTU*–Remote terminal units are devices in the substation that are responsible for responding to requests from the control station.
- *SCADA*–Supervisory control and data acquisition: A generic term for industrial control systems.
- *Control Station*–Central locations used to coordinate control decisions for regions of the power grid.
- *TCP/IP*–The Transmission Control Protocol and the Internet Protocol: Commonly used protocol combination in the Internet to address packets to processes and route segments to hosts.
- *CRC*–Cyclic redundancy check: The CRC is used to detect bit errors that occur during communication.
- *Relay*–A digital device that measures circuit status and can open/close a physical breaker.
- *PowerWorld*–A steady-state power simulator.

14.1 INTRODUCTION*

The power grid is an important part of the infrastructure that helps us communicate, enjoy most modern comforts, and complete our day-to-day work. Essentially, without the power delivered in a reliable and safe manner, nothing would work. The United States is currently going through a power grid upgrade, termed the "smart grid." While the existing architecture is by no means "dumb," it is a heterogeneous mixture of old machines, new machines, old protocols, new protocols, and vastly different networks across various domains.

---

* Large portions of this work have been previously published as part of David Bergman's M.S. thesis [1].

The power grid is an incredibly large, incredibly complicated network that combines both physical and cyber devices. The power grid consists of three main parts: generation, transmission, and distribution. The generation occurs at power plants and can derive from various sources such as nuclear, coal, or wind. The transmission lines are the high-voltage lines that run across the country, transporting energy from the generation facilities to the areas of consumption. Distribution occurs at a local level, serving as a step-down from the high-voltage transmission lines to the lower-voltage energy that is fit for the consumers. All three of these domains have their own issues, security, and so on. For instance, security at the generation sites involves armed guards, hardened firewalls, and so on. However, the larger concern is safety because a plant operating outside normal operating parameters poses a risk to personnel and machinery. In the distribution regime, a concern regarding the new smart grid is that of securing private data against disclosure. The transmission section of the grid is smaller in scale than the distribution side, but it has its own unique concerns. Much of the control systems that regulate the transmission lines are now highly connected, both to the control systems' subsystems and to the Internet. This is an issue that must be addressed, and the research documented in this chapter has been a step forward in evaluating potential solutions to potential problems.

Traditionally, the grid has used Modbus over serial connections to communicate between devices. More recently, protocols such as DNP3 in the United States and IEC 60870-5 in Europe have been used to communicate over the long distances between substations and control stations. These protocols are also used to communicate between the substations and the remote terminal units (RTUs) in the field. Both in the United States and in Europe, the current trend in the substation itself is to use IEC 61850 to communicate. These current and future trends are discussed in the work by Mohagheghi, Stoupis, and Wang [2].

Changes to these protocols are part of an emerging awareness of the need for security in supervisory control and data acquisition (SCADA) systems, and the Department of Energy (DOE) is giving considerable attention to modernizing one of our country's critical infrastructures (CI), the power grid. We have seen the effects of a power outage in many areas [3–5], and the disastrous effects are exponentially worse at even larger scales. Blackouts have occurred for various reasons, be they mechanical failures, control failures, network failure, operator failure, or a combination. There are also realistic attack vectors—both cyber [6–8] and physical [9]—that must be better understood before a potential attempt at exploiting them. Investigation of all of these failure modes will result in remediation techniques that can provide protection. Both independent researchers and government officials have formed workgroups and task forces [10,11] to investigate and offer their advice [12], but their suggestions must be tested and refined before they are implemented.

However, working directly with the power grid to conduct these investigations is a poor decision for multiple reasons. For the very same reason that the power grid must be made secure—that it is critical to the operation of the United States—modifications to the grid must not be done until adequate testing has taken place. As such, an important approach to studying cyber-security of the power grid is through test systems, so as not to interfere with operational systems. These test systems may use actual equipment to properly understand how the equipment will react in various situations.

However, the scope of the power grid makes it infeasible to create a physical test system anywhere near its full scale.

Parts of this problem can be solved piece by piece, but to obtain a good understanding of how technologies interact in the grid, test beds play a crucial role. The Virtual Power System TestBed (VPST) aims to provide a national resource, whereby new technologies can be tested in a realistic environment, and clear feedback can be given about their efficacies. VPST is part of the Trustworthy Cyber Infrastructure for the Power Grid (TCIP) [13] project and supported by the National Science Foundation (NSF), the DOE, and the Department of Homeland Security (DHS).

Certain technologies, such as cryptography, would stand to gain a lot from detailed simulation. Cryptography research can take different forms in the SCADA environment. It can be used to investigate the practicality of retrofitting bump-in-the-wire devices between legacy devices [14], research various forms of key management [15], test the efficacy of using puzzles to confirm identity in a large-scale network [16], or examine protocols such as DNP3 with Secure Authentication [17] or DNPSec [18].

Unique requirements of the power grid motivate the work documented in this chapter. The massive scale of the power grid necessitates the use of a virtual environment that can match its scope. Other test beds such as Defense Technology Experimental Research (DETER) [19] will not scale well enough because of the dependence on physical SCADA hardware (as opposed to simulated SCADA hardware). OPNET-based simulators will not scale, because, unlike Real-Time Immersive Network Simulation Environment (RINSE), they do not support multiresolution simulation—a technique that allows highly scalable frameworks. Certainly, these test beds provide essential services and are not to be discounted. However, they do present a gap that we intend to fill with VPST.

The beginning of this chapter focuses on describing the SCADA simulation portion of VPST. First, we provide a brief background on the network communications simulator, RINSE, which provides the basis for our SCADA models discussed in Section 14.2 as well as the unique characteristics of the power grid. In Section 14.2, we also describe DNP3, one of the primary protocols used in SCADA, as well as attacks against DNP3 and the approach used to model it. Then, Section 14.3 discusses virtual hosts and the control flows for each of the SCADA-specific virtual hosts. Section 14.4 details the local lab enironment and what each portion is capable of providing. Section 14.4.1 describes the steps we have taken to develop and validate our efforts to create the virtual test bed. Next, Section 14.5 describes a potential workflow for testing a new technology in the local test bed. Finally, we conclude in Section 14.6 while detailing some of our ongoing work.

14.2 SIMULATING DNP3

14.2.1 RINSE

The Scalable Simulation Framework (SSF) [20] is a framework that can be extended to support complex systems such as fluid dynamics, raytracing, and computer networks. Based on SSF, RINSE is a network simulator that serves as the basis for this research and has been through several iterations. RINSE is currently maintained by a team of developers working on the TCIP project at the University of Illinois. Developers are working on various topics including wireless communications, switch modeling, and intrusion prevention through game theory. RINSE has traditionally been used as a wireline network communications simulator to explore malicious behavior in the Internet—namely, worms, botnets, and denial-of-service attacks. The scalability that RINSE allows and the similarity of the Internet to the SCADA infrastructure make it a good candidate for simulating SCADA traffic.

RINSE has a number of properties that make it amenable to large-scale simulation. SSF enables highly parallelizable models by partitioning graphs into submodels. These submodels, which compose the main model, are divided such that communication between them is kept to a minimum, which allows the maximal advantage of multicore systems. Also, RINSE supports multiple resolutions. That is, RINSE can calibrate the fidelity of a simulation to ensure that it runs in real time. It does this by providing a fluid model [21] for traffic and allowing both full-fidelity traffic and fluid models to exist within the same simulation. These fluid models exist for various components of RINSE such as transport protocols, routers, and links. These models also exist for modeling network topologies, and by utilizing these models, RINSE achieves a significant speedup over using a full-resolution model.

RINSE also supports simulation speeds that are faster than real time such that the simulator is not dependent on wall-clock time to advance the simulator time. This is important when interfacing with real devices through emulation. Instead of synchronizing with the wall clock, the simulator keeps track of its own timeline and, where possible, computes traffic ahead of when it will be needed. Combined with prioritizing emulated packets, RINSE can provide a large-scale simulation environment that is not slowed down by interacting with physical devices.

Additionally, RINSE is modular in such a way that new protocols and new models can be developed. By implementing new protocols through extending the base *ProtocolMessage*, RINSE can support the integration of existing or nascent protocols. Likewise, the *ProtocolSession* class allows us to develop new layers in the protocol stack. In the case of DNP3, for instance, the relays are derived from the *ProtocolSession* class and communicate with the data aggregators through messages derived from the *ProtocolMessage* class. The *ProtocolSessions* that comprise a host are indicated in a file by using the Domain Modeling Language (DML). DML describes a model as a tree of key-value pairs, and a file containing a DML model is passed to the simulation engine at runtime. These key-value pairs are then interpreted and used to derive the network topology, hosts, protocols, and traffic patterns. This allows RINSE to simulate any number of models without having to recompile, thus increasing the turnaround time between testing iterations of the same model.

To use RINSE to model the power grid, we must first define the protocol that virtual SCADA devices will use to communicate with one another. When simulating a new protocol or device, there always exists the question of how accurately we must model the proposed design. The option of tweaking parameters of a different model to estimate the new model rarely works because often there are fundamental features that do not lend themselves to be portrayed by a different model. For instance, we could hypothetically model the power grid as a purely Transmission Control

Protocol/Internet Protocol (TCP/IP)-oriented network, with parameters such as poll interval, packet size, latency, and bandwidth set to portray the parameters present in the power grid. However, this would completely ignore the idiosyncrasies of the grid itself. For instance, if there is a vulnerability in the protocol specification or implementation, it would be agnostic to the network layout and configuration. In this manner, we can say that to capture the quirks and security assumptions of a protocol, we must model that protocol as accurately as the scale of the network permits. Here, we provide a brief overview of the three DNP3 layers.

14.2.2 DNP3 OVERVIEW

Designed to provide interoperability and an open standard to device manufacturers, DNP v3.0[22] has gained prominence in the U.S. electrical grid as the communication protocol among power grid equipment. Versions 1.0 and 2.0 were never released to the public. So as to not confuse this protocol with previous implementations, this chapter will refer to DNP v3.0 as DNP3. DNP3 is designed to operate in environments with a high electronics density. These environments can be fairly noisy, and therefore, DNP3 is designed to be as robust as possible with respect to detecting and recovering from error. Cyclic redundancy check (CRC) bytes are extensively used to detect when bits have been flipped and a small frame size (292 bytes) is used to localize errors and reduce the overhead imposed by resending frames.

The protocol was designed as a stack of three layers: the Data Link Layer, the Pseudo Transport Layer, and the Application Layer. The physical medium is generally either Ethernet or RS-485. Since these standards are common, DNP3 can be run over existing networks or networks can be built from the ground up to support DNP3 SCADA networks. The Data Link Layer can either be run directly on the physical medium or it can be encapsulated by other protocols as seen in Figure 14.1.

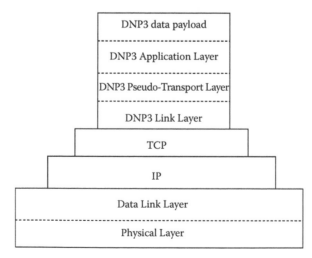

FIGURE 14.1 DNP3 Protocol stack encapsulated by Transmission Control Protocol/Internet Protocol (TCP/IP).

This layer provides framing information and reliability. On top of the Data Link Layer is the Pseudo-Transport Layer. This simple layer is used to support fragmentation. Finally, the top layer is the Application Layer that acts on behalf of the user for requesting/confirming/sending/receiving requests and data.

14.2.2.1 Data Link Layer

The DNP3 Data Link Layer is responsible for point-to-point communication. Essentially, what is contained in this layer is addressing. More specifically, the individual fields for this layer can be seen in the top portion of Figure 14.2. From this layer, we obtain information about the direction of travel, which party initiated the communication, whether the stream is on an odd- or even-numbered frame, and for what function the frame is going to be used. Finally, there is one byte for the length

Data Link Header

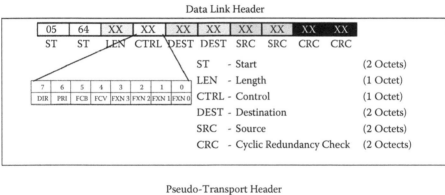

Pseudo-Transport Header

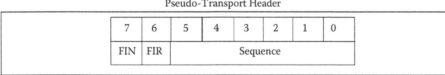

Application Layer

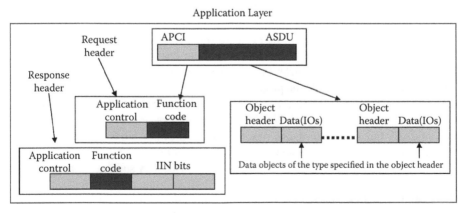

FIGURE 14.2 Three layers of DNP3. (Form DNP Users Group, *DNP v3.0 Guide: A Protocol Primer*, 2008. With permission.)

of the packet. This length is the number of bytes in the message *disregarding* the CRC bytes, which are computed against the first eight bytes.

14.2.2.2 Psuedo-Transport Layer

This layer is responsible for the segmentation of the Application Layer into lower-level frames. It is a rather simple layer that consists of only one byte. As seen in the middle box in Figure 14.2, this byte is divided into three fields. The most significant bit, *FIN*, indicates whether this frame is the last in a sequence of frames that correspond to one Application Layer fragment. The second bit, *FIR*, indicates whether this frame is the first in a sequence of frames that correspond to one Application Layer fragment. For a message that can fit into one frame, both of these bits would be set.

The lower six bits act as the sequence number, which serves to detect dropped frames, out-of-order frames, and other such errors. Once the counter reaches 0x3*F*, it simply resets to 0. Further information can be found in the DNP3 specification for the transport function, volume 3 [24].

14.2.2.3 Application Layer

The bottom box of Figure 14.2 shows the Application Layer. The Application Layer is composed of two sections—the Application Protocol Control Information (APCI) and the Application Service Data Unit (ASDU). The APCI can either be a response header or a request header depending on the purpose of the packet. The fields in the APCI contain fragmentation information as well as a function code describing the purpose of the message. A response header also contains Internal Indications that are spread across 2 bytes as 16 separate bit fields. Their values are described here, but their general purpose is to indicate device status or provide an error message. Following this identification code in a response is any data that the master may have requested. Information can be found in the DNP3 specification, volume 2 [25].

14.2.3 Attacking DNP3

As one of the primary protocols used to transmit information in the power grid, it is important that we model DNP3 as accurately as possible. One reason for this is that there exist deficiencies in the protocol that allow it to be compromised. For instance, related work has developed a taxonomy to describe attacks against DNP3 that comprises 28 generic attacks (of which there are 91 specific instances) [26], thus showing how vulnerable the protocol is. Countermeasures for these holes could be tested in a simulation environment before becoming part of the standard. Other work has compiled a survey of SCADA-related attacks and discusses techniques such as attack trees, fault trees, and risk analysis as they pertain to CIs [27]. Indeed, much research has been done on both SCADA security gaps [12,28,29,30] and their countermeasures, including data set security [31], SCADA-specific intrusion prevention systems [32], and encapsulating DNP3 in another protocol such as SSL/TLS (commonly used for secure internet communications, operating at layer 6 of the network stack) or IPSec (a less common protocol used for securing internet connections at network layer 3) [33]. In general, the power grid is susceptible to attack, and while the exact nature of these vulnerabilities is out of the scope of this chapter, this knowledge provides the impetus to model DNP3 accurately enough to reproduce the vulnerabilities in our virtual test bed.

14.2.4   Modeling DNP3

No matter the actual implementation of the protocol in the simulator, the protocol must be able to handle all three layers of the stack (Data Link, Transport, and Application). Inside the simulator, we treat the three layers as a combined payload to be transported by TCP/IP, and packets are routed using the IP header as opposed to the Data Link Layer header. However, when dealing with emulated packets, the Data Link source and destination fields are used to direct packets to and from the proper hosts. More information about this can be found in Section 14.2.3, which discusses how translation is done between the IDs of the virtual hosts and their emulated DNP3 addresses.

Dealing with emulated packets is an important concern for our use of this virtual DNP3 model. Being compliant with physical devices enables many potential use cases. Without external communications, the RINSE model would provide limited usefulness. It would provide background traffic, metrics regarding correctness and scaling of technologies, and insight into large-scale SCADA networks. However, by being interoperable with physical equipment, more use cases are available that involve a control station. This provides benefit by potentiating training with human-in-the-loop event analysis, incremental deployment analysis, and attack robustness analysis [34].

14.2.4.1   Approach

Instead of using a full-fledged implementation of the DNP3 stack, we model our own slightly abstract view because of a number of different reasons. The main reason is scalability. With the potential of modeling hundreds of thousands of relays, it would be intractable to model the full functionality of DNP3. Instead, by focusing on supporting two classes of reads, with only a few object types, and one type of command, we can simplify the control flow to enable quick computation and low-latency replies. However, if requested by a collaborator, the structure to extend the models to support extra function codes does exist.

Since the IP layer of our simulator provides routing, it would seem that we do not necessarily have to rely on the DNP3 Data Link Layer to route information from one device to another. We justify this by noting that industry is making a move toward encapsulating DNP3 to take advantage of its routing strengths. Some reasons that we choose to still include the Data Link Layer are that there may be unknown interactions between layers. For instance, if an adversary tampers with a field in this layer, the Application Layer may not function properly. Since some of the attacks mentioned in Section 14.2.3 directly attack the Data Link Layer, without modeling this layer, we would not know whether defenses properly address the vulnerabilities. Also, since our simulator has the capability to emulate nodes (i.e., representing a real host as a stub in the simulation), we must support communication with real hosts. All the pieces of physical equipment in our lab require a well-formed packet to function properly. If we fail to deliver that, then our simulation will not work.

14.2.4.2 Trade-offs

We have also made some decisions about the implementation of the protocol. One such decision that is left up to the vendor is how to implement the CRC function. We chose to implement it as a shift/accumulator, since this method has a constant calculation time. The table lookup method, on the contrary, is faster in the best case scenario but endures a penalty if the table is ever evicted from the cache. The RINSE implementation for this function was based on an algorithm released by the DNP3 User's Group [35], with some modifications to fit within the RINSE framework. In general, when we had to make a decision like this, we decided to follow the industry norms.

While modeling the three layers as one layer can accelerate simulation time, it also means that DNP3 cannot be used on its own to provide any of its functionality. Currently, this is not a problem, but if there is a need to model DNP3 directly on top of the physical link, it would require a reworking of the DNP3 implementation. Additionally, it would require a rewrite to the way RINSE handles routing as it currently routes based on IP address. In illuminating these drawbacks, it is our hope that we have further illustrated the trade-offs of our design decisions. Where possible, speed and scalability have been optimized over other characteristics.

14.3 VIRTUALIZING HOSTS

To utilize the virtual DNP3 protocol, we now discuss the virtual *hosts* that use DNP3 to communicate with one another. A *host* is the term we give to nodes in the simulator that have some sort of computational power. That is, a host models some physical device. Namely, we have chosen to simulate data aggregators and relays. On the contrary, we have chosen *not* to simulate control stations because of a number of factors including their complex structure, their proprietary nature, and heavy customizability. Instead, we rely on our physical control station, which provides the added benefit of allowing a human to view the simulated network as part of an operational network. These virtual hosts represent a portion of the network that is purely cyber. The underlying electrical properties of this portion of the grid is simulated by PowerWorld—a commercially available power flow simulator, which we use due to its prevalent use in industry. To bridge the gap between PowerWorld and these hosts, we have implemented the State Server, which is discussed in Section 14.3.3.

14.3.1 VIRTUAL RELAYS

Relays are responsible for control of physical lines as well as gathering data pertaining to their operation. Relays must determine various characteristics of these lines, such as phase angle, voltage, real and reactive power, and other such values. Additionally, relays must provide information about their own operation such as status values, counters, and synchronization efforts. It is unrealistic to support all of these features in a virtual relay, as that would severely hamper scalability. As such, it is important to determine which subset of features must be supported to provide the largest functional coverage at the lowest computational cost. The functionality that covers a large majority of typical requests is as follows:

- *Read Class X data*—*X* can be 0, 1, 2, or 3. Different values of *X* represent different priorities, where 1 refers to data that changes the most often, and 3 the least. Class 0 data includes all data points. Since PowerWorld provides some of these values either directly or indirectly (through simple mathematical operations), this sort of data can be provided to the data aggregators at the substation level.
- *Turn breaker on/off*—A request sent to the relay to turn on or off can then be passed on to the State Server.

When not processing a request, the relays are sitting idle. During this time, they are waiting for a data aggregator to issue a request. Currently, the relays support a limited subset of what their physical analogues support. The relay will thus determine if it can answer the request, and if it is either a data read request or a command, it will perform the required actions. If the request is a read, then the relay will retrieve data from the State Server through its own designated shared memory. If the data aggregator request is a command, the relay takes a corresponding action by writing to shared memory. The State Server is then responsible for forwarding this command to PowerWorld. Once the appropriate action is taken, the relay prepares a response. The response contains either the data or a confirmation, depending on whether the request was a data request or a command, respectively. Once sent, the relay reenters the idle mode and waits for the next request. Some of the difficulties of detecting events and issuing unsolicited responses are discussed in Section 14.6.2.

14.3.2 VIRTUAL DATA AGGREGATORS

The virtual data aggregators have a control flow similar to that of the virtual relays. However, since they require data from the relays instead of PowerWorld, their polling patterns are somewhat different. Instead of sending out one data request per window, an aggregator will send out *n* requests per window, where *n* is the number of relays reporting to it.

The control flow for the master thread, seen in Figure 14.3, describes how the virtual data aggregator polls the relays for which it is responsible. First, the data aggregator waits *t* seconds before beginning the round of requests. Starting with relay 0, the data aggregator prepares a request and waits the appropriate amount of time that its physical analogue would take to produce the request. Then it sends the request and waits for the reply. Once it receives the response, the data aggregator then begins processing the command. Processing the command depends on which data were requested and how the response is formatted. In the basic scenario, the data supplied by the response is simply entered as a corresponding entry into a table, which can be queried by the control station. Once the response is processed, the data aggregator moves on to the next relay and starts the request flow again. Once the data aggregator has polled all of the relays in its list, it waits until the next polling period and resets the current relay to 0.

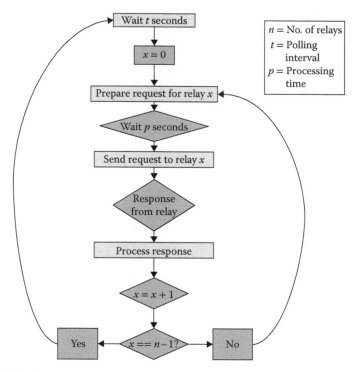

FIGURE 14.3 Data aggregator control flow (master thread).

The data aggregator also acts as a source of data for the control station. This responsibility is handled by the virtual data aggregator's slave thread. The slave thread essentially acts as a server, remaining idle until it receives an incoming request from the control station. When it receives a request, it begins to process it. The data aggregator then makes a decision based on the function code in the request. If the function corresponds to a read, then the data aggregator will provide the requested data out of its table. If the function is a command, then the data aggregator passes the command to the correct relay and waits for a response. Once either the relay responds or the data is ready (depending on which function code was sent to the data aggregator), the data aggregator prepares a response to the control station. After this, the data aggregator waits for an appropriate delay, corresponding to the delay its physical analogue would take for preparing the packet. Finally, the data aggregator sends the response and goes idle.

14.3.3 State Server

The State Server acts as a single point of contact between the RINSE and the PowerWorld simulation environments. The same procedure that generates the DML file also generates a mapping between PowerWorld entities and the virtual relays that are assigned to monitoring them. Using this mapping, the State Server generates requests according to the PowerWorld Application Programming Interface (API). The requests are split into four different groups—lines, generators, loads, and shunts—which correspond to power grid components. After PowerWorld processes these commands, it sends back a response with the appropriate values.

PowerWorld was designed with a server thread, which can serve data to external applications, for which no DNP3 converter had been developed. Hence, we have created this State Server, which consists of four similar tasks being carried out in succession. Polling periods start by preparing a request for shunt values from PowerWorld. After waiting for a response, it processes the response and moves on to requesting generator values. The State Server then repeats the process for load and line values. Finally, the State Server waits *t* seconds before starting the process over again.

Once the data is transferred from PowerWorld to RINSE, it is the responsibility of the State Server to partition the data. Depending on which values were requested, the amount of data a relay will receive can vary from five to nine bytes total. Then, the State Server splits up the response into appropriately sized portions and shares them with the relays through shared memory.

This choice of using shared memory as opposed to explicitly passing a message with the data was made to better represent the practical circumstances. Its sole role is to deliver data from PowerWorld to the virtual relays, and since the State Server has no physical analogue, its latency should be no greater than the time it takes for a relay to measure the line it is connected to—essentially zero. Transmitting this information over the routing network would introduce latency into the system that has no real world analogue.

14.4 DEVELOPMENT AND VALIDATION

14.4.1 LAB SETUP

To connect our complete virtual SCADA system to physical devices, we rely on the generosity of several donors, who have donated a number of SCADA devices, software, and technical support. VPST currently utilizes two SEL-421 relays, an SEL-3351 data aggregator, and a workstation running the OSI Monarch OpenView Energy Management System (EMS) software suite [36]. The relays are each attached to an Adaptive Multichannel Source (AMS) that supports lab testing of Schweitzer Engineering Laboratories (SEL) devices by providing sinusoidal waveforms that feed into the relays.

Our communications simulator, RINSE, provides the SCADA simulation for our local lab environment. Its components have been described in Sections 14.2.1 and 14.3. Using these components, we can generate any desired topology. Shown in Figure 14.4 is a simple topology that includes a virtual control station proxy, multiple substations with one data aggregator each and multiple relays per substation, and a State Server that communicates with a virtual PowerWorld proxy. The PowerWorld proxy is directed through emulation to a computer running PowerWorld, which runs a steady-state power simulation. PowerWorld has an API that can interface with external devices using its proprietary protocol. In this figure, we do not show

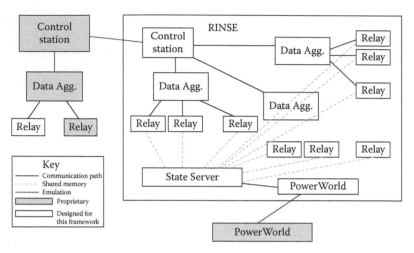

FIGURE 14.4 Simulation framework.

PowerWorld interfacing with anything besides RINSE, although the capability is there, and may be beneficial to certain use cases.

To provide emulation support, we also utilize proxies hosted on physical machines. Although they are not shown in Figure 14.4, our setup would not function without them. Since RINSE operates on its own virtual private network, to direct traffic toward one of the virtual nodes, the traffic must originate from within that virtual private network. To allow external traffic passing, RINSE comes equipped with a gateway that supports OpenVPN connections. From a workstation, we connect using an OpenVPN client to the OpenVPN server on RINSE's host machine. This allows a physical machine to be in the same private address space as the virtual machines. Once this connection is complete, we can redirect this proxy machine's incoming traffic to any destination inside RINSE. We also can redirect any traffic that arrives on the private network address to physical computers. In this manner, the State Server can communicate with PowerWorld in both directions. Also, by redirecting different ports to different addresses, we can allow the Control Station to poll any device in RINSE.

The lab is on a switched network, which provides two useful capabilities. The first is that we can use WireShark to examine traffic between all of these hosts. This allows us to troubleshoot communications between the devices, which is important since the relays only support maintenance through *telnet*. For instance, if we notice that the control station is no long updating, we can pinpoint the break in communication. In addition, it also allows us to inspect the packets themselves. This has proved helpful for meeting timing requirements as well as modeling both the DNP3 and the virtual hosts.

When building a system such as this test bed, fidelity is an important concern. That is, the system must perform in the same manner as the operational analogue. There are varying degrees of fidelity, and development of this system has gone through phases addressing different levels of fidelity. First, the system must interact with physical equipment in such a way that the physical equipment does not realize that it is operating inside a partially virtualized environment. This is termed interoperability. Second, the virtual hosts must accurately reflect characteristics such as response times and accuracy of data. This can be achieved by tuning the corresponding parameters. Third, verification must be run on the virtual system itself to ensure that no race conditions exist in the modeling code. Finally, since SCADA network information is closely guarded, an expert must be brought in to verify that the virtual network reflects a practical network. The following process should serve as a guideline for incorporating new virtual hosts into our framework.

14.4.2 INTEROPERABILITY

The first goal in creating a virtual host is allowing it to communicate with an external device. Reaching this goal requires several steps. First, we create a new virtual host that receives a transmission and responds with a hard-coded message. This hard-coded message can be devised by examining traffic between physical devices. For instance, when devising responses for virtual data aggregators, we look at requests from the actual control station to the actual data aggregator and copy the response. With this model, it allows us to test the communication channel. The communication channel, which is made up of the virtual networking stack, a virtual proxy, a physical proxy, a physical routing network, and a physical end-host, can have some robustness issues that must be resolved before experimenting with a more dynamic response. Once the physical end-host receives and successfully decodes the hard-coded message, we consider this stage a success.

The next step is to generate responses dynamically. To do this, there are two steps. The first step is to create a virtual analogue of the physical protocol. In this case, the protocol is DNP3.0 encapsulated by TCP/IP. Within RINSE, this is represented as a *ProtocolMessage* with a corresponding *ProtocolSession*. The *ProtocolMessage* class contains member data to correspond to each of the fields specified in Section 14.2.2. When communicating with a physical host, the class structure must be converted to a byte stream and there are functions that perform this for both directions. Additionally, there is a helper function to compute and interpose the CRCs when converting to a byte stream.

After the protocol is implemented, the next step is to produce the dynamically generated message with static data. To do this, we extract the DNP3.0 request, convert it to a DNP3.0 *ProtocolMessage*, and process it. The control flow has been discussed in Section 14.3.2.

Once we ensure that the virtual data aggregator is able to correctly generate a response with static data, it is time to generate a response with dynamic data. For the control station to observe a changing environment, responses from the data aggregators must reflect a changing environment. Therefore, each of the relays must also derive its state from a changing environment. Each could poll PowerWorld independently, but this would not scale well. Instead, we chose to implement a single point of contact—namely, the state server. In this manner, we eliminate costly overhead by compressing multiple requests and responses into one.

14.4.3 Tuning Parameters

Once we have a generalized virtual model, it can be tuned to represent any number of physical analogues. By tuning parameters such as polling interval, polling pattern, and response time, we can model various types of data aggregator computers. New features could also be added by specifying them in the DML without designing brand new virtual data aggregators. Virtual relays do not have any polling interval (the data is already available upon receiving a request); however, they can be matched to their physical analogues by changing their response time. Were a virtual Modbus protocol created, another parameter in DML could be specified for protocol support. If a new physical device cannot be modeled by changing some of these parameters, then either altering the model or creating a new device class would provide the necessary functionality.

14.4.4 Expert Analysis

Because of the critical nature of the power grid operations, it is difficult to acquire network topologies. It is also difficult to acquire the designs for SCADA devices from their vendors. In addition to creating models based on observation of the lab devices, some power experts in TCIP have offered their expertise. According to them, aspects such as polling patterns and the parameterization of virtual models appear to be sufficient. Network topologies can be managed through DML files and do not require RINSE development access. The topology described in Figure 14.4 is simplified, but accurate enough to start security analysis. The two suggested topics to be tackled next are expanded functionality of the meters and extra virtual models for other devices. However, this ought to be done on a case-by-case basis as model development consumes time and resources. For current models, there exist enough parameters to allow an experimenter to model a device without delving into the RINSE source code, but future models may require extensive familiarity with the simulation engine.

14.5 SAMPLE WORKFLOW

Now that we have demonstrated a fully functional test bed, when a third party wishes to collaborate with us, we will use a workflow similar to the following.

14.5.1 Create the Basic Network

The first step in the process is to design a power network that exhibits traits that may be of interest. Someone who is knowledgeable of a typical power grid network will design this, as their knowledge should translate into a realistic representation. One such power network can be seen in Figure 14.5, which shows a power transmission network system for a large city, as displayed by PowerWorld. The long lines are transmissions lines that transmit power between buses, which are the short lines with dots in the figure. Connected to the buses are generators and loads, represented by the circled arrows and ground symbols, respectively. PowerWorld itself offers many

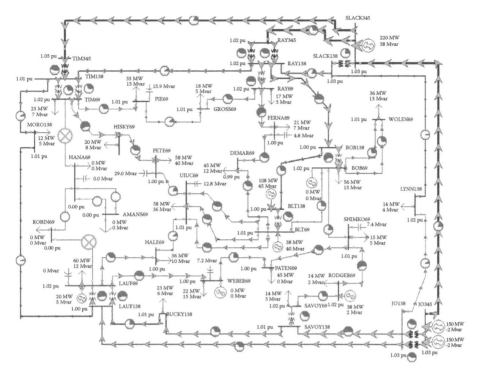

FIGURE 14.5 PowerWorld sample design.

tools for designing, interacting, and assessing power flow models. For the purpose of VPST, the important features that it offers are the abilities to manually create and export a design as well as support a real-time simulation.

From the PowerWorld simulation, key information will be extracted. Typically, these are the IDs of lines, buses, generators, shunts, and loads. From here, we generate the DML and EMS configuration as well as set up the proxies. The corresponding RINSE model is shown in Figure 14.6. From the PowerWorld view, 36 buses and 196 distinct connections were found. These distinct connections have been categorized into lines, loads, generators, and shunts and are then represented by relays in the DML. As can be seen in the zoomed-in box in Figure 14.6, the relays are then annotated to provide a reference point in the PowerWorld model. Likewise, the corresponding EMS configuration in Figure 14.7 shows a list of RTUs (i.e., data aggregators) that the EMS is aware of and is able to poll for data. Currently, configuring the EMS is done by hand, but efforts are under way to automate this process as well.

This portion of the workflow is set up to be as automatic as possible. There are scripts to extract the information from PowerWorld, to create the DML model and its corresponding routing information, to log in the proxy machines and set up the redirects, and to produce the control station configuration file. Future efforts are ongoing to reproduce the entire lab setup in another location, provided similar resources exist there as well.

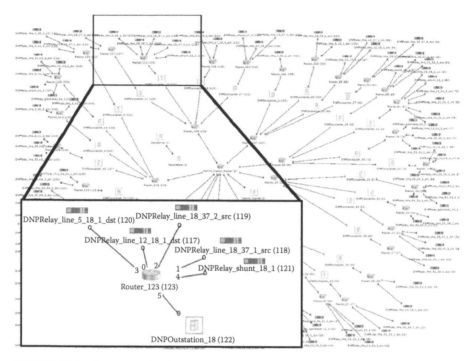

FIGURE 14.6 Corresponding Real-Time Immersive Network Simulation Environment view.

RTU Data

RTU	Name	Sta Addr.	Mode	State	Protocol	Comm Svr On-line	Comm Svr Backup	Channel On-line	Channel Backup	Scans	Errors
1	SubstationA	2	ON	ON	DNP	130.126.140.164		1	0	0	0
2	SubstationA	1	ON	ON	DNP	130.126.140.164		1	0	0	0
3	SubstationA	3	ON	ON	DNP	130.126.140.164		1	0	0	0
4	SubstationA	4	ON	ON	DNP	130.126.140.164		1	0	0	0
5	SubstationA	5	ON	ON	DNP	130.126.140.164		1	0	0	0
6	SubstationA	6	ON	ON	DNP	130.126.140.164		1	0	0	0
7	SubstationB	1	ON	ON	DNP	130.126.140.164		2	0	0	0
8	SubstationB	2	ON	ON	DNP	130.126.140.164		2	0	0	0
9	SubstationB	3	ON	ON	DNP	130.126.140.164		2	0	0	0
10	SubstationB	4	ON	ON	DNP	130.126.140.164		2	0	0	0
11	SubstationC	1	ON	ON	DNP	130.126.140.164		3	0	0	0
12	SubstationC	2	ON	ON	DNP	130.126.140.164		3	0	0	0
13	SubstationC	3	ON	ON	DNP	130.126.140.164		3	0	0	0
14	SubstationC	4	ON	ON	DNP	130.126.140.164		3	0	0	0
15	SubstationD	1	ON	ON	DNP	130.126.140.164		4	0	0	0
16	SubstationD	2	ON	ON	DNP	130.126.140.164		4	0	0	0
17	SubstationD	3	ON	ON	DNP	130.126.140.164		4	0	0	0
18	SubstationD	4	ON	ON	DNP	130.126.140.164		4	0	0	0
19	SubstationD	5	ON	ON	DNP	130.126.140.164		4	0	0	0
20	SubstationD	6	ON	ON	DNP	130.126.140.164		4	0	0	0

FEP Menu | RTU Data | RTU Errors | RTU Error Bars | RTU Config

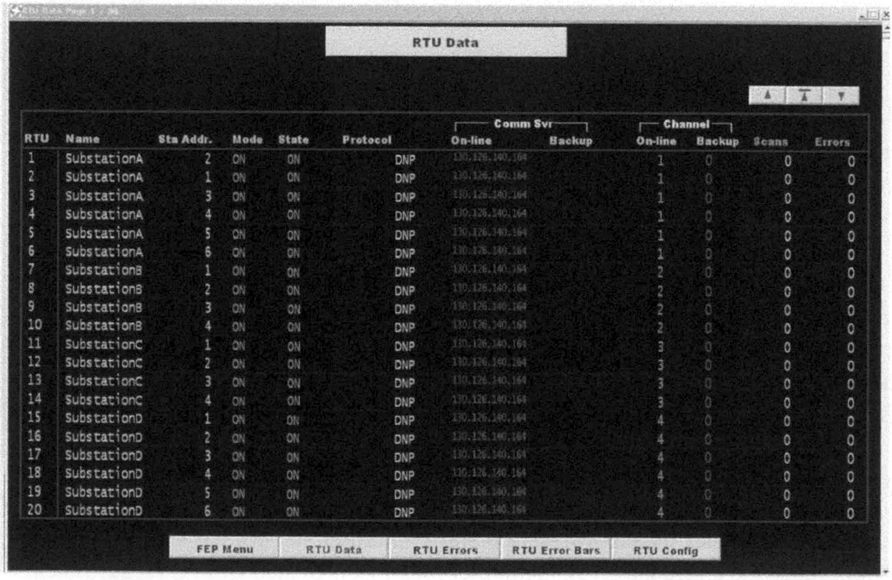

FIGURE 14.7 Corresponding Energy Management System (EMS) view.

14.5.2   INTEGRATE NEW TECHNOLOGY

To integrate a new technology into the framework, there exist two options. The first option is to provide a stand-alone device or model that can be accessed through a network device. By doing this, the actual technology is tested, as opposed to a recreation. The way in which this is done is similar to how the State Server receives its values from PowerWorld (see Section 14.3.3).

The other option is to incorporate a model of the technology directly into RINSE. When using this option, the two parties will develop an accurate model of the new technology, whether that is a protocol or a bump-in-the-wire device or anything else. RINSE is written in C++ and as such can accept a C++ implementation of the new technology. The original implementation must be ported to C++, allowing some modifications to provide networking support inside RINSE. Further efforts are planned to make RINSE as modular as possible, to ease extensibility for cases such as this.

There are trade-offs to consider here, which must be evaluated on a case-by-case basis. For instance, in the C++ implementation case, there are issues of fidelity and time of development. It takes time to develop a model and ensure that it corresponds to the actual technology. There are also issues of ensuring that the C++ model is in lockstep with the current iteration of the product. In the emulation case, there may be issues of scalability, latency, and availability of measurement data. If time allows, implementing the technology directly inside of RINSE will usually allow for better results, as the full benefit of scalable simulation can be realized.

14.5.3   EVALUATE EXPERIMENT

Once all of the connections are set up and the entire communication flow is shown to be working, it is time to test the technology in silico. During the course of the experiment, it is possible to interact with the model and test how it reacts to certain commands. For instance, the EMS software can provide a place for an operator to monitor the grid. Depending on the new technology, it may become easier or more difficult for the operator to understand what is happening in the system. For instance, new technology implemented in RINSE may simplify the job of the operator so that fewer errors occur. On the contrary, it may accidentally drop communication under certain conditions. PowerWorld can offer information about the performance of a new technology. For instance, if a command sent from the EMS never produces a change in PowerWorld, this can be seen quickly in PowerWorld and traced to its root cause in RINSE by looking at metrics such as bandwidth usage, dropped packets, and average latency, or by using *tcpdump* for postmortem evaluation. From there, RINSE supports quick turnaround time through network creation at runtime, easily configurable network topologies, and immediate feedback.

14.6 CONCLUDING REMARKS

14.6.1 CONCLUSION

In this chapter, we have discussed the development of virtual models for DNP3.0, data aggregators, and relays within the RINSE framework. These models are important as they provide a realistic environment in which experimentation on emerging power technologies can take place. The smart grid and its related technology are being integrated into the existing grid at a rapid pace and with little validation. We must understand how this new technology affects existing technologies and infrastructure. By utilizing the models discussed above, we provide such an environment to test these technologies. To provide a realistic simulation, the virtual relays acquire data from PowerWorld, a power flow simulator, through a virtual host termed the State Server. This State Server captures snapshots of the electrical simulation and provides them to the virtual relays through shared memory. By enabling real-time information to be passed through our SCADA simulation, RINSE provides a platform for testing new technologies in a scalable, high-fidelity manner. A sample workflow that shows the steps that are required to verify the functionality of a new technology was presented.

14.6.2 FUTURE WORK

One area that must be modeled is *report-by-exception* polling. This can come in two flavors. The first is a *polled report-by-exception*, which is a poll that only asks for data that has changed. Another form is *unsolicited report-by-exception*. This allows a relay to send a message if something has changed state, without being asked to do so. This type of reporting is also currently unimplemented, but would require greater effort because of the nature in which the State Server is implemented. Without having access to the actual grid, but through the advice of experts, it seems as though relays support *report-by-exception*, but often do not use it as resources may be better spent on real-time monitoring. In any case, this seems like a natural feature to implement.

The DNP Users Group has published test procedures for Levels 1 and 2, the simplest implementations. Future work should ensure that the virtual models developed in RINSE pass these tests, or at least that subportions of these tests are deemed relevant to the demands of the experiment. Models may be introduced that conform to varying levels of these tests, so as to provide certain guarantees regarding both fidelity and performance.

One area of concern is that of clock skew among the relays. As implemented, the State Server receives values in chunks. All data within a given chunk has occurred during the same time slice. However, in the physical grid, data from separate relays are never guaranteed to occur during the same time slice. This is the type of problem that synchrophasor technology [37] is attempting to solve. However, the question of how, or even whether, to simulate this effect will require some examination. Again, this raises the question regarding fidelity versus performance. To simulate this clock skew, the state server must query each data point separately and provide randomization in the timing as well. This overhead may impact the performance of the simulation as well as the freshness of the data.

REFERENCES

1. Bergman, D. C. 2010. "Power Grid Simulation, Evaluation, and Test Framework." Master's thesis, University of Illinois at Urbana-Champaign, Champaign, IL.
2. Mohagheghi, S., J. Stoupis, and Z. Wang. 2009. "Communication Protocols and Networks for Power Systems—Current Status and Future Trends." In *Power Systems Conference and Exposition, 2009. PSCE '09. IEEE/PES*, IEEE, pp. 1–9, March 2009.
3. Pacific Northwest National Laboratory (PNNL). Looking Back at the August 2003 Blackout. Available at http://eioc.pnnl.gov/research/2003blackout.stm. Accessed February 2010
4. Andersson, G. et al. 2005. "Causes of the 2003 Major Grid Blackouts in North America and Europe, and Recommended Means to Improve System Dynamic Performance." *IEEE Transactions on Power Systems* 20 (4): 1922–28.
5. Corsi, S., and C. Sabelli. 2004. "General Blackout in Italy Sunday September 28, 2003, h. 03: 28: 00." In *IEEE Power Engineering Society General Meeting, 2004*, IEEE, pp. 1691–702.
6. Bompard, E., C. Gao, R. Napoli, A. Russo, M. Masera, and A. Stefanini. 2009. "Risk Assessment of Malicious Attacks against Power Systems." *Transactions on Systems Man and Cybernetics. Part A* 39 (5): 1074–85.
7. Fernandez, J. D., and A. E. Fernandez. 2005. "SCADA Systems: Vulnerabilities and Remediation." *Journal* of *Computing Sciences in Colleges* 20 (4): 160–68.
8. Igure, V. M., S. A. Laughter, and R. D. Williams. 2006. "Security Issues in SCADA Networks." *Computers & Security* 25 (7): 498–506.
9. Salmeron, J. et al. 2004. "Analysis of Electric Grid Security Under Terrorist Threat." *IEEE Transactions on Power Systems* 19 (2): 905–12.
10. Department of Energy Smart Grid Task Force. 2007. Available at http://energy.gov/oe/technology-development/smart-grid/federal-smart-grid-task-force.
11. U.S. National Institute of Standards and Technology (NIST). 2010. *Smart Grid Interoperability Standards Project*. Available at http://www.nist.gov/smartgrid/
12. U.S. National Institute of Standards and Technology (NIST). 2009. *Smart http://www.nist.gov/smartgrid/ Grid Cybersecurity Strategy and Requirements*. Available at http://www.naseo.org/eaguidelines/documents/cybersecurity/NistIr-7628%20smart%20grid.pdf
13. UIUC. 2010. *Trustworthy Cyber Infrastructure for the Power Grid*. Available at http://tcipg.org/
14. Tsang, P., and S. Smith. 2008. "YASIR: A Low-Latency, High-Integrity Security Retrofit http://tcipg.org/ for Legacy SCADA Systems." In *Proceedings of The IFIP TC 11 23rd International Information Security Conference*, Milano, Italy pp. 445–59.
15. Piètre-Cambacédès, L., and P. Sitbon. 2008. "Cryptographic Key Management for SCADA Systems—Issues and Perspectives." In *ISA '08: Proceedings of the 2008 International Conference on Information Security and Assurance (isa 2008)*. Washington, DC pp. 156–61.
16. Bowen III, C. L.,T. K. Buennemeyer, and R. W. Thomas. 2005. *A Plan for SCADA Security Employing Best Practices and Client Puzzles to Deter DoS Attacks*. Presented at Working Together: R&D Partnerships in Homeland Security, Boston, Massachusetts.
17. DNP Users Group. 2010. *DNP3 Specification, Secure Authentication, Supplement to* www.dnp.org, *Volume 2*. Available at www.dnp.org

18. Majdalawieh, M., F. Parisi-Presicce, and D. Wijesekera. 2006. "DNPSec: Distributed Network Protocol Version 3 (DNP3) Security Framework." In *Advances in Computer, Information, and Systems Sciences, and Engineering: Proceedings of IETA 2005, TeNe 2005, and EIAE 2005*. Springer, Netherlands, pp. 227–34.

19. Benzel, T., R. Braden, D. Kim, C. Neuman, A. Joseph, K. Sklower, R. Ostrenga, and S. Schwab. 2006. "Experience with DETER: A Testbed for Security Research." In *2nd International Conference on Testbeds and Research Infrastructures for the Development of Networks and Communities, 2006. TRIDENTCOM 2006*, IEEE, pp. 10.

20. SSF. Scalable Simulation Framework. 2004. Available at http://www.ssfnet.org/home-Page.html. Accessed January 2010.

21. Nicol, D., M. Goldsby, and M. Johnson. 1999. "Fluid-Based Simulation of Communication Networks Using SSF." In *Proceedings of the 1999 European Simulation Symposium*, vol. 2, Erlanger, Germany.

22. DNP Users Group. 2010. *DNP: Distributed Network Protocol*. Available at www.dnp.org

23. DNP Users Group. 2008. *DNP v3.0 Guide: A Protocol Primer*. Available at www.dnp.org

24. DNP Users Group. 2002. *DNP3 Specification Volume 3: Transport Function*. Available at www.dnp.org

25. DNP Users Group. 2005. *DNP3 Specification Volume 2: Application Layer*. Available at www.dnp.org

26. East, S., J. Butts, M. Papa, and S. Shenoi. 2009. "A Taxonomy of Attacks on the DNP3 Protocol." In *Critical Infrastructure Protection III, IFIP Advances in Information and Communication Technology*, vol. 311, Springer pp. 67–81.

27. Ralston, P. A. S., J. H Graham, and J. L Hieb. 2007. "Cyber Security Risk Assessment for SCADA and DCS Networks." *ISA Transactions* 46 (4): 583–94.

28. Patel, S. C., and Y. Yu. 2007. "Analysis of SCADA Security Models." *International Management Review* 3 (2): 68–76.

29. Faruk, A. B. M. O. 2008. "Testing & Exploring Vulnerabilities of the Applications Implementing DNP3 Protocol." Master's thesis, Kungliga Tekniska högskolan Stockholm, Sweden.

30. Hong, S., and S. J. Lee. 2008. "Challenges and Pespectives in Security Measures for the SCADA System." In *Proceedings of 5th Myongji-Tsinghua University Joint Seminar on Prototection & Automation*. IEEE

31. Mander, T. et al. 2009."Power System DNP3 Data Object Security Using Data Sets." *Computers Security* 29 (4): 487–500.

32. Rrushi, D. J. L., and U.S. di Milano. 2006. "SCADA Intrusion Prevention System." In *Proceedings of 1st CI2RCO Critical Information Infrastructure Protection Conference*. Springer-Verlag Berlin, Heidelberg.

33. Graham, J. H., and S. C. Patel. 2004. "Security Considerations in SCADA Communication Protocols." Technical Report TR-ISRL-04-01, Intelligent Systems Research Laboratory, Department of Computer Engineering and Computer Science, University of Louisville, Louisville, Kentucky, September 2004.

34. Bergman, D. C., D. Jin, D. M. Nicol, and T. Yardley. 2009. The Virtual Power System Testbed and Inter-Testbed Integration. In *Proceedings of the Conference on Cyber Security Experimentation and Test (CSET)*, pp. 1–6. Berkeley, CA: USENIX Association.

35. DNP Users Group. 2002. *Guide to Calculate DNP CRC*. Available at www.dnp.org

36. OpenView: Graphical User Interface. 2009. Available at http://www.osii.com/pdf/scada-ui/OpenView_PS.pdf

37. Schweitzer, E. O., and D. E. Whitehead. 2007. *Real-Time Power System Control Using Synchrophasors*. Presented at 34th Annual Western Protective Relay Conference, IEEE, October 2007.

15 System Approach to Simulations for Training
Instruction, Technology, and Process Engineering

Sae Schatz, Denise Nicholson,
and Rhianon Dolletski

CONTENTS

15.1 Introduction 371
    15.1.1 What Is Distributed Real-Time SBT? 372
15.2 Brief History of Distributed SBT 373
15.3 Persistent Challenges 375
    15.3.1 Interoperability 375
        15.3.1.1 Software Protocols 375
        15.3.1.2 Domain Architectures 376
        15.3.1.3 Fair Play 377
    15.3.2 Fidelity 377
        15.3.2.1 Expanding Fidelity into New Domains 378
    15.3.3 Instructional Strategies 379
    15.3.4 Instructor Workload 380
        15.3.4.1 Human Effort Required for Distributed AAR 380
    15.3.5 Lack of Effectiveness Assessment 381
    15.3.6 Lack of Use Outside of the Defense Sector 382
15.4 Conclusion 383
Acknowledgments 384
References 384

15.1 INTRODUCTION

Parallel and distributed simulation (PADS) emerged in the early 1980s, and over the past three decades, PADS systems have matured into vital technologies, particularly for military applications. This chapter provides a basic overview of one specific PADS application area: *distributed real-time simulation-based training* (SBT). The chapter opens with a brief description of distributed real-time SBT and its development. It then lists the recurrent challenges faced by such systems with respect to instructional best practices, technology, and use.

15.1.1 What Is Distributed Real-Time SBT?

We begin by clarifying the phrase "distributed real-time SBT." Consider each key word, in turn. First, *distributed* simulations are networks of geographically dispersed simulators (often called "federates") that execute a single overall model, or more colloquially, that share a single "place." Next, *real-time* simulations are those that advance at the same rate as actual time [1]. Finally, *training* simulations are instructional tools that employ training strategies (such as scenario-based training) and represent a problem-solving context to facilitate guided practice [2,3]. Thus, for this chapter, consider the following definition (see Figure 15.1 for an illustration):

> Distributed real-time SBT involves two or more geographically distributed computer-based simulations that are interconnected through a network and track the passing of real time; they are used concurrently by multiple trainees, whose individual inputs affect the overall shared environment, and they include instructionally supported, sufficiently realistic problem-solving environments in which trainees can learn or practice their knowledge, skills, and attitudes.

Distributed real-time SBT is considered a primary enabling technology for the U.S. military. It supports Joint (i.e., across military branches) and interagency training and mission rehearsal, development and testing of new tactics and techniques, and assessment of personnel skills [4]. In turn, these capabilities reduce risks, save lives and money, and engender increased operational capabilities. Section 15.2 offers

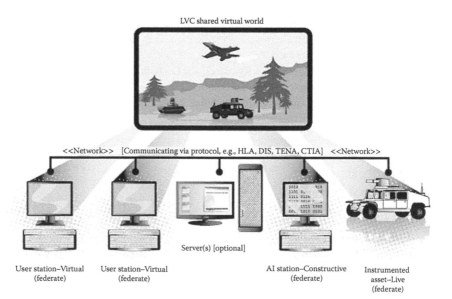

FIGURE 15.1 Illustration of a generic live, virtual, and constructive (LVC) distributed real-time simulation-based training system.

a sketch of the historic milestones that have led to the development of distributed real-time SBT and contributed to its widespread use within the Defense sector.

15.2 BRIEF HISTORY OF DISTRIBUTED SBT

Simulation has existed for centuries. As early as 2500 BC, ancient Egyptians used figurines to simulate warring factions. However, "modern" simulation only began around the turn of the past century with the invention of mechanical flight simulators. These started appearing around 1910 and are often considered the forerunners of contemporary simulation, in general. The best known early flight simulator, called the Link Trainer (or colloquially, the "Blue Box"), was patented in 1930 [5]. In the years that followed, thousands of Blue Boxes were put into service, particularly by the U.S. government in their effort to train military aviators during World War II.

In general, World War II served as a major catalyst for modern modeling and simulation (M&S). In the United States, the war encouraged the federal government to release unprecedented funding for research, a significant portion of which went to advancing computer development as well as early M&S. The war effort also created the need to analyze large-volume data sets and to perform real-time operations. These computational needs, coupled with the flood of federal money, created a frenzy of competition in computer, modeling, and simulation sciences [6].

The research led to a series of technology breakthroughs throughout the 1940s and 1950s [7], and use of simulation swelled. NASA and the U.S. military, in particular, developed large complex simulators for training purposes [8]. The growth of M&S continued throughout the 1960s, and, by the 1970s, the combination of less-expensive computing technology and its improved effectiveness ushered in an era of widespread simulation use—particularly within the Defense training community.

A new outlook on training also began to surface. Before the 1970s, SBT had been primarily applied to individual tasks and viewed as substitution training (i.e., a substitution for real-life training used for cost, safety, or other purely logistical reasons). However, during the 1970s, the training community began to value instructional simulation beyond mere substitution [9]. New efforts focused on collective training, the use of realistic and measurable training requirements, and the development of expertise.

Spurred, in part, by the demand for collective and improved training, military investigators began exploring distributed simulation technology. In 1983, the Defense Advanced Research Projects Agency (DARPA) started development of the futuristic Simulation Network (SIMNET). This effort, championed by Jack A. Thorpe, promised to provide affordable, networked SBT for U.S. warfighters. In the 1980s, the idea of networked simulators was viewed, at best, as high-risk research—or as an impossible pipedream, at worse! However, history now reveals that SIMNET, and the notion of distributed simulation that it represented, were major leaps forward for SBT [10].

By the 1990s, SIMNET had given birth to the era of networked real-time SBT, and although still considered highly experimental, the military started sponsoring more distributed SBT efforts. In addition to SIMNET, the Aggregate-Level Simulation Protocol (ALSP) emerged as another early distributed SBT success [11], and around the same time, software developers released the Distributed Interactive Simulation

(DIS), a more mature version of the original SIMNET design [12]. Both ALSP and DIS were intended to support flexible exchange of data among federated military simulation systems. However, these protocols failed to meet emerging requirements.

Therefore, to address a wider set of needs, military M&S leaders developed the High Level Architecture (HLA) protocol. Released in 1996, HLA combined the best features of ALSP and DIS, while also including support for the analysis and acquisition communities [13]. Yet, technologists began to "perceive HLA as a 'jack of all trades, but master of none' " [12, p. 8], and various user communities started developing their own distinct protocols, including the Extensible Modeling and Simulation Framework (XMSF), Test and Training Enabling Architecture (TENA), Virtual Reality Transfer Protocol (VRTP), Common Training Instrumentation Architecture (CTIA), Distributed Worlds Transfer and Communication Protocol (DWTP), and Multi-user 3D Protocol (Mu3D) (for an overview of these middleware technologies, see the work by Richbourg and Lutz [12]).

Meanwhile, as the development community advanced the technological capabilities of distributed simulation, the training community was identifying new opportunities to improve collective training. Throughout the 1980s, behavioral scientists worked to systemically enhance SBT (in general) by developing Instructional Systems Design (ISD) approaches to simulation content creation, methodically pairing simulation features with desired training outcomes, analyzing instructional strategies for simulation, creating performance assessment techniques, and designing feedback delivery mechanisms—just to name a few topics [14].

Naturally, these instructional advances were also applied to *distributed* SBT. Plus, distributed systems offered their own challenges for the training research community, such as how to build team cohesion among distributed personnel or deliver feedback to geographically distributed trainees.

As the technological and instructional R&D communities created leap-ahead advancements in distributed SBT throughout the 1980s and 1990s, the administrative communities debated and refined the employment of distributed system. Notably, the Defense Modeling and Simulation Office (DMSO; later renamed the Modeling and Simulation Coordination Office [MSCO]) was established in late 1991 to solve coordination gaps among the Services. Before DMSO, military M&S efforts were described as "out of control" with "no identified POC (point of contact) for M&S nor any real focus to the dollars being spent in this area" [15, para. 4]. As Col (ret.) Fitzsimmons, the first DMSO Director, illustrates,

> ... I placed four toilet paper cardboard rolls on [the] table and said that if you look down each roll you see what the individual Services are doing ... basically good stuff. However there is no way to work across the Services to coordinate activities (such as reduce development costs, share ideas etc). Duplication of efforts was obvious from even our short look into all this. I then pointed out that the main mission of the Services was to provide "interoperable" force packages ... which meant there needed to be a way to do joint training, which also meant there needed to be interoperable training systems. Bottom line was that we felt there needed to be someone guiding development of M&S from the perspective of joint interoperable training and helping coordinate reuse...
>
> **Fitzsimmons [15, para. 7]**

Thus, DMSO developed into a policy-level office focused on reuse and Joint interoperability for training. Early in DMSO's existence, its leaders recognized the benefits of integrating live, virtual, and constructive (LVC) simulations through distributed technologies, and LVC became a cornerstone of DMSO efforts [15]. DMSO also championed distributed simulation verification, validation, and accreditation (VV&A) efforts, publishing standards and policies for the Defense community, as well as guiding M&S policy through return-on-investment analyses.

Looking back upon the emergence of distributed real-time SBT, it becomes clear that the military community pioneered this discipline. More specifically, military technologists, training scientists, and visionary policy leaders together made distributed SBT a functional, effective, and usable tool for the Defense community. Yet, despite widespread use of distributed SBT for military training, it remains a fairly nascent capability. In Section 15.3, we outline various issues commonly associated with distributed real-time SBT.

15.3 PERSISTENT CHALLENGES

Over the past 30 years, the pipedream of distributed real-time simulation has transformed into a fundamental training technology; still, challenges persist. These include long-standing technology issues, questions regarding how to best attain distributed simulation instructional goals, and discussions on how to best employ this distributed real-time SBT.

15.3.1 INTEROPERABILITY

Interoperability concerns the interlinking of the many diverse systems involved with distributed simulation. In other words, an interoperable simulation must be able to both pass and interpret data among its federates [16]. Naturally, interoperability was among the first issues distributed SBT developers confronted, and, although great advances have been made, significant technological gaps remain.

15.3.1.1 Software Protocols

Generally speaking, many interoperability gaps result from limitations found in the software protocols used to communicate data among interlinked simulations. Common protocol issues include the following:

- Scalability limitations
- Challenges regarding time synchronization
- Lack of interoperability among different protocols
- Lack of true plug-and-play capabilities
- Lack of support for semantic interoperability

DIS and HLA remain the most common interoperability protocols; yet, they still only achieve moderate ratings of practical relevance: "(3.5 and 3.4 respectively [out of 5.0]), a value which is relatively high, but might be expected to be even higher considering that both standards have been on the market for more than 10 years

(HLA) or 15 years (DIS)" [17]. Both DIS and HLA are most notability affected by scalability, plug-and-play, and semantic interoperability problems [12].

More recent protocols attempt to address the difficulties of DIS and HLA; however, the introduction of new protocols, ironically, contributes to another significant challenge—lack of interoperability among protocols. To use simulations that employ different protocols, developers must use special bridges or gateways, which introduce "increased risk, complexity, cost, level of effort, and preparation time into the event" [12, pp. 8–9]. Plus, the ability to reuse models and applications across different protocols is limited.

To help address these issues, the MSCO recently completed a large-scale effort to identify the best way forward for interoperability. The Live Virtual Constructive Architecture Roadmap (LVCAR) study examined technical architectures, business models, and standards and then considered various strategies for improving the state of simulation interoperability [18]. The study produced 13 documents, including an extensive main report and several companion papers that were delivered in 2008. Although formal next steps have not yet been announced, the results of this effort will undoubtedly shape MSCO policy in the years to come, directly affecting which protocols are developed and/or used.

15.3.1.2   Domain Architectures

At the component level, various domain architectures have been established to maximize the potential for interoperability of simulations. Such frameworks attempt to create common software platforms for use across the (mainly government) simulation community. For example, created in the early 1990s, the Joint Simulation System (JSIMS) and One Semi-Automated Forces (OneSAF) systems aim to provide common architectures for synthetic environments and computer generated forces, respectively.

New components are commonly added to government domain architectures, and as a result, these packages are routinely upgraded. Regrettably, many simulation practitioners fail to upgrade their simulations, and now a hodgepodge of versions are found across the user community. Moreover, different versions of an architecture do not necessarily support federation. In other words, the variety of versions actually creates a hindrance to interoperability [19].

No clear solution yet exists to this challenge. However, unifying agencies, such as Simulation Interoperability Standards Organization (SISO), attempt to encourage more standardized architecture use throughout the M&S community. SISO, whose roots can be traced back to SIMNET, is an international organization that hosts regular meetings as well as annual Simulation Interoperability Workshops. SISO also maintains interoperability standards and common components packages, such as SEDRIS, an enabling technology that supports the representation of environmental data and the interchange of environmental data sets among distributed simulation systems [20].

Agencies such as SISO can help remove some barriers to interoperability and promote more homogeneous use of common domain architectures. However, for change to truly occur, user communities must participate with standard activities and commit to building a more cohesive sense of community among distributed SBT practitioners [18].

15.3.1.3 Fair Play

Finally, a range of other interoperability issues can be roughly grouped under the concept of "fair play." *Fair play* means ensuring that no trainee has an unfair advantage because of technical issues outside of the training, such as improved graphics giving an undue visual advantage. Fair play issues often involve *time synchronization* or *model composability*.

Achieving optimal time synchronization (sometimes called "time coherence") is a complicated, ongoing struggle for software developers. In brief, time synchronization concerns controlling the timing of simulation events so that they are reflected in the proper order at each participating federate [21]. As of the penning of this chapter, no clear mitigation strategy has been established to overcome distributed synchronization issues; however, investigators continue to search for a solution.

Model composability, another as-yet unsolved technological conundrum, is concerned with using the data models of a simulation to effectively represent the common virtual environment in which the distributed participants interact [22]. When simply stated, this challenge seems manageable, but the complexity becomes apparent once one considers the array of intricate, heterogeneous systems—often LVC—that must be capable of expressing the same environment, agents, and behaviors. As with synchronization challenges, no clear resolution has yet emerged, and researchers continue to publish extensively, seeking a more optimal solution.

15.3.2 FIDELITY

Since the early days of simulation, the M&S community has debated the necessary degree of fidelity. Originally, many practitioners believed that the more physical fidelity a simulator offered, the better its learning environment became. Toward that end, early SBT efforts often focused on perfecting the realism of the simulator, so as to mimic the real world as closely as possible. Consider this quotation describing (but not necessarily condoning) early views on simulation:

> The more like the real-world counterpart, the greater is the confidence that performance in the simulator will be equivalent to operational performance and, in the case of training, the greater is the assurance that the simulator will be capable of supporting the learning of the relevant skills ... Designing a simulator to realistically and comprehensively duplicate a real-world item of equipment or system is a matter of achieving physical and functional correspondence. The characteristics of the human participant can be largely ignored.
>
> **National Research Council [23, p. 27]**

The idea of *selected fidelity* was articulated in the late 1980s, and it suggests that individual stimuli within a simulation can be more-or-less realistic (depending on the specific task) and still support effective training [24]. Later research led to the notion that *psychological fidelity* was more important than physical fidelity. With psychological fidelity, simulations could be successful without high physical fidelity—so long as the overall fidelity configuration supported the educational goals of the system (e.g., [2,23,25,26]).

Today, effective simulators come in an array of different fidelity levels, and social scientists better understand (although not completely) how to use more limited degrees of fidelity appropriately. Still, the idea that "more fidelity is better" remains entrenched among many practitioners and simulation users, as a recent U.S. Army Research Institute survey discovered [27].

15.3.2.1 Expanding Fidelity into New Domains

In addition to meeting basic fidelity constraints, new requirements call for M&S capabilities in novel application domains. For instance, the U.S. Marine Corps has recently embarked on an effort to deliver pilot-quality SBT to infantry personnel:

> Today, there are cockpit trainers that are so immersive—for both pilot training and evaluation—that the Services and the Federal Aviation Administration (FAA) allow their substitution for much of the actual flying syllabus. Unfortunately, this level of maturity has not been reached for immersive small unit infantry training which necessarily includes an almost limitless variety of localities, environments, and threats.

> **NRAC [28, p. 58]**

Other Defense simulation goals are concerned with replicating the stress, cognitive load, and emotional fatigue found in the real-life combat situations, or refining human social, cultural, and behavioral models to the point where agents act realistically—down to their body language [29]. These new requirements challenge developers, who must extend the supporting technologies. Even more, they challenge instructional and behavioral scientists, who must define the features of each new domain and determine what levels of fidelity support their instruction.

The Defense sector has apportioned substantial resources to meet these challenges, from both the technological and the behavioral science perspectives. For example, U.S. Joint Forces Command (USJFCOM) recently began the Future Immersive Training Environment (FITE) effort, which seeks to deliver highly realistic infantry training to ensure that the first combat action of personnel is no more difficult or stressful than their last simulated mission. As of mid-2010, FITE investigators had successfully demonstrated a wearable, dismounted immersive simulation system (see Figure 15.2). The body-worn system provides interconnectivity among users, who interact with a virtual environment through synchronized visual, auditory, olfactory, and even haptic events. If shot, the participant even feels a slight electrical charge so that they know they are "injured" or "dead" [30].

Other military programs seek to more faithfully model humans. For instance, the Human Social Culture Behavior (HSCB) initiative is a Defense-wide effort to "develop a military science base and field technologies that support socio-cultural understanding and human terrain forecasting in intelligence analysis, operations analysis/planning, training, and Joint experimentation" [29, p. 4]. Under this venture, numerous projects are improving the fidelity of human models, cognitive/behavioral models, and the interaction among the behaviors of agents. By increasing the fidelity of such components, next-generation simulations will be able to support unique "nonkinetic" (i.e., noncombat) training, such as the acquisition of cultural competence, which is not presently well supported in SBT.

FIGURE 15.2 Camp Lejeune, North Carolina (February 24, 2010). Marines from the 2d Battalion, 8th Marine Regiment, train with the Future Immersive Training Environment (FITE) Joint Capabilities Technology Demonstration (JCTD) virtual reality system in the simulation center at Camp Lejeune, North Carolina. Sponsored by the U.S. Joint Forces Command, with technical management provided by the Office of Naval Research, the FITE JCTD allows an individual wearing a self-contained virtual reality system, with no external tethers and a small joystick mounted on the weapon, to operate in a realistic virtual world displayed in a helmet mounted display. (U.S. Navy photo by John F. Williams/Released.)

These are just two of many Defense efforts designed to expand the range of high-fidelity SBT. These two instances exemplify a trend with respect to SBT fidelity endeavors. Specifically, much of the "low-hanging fruit" has been plucked, leaving instead requirements for nuanced capability improvements (such as precision tuning of character models), expansion into new content domains (such as cultural training), and expansion into new delivery modalities (such as wearable, immersive systems).

15.3.3 INSTRUCTIONAL STRATEGIES

As discussed in Section 15.2, a number of factors converged through the 1970s to reshape the general approach of practitioners to SBT. The cognitive revolution had finally taken hold in the United States, and constructivist theories were gaining attention. Instructional designers were beginning to apply ISD approaches to simulation curricula, and behavioral researchers were finally invited to help address the negative training potential of simulations.

These factors contributed to the development and widespread use of simulation-specific instructional strategies such as the event-based [31] or scenario-based training approach [3]. Proponents of such event-based strategies first debunked "the widely held belief that just more practice automatically leads to better skills" by demonstrating that unstructured simulation-based practice engenders "rote behaviors and an inflexibility to recognize errors" [32, p. 15]. They then called for the scenarios to be systematically organized around predictable training objectives and employ guided-practice principles [31].

Thanks to these efforts, contemporary SBT is more effective. However, three areas of concern remain with respect to instructional strategies. First, while current practice provides some general recommendations for instructional design, few, if any, specific instructional strategies have been identified that span the entire simulation-based learning process (from design to execution, feedback, and remediation). Second, instructional strategies for unique problem domains, such as improving infantry personnel's perceptual abilities or more rapidly imbuing novice trainees with sophisticated higher-order cognitive abilities have not been well developed. Third, the impact of many uncommon and/or combined strategies has not yet been well documented [33]. Consequently, SBT instructors are often left to choose their own instructional approach, which contributes to wide variations in training effectiveness and puts a greater task burden on SBT facilitators [34].

The military has funded several projects intended to identify theoretical frameworks for improving the pedagogy (or more accurately, "andragogy") of its SBT. For example, the Office of Naval Research (ONR) recently sponsored the Next-generation Expeditionary Warfare Intelligent Training (NEW-IT) initiative, which seeks to improve SBT instructional effectiveness, in part, by including automated instructional strategies within distributed SBT software. NEW-IT is scheduled to complete in 2011; however, its investigators already report training performance gains of 26–50% in their empirical field testing [35]. Another ONR effort, called Algorithms Physiologically Derived to Promote Learning Efficiency (APPLE), aims to systematically inform the selection of instructional strategies across a wide range of domains, including distributed SBT [36]. The APPLE effort just began, but it has potential to greatly improve our collective understanding of optimum instructional methods. Programs such as NEW-IT and APPLE promise to improve the state of instructional strategy use within the military SBT community, but many more investigations into methods for improving the effectiveness of SBT are required before a full understanding of SBT instructional best practices is achieved.

15.3.4 INSTRUCTOR WORKLOAD

To be effective, SBT systems rely on significant involvement from expert instructors, who often have all-encompassing duties. The workload of instructors becomes even more pronounced in distributed training contexts, where additional human effort is required to configure and initialize system setup, monitor distributed trainees and LVC entities during the exercise, and manage the delivery of distributed postexercise feedback (e.g., [37,38]). This has led some to argue that a good instructor is the primary determinant of the effectiveness of SBT (e.g., [26,23]) or that "simulators without instructors are virtually useless for training" [39, p. 5]. For distributed real-time SBT, these workload dilemmas become especially pronounced during After Action Review (AAR).

15.3.4.1 Human Effort Required for Distributed AAR

AAR became a formalized military training technique in the 1940s, and it is intended to provide a structured, nonpunitive approach to feedback delivery. Naturally, once the capacity existed, the objective data generated by simulations began to inform

AARs, and as simulators have grown in complexity, greater and greater amounts of objective data have became available. With the introduction of SIMNET in the 1980s, the amount of information that could be utilized in the AAR increased dramatically and so too did the workload on instructors [40].

Typically, an instructor must attend to each distributed training "cell." This means that the instructors, like the trainees, are geographically dispersed and cannot readily respond to all trainees or interact with one another. During a training exercise, instructors are responsible for observing actions of trainees to identify critical details not objectively captured by the simulation. These details may include radio communications that occur among trainees (outside of the simulation itself) or strategic behaviors too subtle for the simulation to analyze [41]. After observing an exercise, instructors typically facilitate the AAR delivery; they assemble the AAR material, find and discuss key points, explain performance outcomes, and lead talks among the trainees [42].

Heavy reliance on instructors in distributed AAR is not merely an issue of individuals' tasking. It affects the cost of deployment and, depending on the quantity and ability of the facilitators, can limit training effectiveness. An apparent mitigation for these issues involves supporting AAR through improved automated performance capture, analysis, and debrief. Such systems could also serve as collaborative tools for the distributed instructors.

The Services have been working toward such solutions for decades. For instance, the Army Research Institute funded development of the Dismounted Infantry Virtual After Action Review System (DIVAARS), which supports "DVD-like" replay of simulation and provides some data analysis support. A similarly focused effort, called Debriefing Distributed Simulation-Based Exercises (DDSBE), was sponsored by ONR.

Other, more particular, AAR systems also exist—from AAR support for analyzing the physical positioning of infantrymen on a live range (e.g., BASE-IT, sponsored by ONR [43]) to interpretation of trainees' cognitive states through neurophysiological sensors (e.g., AITE, sponsored by ONR [44]).

A number of other AAR tools, both general and specifically targeted, can be readily discovered. On the one hand, the variety of tools gives practitioners a range of options; however, on the other hand, many of these tools operate within narrow domains and software specifications, have been mainly demonstrated in carefully scripted use-case settings, and fail to interoperate with one another. In practice, current AAR technologies rarely provide adequate support for trainers and fail to provide deep diagnosis of performance [41]. Nonetheless, ongoing efforts continue to attempt to remedy these problems.

15.3.5 LACK OF EFFECTIVENESS ASSESSMENT

In contrast to the AAR challenge (upon which many researchers and developers are concentrating), few investigators have addressed the question of effectiveness assessment. A recent Army Research Institute report explains:

> More often than not, the simulators are acquired without knowing their training effectiveness, because no empirical research has been done. The vendors, who manufacture and integrate these devices, do not conduct such research because they are

in the business of selling simulators, not research. Occasionally training effectiveness research is conducted after the simulators have been acquired and integrated, but this is narrowly focused on these specific simulators, training specific tasks, in this specific training environment. The research tends to produce no general guidance to the training developer, because of its narrow focus, and because it is conducted on a noninterference basis, making experimental control difficult if not impossible.

Stewart, Johnson, and Howse [27, p. 3]

Effectiveness assessments fall roughly into two categories. They may consider specific instances or approaches, as the preceding quotation suggests, or they may be carried out at the enterprise level (i.e., strategy-wide impact assessment). Such strategic impact assessments involve identifying future consequences of a course of action, such as the expected return-on-investment from pursuing large-scale SBT curriculum.

Assessing the impact and value of distributed SBT—at either the project or the enterprise levels—is challenging and costly. Many benefits of distributed SBT, such as avoidance of future errors, are difficult to measure, and simulation analysts lack the ability to conduct upfront impact assessments or communicate their results [45]. In addition, developers and sponsors are often reluctant to spend resources on verifying the effectiveness of a system that they already expect (or at least very much hope) to work.

Although Defense organizations encourage researchers and developers to conduct effectiveness evaluations, such efforts are carried out intermittently. Fortunately, MSCO is leading attempts to increase such testing. MSCO publishes standards for project-level VV&A evaluation, and the office recently completed a strategy-level evaluation of return-on-investment for Defense M&S [46]. Nonetheless, many more researchers will have to follow the example set by MSCO before this gap is mitigated.

15.3.6 LACK OF USE OUTSIDE OF THE DEFENSE SECTOR

The roots of distributed simulation can be traced back primarily to the Defense sector [47], and unfortunately, its use has remained largely confined to that community [46,48]. Around 2002, academic papers began routinely appearing asking why commercial industry has yet to embrace distributed simulation (e.g., Ref. [49]). This publishing trend continues to date, and the obstacles to adoption outside of the military sector include the following (for more details see Refs. [48] and [50]):

- Insufficient integration with commercial off-the-shelf (COTS) simulation packages
- Technical difficultly (or perceived technical difficulty) of federating systems
- Inefficiency of synchronization algorithms
- Bugs and lack of verification in distributed models
- Overly complex runtime management
- Perceived lack of practical return-on-investment
- Too much functionality in existing distributed packages not relevant for industry

The dearth of effectiveness testing also contributes to the reluctance of some to embrace distributed SBT. As Randall Gibson explains:

> I would have to conclude that the simulation community is not doing an adequate job of selling simulation successes. Too often I see management deciding *not* to commit to simulation for projects where it could be a significant benefit.
>
> **Gibson et al. [45, p. 2027]**

In short, money is the major driver in business, and most businesses fail to see sufficient return-on-investment for their use of distributed SBT. Additionally, most commercial practitioners claim to be interested in low-cost, throw-away COTS tools, instead of more robust enterprise systems. However, current COTS simulation tools do not adequately support distributed simulation applications, and most industry practitioners are not willing to pay much more than a 10% increase in cost to upgrade current technologies [48].

Recently, several pilot projects investigated the benefits of distributed SBT for businesses such as car manufacturing (see the work by Boer [51]). These studies help advance the cause of distributed SBT for business. However, greater numbers of use cases and return-on-investment analyses are required. A 2008 survey of simulation practitioners also suggests that more "success stories" are necessary to overcome industry representatives' "psychological barriers" to accepting distributed SBT. Other responses in this survey indicate that "ready and robust solutions" and "technological advances" must also be made before the commercial sector fully embraces distributed SBT [17].

15.4 CONCLUSION

Distributed real-time SBT fills a vital training gap, and it is actively used by various organizations. However, as this chapter has illustrated, employment of distributed real-time SBT still involves challenges, including technological, instructional, and logistical obstacles. Investigators continue to address these gaps; however, practitioners must make trade-offs due to these limitations.

This chapter was intended to provide readers with a general overview of the history and challenges of distributed real-time SBT. The chapter did not contain a comprehensive listing of all gaps or related projects. Instead, we attempted to identify a general set of challenges that are unique to, or otherwise exacerbated by, distributed systems, as well as some formal initiatives attempting to address these issues. We hope that this chapter has provided a broad perspective of the field and that readers have noticed the integrated nature of the distributed SBT challenges discussed herein. Distributed real-time SBT involves technology, people, and policies; thus, by its very nature, this field is a systems engineering challenge. As innovations continue to be made in this domain, it is critical that technologists, researchers, and policy leaders maintain a broad perspective—carefully balancing the resources and outcomes of individual efforts with the impact to the overall system.

ACKNOWLEDGMENTS

This work is supported by the Office of Naval Research Grant N0001408C0186, the Next-generation Expeditionary Warfare Intelligent Training (NEW-IT) program. The views and conclusions contained in this document are those of the authors and should not be interpreted as representing the official policies, either expressed or implied, of the ONR or the U.S. Government. The U.S. Government is authorized to reproduce and distribute reprints for Government purposes notwithstanding any copyright notation hereon.

REFERENCES

1. MSCO. 2010. *DoD Modeling and Simulation (M&S) Glossary.* Alexandria, VA: Modeling and Simulation Coordination Office (MSCO).
2. Salas, E., L. Rhodenizer, and C. A. Bowers. 2000. "The Design and Delivery of Crew Resource Management Training: Exploiting Available Resources." *Human Factors* 42 (3): 490–511.
3. Oser, R. L., J. A. Cannon-Bowers, D. J. Dwyer, and H. Miller. 1997. *An Event Based Approach for Training: Enhancing the Utility of Joint Service Simulations.* Paper Presented at the 65th Military Operations Research Society Symposium, Quantico, VA, June 1997.
4. Joint Staff/J-7. 2008. *The Joint Training System: A Primer for Senior Leaders (CJCS Guide 3501).* Washington, DC: Department of Defense.
5. Link, E. A. 1930. Jnr U.S. Patent 1,825,462, filed 1930.
6. Owens, L. 1986. "Vannevar Bush and the Differential Analyzer: The Text and Context of an Early Computer." *Technology and Culture* 27 (1): 63–95.
7. Dutton, J. M. 1978. "Information Systems for Transfer of Simulation Technology." In *Proceedings of the 10th Conference on Winter Simulation*, 995–99. New York: ACM.
8. Nance, R. E. 1996. "A History of Discrete Event Simulation Programming Languages." In *History of Programming Languages*, vol. 2, edited by T. J. Bergin and R. G. Gibson, 369–427. New York: ACM.
9. Miller, D. C., and J. A. Thorpe. 1995. "SIMNET: The Advent of Simulator Networking." *Proceedings of the IEEE* 83 (8): 1114–23.
10. Cosby, L. N. 1999. "SIMNET: An Insider's Perspective." *Simulation Technology* 2 (1).
11. Miller, G., A. Adams, and D. Seidel. 1993. *Aggregate Level Simulation Protocol (ALSP)* (1993 Confederation Annual Report; Contract No. DAAB07-94-C-H601). McLean, VA: MITRE Corporation.
12. Richbourg, R., and Lutz, R. 2008. *Live Virtual Constructive Architecture Roadmap (LVCAR) Comparative Analysis of the Middlewares* (Interim Report Appendix D; Case No. 09-S-2412 / M&S CO Project No. 06OC-TR-001). Alexandria, VA: Institute for Defense Analyses.
13. DMSO. 1998. *High-Level Architecture (HLA) Transition Report* (ADA402321). Alexandria, VA: Defense Modeling and Simulation Office.
14. Stout, R. J., C. Bowers, and D. Nicholson. 2008. "Guidelines for Using Simulations to Train Higher Level Cognitive and Teamwork Skills." In *The PSI Handbook of Virtual Environments for Training and Education*, edited by D. Nicholson, D. Schmorrow, and J. Cohn, 270–96. Westport, CT: PSI.
15. Fitzsimmons, E. 2000. "The Defense Modeling and Simulation Office: How It Started." *Simulation Technology Magazine* 2 (2a). http://www.stanford.edu/dept/HPS/TimLenoir/MilitaryEntertainment/Simnet/SISO%20News%20-%20Preview%20Article.htm, accessed: 10 April, 2011.

16. Searle, J., and J. Brennan. 2006. "General Interoperability Concepts." In *Integration of Modelling and Simulation, Educational Notes* (RTO-EN-MSG-043, Paper 3), 3-1–3-8. Neuilly-sur-Seine, France: RTO. http://www.dtic.mil/cgi-bin/GetTRDoc?AD=ADA470923

17. Straßburger, S., T. Schulze, and R. Fujimoto. 2008. *Future Trends in Distributed Simulation and Distributed Virtual Environments: Peer Study Final Report, Version 1.0.* Ilmenau, Magdeburg, Atlanta: Ilmenau University of Technology, University of Magdeburg, Georgia Institute of Technology.

18. Henninger, A. E., D. Cutts, M. Loper, R. Lutz, R. Richbourg, R. Saunders, and S. Swenson. 2008. *Live Virtual Constructive Architecture Roadmap (LVCAR) Final Report* (Case No. 09-S-2412 / M&S CO Project No. 06OC-TR-001). Alexandria, VA: Institute for Defense Analyses.

19. Hassaine, F., N. Abdellaoui, A. Yavas, P. Hubbard, and A. L. Vallerand. 2006. "Effectiveness of JSAF as an Open Architecture, Open Source Synthetic Environment in Defence Experimentation." In *Transforming Training and Experimentation through Modelling and Simulation* (Meeting Proceedings RTO-MP-MSG-045, Paper 11; 11-1–11-6). Neuilly-sur-Seine, France: RTO.

20. SISO. 2009. "A Brief Overview of SISO." www.sisostds.org, accessed: 1 April, 2011.

21. Fujimoto, R. M. 2001. "Parallel and Distributed Simulation Systems." In *Proceedings of the 33th Conference on Winter Simulation*, edited by B. A. Peters, J. S. Smith, D. J. Medeiros, and M. W. Rohrer, 147–57. New York: ACM.

22. Sarjoughian, H. S. 2006. "Model Composability." In *Proceedings of the 38th Conference on Winter Simulation*, edited by L. F. Perrone, F. P. Wieland, J. Liu, B. G. Lawson, D. M. Nicol, and R. M. Fujimoto, 149–58. New York: ACM.

23. National Research Council, Committee on Modeling and Simulation for Defense Transformation. 2006. *Defense Modeling, Simulation, and Analysis: Meeting the Challenge.* Washington, DC: The National Academies Press.

24. Andrews, D. A., L. A. Carroll, and H. H. Bell. 1995. *The Future of Selective Fidelity in Training Devices* (AL/HR-TR-1995-0195). Mesa, AZ: Human Resources Directorate Aircrew Training Research Division.

25. Beaubien, J. M., and D. P. Baker. 2004. "The Use of Simulation for Training Teamwork Skills in Health Care: How Low Can You Go?" *Quality and Safety in Health Care* 13 (Suppl 1): i51–i56.

26. Ross, K. G., J. K. Phillips, G. Klein, and J. Cohn. 2005. "Creating Expertise: A Framework to Guide Simulation-Based Training." In *Proceedings of I/ITSEC.* Orlando, FL: NTSA.

27. Stewart, J. E., D. M. Johnson, and W. R. Howse. 2008. *Fidelity Requirements for Army Aviation Training Devices: Issues and Answers* (Research Report 1887). Arlington, VA: U.S. Army Research Institute.

28. NRAC. 2009. *Immersive Simulation for Marine Corps Small Unit Training* (ADA523942). Arlington, VA: Naval Research Advisory Committee.

29. Biggerstaff, S. 2007. *Human Social Culture Behavior Modeling (HSCB) Program.* Paper Presented at the 2007 Disruptive Technologies Conference, Washington, DC, September 2007.

30. Lawlor, M. 2010. "Infusing FITE into Simulations." *SIGNAL Magazine* 64 (9): 45–50.

31. Fowlkes, J., D. J. Dwyer, R. J. Oser, and E. Salas. 1998. "Event-Based Approach to Training (EBAT)." *The International Journal of Aviation Psychology* 8 (3): 209–221.

32. van der Bosch, K., and J. B. J. Riemersma. 2004. "Reflections on Scenario-Based Training in Tactical Command." In *Scaled Worlds: Development, Validation, and Applications*, edited by S. Schiflett, L. R. Elliott, E. Salas, and M. D. Coovert, 22–36. Burlington, VT: Ashgate.

33. Vogel-Walcutt, J. J., T. Marino-Carper, C. A. Bowers, and D. Nicholson. 2010. "Increasing Efficiency in Military Learning: Theoretical Considerations and Practical Applications." *Military Psychology* 22 (3): 311–39.

34. Schatz, S., C.A. Bowers, and D. Nicholson. 2009. "Advanced Situated Tutors: Design, Philosophy, and a Review of Existing Systems." In *Proceedings of the 53rd Annual Conference of the Human Factors and Ergonomics Society*, 1944–48. Santa Monica, CA: Human Factors and Ergonomics Society.

35. Vogel-Walcutt, J. J., Marshall, R., Fowlkes, J., Schatz, S., Dolletski-Lazar, R., & Nicholson, D. (2011). Effects of instructional strategies within an instructional support system to improve knowledge acquisition and application. In *Proceedings of the 55th Annual Meeting of the Human Factors and Ergonomics Society*. Santa Monica, CA: Human Factors and Ergonomics Society. September 19–23, 2011 Las Vegas, Nevada.

36. Nicholson, D., and J. J. Vogel-Walcutt. 2010. *Algorithms Physiologically Derived to Promote Learning Efficiency (APPLE) 6.3*. http://active.ist.ucf.edu.

37. Ross, P. 2008. "Machine Readable Enumerations for Improved Distributed Simulation Initialisation Interoperability." In *Proceedings of the SimTect 2008 Simulation Conference*. http://hdl.handle.net/1947/9101, accessed 5 April, 2011.

38. Weinstock, C. B., and J. B. Goodenough. 2006. *On system scalability* (Technical Note, CMU/SEI-2006-TN-012). Pittsburgh, PA: Carnegie Mellon University.

39. Stottler, R. H., R. Jensen, B. Pike, and R. Bingham. 2006. *Adding an Intelligent Tutoring System to an Existing Training Simulation* (ADA454493). San Mateo, CA: Stottler Henke Associates Inc.

40. Morrison, J. E., and L. L. Meliza. 1999. *Foundations of the After Action Review Process* (ADA368651). Alexandria, VA: Institute for Defense Analyses.

41. Wiese, E., J. Freeman, W. Salter, E. Stelzer, and C. Jackson. 2008. "Distributed After Action Review for Simulation-Based Training." In *Human Factors in Simulation and Training*, edited by D. A. Vincenzi, J. A. Wise, M. Mouloua and P. A. Hancock. Boca Raton, FL: CRC Press.

42. Freeman, J., W. J. Salter, and S. Hoch. 2004. "The Users and Functions of Debriefing in Distributed, Simulation-Based Training." In *Proceedings of the Human Factors and Ergonomics Society 48th Annual Meeting*, 2577–81. Santa Monica, CA: Human Factors and Ergonomics Society.

43. Rowe, N. C., J. P. Houde, M. N. Kolsch, C. J. Darken, E. R. Heine, A. Sadagic, C. Basu, F. Han. 2010. *Automated Assessment of Physical-Motion Tasks for Military Integrative Training*. Paper Presented at the 2nd International Conference on Computer Supported Education (CSEDU 2010), Noordwijkerhout, The Netherlands, May 2010.

44. Vogel-Walcutt, J. J., D. Nicholson, and C. Bowers. 2009. *Translating Learning Theories into Physiological Hypotheses*. Paper Presented at the Human Computer Interaction Conference, San Diego, CA, July 19–24, 2009.

45. Gibson, R., D. J. Medeiros, A. Sudar, B. Waite, and M. W. Rohrer. 2003. "Increasing Return on Investment from Simulation (Panel)." In *Proceedings of the 35th Conference on Winter Simulation*, edited by S. Chick, P. J. Sánchez, D. Ferrin, and D. J. Morrice, 2027–32. New York: ACM.

46. MSCO. 2009. *Metrics for M&S Investment* (REPORT No. TJ-042608-RP013). Alexandria, VA: Modeling and Simulation Coordination Office (MSCO).

47. Fujimoto, R. M. 2003. "Distributed Simulation Systems." In *Proceedings of the 35th Conference on Winter Simulation*, edited by S. Chick, P. J. Sánchez, D. Ferrin, and D. J. Morrice, 2027–32. New York: ACM.

48. Boer, C. A., A. de Bruin, and A. Verbraeck. 2009. "A Survey on Distributed Simulation in Industry." *Journal of Simulation* 3: 3–16.

49. Taylor, S. J. E., A. Bruzzone, R. Fujimoto, B. P. Gan, S. Straßburger, and R. J. Paul. 2002. "Distributed Simulation and Industry: Potentials and Pitfalls." In *Proceedings of the 34th Conference on Winter Simulation*, edited by E. Yücesan, C.-H.Chen, J. L. Snowdon, and J. M. Charnes, 688–94. New York: ACM.

50. Lendermann, P., M. U. Heinicke, L. F. McGinnis, C. McLean, S. Straßburger, and S. J. E. Taylor. 2007. "Panel: Distributed Simulation in Industry—A Real-World Necessity or Ivory Tower Fancy?" In *Proceedings of the 39th Conference on Winter Simulation*, edited by S. G. Henderson, B. Biller, M.-H. Hsieh, J. Shortle, J. D. Tew, and R. R. Barton, 1053–62. New York: ACM.
51. Boer, C. A. 2005. *Distributed Simulation in Industry*. Rotterdam, The Netherlands: Erasmus University Rotterdam.

16 Concurrent Simulation for Online Optimization of Discrete Event Systems

Christos G. Cassandras and Christos G. Panayiotou

CONTENTS

16.1 Introduction 389
16.2 Modeling and Simulating Discrete Event Systems 392
16.3 Sample Path Constructability 395
    16.3.1 Standard Clock Approach 398
    16.3.2 Augmented System Analysis 399
16.4 Concurrent Simulation Approach for Arbitrary Event Processes 401
    16.4.1 Notation and Definitions 402
    16.4.2 Observed and Constructed Sample Path Coupling Dynamics 404
    16.4.3 Speedup Factor 408
    16.4.4 Extensions of the TWA 409
16.5 Use of TWA for Real-Time Optimization 410
16.6 Conclusions 415
Acknowledgments 415
References 415

16.1 INTRODUCTION

The path to fast and accurate computer simulation depends on the objective one pursues. In the case of offline applications, the goal is typically to simulate a highly complex system to study its behavior in detail and to ascertain its performance. An example might involve the design of an expensive large-scale system to be built, in which case it is essential to use simulation for the purpose of exploring whether it meets performance specifications and cost constraints under alternative designs and control mechanisms. In such a realm, reasonably fast simulation time is desirable but not indispensable, and one may be more likely to trade off execution speed for higher fidelity. On the contrary, for online applications, we often refer to *real-time simulation* as a tool for generating efficient models providing predictive capability while an actual system is operating as well as the means to explore alternatives that may be implemented in real time. An example might arise when a system is about to operate in an environment it was never originally designed for; one is then interested in simulating how the system would behave in such an environment, possibly under various implementable alternatives (e.g., adding resources to it or modifying the parametric settings of a controller). On the basis of the results, one might proceed in the new setting, abort, or choose the best possible modification before proceeding. In such cases, obtaining information fast from simulation is not just desirable, but an absolute requirement.

Speeding up the process of information extraction from simulation may be attained in various ways. First, one can use *abstraction* to replace a detailed model by a less detailed one that is still sufficiently accurate to provide useful information in a real-time frame for making decisions such as choosing between two or more alternatives to be implemented. Another approach is to use multiple processors operating in parallel. This is usually referred to as *distributed simulation*, where different parts of the entire simulation code are assigned to different processors. By operating in parallel, N such processors have the potential of reducing the time to complete a simulation run by a factor of N. However, this is rarely feasible since the execution must be synchronized. Thus, one slow processor may delay the rest either because its output is necessary for them to continue execution or because of inefficiencies in the way the code was distributed. Nonetheless, this approach enables interoperability as well as allowing for geographic distribution of processors and some degree of fault tolerance.[1,2]

The goal of *concurrent simulation* is to provide, at the completion of a single run, information that would normally require M separate simulation runs and to do so with minimal overhead. Thus, suppose that the goal of a real-time simulation application is to determine how a given performance measure $J(\theta)$ varies as a function of some parameter θ in the absence of any analytical expression for $J(\theta)$. This requires the generation of a simulation run at least once for every value of θ of interest. It is easy to see how demanding this process becomes: if θ represents options from a discrete set $\{\theta_1, \dots, \theta_M\}$, we need M simulation runs under each of $\theta_1, \dots, \theta_M$. It is even more demanding when the system is stochastic, requiring a large number of simulation runs under a fixed θ_i to achieve desired levels of statistical accuracy. Concurrent simulation techniques are designed to accomplish the task of providing estimates of all $J(\theta_1), \dots, J(\theta_M)$ at the end of a *single* simulation run instead of M runs. Moreover, if a simulation run under some θ_i requires T time units, then instead of $M \cdot T$ total simulation time, a concurrent simulation algorithm is intended to deliver all M estimates in $T + \tau_M$ time units where often $\tau_M < T$ and certainly $T + \tau_M \ll M \cdot T$. In general, $\theta = [\theta^1, \dots, \theta^N]$ is a vector of parameters, in which case concurrent simulation is applied to each element θ^i over all M^i values of θ^i that are of interest. Thus, at the end of a single simulation run, we obtain estimates of $J(\theta^i_1), \dots, J(\theta^i_{M^i})$ for all $i = 1, \dots, N$ at the same time. To keep notation manageable, however, in most of the discussion that follows we consider a scalar θ.

It is important to emphasize the difference between distributed and concurrent simulation. In distributed simulation, the code for a single simulation run is shared over M processors. Thus, at the end of the run, a single set of input–output data (equivalently, one estimate of the relationship $J(\theta)$) is obtained. In concurrent simulation, a single processor simulates a system but also concurrently simulates it under $M - 1$ additional settings. Thus, at the end of the run, we obtain a total of M input–output data (equivalently, estimates of $J(\theta_1), \ldots, J(\theta_M)$). If, in addition, M processors are available for concurrent simulation, then the speedup resulting from the concurrent construction is further increased by naturally assigning a processor to each of the M threads estimating $J(\theta_i)$ for each $i = 1, \ldots, M$. The speedup benefit in concurrent simulation comes not from the presence of extra processing capacity (which also implies a higher cost) but from the efficient sharing of input data over multiple simulations of the same system under different settings. Another way of interpreting it is as a method that addresses the question "What if a given system operating under a nominal value θ_1 were to operate under $\theta_2, \ldots, \theta_M$ with the same input conditions?" The answer to these $M - 1$ "what if" questions can be obtained without having to reproduce $M - 1$ distinct simulations by "brute force"; instead, it can be deduced by processing data from the nominal simulation so as to track the differences caused by replacing θ_1 by θ_i, $i = 2, \ldots, M$. In Monte Carlo simulation for instance, the computational cost of generating random variates from several distributions is much higher than the cost of performing this difference tracking and constructing state trajectories under multiple settings captured by $\theta_1, \ldots, \theta_M$.

In this chapter, we will first review the formalism and theoretical foundations of concurrent simulation as it applies to the class of discrete event systems (DES), also developed in the works by Cassandras and Lafortune[3] and Cassandras and Panayiotou.[4] The state space of a typical DES involves at least some discrete components, giving rise to piecewise constant state trajectories. This can be exploited to predict changes in an observed state trajectory under some setting θ_i when the system operates under $\theta_j \neq \theta_i$. Formally, this is studied as a general *sample path constructability* problem. In particular, given a DES sample path under a parameter value θ, the problem is to construct multiple sample paths of the system under different values *using only information available along the given sample path*. A solution to this problem can be obtained when the system under consideration satisfies a *constructability* condition. Unfortunately, this condition is not easily satisfied. However, it is possible to enforce it at some expense. For a large class of DES modeled as Markov chains, there are two techniques known as the *Standard Clock* (SC) method and the *Augmented System Analysis* (ASA), which provide solutions to the constructability problem. In a more general setting (where one cannot impose any assumptions on the stochastic characteristics of the event processes), we will describe the process through which the problem is solved by coupling an observed sample path to multiple concurrently generated sample paths under different settings. This ultimately leads to a detailed procedure known as the *Time Warping Algorithm* (TWA).

At this point, it is worth emphasizing that concurrent sample path constructability techniques can be used in two different modes. First, they can be used offline to obtain estimates of the system performance under different parameters $\theta_1, \ldots, \theta_M$ from a single simulation run under θ_0. A main objective in this case is to reduce the overall simulation time to achieve high speedup. This approach was investigated in the work by Cassandras and Panayiotou.[4] The emphasis of this chapter is on the online use of sample path constructability techniques where the nominal sample path θ_0 is simply the one observed during the operation of the *actual system*. In this case, based on the observed data, the goal is to estimate the system's performance again under a set of hypothetical parameter values $\theta_1, \ldots, \theta_M$. In this case, a controller can monitor the performance of the system under the different parameters $J(\theta_0), \ldots, J(\theta_M)$, and at the end of each epoch, it can switch to the parameter that achieves the best performance. In this setting, another important advantage of concurrent sample path constructability techniques is that they use *common random numbers* (CRN), and as a result, the probability of selecting the parameter that optimizes the performance $J(\theta)$ is increased. Another important advantage of the online use of concurrent sample path constructability techniques is that it does not require any prior knowledge of the underlying distributions of the various events since those are already embedded in the occurrence times of the observed events. This makes the approach applicable even when the distributions are time varying.

We begin the chapter by reviewing the *Stochastic Timed Automaton* (STA) modeling framework for DES in Section 16.2. This is the framework used by most commercial discrete event simulators, thus facilitating the implementation of the concurrent simulation techniques to be described. In Section 16.3, we define the general sample path constructability problem. We also review the SC method and the ASA for DES with Markovian structure in their event processes. Then, in Section 16.4, we describe the procedure through which the general constructability problem is solved and derive the aforementioned TWA. In Section 16.5, we describe how the TWA can be used together with optimization schemes (e.g.,[5-8] to solve dynamic resource allocation problems (see also works by Panayiotou and Cassandras[9,10]). The chapter ends with some final thoughts and conclusions.

16.2 MODELING AND SIMULATING DISCRETE EVENT SYSTEMS

We begin with a DES modeled as an automaton $(X, \mathcal{E}, f, \Gamma, x_0)$, where X is a countable state space, $\mathcal{E}$ is a countable event set, $f:X \times \mathcal{E} \rightarrow X$ is a state transition function, $\Gamma:X \rightarrow 2^{\mathcal{E}}$ is the active (or feasible) event function so that $\Gamma(x)$ is the set of all events $e \in \mathcal{E}$ for which $f(x, e)$ is defined, and it is called the active event set (or feasible event set), and finally, x_0 is the initial state. The automaton is easily modified to $(X, \mathcal{E}, \Gamma, x_0, p, p_0)$ to include probabilistic state transition mechanisms: the state transition probability $p(x'; x, e')$ is defined for all $x, x' \in X$, $e' \in \mathcal{E}$ and is such that $p(x'; x, e') = 0$ for all $e' \notin \Gamma(x)$; in addition, $p_0(x)$ is the probability mass function (pmf) $P[x_0 = x]$, $x \in X$, of the initial state x_0. For simplicity, we assume that the DES satisfies the "noninterruption condition," that is, once an event is enabled, it cannot be disabled; however, this is not essential to the rest of our discussion.

This model is referred to as an *untimed* automaton,[3] since it provides no information as to which among all events feasible at some state x will occur next. To resolve this issue, we define a *clock structure* associated with an event set $\mathcal{E}$ (which we will hencefoth assume finite with cardinality N) to be a set $V = \{V_1, \ldots, V_N\}$ of *event lifetime sequences* $V_i = \{v_{i,1}, v_{i,2}, \ldots\}$, one for each event $i \in \mathcal{E}$, with $v_{i,k} \in \mathbb{R}^+$. This leads to the definition of a *timed* automaton $(X, \mathcal{E}, f, \Gamma, x_0, V)$, where V is a clock structure, and $(X, \mathcal{E}, f, \Gamma, x_0)$ is an (untimed) automaton. The timed automaton generates a state sequence

$$x' = f(x, e') \tag{16.1}$$

driven by an event sequence $\{e_1, e_2, \ldots\}$ generated through

$$e' = \arg\min_{i \in \Gamma(x)} \{y_i\} \tag{16.2}$$

with the *clock values* y_i, $i \in \mathcal{E}$, defined by

$$y_i' = \begin{cases} y_i - y^* & \text{if} \quad i \neq e' \quad \text{and} \quad i \in \Gamma(x) \\ v_{i,N_{i+1}} & \text{if} \quad i = e' \quad \text{or} \quad i \notin \Gamma(x) \end{cases} \quad i \in \Gamma(x') \tag{16.3}$$

where the *interevent time* y^* is defined as

$$y^* = \min_{i \in \Gamma(x)} \{y_i\} \tag{16.4}$$

and the *event scores* N_i, $i \in \mathcal{E}$, are defined by

$$N_i' = \begin{cases} N_{i+1} & \text{if} \quad i = e' \quad \text{or} \quad i \notin \Gamma(x) \\ N_i & \text{otherwise} \end{cases} \quad i \in \Gamma(x') \tag{16.5}$$

In addition, initial conditions are $y_i = v_{i,1}$ and $N_i = 1$ for all $i \in \Gamma(x_0)$. If $i \notin \Gamma(x_0)$, then y_i is undefined and $N_i = 0$.

Note that this is precisely how a discrete event simulator generates state trajectories of DES using the event lifetime sequences $\mathbf{V}_i = \{v_{i,1}, v_{i,2}, \ldots\}$ as input. The simulator maintains the *Clock*, the *State x*, and the *Event Calendar* where all feasible events at state x along with their clock values are maintained. The entries of the Event Calendar are ordered so as to determine y^* through Equation 16.4 and hence determine the "triggering event" e' in Equation 16.2. Once this is accomplished, the simulator updates the State through Equation 16.1 and the Clock by setting its new value to

$$t' = t + y^* \tag{16.6}$$

Then, since x' is available, $\Gamma(x')$ is determined, and all Clock values for $i \in \Gamma(x')$ are updated through Equation 16.3. This results in an updated Event Calendar and the process repeats. It is important to note that this state trajectory construction is entirely *event driven* and *not time driven*. In other words, it is the Event Calendar that determines the next State of the system as well as the next value of the Clock. If the Event Calendar ever becomes empty, the process cannot continue, a situation that we often identify as a "deadlock" in the system. This event-driven mechanism is to be contrasted to a time-driven approach where the clock is updated through $t' = t + \Delta$, where Δ is a fixed time step. This is clearly inefficient, since often there is a large interval between the current event and the next event; during this interval, the simulator needlessly updates the clock. What is worse, however, is that one or more events may in fact occur within an interval $[t, t + \Delta]$ in which case such a time-driven procedure fails to update the state until $t + \Delta$. To counteract that, one might use a smaller value of Δ, which in turn forces more needless clock updates.

The final step is to incorporate randomness into a timed automaton by allowing the elements of an event lifetime sequence to be random variables. Furthermore, in a more general setting, the state transition mechanisms can be assumed probabilistic. Thus, we define a *stochastic clock structure* associated with an event set $\mathcal{E}$ to be a set of distribution functions $G = \{G_i : i \in \mathcal{E}\}$ characterizing the stochastic clock sequences. This leads to the definition of a STA $(\mathcal{E}, \mathbf{X}, \Gamma, p, p_0, G)$, where $(\mathcal{E}, \mathbf{X}, \Gamma, p, p_0)$ is an automaton with a probabilistic state transition mechanism and G is a stochastic clock structure. The STA generates a stochastic state sequence $\{X_0, X_1, ...\}$ (i.e., a sample path) through a transition mechanism (based on observations $X = x, E' = e'$):

$$X' = x' \text{ with probability } p(x'; x, e')$$

and it is driven by a stochastic event sequence $\{E_1, E_2, ...\}$ generated through the same process as Equations 16.2 through 16.5 with E replacing e, Y_i replacing y_i, and $V_{i,k}$ replacing $v_{i,k}$. In addition,

$$\{V_{i,k}\} \sim G_i$$

where the tilde ($\sim$) notation denotes "with distribution" and initial conditions are $X_0 \sim p_0(x)$, and $Y_i = V_{i,1}$ and $N_i = 1$ if $i \in \Gamma(X_0)$. If $i \notin \Gamma(X_0)$, Y_i is undefined and $N_i = 0$. For simplicity, for the remainder of the chapter, we will assume a STA with deterministic state transition functions $(\mathcal{E}, \mathbf{X}, f, \Gamma, G)$ where the source of randomness is the stochastic nature of event lifetimes.

Example

To illustrate the definition of a STA, let us apply it to a simple single-server queueing system with a sequence of arriving tasks waiting in a First-In-First-Out (FIFO) queue with infinite capacity. This is normally written as "$G/G/1$," where the first two Gs denote the probability distribution characterizing the arrival process and the service process, respectively: the letter G stands for "General" to indicate that neither distribution is assumed known. The number 1 indicates that there is a single server processing arriving task requests. In this case, we set $\mathbf{X} = \{0, 1, ...\}$ to be the state space, so that $x \in \mathbf{X}$ is the number of tasks in the system (including one in process), and $\mathcal{E} = \{a, d\}$ to denote an arrival and a departure event, respectively. Clearly, $\Gamma(x) = \{a, d\}$ for all $x \in \mathbf{X}$ except for $\Gamma(0) = \{a\}$, since a departure from an empty system is infeasible. Since all state transitions are deterministic, we revert to the usual state transition mechanism $x' = f(x, e')$, and we have $x' = x + 1$ if $e' = a$ and $x' = x - 1$ if $e' = d$ and $x > 0$.

A $G/G/1/K$ queueing system is one where the queueing capacity is limited to K tasks. In this case, the model above is modified so that $\mathbf{X} = \{0, 1, ..., K\}$ and $x' = x + 1$ if $e' = a, x < K$, while $x' = x$ if $e' = a, x = K$. If, in addition, there is some information regarding the distribution functions of the arrival and service processes, then these are specified through stochastic clock structure $G = \{G_a, G_d\}$. For example, if the arrival process is Poisson with rate λ, we have $G_a(t) = 1 - e^{-\lambda t}, t \geq 0$ with $G_a(t) = 0$ for $t < 0$, and if the service process is uniformly distributed over $[0, 1]$, we have $G_d(t) = t$ for $t \in [0, 1]$, $G_d(t) = 0$ for $t < 0$, and $G_d(t) = 1$ for $t > 1$. In this case, the queueing system is represented by the notation $M/U/1/K$.

16.3   SAMPLE PATH CONSTRUCTABILITY

Let us consider a DES modeled as a STA and let $\Theta = \{\theta_0, \theta_1, \ldots, \theta_M\}$ be a finite set containing all possible values that a parameter θ in this model may take. As mentioned earlier, in general, $\boldsymbol{\theta}$ is a vector, so if $\boldsymbol{\theta} = [\theta^1, \ldots, \theta^N]$ and θ^i can take M^i values, then we can write $\Theta = \{\theta_1^1, \ldots, \theta_{M^0}^1, \ldots, \theta_1^N, \ldots, \theta_{M^N}^N\}$. Moreover, it is possible that Θ is a set of feasible operating policies or design choices that affect the dynamics of the DES. For example, in a G/G/1/θ queueing system, we may consider the capacity θ as a model parameter taking integer values for some set $\Theta_1 = \{1, \ldots, K\}$. Alternatively, we may want to study the performance of this DES as a function of different queueing disciplines such as FIFO, Last-In-First-Out (LIFO), and Random; in this case, $\Theta_2 = \{\text{FIFO, LIFO, Random}\}$. Yet another example is the simultaneous investigation of the queue capacity and the queueing discipline, and in this case, $\Theta_3 = \Theta_1 \times \Theta_2$. Let $J(\theta)$ denote a performance metric of the system under θ. Since the DES is in general stochastic, let

$$\omega(\theta) = \{(e_k, t_k), \quad k = 1, 2, \ldots\}$$

be a sample path of the system when it is operating under θ, where $e_k \in \mathcal{E}$ is the kth event observed in this sample path and t_k is the associated occurrence time. The corresponding sample performance metric is denoted by $L[\omega(\theta)]$, and we are interested in estimating performance metrics of the form

$$J(\theta) = E\big[L[\omega(\theta)]\big]$$

Our premise is that there is no closed-form expression for $J(\theta)$; therefore, we rely on simulation to obtain sample paths $\omega(\theta)$ from which we can compute $L[\omega(\theta)]$ and ultimately estimate the expected value $E[L[\omega(\theta)]]$, typically through

$$\hat{J}(\theta) = \frac{1}{R} \sum_{i=1}^{R} L[\omega_i(\theta)]$$

where $\omega_i(\theta)$ is the ith sample path (equivalently, simulation run or, for online applications, a finite length observation interval). In applications where the objective is to select the value of θ in $\Theta = \{\theta_0, \theta_1, \ldots, \theta_M\}$ that minimizes $J(\theta)$, then the "brute force" trial-and-error approach involves a total of $(M + 1) \cdot R$ simulation runs to estimate performance for each of the $M + 1$ settings and then select the optimal one. For offline optimization, assuming all event lifetime distributions are known, then this is feasible, although possibly computationally expensive. For online applications, however, where event lifetime distributions may not be known exactly or where event lifetime distributions may be time varying, the brute force approach is infeasible. The goal of concurrent simulation methods is to extract estimates $\hat{J}(\theta_i)$, $i = 1, \ldots, M$ by observing only one sample path under parameter θ_0. To study whether this is possible and under what conditions, we pose the sample path constructability problem as follows:

> For a DES under θ_0, construct all sample paths $\omega(\theta_1), \ldots, \omega(\theta_M)$ given a realization of lifetime sequences $V_1, \ldots, V_N$ and the sample path $\omega(\theta_0)$.

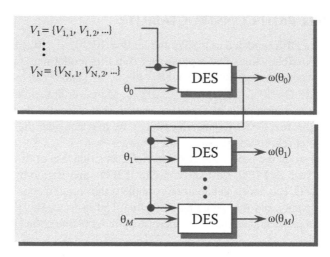

FIGURE 16.1   The sample path constructability problem.

This is illustrated in Figure 16.1. There is a simple sufficient condition under which the constructability problem can be solved. This condition consists of two parts. First, let $\{X_k(\theta)\}$, $k = 1, 2, \ldots$, be the state sequence observed in a sample path $\omega(\theta)$ under θ in which the lifetime sequences are $V_i = \{V_{i,1}, V_{i,2}, \ldots\}$, $i = 1, \ldots, N$. We refer to this as the *nominal* sample path. Then, for $\theta_m \neq \theta$, suppose the DES is driven by the same event sequence generated under θ, giving rise to a new state sequence $\{X_k(\theta_m)\}$, $k = 1, 2, \ldots$, and sample path $\omega(\theta_m)$. We say that $\omega(\theta_m)$ is *observable* with respect to $\omega(\theta)$ if $\Gamma(X_k(\theta_m)) \subseteq \Gamma(X_k(\theta))$ for all $k = 1, 2, \ldots$. Thus, we have the following *observability condition* (**OB**):

> (**OB**) Let $\omega(\theta)$ be a sample path under θ, and $\{X_k(\theta)\}$, $k = 0, 1, \ldots$, the corresponding state sequence. For $\theta_m \neq \theta$, let $\omega(\theta_m)$ be the sample path generated by the event sequence in $\omega(\theta)$. Then, for all $k = 0, 1, \ldots$, $\Gamma(X_k(\theta_m)) \subseteq \Gamma(X_k(\theta))$.

By construction, the two sample paths are "coupled" so that the same event sequence drives them both. As the two state sequences subsequently unfold, condition (**OB**) states that every state observed in the nominal sample path is always "richer" in terms of feasible events. In other words, from the point of view of $\omega(\theta_m)$, all feasible events required at state $X_k(\theta_m)$ are observable in the corresponding state $X_k(\theta)$.

The second part of the constructability condition involves the clock values Y_i of events $i \in \mathcal{E}$. Let $H(t, z)$ be the conditional distribution of Y_i defined as follows:

$$H(t,z) = P\big[Y_i \leq t \mid V_i > z\big]$$

where z is the observed age of the event at the time $H(t, z)$ is evaluated (i.e., the time elapsed since event i was last triggered). Given the lifetime distribution $G_i(t)$, and since $V_i = Y_i + z$, we have

$$H(t,z) = P\big[V_i \leq z + t \mid V_i > z\big] = \frac{G_i(z+t) - G_i(z)}{1 - G_i(z)}$$

Since both the distribution $H(t, z)$ and the event age generally depend on the parameter θ, we write $H(t, z_{i,k}(\theta); \theta)$ to denote the cumulative distribution function (cdf) of the clock value of event $i \in \mathcal{E}$, given its age $z_{i,k}(\theta)$. Then, the *constructability condition* (**CO**) is as follows:

(**CO**) Let $\omega(\theta)$ be a sample path under θ, and $\{X_k(\theta)\}$, $k = 0, 1, \dots$, the corresponding state sequence. For $\theta_m \neq \theta$, let $\omega(\theta_m)$ be the sample path generated by the event sequence in $\omega(\theta)$. Then,

$$\Gamma\big(X_k(\theta_m)\big) \subseteq \Gamma\big(X_k(\theta)\big) \quad \text{for all} \quad k = 0, 1, \dots$$

and

$$H\big(t, z_{i,k}(\theta_m); \theta_m\big) = H\big(t, z_{i,k}(\theta); \theta\big) \quad \text{for all} \quad i \in \Gamma\big(X_k(\theta)\big), k = 0, 1, \dots$$

The first part is simply (**OB**), while the second part imposes a requirement on the event clock distributions. Details on (**CO**) are found in the work by Cassandras and Lafortune,[3] but we note here two immediate implications: (1) if the STA has a Poisson clock structure, then (**OB**) implies (**CO**), and (2) if $\Gamma(X_k(\theta_m)) = \Gamma(X_k(\theta))$ for all $k = 0, 1, \dots$, then (**CO**) is satisfied. Clearly, (1) simply follows from the memory-less property of the exponential distribution, and it reduces the test of constructability to the purely structural condition (**OB**). Regarding (2), it follows from the fact that if feasible event sets are always the same under θ and θ_m, then all event clock values are also the same.

Example

Consider a G/G/1/K queueing system, where $K = 1, 2, \dots$ is the parameter of interest. Under $K = 3$, a sample path ω_3 is obtained. We then pose the question whether a sample path ω_2, constructed under $K = 2$, is observable with respect to ω_3. In what follows, $\{X_k(K)\}$, $k = 1, 2, \dots$, denotes the state sequence in a sample path generated with queueing capacity $K = 1, 2, \dots$. The two state transition diagrams are shown in Figure 16.2. Since condition (**OB**) must be tested under the assumption that both sample paths are generated by the exact same input, it is convenient to construct a new system whose state consists of the joint queue lengths of the two original systems. This is referred to as an *augmented system*. As shown in Figure 16.2, we assume both sample paths start from a common state (say 0, for simplicity). On the left part of the augmented system state transition diagram, the two states remain the same. In fact, a difference arises only when (1) $X_k(3) = X_k(2) = 2$ for some k, and (2) event a occurs. At this point, the state transition gives $X_{k+1}(3) = 3$, but $X_{k+1}(2) = 2$. From this point on, the two states satisfy $X_k(3) = X_k(2) + 1$, as seen in Figure 16.2, until state $(1, 0)$ is visited. At this point, the only feasible event is a departure d, bringing the state back to $(0, 0)$.

By inspection of the augmented system state transition diagram, we see that (**OB**) is trivially satisfied for the three states such that $X_k(3) = X_k(2)$. It is also easily seen to be satisfied at $(3, 2)$ and $(2, 1)$, since $\Gamma(x) = \{a, d\}$ for all $x > 0$. The only remaining state is $(1, 0)$, where we see that $\Gamma(X_k(2)) = \Gamma(0) = \{a\} \subset \{a, d\} = \Gamma(1) = \Gamma(X_k(3))$, and (**OB**) is satisfied. Thus, sample paths of this system under $K = 2$ are indeed observable with respect to sample paths under $K = 3$.

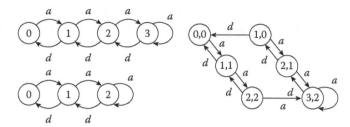

FIGURE 16.2 State transition diagrams for G/G/1/3 and G/G/1/2 along with the corresponding augmented system.

Let us now reverse the roles of the two parameter values and corresponding sample paths. Thus, the question is whether a sample path constructed under $K = 3$ is observable with respect to a sample path under $K = 2$. Returning to Figure 16.2, and state (1, 0) in particular, note that $\Gamma(X_k(3)) = \Gamma(1) = \{a, d\} \supset \{a\} = \Gamma(0) = \Gamma(X_k(2))$. Thus, the feasible event d, needed at state $X_k(3) = 1$ is *unobservable* at the corresponding state $X_k(2) = 0$. Therefore, it is interesting to note that condition (**OB**) is not symmetric with respect to the role of the parameter chosen as the "nominal" one.

As the previous example illustrates, even for DES modeled as Markov chains, condition (**CO**) is not generally satisfied because (**OB**) may fail. However, it is possible to enforce (**OB**) at the expense of some computational cost or the cost of temporarily "suspending" the construction of the sample paths under some values in $\{\theta_1, \ldots, \theta_M\}$. We will review next two methods that accomplish this goal, the SC method and ASA, before addressing the more general question of solving the constructability problem for arbitrary DES.

16.3.1 STANDARD CLOCK APPROACH

The SC approach[11] applies to stochastic timed automata with a Poisson clock structure so that all event lifetime distributions are exponential, and we have $G_i(t) = 1 - e^{-\lambda_i t}$ for all $i \in \mathcal{E}$. Thus, every event process in such systems is characterized by a single Poisson parameter λ_i. As already pointed out, the distributional part of (**CO**) is immediately satisfied by the memoryless property of the exponentially distributed event lifetimes. The observability condition (**OB**) will be forced to be satisfied by choosing a nominal sample path to simulate such that $\Gamma(x) = \mathcal{E}$ for all states x. To accomplish this, let

$$\Lambda(x) = \sum_{i \in \Gamma(x)} \lambda_i \tag{16.7}$$

A property of Markov chains is that the distribution of the triggering event taking place at state x is given by

$$p_i(x) = \frac{\lambda_i}{\Lambda(x)}, \quad i \in \Gamma(x) \tag{16.8}$$

Moreover, exploiting the uniformization of Markov chains, we replace $\Lambda(x)$ by a uniform rate $\gamma \geq \Lambda(x)$ common to all x so that the additional probability flow

$[\gamma - \Lambda(x)]$ at state x corresponds to "fictitious events" that leave the state unchanged. In our case, we choose this uniform rate to be the maximal rate over all events:

$$\Lambda = \sum_{i \in \mathcal{E}} \lambda_i \qquad (16.9)$$

In the uniformized model, every event is always feasible. However, events i such that $i \notin \Gamma(x)$ for the original model leave the state unchanged, and they are ignored. Moreover, the mechanism through which the triggering event at any state is determined becomes independent of the state. Specifically, in Equation 16.7, we can now set $\Lambda(x) = \Lambda$ for all x. Therefore, the triggering event distribution at any state is simply $p_i = \lambda_i/\Lambda$. The fact that all events are permanently feasible in the uniformized model, combined with the memoryless property of the exponential event lifetime distribution, immediately implies the constructability condition (**CO**). In short, the way the SC approach gets around the problem of potentially unobservable events is by simulating a system in which *all* events are forced to occur at *all* states, and hence satisfy (**OB**). Although this results in several fictitious events, that is, events i such that $i \notin \Gamma(x)$, it also makes for a general-purpose methodology within the realm of Poisson event processes.

There is one additional simplification we can make to this scheme. We can generate all interevent times $\{V_1, V_2, ...\}$ from the cdf $G(t) = 1 - e^{-t}$ and then rescale any specific V_k through $V_k(\Lambda) = V_k/\Lambda$ for any given Λ. This allows us to define all event times: assuming the system starts at time zero, we have $t_1 = V_1$, $t_2 = t_1 + V_2$, and so on. The resulting sequence is referred to as the SC and can be generated in advance of any simulation. The specific SC algorithm for constructing sample paths under $\{\theta_0, ... , \theta_M\}$ concurrently with the simulation of a nominal system constructed in this manner is as follows:

INITIALIZATION: Construct a SC: $\{V_1, V_2, ...\}$, $V_k \sim 1 - e^{-t}$, $t > 0$.
FOR EVERY CONSTRUCTED SAMPLE PATH $m = 0, 1, ... , M$:
1. Determine the triggering event E_m (by sampling from the pmf $p_i = \lambda_i/\Lambda$).
2. Update the state: $X'_m = f_m(X_m, E_m)$
3. Rescale the observed interevent time V: $V_m = V/\Lambda_m$. Return to step 1 for the next event.

The SC concurrent simulation scheme is extremely simple and applies to any Markov chain regardless of its state transition mechanism. Its main drawback is that it is not applicable for online applications where some of the rates of the event lifetime distributions are not known a priori since fictitious events are not observable. Futhermore, for offline applications, several of the events generated at step 1 of the algorithm may be wasted in the sense that they are fictitious and cause no change in the state of one or more of the constructed sample paths.

16.3.2 Augmented System Analysis

Unlike the SC method, ASA[12,13] is based on observing a real sample path under some setting θ_0 and accepting the fact that constructability must be satisfied based on whatever events this observed sample path provides. Let $\omega(\theta_0)$ denote the observed sample path, and we need to check whether we can use information extracted from $\omega(\theta_0)$ to construct $\omega(\theta_m)$ for some $\theta_m \neq \theta_0$. As the observed state sequence $\{X_k(\theta_0)\}$ unfolds, we can construct the sequence $\{X_k(\theta_m)\}$ as long as $\Gamma(X_k(\theta_m)) \subseteq \Gamma(X_k(\theta_0))$. Now, suppose that we encounter a state pair $(X(\theta_0), X(\theta_m))$ such that $\Gamma(X(\theta_m)) \supset \Gamma(X(\theta_0))$. This means that there is at least one event, say $i \in \Gamma(X(\theta_m))$, which is unobservable at state $X(\theta_0)$. At this point, the construction of $\omega(\theta_m)$ must be suspended, since the triggering event at state $X(\theta_m)$ cannot be determined. Then, the key idea in ASA is the following: keep the state in $\omega(\theta_m)$ fixed at $X(\theta_m)$ and keep on observing $\omega(\theta_0)$ *until some new state $X'(\theta_0)$ is entered that satisfies* $\Gamma(X(\theta_m)) \subseteq \Gamma(X'(\theta_0))$. The only remaining question is how long $\omega(\theta_m)$ must wait in suspension. Note that we do not require "state matching," that is, we do not require that $X'(\theta_0) = X(\theta_m)$ to continue the construction, but the weaker condition of "event matching": $\Gamma(X(\theta_m)) \subseteq \Gamma(X'(\theta))$.

We will use $\mathcal{M}_m$ to denote the *mode* of $\omega(\theta_m)$, which can take on either the special value ACTIVE or a value from the state space X_m. This gives rise to the following *Event Matching* algorithm:

INITIALIZATION: Set $\mathcal{M}_m$ = ACTIVE and $X_m = X_0$.
WHEN ANY EVENT (say α) IS OBSERVED AND THE STATE IN $\omega(\theta_0)$
    BECOMES X, FOR $m = 1, \ldots, M$:
    1. If $\mathcal{M}_m$ = ACTIVE, update the state: $X'_m = f_m(X_m, \alpha)$ and if $\Gamma(X'_m) \supset \Gamma(X)$ (i.e., observability is violated), set $\mathcal{M}_m = X'_m$.
    2. If $\mathcal{M}_m \neq$ ACTIVE and $\Gamma(\mathcal{M}_m) \subseteq \Gamma(X)$, set $\mathcal{M}_m$ = ACTIVE; if $\mathcal{M}_m \neq$ ACTIVE, and $\Gamma(\mathcal{M}_m) \supset \Gamma(X)$, then continue.

In step 2, the condition $M_m \neq$ ACTIVE and $\Gamma(M_m) \subseteq \Gamma(X)$ implies that $\omega(\theta_m)$ was suspended at the time this event occurs, but since the observability condition $\Gamma(M_m) \subseteq \Gamma(X)$ is satisfied, the sample path construction can resume. The validity of the event matching scheme rests on basic properties of Markov processes, which enable us to "cut and paste" segments of a given sample path to produce a new sample path that is stochastically equivalent to the original one (for more details, see the work by Cassandras and Lafortune[3]). Note that no stopping condition is specified, since this is generally application dependent. Also, as in the case of the SC construction, the result of the Event Matching algorithm is a complete sample path; any desired sample function can subsequently be evaluated from it for the purpose of estimating system performance under all $\{\theta_0, \theta_1, \ldots, \theta_M\}$ of interest. The main drawback of ASA is the need to suspend a sample path construction for a potentially long time interval. This problem is the counterpart of the drawback we saw for the SC approach, which consists of "wasted random number generations," however, unlike SC, ASA can construct sample paths under different parameters using information directly observable from the sample path of an actual system. Finally, it is worth pointing out that the Event Matching algorithm can be modified so as to allow for at most one event process that is not Poisson; details of this extension are given in the work by Cassandras and Lafortune.[3]

16.4 CONCURRENT SIMULATION APPROACH FOR ARBITRARY EVENT PROCESSES

In this section, we describe a method for solving the constructability problem without having to rely on the Markovian structure of the event processes in the DES model, based on the sample path coupling approach first presented in the work by Cassandras and Panayiotou.[4] We start by making some assumptions, followed by three subsections. Section 16.4.1 introduces some notation and definitions; Section 16.4.2 presents the *Time Warping Algorithm* (TWA) for implementing the general solution to the constructability problem; Section 16.4.3 quantifies the speedup realized through the TWA; and Section 16.4.4 discusses extensions resulting from relaxing the assumptions presented next.

The four assumptions that follow simplify our analysis and apply to a large class of DES. However, they can also be eventually relaxed (see the work by Cassandras and Panayiotou[4]).

(**A1**) *Feasibility Assumption*: Let x_n be the state of the DES after the occurrence of the nth event. Then, for any n, there exists at least one $r > n$ such that $e \in \Gamma(x_r)$ for any $e \in \mathcal{E}$.

(**A2**) *Invariability Assumption*: Let $\mathcal{E}$ be the event set under the nominal parameter θ_0 and let $\mathcal{E}_m$ be the event set under $\theta_m \neq \theta_0$. Then, $\mathcal{E}_m = \mathcal{E}$.

(**A3**) *Similarity Assumption*: Let $G_i(\theta_0)$, $i \in \mathcal{E}$ be the event lifetime distribution for the event i under θ_0 and let $G_i(\theta_m)$, $i \in \mathcal{E}$ be the corresponding event lifetime distribution under θ_m. Then, $G_i(\theta_0) = G_i(\theta_m)$ for all $i \in \mathcal{E}$.

(**A4**) *State Invariability Assumption*: Let $G_i(x_k)$, $i \in \mathcal{E}$ and $x_k \in X$ be the event lifetime distribution of event i if i is activated when the system state is x_k. Then, $G_i(x_k) = G_i(x_l)$ for all $x_k, x_l \in X$ and all $i \in \mathcal{E}$.

Assumption **A1** guarantees that in the evolution of any sample path, all events in $\mathcal{E}$ will always become feasible at some point in the future. If, for some DES, assumption **A1** is not satisfied, that is, there exists an event α that never gets activated after some point in time, then, as we will see, it is possible that the construction of some sample path will remain suspended forever waiting for α to happen. Note that a DES with an irreducible state space immediately satisfies this condition.

Assumption **A2** states that changing a parameter from θ_0 to some $\theta_m \neq \theta_0$ does not alter the event set $\mathcal{E}$. More importantly, **A2** guarantees that changing to θ_m does not introduce any new events so that all event lifetimes for all events can be observed from the nominal sample path (the converse, i.e., fewer events in $\mathcal{E}_m$, would still make it possible to satisfy (**OB**)).

Assumption **A3** guarantees that changing a parameter from θ_0 to some $\theta_m \neq \theta_0$ does not affect the distribution of one or more event lifetime sequences. This allows us to use exactly the same lifetimes that we observe in the nominal sample path to construct the perturbed sample path. In other words, our analysis focuses on *structural* system parameters rather than *distributional* parameters. As we will see, however, it is straightforward to handle the latter at the expense of some computational cost.

Finally, assumption **A4** guarantees that the observed lifetimes do not depend on the current state of the system. This allows us to use any event lifetime of an event $e \in \mathcal{E}$ that has been observed irrespective of the current state of the system and, therefore, to employ "event matching" as opposed to "state matching" when a constructed sample path becomes active after being suspended for violating the (**OB**) condition.

16.4.1 NOTATION AND DEFINITIONS

Let

$$\xi(n,\theta) = \left\{ e_j : j = 1, \ldots, n \right\}$$

with $e_j \in \mathcal{E}$, be the sequence of events that constitute an observed sample path $\omega(\theta)$ up to n total events. Although $\xi(n,\theta)$ is clearly a function of the parameter θ, we will write $\xi(n)$ and adopt the notation

$$\hat{\xi}(k) = \left\{ \hat{e}_j : j = 1, \ldots, k \right\}$$

for any constructed sample path under a different value of the parameter up to k events in that path. It is important to realize that k is actually a function of n, since the constructed sample path is coupled to the observed sample path through the observed event lifetimes. However, again for the sake of notational simplicity, we will refrain from continuously indicating this dependence.

Next, we define the *score* of an event $i \in \mathcal{E}$ in a sequence $\xi(n)$, denoted by $s_i^n = [\xi(n)]i$, to be the nonnegative integer that counts the number of instances of event i in this sequence. The corresponding score of i in a constructed sample path is denoted by $\hat{s}_i^k = \left[\hat{\xi}(k) \right]_i$. In what follows, all quantities with the symbol "∧" above them refer to a typical constructed sample path.

Associated with every event type $i \in \mathcal{E}$ in $\xi(n)$ is a sequence of s_n event lifetimes

$$\mathbf{V}_i(n) = \left\{ v_i(1), \ldots, v_i\left(s_i^n\right) \right\} \quad \text{for all} \quad i \in \mathcal{E}$$

The corresponding set of sequences in the constructed sample path is as follows:

$$\hat{\mathbf{V}}_i(k) = \left\{ v_i(1), \ldots, v_i\left(\hat{s}_i^k\right) \right\} \quad \text{for all} \quad i \in \mathcal{E}$$

which is a subsequence of $\mathbf{V}_i(n)$ with $k \leq n$. In addition, we define the following sequence of lifetimes:

$$\tilde{\mathbf{V}}_i(n,k) = \left\{ v_i\left(\hat{s}_i^k + 1\right), \ldots, v_i\left(s_i^n\right) \right\} \quad \text{for all} \quad i \in \mathcal{E}$$

which consists of all event lifetimes that are in $\mathbf{V}_i(n)$ but not in $\hat{\mathbf{V}}_i(k)$. Associated with any one of these sequences are the following operations. Given some $\mathbf{W}_i = \left\{ w_i(j), \ldots, w_i(r) \right\}$,

Suffix Addition: $\mathbf{W}_i + \{w_i(r + 1)\} = \{w_i(j), \ldots, w_i(r), w_i(r + 1)\}$ and
Prefix Subtraction: $\mathbf{W}_i - \{w_i(j)\} = \{w_i(j + 1), \ldots, w_i(r)\}$.

Note that the addition and subtraction operations are defined so that a new element is always added as the *last* element (the *suffix*) of a sequence, whereas subtraction always removes the *first* element (the *prefix*) of the sequence. At this point, it is worth pointing out that to construct the various event lifetime sequences, it is necessary that the event lifetime sequences are observable from the nominal sample path. For offline approaches, this is generally simple since the lifetimes can be directly recorded from the simulator. For online approaches, this is possible for *invertible* systems, that is, systems for which event lifetimes can be recovered from the output of a system (i.e., sequence of events, states, and transition epochs) (see the work by Park and Chong[14] for invertibility conditions as well as an appropriate algorithm). Next, define the set

$$A(n,k) = \left\{ i : i \in \mathcal{E}, s_i^n > \hat{s}_i^k \right\} \tag{16.10}$$

which is associated with $\tilde{\mathbf{V}}_i(n,k)$ and consists of all events i whose corresponding sequence $\tilde{\mathbf{V}}_i(n,k)$ contains at least one element. Thus, every $i \in A(n, k)$ is an event that has been observed in $\xi(n)$ and has at least one lifetime that has yet to be used in the coupled sample path $\hat{\xi}(k)$. Hence, $A(n, k)$ should be thought of as the set of *available* events to be used in the construction of the coupled path.

Finally, we define the following set, which is crucial in our approach:

$$M(n,k) = \Gamma(\hat{x}_k) - \left(\Gamma(\hat{x}_{k-1}) - \{\hat{e}_k\} \right) \tag{16.11}$$

where, clearly, $M(n, k) \subseteq \mathcal{E}$. Note that $\hat{e}_k$ is the triggering event at the $(k - 1)$th state visited in the constructed sample path. Thus, $M(n, k)$ contains all the events that are in the feasible event set $\Gamma(\hat{x}_k)$ but not in $\Gamma(\hat{x}_{k-1})$; in addition, $\hat{e}_k$ also belongs to $M(n, k)$ if it happens that $\hat{e}_k \in \Gamma(\hat{x}_k)$. Intuitively, $M(n, k)$ consists of all *missing* events from the perspective of the constructed sample path when it enters a new state $\hat{x}_k$, that is, those events already in $\Gamma(\hat{x}_{k-1})$ that were not the triggering event remain available to be used in the sample path construction as long as they are still feasible. All other events in the set are "missing" as far as residual lifetime information is concerned.

The concurrent sample path construction process we are interested in consists of two coupled processes, each generated by an STA as detailed in Section 16.2, through Equations 16.1 to 16.6. The observed sample path and the one to be constructed both satisfy this set of equations. Our task is to derive an additional set of equations that captures the coupling between them. In particular, our goal is to enable event lifetimes from the observed $\xi(n)$ to be used to construct a sequence $\hat{\xi}(k)$. First, observe that the process described by Equations 16.1 through 16.6 and applied to $\hat{\xi}(k)$ hinges on the availability of residual lifetimes $\hat{y}_i(k)$ for all $i \in \Gamma(\hat{x}_k)$. Thus, the constructed sample path can only be "active" at state $\hat{x}_k$ if every $i \in \Gamma(\hat{x}_k)$ is such that either $i \in (\Gamma(\hat{x}_{k-1}) - \{\hat{e}_k\})$ (in which case $\hat{y}_i(k)$ is a residual lifetime of an event available from the previous state transition) or $i \in A(n, k)$ (in which case a full lifetime of i is available from the observed sample path). This motivates the following:

Definition 16.1

A constructed sample path is *active* at state $\hat{x}_k$ after the occurrence of an observed event e_n if for every $i \in \Gamma(\hat{x}_k), i \in \left(\Gamma(\hat{x}_{k-1}) - \{\hat{e}_k\}\right) \cup A(n,k)$.

Thus, the start/stop conditions for the construction of a sample path are determined by whether it is active at the current state or not.

16.4.2 OBSERVED AND CONSTRUCTED SAMPLE PATH COUPLING DYNAMICS

Upon occurrence of the $(n + 1)$th observed event, e_{n+1}, the first step is to update the event lifetime sequences $\tilde{V}_i(n,k)$ as follows:

$$\tilde{V}_i(n+1,k) = \begin{cases} \tilde{V}_i(n,k) + v_i\left(\hat{s}_i^n + 1\right) & \text{if } i = e_{n+1} \\ \tilde{V}_i(n,k) & \text{otherwise} \end{cases} \quad (16.12)$$

The addition of a new event lifetime implies that the available event set $A(n, k)$ defined in Equation 16.10 may be affected. Therefore, it is updated as follows:

$$A(n+1,k) = A(n,k) \cup \{e_{n+1}\} \quad (16.13)$$

Finally, note that the missing event set $M(n, k)$ defined in Equation 16.11 remains unaffected by the occurrence of observed events:

$$M(n+1,k) = M(n,k) \quad (16.14)$$

At this point, we are able to decide whether all lifetime information to proceed with a state transition in the constructed sample path is available or not. In particular, the condition

$$M(n+1,k) \subseteq A(n+1,k) \quad (16.15)$$

may be used to determine whether the constructed sample path is active at the current state $\hat{x}_k$ (in the sense of Definition 16.1). The following is a formal statement of this fact and is proved in the work by Cassandras and Panayiotou.[4]

Lemma 16.1

A constructed sample path is active at state $\hat{x}_k$ after an observed event e_{n+1} if and only if $M(n + 1, k) \subseteq A(n + 1, k)$.

Assuming Equation 16.15 is satisfied, Equations 16.1 through 16.6 may be used to update the state $\hat{x}_k$ of the constructed sample path. In so doing, lifetimes $v_i\left(s_i^k + 1\right)$ for all $i \in M(n + 1, k)$ are used from the corresponding sequences $\tilde{V}_i(n+1,k)$. Thus, upon completion of the state update steps, all three variables associated with the coupling process, that is, $\tilde{V}_i(n,k)$, $A(n, k)$, and $M(n, k)$ must be updated. In particular,

$$\tilde{V}_i(n+1,k+1) = \begin{cases} \tilde{V}_i(n+1,k) - v_i\left(\hat{s}_i^k + 1\right) & \text{for all} \quad i \in M(n+1,k) \\ \tilde{V}_i(n+1,k) & \text{otherwise} \end{cases}$$

This operation immediately affects the set $A(n + 1, k)$, which is updated as follows:

$$A(n+1,k+1) = A(n+1,k) - \left\{ i : i \in M(n+1,k), \hat{s}_i^k + 1 = s_i^n + 1 \right\}$$

Finally, applying Equation 16.11 to the new state $\hat{x}_{k+1}$,

$$M(n+1,k+1) = \left(\left(\hat{x}_{k+1}\right) - \left(\Gamma\left(\hat{x}_k\right) - \left\{\hat{e}_{k+1}\right\}\right)\right)$$

Therefore, we are again in a position to check Equation 16.15 for the new sets $M(n + 1, k + 1)$ and $A(n + 1, k + 1)$. If it is satisfied, then we can proceed with one more state update on the constructed sample path; otherwise, we wait for the next event on the observed sample path until Equation 16.15 is again satisfied. Similar to Lemma 16.1, we have the following:

Lemma 16.2

A constructed sample path is active at state $\hat{x}_{k+1}$ after event $\hat{e}_{k+1}$ if and only if $M(n + 1, k + 1) \subseteq A(n + 1, k + 1)$.

The analysis above can be put in the form of an algorithm termed *TWA*, which is described next.

Time Warping Algorithm (TWA):

The TWA consists of three parts: an initialization, an update of the observed system state and of sample path coupling variables, and the "time warping" operation.

1. Initialization

The event counts, event scores, clocks, and states of the observed and constructed sample paths are initialized in the usual way:

$$n = 0, \ k = 0, \ s_i^n = 0, \ \hat{s}_i^k = 0 \quad \text{for all} \quad i \in \mathcal{E},$$

$$t_n = 0, \ \hat{t}_k = 0, \ x_n = x_0, \ \hat{x}_k = \hat{x}_0$$

along with the lifetimes of the events feasible in the initial states:

$$y_i(n) = v_i(1) \quad \text{for all} \quad i \in \Gamma(x_n)$$

and the missing and available event sets:

$$M(0,0) = \Gamma(\hat{x}_0), \quad A(0,0) = \varnothing$$

2. When Event e_n Is Observed
   2.1. Use the STA model dynamics (Equations 16.1 through 16.6) to determine $e_{n+1}, x_{n+1}, t_{n+1}, y_i(n+1)$ for all $i \in \Gamma(x_{n+1})$, s_i^{n+1} for all $i \in \mathcal{E}$.
   2.2. Add the e_{n+1} event lifetime to $\tilde{\mathbf{V}}_i(n+1,k)$:

$$\tilde{\mathbf{V}}_i(n+1,k) = \begin{cases} \tilde{\mathbf{V}}_i(n,k) + v_i(\tilde{s}_i^n + 1) & \text{if} \quad i = e_{n+1} \\ \tilde{\mathbf{V}}_i(n,k) & \text{otherwise} \end{cases}$$

   2.3. Update the available event set $A(n, k)$:

$$A(n+1,k) = A(n,k) \cup \{e_{n+1}\}$$

   2.4. Update the missing event set $M(n, k)$:

$$M(n+1,k) = M(n,k)$$

   2.5. If $M(n+1, k) \subseteq A(n+1, k)$, then go to the time warping operation in step 3; otherwise, set $n \leftarrow n+1$ and go to step 2.1.

3. Time Warping Operation
   3.1. Obtain all missing event lifetimes to resume sample path construction at state $\hat{x}_k$

$$\hat{y}_i(k) = \begin{cases} v_i(\hat{s}_i^k + 1) & \text{for} \quad i \in M(n+1,k) \\ \hat{y}_i(k-1) & \text{otherwise} \end{cases}$$

   3.2. Use the STA model dynamics (Equations 16.1 through 16.6) to determine $\hat{e}_{k+1}, \hat{x}_{k+1}, \hat{t}_{k+1}, \hat{y}_i(k+1)$ for all $i \in \Gamma(\hat{x}_{k+1}) \cap (\Gamma(\hat{x}_k) - \{\hat{e}_{k+1}\}), \hat{s}_i^{k+1}$ for all $i \in \mathcal{E}$.
   3.3. Discard all used event lifetimes:

$$\tilde{\mathbf{V}}_i(n+1,k+1) = \tilde{\mathbf{V}}_i(n+1,k) - v_i(\hat{s}_i^k + 1) \text{ for all } i \in M(n+1,k)$$

   3.4. Update the available event set $A(n+1, k)$:

$$A(n+1,k+1) = A(n+1,k) - \{i : i \in M(n+1,k), \hat{s}_i^{k+1} = s_i^{n+1}\}$$

   3.5. Update the missing event set $M(n+1, k)$:

$$M(n+1,k+1) = \Gamma(\hat{x}_{k+1}) - (\Gamma(\hat{x}_k) - \{\hat{e}_{k+1}\})$$

3.6. If $M(n + 1, k + 1) \subseteq A(n + 1, k + 1)$, then set $k \leftarrow k + 1$ and go to step 3.1; otherwise, set $k \leftarrow k + 1$, $n \leftarrow n + 1$ and go to step 2.1.

The computational requirements of the TWA are minimal: adding and subtracting elements to sequences, simple arithmetic, and checking conditions (Equation 16.15). It is the storage of additional information that constitutes the major cost of the algorithm.

Example

Let us consider once again a G/G/1/K queueing system as in previous examples, where the event set is $\mathcal{E} = \{a, d\}$ and the state space is $\mathbf{X} = \{0, 1, \dots, K\}$. Let the observed sample path be one with queue capacity $K = 2$, and let us try to construct a sample path under $K = 3$ in the framework of Figure 16.1. Let $\Gamma(x[K])$ be the feasible event set at state x for the system under K and assume that both systems are initially empty. Unlike the SC and ASA methods, we can no longer maintain between the two sample paths (the observed one and the one to be constructed) a coupling that preserves full synchronization of events. This is because of the absence of Markovian event processes that allow us to exploit the memoryless property. Thus, we must maintain each feasible event set, $\Gamma(x[2])$ and $\Gamma(x[3])$, separately for each observed state $x[2]$ and constructed state $x[3]$. Whenever an event is observed, its lifetime is assumed to become available (i.e., the time when this event was activated is known). Each such lifetime is subsequently used in the construction of the sample path under $K = 3$.

To see precisely how this can be done, we start out with a state $x[3] = 0$ for the constructed sample path so that $\Gamma(x[3]) = \{a\}$. Since no event lifetimes are initially available, we consider the sample path of this system as "suspended" with a missing event set $M(0, 0) = \{a\}$. The initial state of the observed sample path is $x[2] = 0$ so that $\Gamma(x[2]) = \{a\}$. Therefore, the first observed event is a. At this point, the constructed sample path may be "resumed," since all lifetimes of the events in $\Gamma(x[3])$ are now available, namely, the lifetime of a, so we have the available event set $A(1, 1) = \{a\}$. This amounts to verifying the condition $M(1, 1) \subseteq A(1, 1)$ in step 3.6 of the TWA above. The constructed sample path advances time and updates its state to $x[3] = 1$. Now $\Gamma(x[3]) = \{a, d\}$, but neither event has been observed yet, and therefore, the constructed sample path is suspended again until at least one a and one d event occur at the observed sample path. This start/ stop (or suspend/resume) process goes on until a sample path under $K = 3$ is constructed up to a desired number of events or some specified time, that is, we have $M(2, 2) = \{a, d\}$. In this example, assuming that both arrival and service processes have positive rates, it is clear that eventually $M(n + 1, k + 1) \subseteq A(n + 1, k + 1)$ will be satisfied, after both an arrival event and a departure event are observed. Note that it is possible that when an event occurs causing the condition to be satisfied, a series of events on the constructed sample path is triggered, hence a sequence of state transitions and time updates as well. For instance, if $\Gamma(x[3]) = \{a, d\}$, a sequence of events $\{a, a, a, d\}$ will cause four state transitions in a row as soon as d is observed. The fact that in this process we move backward in time to revisit a suspended sample path and then forward by one or more event occurrences lends itself to the term "time warping" in the TWA.

Regarding the *scalability* of the TWA, as in the case of the SC and ASA methods, it should be clear that the computational effort involved scales with the number of parameters considered, N, and the number of values to be explored for each parameter, M_i, $i = 1, \ldots, N$. Setting $M_i = M$ for all $i = 1, \ldots, N$ for simplicity, it follows that $N \cdot M$ concurrent estimators are active for each simulation run. This number is a worst case scenario. When a specific optimization problem is considered, where one can generally exploit the structure of the DES (as will be seen in the example mentioned in Section 16.5), it is possible to significantly reduce the number of active concurrent sample paths.

16.4.3 SPEEDUP FACTOR

In a simulation setting, it makes sense to use concurrent simulation methods only if they can generate the required results faster than brute-force simulation. Thus, to define a speedup factor associated with a particular concurrent simulation method, suppose that the sample path constructed through such a method were instead generated by a separate simulation whose length is defined by N total events. Let T_N be the time it takes in Central Processor Unit (CPU) units to complete such a simulation run. Furthermore, suppose that when the nominal simulation is executed with a concurrent simulation algorithm as part of it, the total time is given by $T_N^o + \tau_K$, where T_N^o is the simulation time (with no concurrent sample path construction) and τ_K is the additional time involved in the concurrent construction of a sample path with $K \leq N$ events. We then define the *speedup factor* as

$$S = \frac{T_N/N}{\tau_K/K}$$

Thus, when a separate simulation (in addition to the one for the observed sample path) is used to generate a sample path under a new value of the parameter of interest, the computation time per event is T_N/N. If, instead, we use a concurrent simulation method in conjunction with the observed path, no such separate simulation is necessary, but the additional time per event imposed by the approach is τ_K/K, where $K \leq N$ in general. Clearly, $S \geq 1$ is required to justify the use of concurrent simulation. Speedup factors resulting from the use of the TWA for various queueing systems are extensively investigated in the work by Cassandras and Panayiotou[4] where the following upper bound is also derived:

$$S \leq \frac{1}{1 + \beta - \alpha}$$

where α is the fraction of time used for generating random numbers and variates ($0 \leq \alpha \leq 1$) in the nominal sample path and $\beta = r_N/T_N$ with r_N being the time taken to write to and read from memory in the TWA, which depends on the number of random variates observed in the nominal sample path and used in the constructed sample path. As a rule of thumb, when we build and execute a simulation model, we can readily measure α; if $(1 - \alpha)$ is relatively small, we can immediately deduce a potential speedup benefit through the TWA (the final result will depend on β as well). Furthermore, the work by Cassandras and Panayiotou[4] has defined the class of *Regular DES* for which the cardinality of the missing event set $|M(n, k)|$

in Equation 16.11 is bounded by 2 with the potential to achieve higher speedups. This class includes a large family of common systems such as all open and closed Jackson-like queueing networks.[3]

16.4.4 EXTENSIONS OF THE TWA

To simplify the notation and presentation of the TWA, at the beginning of Section 16.4, we made four assumptions that can be relaxed and still make TWA applicable.

In **A1**, we assumed that that the nominal sample path will never go into an absorbing state, or a set of states, for which an event $e \in \mathcal{E}$ will never become feasible at any time in the future. This may freeze indefinitely the construction of one or more concurrent sample paths since such a sample path may have e in its missing event set; however, e will never occur in the future. In situations such as this, assuming some knowledge of the event lifetime distributions, one can use a random number generator to generate the required event lifetimes, thus allowing the construction of all constructed sample paths. A similar situation may arise when an event occurs rarely, which may force the construction of all concurrent sample paths to be suspended waiting for the occurrence of the rare event. In this case, a possible policy would be to use a random number generator to obtain the required lifetime if a constructed sample path is suspended for more than a certain number of observed events.

In **A2**, we assumed that changing a parameter from θ_0 to some $\theta \neq \theta_0$ does not alter the event set $\mathcal{E}$. Clearly, if the new event set $\mathcal{E}_m$ is such that $\mathcal{E}_m \subseteq \mathcal{E}$, the development and analysis of TWA is not affected. If, on the contrary, $\mathcal{E} \subset \mathcal{E}_m$, this implies that events required to cause state transitions under θ_m are unavailable in the observed sample path, which make the application of our algorithm impossible. Notice the similarity of this problem with the one discussed above. In this case, one can introduce *phantom* event sources that generate all the unavailable events as described, for example, in the work by Cassandras and Shi,[15] provided that the lifetime distributions of these events are known.

In **A3**, we assumed that changing a parameter from θ_0 to some $\theta_m \neq \theta_0$ does not affect the distribution of one or more event lifetime sequences. This assumption is used in Equation 16.12 where the observed lifetime $v_i\left(s_i^n + 1\right)$ is directly suffix-added to the sequence $\tilde{V}_i(n+1, k)$. Note that this problem can be overcome by transforming observed lifetimes $V_i = \{v_i(1), v_i(2), \ldots\}$ with an underlying distribution $G_i(\theta_0)$ into samples of a similar sequence corresponding to the new distribution $G_i(\theta_m)$ and then suffix-add them in $\tilde{V}(n+1, k)$. This is indeed possible, if $G_i(\theta_0)$, $G_i(\theta_m)$ are known, at the expense of some additional computational cost for this transformation (e.g., see the work by Cassandras and Lafortune[3]). One interesting special case arises when the parameter of interest is a scale parameter of some event lifetime distribution (e.g., it is the mean of a distribution in the Erlang family). Then, simple rescaling suffices to transform an observed lifetime v_i under θ_0 into a new lifetime $\tilde{v}_i$ under θ_m:

$$\tilde{v}_i = \left(\theta_m / \theta_0\right) v_i$$

Finally, in principle, one can also relax assumption **A4** and record all observed lifetimes not only based on the associated event but also based on the state in which they have been activated. Computationally, this is feasible. However, depending on the state spaces of the nominal and constructed sample paths, it is likely that the aforementioned "tricks" with *phantom* event sources will be required more frequently. A similar situation may arise if the underlying STA allows probabilistic state transition mechanisms. In this case, the state transitions should also be recorded on a per-state basis unless the state transition mechanism has a structure that can be exploited such that "event matching" is again adequate. For example, consider a system with two parallel FIFO queues with infinite capacity where an arriving customer enters queue 1 with probability p or queue 2 with probability $1 - p$. Let (x_1, x_2) denote the state of the system where x_i, $i \in \{1, 2\}$ is the length of each queue. In the event of a customer arrival, the new state will become $(x_1 + 1, x_2)$ with probability p and $(x_1, x_2 + 1)$ with probability $1 - p$. In this case, the state transition probabilities have a structure that is independent of the current state; thus, one can exploit this and use the observed outcomes irrespective of the current state. Consequently, event matching is no longer necessary.

16.5   USE OF TWA FOR REAL-TIME OPTIMIZATION

In this section, we present an example where the TWA is used together with a real-time controller that controls the buffers allocated to different processes. Specifically, we consider the system shown in Figure 16.3 where customers arrive according to some distribution $G_a(\cdot)$ (generally unknown and possibly time varying) and are probabilistically routed to one of the S available servers. The routing probabilities $p_i(t)$, $i = 1, \ldots, S$, may also be unknown and time varying. Each customer, once assigned to a server, immediately proceeds to the corresponding finite capacity queue where, if there is available space, it waits using a FIFO discipline to receive service from the server. If no buffering space is available, the customer is considered blocked. Furthermore, the time taken to service a customer at server i is according to some

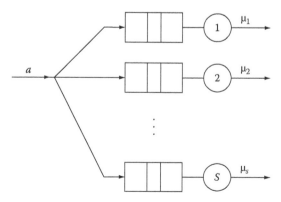

FIGURE 16.3   Queueing system with S parallel servers.

distribution $G_{d,i}(\cdot)$, $i = 1, \ldots, S$. In this context, we assume that there are K available buffers that must be allocated to the S servers (and can be freely reallocated to any server) to minimize some metric such as the blocking probability (the probability that a customer that is assigned to a server will find the corresponding buffer full). In other words, we are interested in solving the following optimization problem:

$$\min_{\theta \in \Theta} \sum_{i=1}^{S} \beta_i J_i(n_i) \quad \text{s.t.} \quad \sum_{i=1}^{S} n_i = K \qquad (16.16)$$

where $1 \leq n_i \leq K - S + 1$ is the number of buffer units assigned to server i, $J_i(n_i)$ is the blocking probability at queue i when allocated n_i buffer units, $\theta = [n_1, \ldots, n_s]$ is an allocation vector and Θ is the set of all feasible allocation vectors. Finally, β_i is a weight associated with server i, but for simplicity, we will assume $\beta_i = 1$ for all i.

Typically, one is interested in solving Equation 16.16 when $J_i(n_i)$ is of the form of an expectation $J_i(n_i) = E[L_i(n_i, \omega(n_i, [T_1, T_2]))]$ where $L_i(n_i, \omega(n_i, [T_1, T_2]))$ is a sample blocking probability observed at the system during the interval $[T_1, T_2]$. In general, there exists no closed-form solution for $J_i(n_i)$ unless the system is driven by Poisson arrivals and exponential service times, all with known rates as well as known routing probabilities. Thus, a possible approach to solve this problem is to make an initial allocation θ_0 and observe the sample path of the system over the interval $[0, T]$ (where T is a predefined period). Using the information extracted from the observed sample path, one can compute the sample performance metrics $L_i(\cdot)$, $i = 1, \ldots, S$ for the allocation θ_0 and use TWA to construct the sample paths for all other allocations $\theta \in \Theta - \{\theta_0\}$ and from them compute the corresponding sample performance functions. Consequently, at $t = T$, the controller can determine the allocation

$$\theta^* = \arg\min_{\theta \in \Theta} L(\theta)$$

where

$$L(\theta) = \sum_{i=1}^{S} L_i(n_i, \omega(n_i, [0,T])), \quad \theta = [n_1, \ldots, n_S] \in \Theta$$

Before proceeding any further, it is instructive to comment on an underlying assumption and two important parameters of the proposed optimization approach, namely, the length of the observation interval T and the cardinality of the set Θ. Note that the optimization approach uses data collected in the interval $[0, T]$ to make buffer assignments that will be valid for $t > T$ or, more precisely, for the interval $[T, 2T]$. Thus, the underlying assumption is that the future behavior of the system will be very similar to its past behavior. In other words, this approach is applicable to systems where the input stochastic processes that drive the system dynamics (in this example, the routing probabilities $p_i(t)$, $i = 1, \ldots, S$) change slowly compared to the observation interval T. Regarding the observation interval, its actual value depends on the variance of the input and output stochastic processes. If set too short, then the obtained performance metrics may be too noisy and, as a result, the controller may pick random allocations. On the contrary, T should not be set too large because the underlying assumption mentioned earlier may not be valid: if T is set too large and the behavior of the system has changed over time, then this will not be detected fast enough, and thus, the system may continue to operate at suboptimal allocations.

Regarding the cardinality of Θ, one can easily notice that it can become very large, which may cause computational problems due to the large number of sample paths that must be constructed. For the above example, the number of feasible allocations is $O\left(\dfrac{(K+S-1)!}{K!(S-1)!}\right)$. To alleviate the computational problem, one may resort to various optimization approaches taking advantage of some properties of the cost function. For the above example, we know that $J_i(n_i)$ is decreasing and convex; thus, one may use "gradient"-based optimization approaches. In this case, we use finite differences together with a discrete resource allocation approach,[16] which is briefly summarized below to reduce the number of constructed sample paths from $O\left(\dfrac{(K+S-1)!}{K!(S-1)!}\right)$ to only $S + 2$. The idea of the discrete optimization approach is rather simple: at every step, it reallocates a resource (buffer) from the least "sensitive" server to the most sensitive one where sensitivity is defined as

$$\Delta L_i(n_i) = L_i(n_i) - L_i(n_i - 1), \quad n_i = 1, \dots, K \tag{16.17}$$

Below are the detailed steps of the algorithm:
Algorithm: *Dynamic Resource Allocation*

1.0 Initialize: $\theta_0 = \left[n_1^{(o)}, \dots, n_S^{(0)}\right]; C^{(0)} = \{1, \dots, S\}; k = 0$

1.1 Evaluate $\mathbf{D}^{(k)}\left(n_1^{(k)}, \dots, n_S^{(k)}\right) \equiv \left[\Delta L_1\left(n_1^{(k)}\right), \dots, \Delta L_S\left(n_S^{(k)}\right)\right]$

2.1 Set $i^* = \arg\max_{i=1,\dots,C^{(k)}}\left[\mathbf{D}^{(k)}\left(n_1^{(k)}, \dots, n_S^{(k)}\right)\right]$

2.2 Set $j^* = \arg\min_{i \in C^{(k)}}\left[\mathbf{D}^{(k)}\left(n_1^{(k)}, \dots, n_S^{(k)}\right)\right]$

2.3 Evaluate $\mathbf{D}^{(k)}\left(n_1^{(k)}, \dots, n_{i^*}^{(k)} + 1, \dots, n_{j^*}^{(k)} - 1, \dots, n_S^{(k)}\right)$

2.4 If $\Delta L_{j^*}\left(n_{j^*}^{(k)} - 1\right) < \Delta L_{i^*}\left(n_{i^*}^{(k)}\right)$ Goto **3.1** ELSE Goto **3.2**

3.1 Update allocation:

$$n_{i^*}^{(k+1)} = n_{i^*}^{(k)} + 1; n_{j^*}^{(k+1)} = n_{j^*}^{(k)} - 1; n_m^{(k+1)} = n_m^{(k)} \text{ for all } m \in C^{(k)} \text{ and}$$
$$m \neq i^*, j^*;$$

Set $k \leftarrow k + 1$

Reset $C^{(k)} = \{1, \dots, S\},$ and Goto **2.1**

3.2 Replace $C^{(k)}$ by $C^{(k)} - \{j^*\}$;
IF $|C^{(k)}| = 1$, Reset $C^{(k)} = \{1, \dots, S\}$, and Goto **2.1**
ELSE Goto **2.2**

When the system operates under a nominal allocation, one can obtain estimates of $L_i(n_i)$ while concurrent estimators can provide estimates for $L_i(n_i - 1)$; thus, the vector $\mathbf{D}^{(k)}$ (·) with all finite differences can be computed in step 1.1. Steps 2.1 and 2.2 determine the servers with the maximum and minimum finite differences i^* and j^*, respectively. Step 2.3 reevaluates the finite differences after a resource is reallocated from the least sensitive j^* to the most sensitive i^*. If the removal of the resource does not make j^* the most sensitive buffer, the reallocation is accepted in step 3.1, otherwise it is rejected in step 3.2.

Next, we consider a numerical example with $S = 4$ and $K = 16$. We assume that the arrival process is Poisson with rate $\lambda = 1.3$, and all service times (at any server) are exponential with rates $\mu_i = 1.0$ for all $i = 1, \dots, 4$. At this point, we emphasize that the controller does not know anything about either the arrival or the service processes. Furthermore, we assume that the routing probabilities are also unknown to the controller and are time varying. Specifically, the routing probabilities change every 50,000 time units as given in Table 16.1.

Figure 16.4 presents the performance (loss probability) of the real-time controller (i.e., the dynamic resource allocation algorithm) in comparison to a static policy. Because of symmetry, the static optimal allocation is [4, 4, 4, 4], which achieves a loss probability between 0.08 and 0.1 for the entire simulation interval. As shown in the figure, the real-time controller can significantly improve the overall system performance since the real-time simulation components can "detect" the overloaded buffer and allocate more resources to it, thus reducing the overall loss probability. Initially, the controller adjusts the buffers such that the loss probability is between 0.02 and 0.04. At time 0.5×10^5, the routing probabilities change abruptly, and thus, the loss probability significantly increases; however, as seen in the figure, the real-time controller quickly reallocates the buffers and the loss probability is again reduced. Similar behavior occurs every time the routing probabilities change.

In Figure 16.5, we also investigate the effect of the observation interval. The scenario investigated is identical to the one presented earlier. The only difference is the length of the observation interval T. In this experiment, we assume that the controller updates the allocation vector every NT observed events (here we use NT instead of T to indicate that the interval is determined by the number of events rather than time units). The figure presents the results for three different values of NT. As indicated in the figure, when NT is too large ($NT = 7000$), the controller does not capture the change in the routing probabilities, and as a result, it leaves the system operating

TABLE 16.1
Routing Probabilities

From	To	Distribution
0	50,000	$\mathbf{p}_1 = [0.7, 0.1, 0.1, 0.1]$
50,000	100,000	$\mathbf{p}_2 = [0.1, 0.7, 0.1, 0.1]$
100,000	150,000	$\mathbf{p}_3 = [0.1, 0.1, 0.7, 0.1]$
150,000	200,000	$\mathbf{p}_4 = [0.1, 0.1, 0.1, 0.7]$

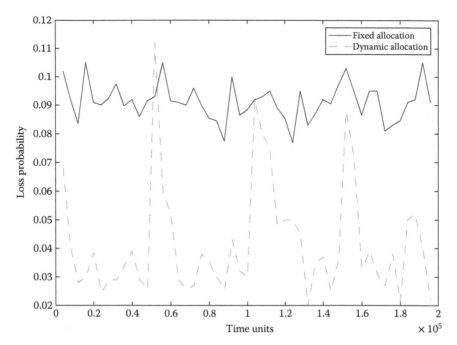

FIGURE 16.4 System performance under the *dynamic* allocation scheme versus a *fixed* resource allocation vector.

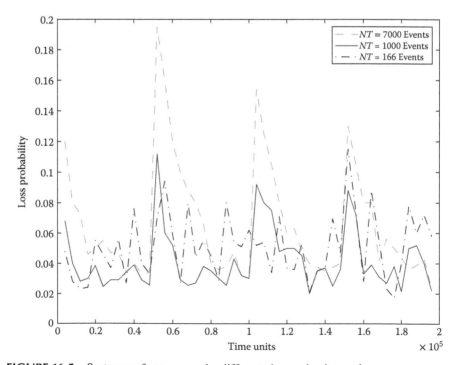

FIGURE 16.5 System performance under different observation intervals.

at suboptimal allocations for long periods of time seriously degrading the overall system performance. On the contrary, when the observation interval is fairly short ($NT = 166$), the obtained estimates are fairly noisy, and as a result, the controller often makes "wrong" decisions, which also leads to poor performance.

16.6 CONCLUSIONS

This chapter has investigated the problem of real-time simulation for the online performance optimization of the policies or parameters of a DES. Given an observed sample path of a DES, the chapter has addressed the general sample path constructability problem and presented three concurrent simulation algorithms for solving the problem. The SC and the ASA approaches solve the problem very efficiently for a class of systems (although SC is more appropriate for offline simulation) with event lifetime distributions that satisfy the memoryless property. For more general systems, the TWA can be used to construct the sample paths of arbitrary DES using a sample path coupling approach. The TWA can be used in two modes, offline and online. In an offline setting, the objective of the TWA is to achieve a speedup factor greater than 1. On the contrary, in an online setting, the approach can be used for real-time control and optimization of DES. In this setting, the proposed concurrent simulation approach provides significant benefits since in general it does not require prior knowledge of the various event lifetime distributions. Furthermore, the approach inherently uses the *CRN* scheme that has been observed experimentally and was proved theoretically in some cases to be effective in variance reduction.[17]

ACKNOWLEDGMENTS

The work of Christos Cassandras was supported in part by NSF under Grant EFRI-0735974, by AFOSR under grants FA9550-07-1-0361 and FA9550-09-1-0095, by DOE under grant DE-FG52-06NA27490, and by ONR under grant N00014-09-1-1051.

The work of Christos G. Panayiotou was supported in part by the Cyprus Research Promotion Foundation, the European Regional Development Fund, the Government of Cyprus, and by the European Project Control for Coordination of distributed systems (CON4COORD - FP7-2007-IST-2-223844).

REFERENCES

1. Fujimoto, M. R. 2000. *Parallel and Distributed Simulation Systems*. New York: Wiley Interscience.
2. Boer, C. A., A. de Bruin, and A. Verbraeck. 2008. "Distributed Simulation In Industry—A Survey, Part 3—The HLA Standard in Industry." In *Proceedings of the 40th Conference on Winter Simulation, WSC '08*, 1094–102.
3. Cassandras, C. G., and S. Lafortune. 2007. *Introduction to Discrete Event Systems*. New York: Springer.
4. Cassandras, C. G., and C. G. Panayiotou. 1999. "Concurrent Sample Path Analysis of Discrete Event Systems." *Journal of Discrete Event Dynamic Systems: Theory and Applications* 9 (2): 171–95.

5. Cassandras, C. G., and V. Julka. 1994. "Descent Algorithms for Discrete Resource Allocation Problems." In *Proceedings of the IEEE 33rd Conference on Decision and Control*, 2639–44.

6. Panayiotou, C. G., and C. G. Cassandras. 1999. "Optimization of Kanban-Based Manufacturing Systems." *Automatica* 35 (9): 1521–33.

7. Yan, D., and H. Mukai. 1992. "Stochastic Discrete Optimization." *SIAM Journal on Control and Optimization*, 30 (3): 594–612.

8. Gong, W.-B., Y. C. Ho, and W. Zhai. 1992. "Stochastic Comparison Algorithm for Discrete Optimization with Estimation." In *Proceedings of 31st IEEE Conference on Decision and Control*, 795–802.

9. Panayiotou, C. G., and C. G. Cassandras. 2001. "On-Line Predictive Techniques for 'Differentiated Services' Networks." In *Proceedings of IEEE Conference on Decision and Control*, 4529–34.

10. Panayiotou, C. G., and C. G. Cassandras. 1997. "Dynamic Resource Allocation in Discrete Event Systems." In *Proceedings of IEEE Mediterranean Conference on Control and Systems*.

11. Vakili, P. 1991. "Using a Standard Clock Technique for Efficient Simulation." *Operations Research Letters* 10 (8): 445–52.

12. Cassandras, C. G., and S. G. Strickland. 1989. "On-Line Sensitivity Analysis of Markov Chains." *IEEE Transactions on Automatic Control* 34 (1): 76–86.

13. Cassandras, C. G., and S. G. Strickland. 1989. "Observable Augmented Systems for Sensitivity Analysis of Markov and Semi-Markov Processes." *IEEE Transactions on Automatic Control* 34 (10): 1026–37.

14. Park, Y., and E. K. P. Chong. 1995. "Distributed Inversion in Timed Discrete Event Systems." *Journal of Discrete Event Dynamic Systems* 5(2/3), 219–41.

15. Cassandras, C. G., and W. Shi. 1996. "Perturbation Analysis of Multiclass Multiobjective Queueing Systems with 'Quality-of-Service' Guarantees." In *Proceedings of the IEEE 35th Conference on Decision and Control*, 3322–7.

16. Cassandras, C. G., L. Dai, and C.G. Panayiotou. 1998. "Ordinal Optimization for a Class of Deterministic and Stochastic Discrete Resource Allocation Problems." *IEEE Transactions on Automatic Control* 43 (7), 881–900.

17. Dai, L., and C. H. Chen. 1997. "Rates of Convergence of Ordinal Comparison for Dependent Discrete Event Dynamic Systems." *Journal of Optimization Theory and Applications* 94 (1): 29–54.

Section IV

Tools and Applications

17 Toward Accurate Simulation of Large-Scale Systems via Time Dilation

James Edmondson and Douglas C. Schmidt

CONTENTS

17.1 Introduction 419
17.2 Background.. 421
    17.2.1 Formal Composition Techniques.. 421
    17.2.2 Simulation Techniques.. 422
17.3 Motivating Scenarios.. 423
    17.3.1 Ethernet Capture Effect.. 424
    17.3.2 Application Services.. 425
    17.3.3 Large-Scale Systems... 426
17.4 Applying Time Dilation with Diecast... 426
    17.4.1 Overview... 426
        17.4.1.1 Time Dilation Factor (TDF) and Scale Factor (SF).......... 426
        17.4.1.2 Paravirtualized vs. Fully Virtualized VMs 428
        17.4.1.3 CPU Scheduling.. 429
        17.4.1.4 Network Emulation 429
    17.4.2 Application to Motivating Scenarios 430
    17.4.3 Future Work... 432
17.5 Addressing Issues of Time Dilation in Physical Memory 432
    17.5.1 Overview... 433
    17.5.2 Solutions 433
        17.5.2.1 More Memory or More Hosts 433
        17.5.2.2 Memory Emulation 434
17.6 Concluding Remarks 435
References.. 436

17.1 INTRODUCTION

Distributed systems, particularly heterogeneous systems, have been historically hard to validate [1]. Even at small scales—and despite significant efforts at planning, modeling, and integrating new and existing systems into a functional system-of-systems—end users often experience unforeseen (and often undesirable) emergent behavior on the target infrastructure. Some types of unexpected emergent behaviors include unwanted synchronization of distributed processes, deadlock and starvation, and race conditions in large-scale integrations or deployments [2].

Deadlock and starvation are not just limited to large-scale systems and can occur when connecting just a few computers or computer systems together. Phenomenon such as the Ethernet Capture Effect [2,3] (which is a type of race condition involving a shared bus and accumulating back-off timers on resending data) once occurred on networks as small as two computers, despite decades of previous protocol use and extensive modeling. If problems like this can occur during small integrations or technology upgrades, the challenges of integrating large-scale systems containing thousands of computers, processing elements, software services, and users are even more daunting. Ideally, all technological upgrades and new protocols could be tested on the actual target infrastructure at full scale and speeds, but developers and system integrators are often limited to testing on smaller-scale testbeds and hoping that the behavior observed in the testbeds translates accurately to the target system.

Consequently, what we need are technologies and methodologies that support representative "at-scale" experiments on target infrastructure or a faithful simulation, including processor time and disk simulation, as well as network simulation. These technologies and methodologies should allow application developers to incorporate their application or infrastructure software into the simulator unmodified, and so they run precisely as expected on the target system. There are many network simulators available for use—some of which we discuss in this chapter—but we also explore a new simulation technology that was introduced by Gupta et al. and is called *time dilation* [4].

The term *time dilation* has roots in the theory of relativity, pioneered by Albert Einstein in the early twentieth century [5]. In physics, time dilation is a set of phenomena that describe how two observers moving through space relative to each other or at different positions relative to objects with gravitational mass will observe each other as having erroneous clocks, even if the clocks are of identical construction. Relative velocity time dilation, the phenomenon described by two bodies observing each other's clocks while moving at different velocities, is the best parallel for the definitions and usage of the terminology in the work on simulation by Gupta et al.

Gupta et al. specifically coin the term *time dilation* in simulation to describe the process of altering time sources, clocks, and disk and network transactions to allow accurate simulation of multiple virtual machines (VM) on a single host, and we use this new definition throughout this chapter. This new usage fits with the original definitions by Einstein since a simulator and the actual operating system running the simulator will see time flowing at different rates due to context switching and other modern operating system techniques, despite both using equivalent clock mechanisms. The time dilation mechanism in the simulation context attempts to correct this clock drift from the simulator and operating system perspective to allow for closer approximation of target behavior by simulated tests.

Simulations based on time dilation allow system integrators and planners to run unmodified executables, services, and processing elements to accurately emulate CPU, network, disk, and other resources for large-scale systems in much smaller testbeds. A prototype of this time dilation technology called DieCast [6] has been implemented by researchers at the University of California at San Diego. This chapter explores the benefits of this technology to date, summarizes what testers must consider when using DieCast, and describes future work necessary to mature time dilation techniques and tools for simulation of large-scale distributed systems. This chapter presents a survey of the work to date by Gupta et al. in the application of time dilation to simulation in DieCast. It also motivates future work on memory management considerations for time dilation and the need for additional improvements to conventional time dilation implementations to address particular types of race conditions (such as the Ethernet Capture Effect) that may not be covered by time dilation.

17.2   BACKGROUND

In general, large-scale distributed systems are difficult to validate. Though some simulators, such as USSF described in Section 17.2, can emulate networks consisting of millions of nodes, simulation technologies that scale to this level often deviate from the performance and characteristics of target large-scale systems. This section describes related work on validating distributed systems and summarizes the pros and cons of current validation techniques with respect to their ability to address the role of time dilation in large-scale system validation. We divide related background material into two main areas: formal composition and simulation techniques.

17.2.1   FORMAL COMPOSITION TECHNIQUES

Formal composition is typically associated with modeling a target system in a computer-aided manner that ensures the distributed system is validated based on validated components and will execute properly on the target system [7]. When constructing mission and safety-critical distributed real time and embedded systems, such as flight avionics computers and nuclear power plant control systems, software developers often use formal composition techniques and tools, such as Step-Wise Refinement [8], Causal Semantics [9], Behavioral Modeling [10], and Object Modeling Technique [7], to validate their software before it goes into production.

Formal composition techniques are often time consuming to validate, however, and can be tightly coupled to a particular development context and domain. Moreover, many formal composition methods require developers to model *everything* (e.g., from processors to operating systems to the application logic) to validate the target system. When composing systems-of-systems with formal composition techniques in this manner, it is difficult to ensure a meaningful composition of heterogeneous components that interoperate correctly [11]. Progress is being made in this area of expertise–including a recent Turing Award awarded to Edmund Clarke, Allen Emerson, and Joseph Sifakis in 2007 for their work in the field of model checking [12], but formal composition of heterogeneous hardware, software, and platforms generally remains an open challenge, especially for large-scale distributed systems.

The main thrust of current development in this area of expertise is the domain-specific modeling language (DSML) [13], which requires developers to tailor a visual modeling language to a specific knowledge domain, allowing business logic programmers to create an application, device driver, etc. for a specific application need (e.g., a device driver for a particular type of hardware). DSMLs can shield developers from many tedious and error-prone hardware and operating system concerns, allowing them to focus on modeling the business application logic and allow further validation by tools developed by researchers and engineers experienced in the target domain.

Though DSMLs can simplify validation of certain software/hardware artifacts (such as device drivers on an airplane) for application developers, it is much more difficult to make a DSML that encompasses all hardware, device drivers, operating systems, etc. for an Internet-connected application or even a small network of application processes. One issue that hinders modeling languages from being able to completely validate arbitrary application and infrastructure software is the sheer variety of orthogonal personal computer architectures and configurations that must be expressed to ensure proper validation.

Validating an application can be simplified somewhat for homogenous configurations (e.g., all personal computers are Dell T310 with a certain number and type of processors, all running a specific version and configuration of Windows, etc.), but complexities remain because of randomized algorithms used in many components of operating systems such as page replacement and queuing. Threading presents additional challenges, even in homogonous hardware configurations, since composing multiple threads has no formally recognized semantic definition that fits all possible compositions [11].

When heterogeneous hardware, operating systems, and software must be supported, validating software with formal composition techniques becomes even harder. Formally composing legacy systems with closed-source implementations into a large-scale system may be aided by solutions that allow descriptions of behaviors by the legacy system, but it is still difficult to ensure that an entire large-scale system is formally composed in a semantically meaningful way. Because of these issues, we do not discuss formal composition techniques in this chapter, but instead focus on a separate vector of validation: simulation.

17.2.2 SIMULATION TECHNIQUES

Simulation is the process of reproducing the conditions of a target platform [14], and because of its flexibility, simulation has been the de facto method for validating many networked and distributed applications. A popular simulation model is discrete event simulation, where business application logic, operating system logic, etc. are treated as distinct, discrete events that are processed by a simulator engine [15,16,17,18]. Though simulation has evolved quite a bit in recent decades, simulators often make approximations that bring testing closer to a target system but do not precisely match what is being simulated, especially when the network connectivity, processing elements, or activities performed that are being simulated experience failures, intermittent behavior, or scarce resources.

Simulation technologies are particularly problematic in highly connected distributed or networked applications where Internet connections with high failure rates or resends are frequent. Although simulating the Internet is generally considered infeasible [19,20], many network emulators exist that attempt to emulate Internet access times, intermittency, network congestion, etc., and local area network testing. Examples of these simulators include Emulab and its derivatives Netlab [21] and ISIS Lab at Vanderbilt University [22], which can simulate dozens to hundreds of processing elements and their interconnections.

Emulab and its derivatives allow swapping in operating system images, applications, and test script setup to enable automated testing. They also provide robust network emulation including bandwidth restriction, packet loss, etc. Accuracy of the simulation is left as an exercise to the developer or user and how they configure operating system images, scripts, etc. Moreover, Emulab does not explicitly support multiple VMs per host, though if a user must scale a small testbed to a larger target system, operating system–specific VM managers, such as Xen [23], may be used.

Other simulators such as Modelnet [24] and USSF [19] enable explicit virtualization of hosts and also include robust network emulators. Modelnet separates a testbed infrastructure into edge nodes and nodes that act as switches between the edge nodes. Vahdat et al. [24] showed that a host emulating Gigabit Ethernet in Modelnet can result in approximations of networked application performance during simulation. The throughput difference between the emulation and real-world performance shown in their results, however, can differ by as much as 20%. Closer performance is possible if more than just networking is emulated, as shown in Section 17.4, via time dilation.

USSF is a simulation technology based on Time Warp and built on top of the WARPED [25] parallel discrete-event simulator that claims to simulate large-scale target systems of over 100,000 nodes [26]. USSF is complicated to use and develop for, requiring the creation of topology models and generation from application model to application code. Developers then tailor their application to the simulator, which may be unacceptable for existing code, particularly complex code that interfaces with undocumented legacy systems.

Working with USSF requires parsing a USSF model into a Topology Specification Language and then code generation via static analysis to reduce memory consumption—a major issue in WARPED libraries. The resulting code then links to WARPED libraries and the USSF kernel, which interfaces to the user application. Although USSF does allow simulation of potentially millions of nodes, there is no guarantee (or even an established working estimate) that the simulation will match a physical target system because USSF development has been prioritized to operate on reduced memory per VM, high cache hit ratios via file-based caching with a least-recently-used policy for cache replacement, etc. and not accuracy of simulation.

17.3 MOTIVATING SCENARIOS

To provide a solid motivation for time dilation and simulation of large-scale systems in general, this section presents several testing scenarios that require scalable and sound validation techniques. These examples focus on system environments where the computer infrastructure is mission-critical. Before we deploy large-scale mission-critical computer systems, it is essential to accurately simulate such systems in a smaller, less critical infrastructure where typical functional and performance problems can be observed, analyzed, and ultimately resolved before production deployment occurs.

We first discuss a smaller situation (connecting two computers together) and specifically look into a phenomenon known as the Ethernet Capture Effect [2,3] that can manifest itself when connecting just two computers together, at least one of which is constantly sending information over an Ethernet connection. To identify the Ethernet Capture Effect during simulation requires the simulator adhering closely to actual performance on a target architecture. We next review Application Services, which are collections of interdependent systems, such as database servers, front end servers, etc. We then expand the second scenario to include collections of such services, which is our targeted large-scale scenario. We refer to these three motivating scenarios when discussing the capacities and capabilities of modeling or simulation techniques.

17.3.1 ETHERNET CAPTURE EFFECT

The Ethernet Capture Effect [2,3] is a form of emergent misbehavior that plagued system integrators during the 1990s and caused unfairness in a networking system once Ethernet hardware became fast enough to support resend frequencies that approached optimal speeds indicated in the Ethernet protocol standard. Though specific to Ethernet, similar problems during integration and scaling can occur with any other system with a shared bus. To replicate this behavior requires just two hosts, both of which send information frequently.

During the 1990s, when a collision occurred between Ethernet connected hosts, all hosts involved in the collision would randomly generate a back-off time to allow the selection of a winner to send information across the Ethernet. The winner of this contest for the shared resource would have its back-off timer reset, while the loser or losers of this contest would essentially accumulate back-off timers until they successfully sent information.

The problem of the Ethernet Capture Effect manifested itself when the winner of these contests had a distinct amount of information to send and the Ethernet hardware was fast enough to allow for the winner to send its next information immediately (see Figure 17.1 for losing hosts increasing their back-off timers).

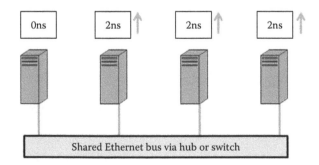

FIGURE 17.1 Example of the Ethernet Capture Effect in a four host system where one host is publishing constantly and three hosts are being forced to increase their back-offs indefinitely to avoid collisions. This race condition occurred despite extensive modeling of the Ethernet protocol and decades of practical use.

Since its timer had been reset, this host or service would be allowed to win each contest, indefinitely, while the losing hosts would be essentially starved until the winning process or service finished. If the winning process never died, the starvation would be indefinite or last until a developer or technician reset the nodes involved.

Ethernet had been modeled extensively before the problem became apparent. Moreover, Ethernet devices had been in service for decades before the Ethernet Capture Effect manifested itself in any type of scale. The protocol had been standardized and validated by thousands of users across the world, but small increases in hardware capabilities caused this emergent misbehavior to cripple previously functioning networks or integrations between seemingly compatible Ethernet networks of services, processes, and hosts.

The Ethernet Capture Effect represents emergent behaviors that should be caught during simulation on a testbed before deployment. If a simulator cannot catch such behaviors, major problems could manifest themselves in the production deployments. Time dilation does not solve this issue. Instead, the Ethernet Capture Effect scenario is presented as a potential motivator of future work.

17.3.2 APPLICATION SERVICES

Amazon, Google, and various other companies maintain thousands of servers providing software services for millions of end users every year. Application Services can range from a dozen to thousands of hosts (see Figure 17.2 for an example of deployment). Though most of these service providers have proprietary networks and systems, there are other open-source auction, e-commerce, and specialty sites for general testing. Many of these proprietary and legacy systems are closed-source, meaning that formal composition methods may not be feasible for most of the

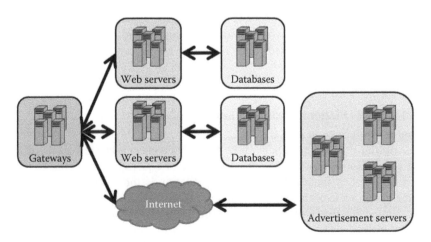

FIGURE 17.2 Example of a medium-scale Application Service with application gateways communicating over local area connections to web servers and databases. Internet connections can be a large part of Application Services, as depicted by the advertisement servers' interactions.

utilized systems (e.g., other than a brief description of what the legacy software does, a developer may not be able to completely model much less formally compose the system without the original developer providing the completed formal model).

Consequently, Modelnet, Emulab, or similar simulators can be used to gauge and validate a target Application Service, though simulation technologies such as USSF (see Section 17.2) may require too much modification for the simulation to be accurate enough for validation before deployment.

17.3.3 LARGE-SCALE SYSTEMS

Large-scale systems are complex systems of systems that evolve over time. For the purposes of this chapter, the motivating scenario here is an integration of a dozen or more Application Services into one large-scale system of 2000 nodes and is comprised of heterogeneous hardware and services. Moreover, we envision a large-scale system to have subsystems linked together via a combination of Internet connections and local area networks.

Consequently, the motivating scenarios presented in this chapter cover simulation needs from a very small network of two to three nodes (e.g., Ethernet Capture Effect) to a medium-sized network of a few dozen nodes (e.g., Application Services) to a large network of thousands of nodes (e.g., large-scale systems). In Section 17.4.2, we show how time dilation can be used to accurately simulate these latter two scenarios by intrinsically maintaining both accuracy and scale. The Ethernet Capture Effect and race conditions like it will require additional refinements to the time dilation simulation process (discussed more in Section 17.4.3).

17.4 APPLYING TIME DILATION WITH DIECAST

Previous sections have examined formal composition and simulation technologies that are being used to approximate and validate large-scale networks. This section expands on these technologies to include descriptions and results of the DieCast simulator system, which is based on the time dilation principle.

17.4.1 OVERVIEW

17.4.1.1 Time Dilation Factor (TDF) and Scale Factor (SF)

As mentioned in Section 17.1, time dilation has roots in the theory of relativity as the description of a phenomenon where two observers might view each other as having erroneous clocks, even if both clocks are of equivalent scale and construction [5]. In work by Gupta et al. [4,6], time dilation is the process of dividing up real time by the scale of the target system that will be emulated on a particular host. The reasoning for this partitioning is simple. Each host machine will potentially emulate numerous other hosts via VMs in an environment called Xen [23], preferably all requiring the same hardware, resources, and timing mechanisms found on the host machine.

A first instinct for emulating nine hosts might be to create nine VMs of the same operating system image and run them with regular system time mechanisms on the testbed hosts. An equivalent to this situation is shown in Figure 17.3. This approach, however, does not emulate the timing of the physical target system because the testing system will be sharing each physical second between the emulated hosts or services, resulting in each VM believing a full second of computational time has passed when the VM was really only able to run for a ninth of that time. This sharing can cause problems in emulation, for example, it can affect throughput to timer firings, sleep statements, etc. Time dilation allows the system developer to adjust the passage of time to more accurately reflect the actual computational time available to each VM. How time dilation affects the simulation of nine VMs on a single host is shown in Figure 17.4.

SF specifically refers to the number of hosts being emulated, while the TDF refers to the scaling of time. Gupta et al. [6] mentioned that these factors may be set to different values, but for the purpose of experimentation in the paper, TDF and SF are both set to the same values.

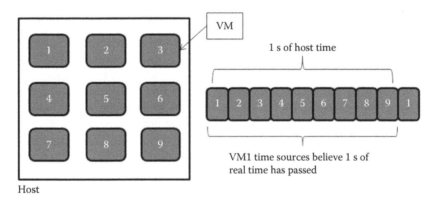

FIGURE 17.3   Most simulators do not modify time sources for virtual machines according to processor time actually used by a virtual machine. This can cause a simulation to drift from actual operation on target hardware because of queuing of timer-related events.

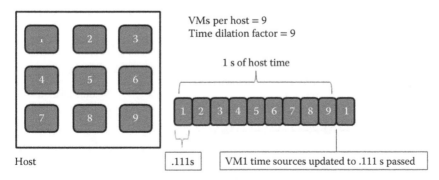

FIGURE 17.4   Basics of time dilation. When running nine virtual machines on a single host, programmable interrupt timers, time sources, and other timer-related events for each individual VM should only increment by the amount of processor time used by the VM.

17.4.1.2 Paravirtualized vs. Fully Virtualized VMs

To accomplish disk input/output emulation, Gupta et al. had to deal with the intricacies of two different types of VMs: paravirtualized and fully virtualized. A paravirtualized VM is a virtualized OS image that was limited to certain flavors of Linux and was soft-emulated on the host. Fully virtualized, in contrast, requires hardware support via Intel Virtualization Technology or AMD Secure VM, but does allow for any operating system image to be emulated directly on the hardware—rather than just a certain type of supported OS.

In their previous work on paravirtualized images [4], Gupta et al. created mechanisms that sat between the disk device driver and the OS, allowing emulation of disk latencies, write times, etc. with time dilation mechanisms. For the fully virtualized model of emulation, Gupta et al. used a disk simulator called DiskSim to emulate a disk drive in memory, which provided more control over buffering of read/write tasks. DiskSim gave the DieCast developer the ability to take the number of VMs into account and partition the read/write queuing accordingly.

The fully virtualized VM implementations give DieCast users the ability to plug in any operating system into the underlying Xen hypervisor and emulate a functional host according to the time slices allocated to each VM via time dilation. Together with changes to the time sources, CPU scheduling, and network emulation, fully virtualized VMs give developers a lot more options and control over what they are going to be testing and amazing scalability (concerning VMs accurately simulated per host). The divergence of a simulation from the target environment (which we call the "overestimation factor") without taking time dilation into account is shown in Figure 17.5, which is based on Figure 17.2c from Gupta [6]. This overestimation

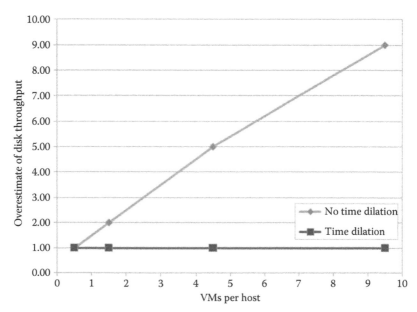

FIGURE 17.5 Analysis of the overestimation of disk throughput without time dilation versus using CPU and disk time dilation scaling.

factor in the simulation is a multiple of the correct target system (i.e., simulating 10 VMs per host results in inaccurate simulation by a factor of 9 on disk throughput). We also experienced this same phenomenon when simulating multiple subscribers and publishers per host in our own testing of the Quality-of-Service (QoS)-Enabled Dissemination (QED) middleware [27].

17.4.1.3 CPU Scheduling

To implement properly scaled CPU scheduling, Gupta et al. had to intercept and scale a number of time sources appropriately. Timer interrupts (e.g., the Programmable Interrupt Timer), specialized counters (e.g., TSC on Intel platforms), and external time sources (e.g., Network Time Protocol) were intercepted and scaled before being handed to VMs.

Timing is more intricate, however, than just allotting 1/(time dilation factor) time to each VM. For IO bound VMs, they may not use their full CPU allocation and this could skew the non-IO-bound VM CPU usage upwards and affect all timing dependent aspects of the emulation. Consequently, Gupta et al. devised a credit-based CPU scheduler in Xen to support a debit/credit scheme, where IO-bound VMs could relinquish CPU, but no VM used more than its exact allocated share of a real time unit. If a VM did not use its share before blocking, the VM received a debit to put towards its next time slice, and the CPU scheduler took into account total usage over a second to make sure that non-IO-bound VMs were not monopolizing the CPU and skewing results.

17.4.1.4 Network Emulation

Network emulation in the DieCast implementation is accomplished via capturing all network traffic and routing it through a network emulator. Though Gupta et al. mention that DieCast has been tested with Dummynet, Modelnet, and Netem, all experiments presented in their work used Modelnet as the network emulator. Since time dilation scales time down, emulating network throughput and latency turns out to be a relatively straightforward task (compared to CPU scheduling). To emulate a 1 Gbps network on a scaled host with time dilation factor of 10, the emulator simply ships 100 Mpbs (1/10 that total) to the host within a second. Latencies are easy to mimic also, since each VM is slowed down to 1/10 speed, and consequently, a system requiring 100 μs latency on the target system could be emulated with data arriving every 1 ms.

Time dilation is a powerful, robust mechanism for network emulation. The first paper on time dilation by Gupta et al. [4] shows how time dilation can even be used to emulate network throughput and speeds that are larger than the network capacity available over a link. As a result, not only can a tester using DieCast simulate ultra high network capacities internal to nodes (e.g., when all VMs are on the same host and the network link is completely emulated), but testers may also scale the number of VMs per host while simultaneously scaling the network capacity between hosts, if required. The key to this ability, once again, is effectively slowing down each VM according to a TDF. If all VMs are operating at that time scale, then the network can be emulated at a factor equal to the TDF.

17.4.2 APPLICATION TO MOTIVATING SCENARIOS

Earlier work has focused on time dilation in two scenarios: scaling network characteristics [4] and validating large-scale applications [6]. In the first paper [4] covering time dilation in networked applications, Gupta et al. scale network bandwidth and response time when needed via time dilation in both CPU and network emulation. This work outlined many results concerning how time dilation brought TCP throughput and response times between a simulated and baseline system into harmony in tests between two machines. These results show that time dilation can accurately simulate traffic characteristics (network capacity, response time, and traffic flows down to the TCP layer). As a first step in reflecting our motivating scenarios, therefore, time dilation can accurately simulate scaling the number of VMs and reflecting a target system when the target system is small—down to 1 to 2 machines if needed.

On the surface, this would appear to signal that the DieCast implementation of time dilation would be able to mimic the Ethernet Capture Effect, presented in Section 17.3, between two or more Ethernet capable hosts. Ethernet is at a lower OSI layer than TCP, however, and it is unlikely that a race condition like the Ethernet Capture Effect, which is essentially a hardware-related phenomenon that requires ultra fast publication to manifest, will occur when a simulator is slowing down publication rates by splitting processor time between multiple VMs. Consequently, the Ethernet Capture Effect and phenomenon like it demonstrate problems in current time dilation implementations and the need for future work to address low level (e.g., hardware level) race conditions that are difficult to emulate in a software environment.

In another paper on time dilation, Gupta et al. [4] performed testing in two different types of scenarios: (1) a baseline that is run without time dilation on the actual hardware—40 machines on their testbed—and (2) a scaled version running on much less hardware—usually 4 machines scaled to 40 machines via a TDF of 10. Several types of distributed systems were tested in the papers on time dilation, but we will be focusing on specific tests that reflect our motivating scenarios: RUBiS (an academic tool that resembles an auction site, along with automatable item auctioning, comments, etc.), Isaac (a home-grown Internet service architecture that features load balancers, front end servers, database servers, and clients and application servers), and PanFS (a commercial file system used in clustering environments).

In Ref. [4], Gupta et al. experimented with the DieCast system with TDFs of 10 (meaning that 10 VMs were running per host) and showed the effects of turning off CPU scheduling, disk IO, etc. to see how the system diverged from the actual baseline when time dilation was not applied in each of the time dilation mechanisms. When CPU and disk IO time dilation were turned off, the graphs diverge drastically for all experiments—often deviating by factors of two or more in perceived network throughput, disk IO, etc. These particular results demonstrate problems with using only network emulation to validate target systems—namely that these testbed systems may misrepresent the target systems by not just small error ratios but factors of error. Consequently, this result extends to experiments using Modelnet, Emulab, or other network emulation frameworks while simulating multiple VMs per node with ns, VMWare, Xen, or other systems without time dilation. This differential in network throughput, disk throughput, or latency could mean developers missing emergent behaviors on the testbed that will occur on the actual target hardware.

The RUBiS [30] completion times for file sharing and auctions on scaled systems closely mirrored the actual baselines for time dilation, while the non-TDF enforced versions of Xen did not. The testing data for RUBiS in particular is abundant with Gupta et al. showing results for CPU usage, memory profiling, throughput, and response times closely mirroring the baseline target system.

As the number of user sessions increased to 5000 or more, the deviation by non-time-dilation-scaled systems in response time grew to 5 to 7 times more than the target systems. Moreover, the average requests-per-second was less than half the corresponding statistic on the target system when not using time dilation with only 10 VMs per host. This inaccuracy in messaging throughput means that without time dilation, the simulation was attempting to put the testbed under less than half of the load that would be experienced on the actual target system. The time dilation tests, however, closely approximated both response time and requests-per-second.

The more complicated Isaac scenario, consisting of load balancers and multiple tiered layers of hosts and services, not only mimicked low-level metrics such as request completion time, but also resembled the base system on high-level application metrics such as time spent at each tier or stage in the transaction. DieCast also matched target systems closely during IO-bound transactions and CPU-bound ones.

The Isaac scenario also demonstrated the ability of time dilation to approximate target systems in fault situations. The DieCast developers caused failures to occur in the database servers in the Isaac scenario at specific times and compared the baseline performance to the time-dilated and non-time-dilated simulations. Without time dilation, the simulated experiments did not follow the baseline experiments. In addition, when requests through the Isaac system were made more CPU-intensive (generating large amounts of SHA-1 hashes) or more database-intensive (each request caused 100 times larger database access), the time dilation simulation was within 10% deviation of the baseline at all times, while non-dilated ended up requiring 3× more time than the baseline to complete the CPU stress tests.

The final tests on the commercial PanFS system showed similar aggregate file throughput to the baseline and also allowed Panasas, the company that makes PanFS, to more accurately test their file system according to target client infrastructures before deployment, which they had not previously been able to do due to their clients having much larger clustering infrastructures than the company had available for testing. While RUBiS and Isaac represent classic application services scenarios, the PanFS results were especially interesting because PanFS is regularly used by clients with thousands of nodes (i.e., large-scale systems). To validate the time dilation work in PanFS, Gupta et al. tested the system on 1000 hosts against time dilation on just 100 hosts and closely mirrored the performance of a deployed system, validating that the PanFS servers were reaching peak theoretical throughput bottlenecks. The PanFS results demonstrate that time dilation may be ready to make the leap to larger testing configurations, perhaps into tens of thousands and hundreds of thousands of target system nodes.

17.4.3 FUTURE WORK

Though DieCast provides an accurate approximation of a large-scale distributed system, time dilation technologies may mask timing-related race conditions (such as the Ethernet Capture Effect and TCP Nagle algorithm problems noted by Mogul [2]) due to DieCast slowing the entire system down and potentially missing boundary conditions that could have happened, for instance, in between real target timer events (which were instead queued up and fired in quick succession in a time dilation system). One of our motivating scenarios, the Ethernet Capture Effect, may not have been caught by systems using time dilation because by slowing down the Ethernet traffic resend rate of an application (by interrupting its resends to run other VMs), we may have allowed the back-off accumulator to reset. The lesson here is that, despite close approximation of the actual target system, DieCast and time dilation may not help catch all types of race conditions and unexpected emergent behavior. Other simulation solutions may be required to catch these types of low-level issues.

Gupta et al. admit that DieCast may not be able to reproduce certain race conditions or timing errors in the original services [6]. The system also has no way to scale memory usage or disk storage, and this can be a large limiting factor when a testbed host system is unable to emulate the time dilation factor of the target system (e.g., 100 hosts on the testbed with 4 GB of RAM trying to emulate 1000 hosts each requiring 1 GB of dedicated RAM a piece). Moreover, Gupta et al. appear to arbitrarily set the TDF to 10 for all experiments, noting that they had empirically found this value was the most accurate for their tests. No formal methods or description appear in their work to instruct how others may find the optimal TDF for target systems or the TDF corresponding to the number of simulated processes, the maximum delays expected from IO operations, or any other metric of the system. For time dilation to gain widespread acceptance and usage, this matter of obtaining an appropriate TDF for experiments should be addressed.

Potential vectors of interest may include augmenting the DiskSim to allow virtualization of memory on disk (possibly by further scaling down of time) to allow for the increased latency of disk drives emulating RAM memory access times. DieCast may also be a good vehicle to implement emergent "signatures" detection algorithms [2] into the testing phases and development cycles of large scale system development.

17.5   ADDRESSING ISSUES OF TIME DILATION IN PHYSICAL MEMORY

Section 17.4 discussed how DieCast can be used to validate large-scale network topologies for production applications. DieCast has built-in support for scaling disk input/output, networking characteristics, and time sources, but it has no mechanisms for scaling memory. This section, therefore, discusses options available to developers when memory scaling is necessary, including how to determine memory requirements to support scaled experiments and custom modifications that may be made to DieCast or VM managers such as Xen to enable scaling memory.

17.5.1 OVERVIEW

One aspect of validation that the DieCast implementation of time dilation does not solve is the situation where a host cannot emulate the physical memory required by the user-specified number of VMs per host. Physical memory is a scarce resource, and if it runs out during emulation of the target infrastructure, VM managers like Xen will begin to emulate physical memory using hard drive memory (virtual memory) on the host. Virtual memory is typically set aside from a hard drive, which has orders of magnitude worse fetch time than physical memory.

Although emulating in virtual memory will not stop the VMs from functioning (unless virtual memory is similarly exhausted), it may result in major timing differences between the testbed system and the target system. If the TDF were set to 10 (e.g., for 10 VMs hosted per host) and we only had enough physical memory to mimic target system performance for 5 of those VMs properly before virtual memory was used, we would likely miss race conditions, deadlock, and other types of emergent misbehavior during testing.

When using time dilation solutions, users should have options to address this issue. We evaluate potential solutions in this section.

17.5.2 SOLUTIONS

17.5.2.1 More Memory or More Hosts

Adding more physical memory may be possible, especially with increased adoption of 64-bit architectures and the ability of modern operating systems to support more than 4 GB of physical memory. This solution is attractive, and though processor speeds appear to have plateaued recently, availability of larger, inexpensive physical memory continues to increase. Users of time dilation systems or any other simulation system need to make sure that the amount of available memory exceeds the memory profiles required by all individual VMs. We discuss a reliable method for doing so here.

One available option is to profile physical memory usage of a single VM using Linux's top utility, the Task Manager of Microsoft Windows, or any other type of monitoring tool. To properly conduct such a memory profile, the VM must not only be up and running when profiling the memory, but also performing in the same type of role that it will be used in on the target system (e.g., serving as a database system during memory profiling). Developers would then have to multiply this maximum physical memory used during the memory profiling session by the number of VMs that the host will be running and add an additional physical memory overhead required by the Xen VM Manager in the case of DieCast, or whatever technology is managing the VMs, and the actual host operating system.

Once these memory overheads are calculated, developers should be able to arrive at the required physical memory for host systems. If implementers or validation testers are unsure of the amount of overhead required by host operating system and the VM manager, it may be best to multiply the amount required by a certain percentage, and remember the following: it is much better to have more physical memory than required than not enough when trying to get an accurate simulation of a target system with time dilation or any simulation system.

Adding more hosts may also be a feasible solution to this scenario if developers can afford to add more hosts to the testbed system. Gupta et al. recommend a TDF of 10 [6], and although there was not much reasoning or testing presented in the work to support this TDF, developers using DieCast may be best served by following this advice and keeping the host to VM ratio at 10 or less (i.e., 10 VMs per host at a maximum).

These two solutions (adding more memory or adding more hosts) are feasible for the vast majority of validation requirements. The next proposed solution tries to cover the other portion of testbed emulation of a target infrastructure.

17.5.2.2 Memory Emulation

This solution requires the most augmentation to a time dilation system like DieCast and is the most likely to deviate from a target system. This solution, however, may be the only option available when obtaining sufficient physical memory is infeasible.

As an example of such a scenario, consider a situation where a testbed system is composed of 10 hosts and a target infrastructure has 1000 nodes. If we were to equally distribute the 1000 VMs required over the 10 hosts, we would require each host to emulate 100 VMs, requiring at least a TDF of 100 to accurately mimic operation of the target system. Assuming that each VM requires a physical memory profile of 4 GB to accurately reflect operation of a target system, a total of 400 GB of physical memory must be installed on each host, before taking into account the memory required by the host operating system and VM manager.

Assuming an overhead of 20% of the VM requirement for the latter (400 GB × 0.2 = 80 GB for a total of 480 GB required per host), if our hosts actually have only 4 GB of installed memory, this situation will result in a simulation that does not accurately reflect timing of target systems, due to virtual memory being much slower than the physical memory used on the target system. A potential solution to this situation is to completely emulate the instruction set for all VMs on the host and run most of the VMs on virtual memory with a TDF that reflects usage of virtual memory instead of physical memory. This solution will result in a significant increase in the amount of time an experiment will require to run.

Figure 17.6 shows the difference between accessing physical memory and a hard drive for memory needs. The difference between access time in physical memory and hard drive data is typically six orders of magnitude. Consequently, emulating all VMs in virtual memory and adjusting the time dilation accordingly to the access time difference could lead to a time dilation simulation taking over 1 million times longer with emulation on hard disks and over 10 thousand times longer with

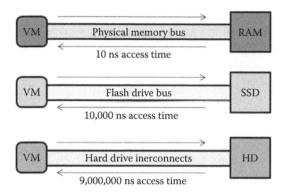

FIGURE 17.6 Memory access time comparison between physical memory (RAM), SSD flash drives, and traditional hard drives. All numbers are approximations to provide scale.

emulation on a flash memory type drive (shown as SSD for Solid-State Drive in Figure 17.6). SSD flash cards or hard drives are currently able to supplement system memory with over 64 GB of flash memory. There has also been recent success with using SSD memory for virtual memory in enterprise database applications and large clusters [28,29]. Obtaining the 480 GB of additional memory for the ten thousand times longer run time system could potentially be possible via USB hubs or similar technologies.

17.6 CONCLUDING REMARKS

Time dilation is a versatile emerging technology that helps developers and testers ease the validation of medium- to large-scale systems. Earlier work on DieCast has provided developers with CPU scheduling, network emulation, and disk emulation informed by time dilation mechanisms, resulting in accurate test runs on reduced hardware. DieCast does not require developers to remodel their business application logic, piggyback unrelated libraries, generate simulation glue code, or do many things required by other simulation frameworks and technologies typically used for large- to ultra large-scale target architectures. Instead, it allows for developers to compile their projects and code as they normally would do, harness the power of the Xen hypervisor—a stable, well supported VM manager for Linux—and run multiple VMs per machine in a way that more accurately reflects the true performance of a target system.

Though time dilation shows promise, it does not identify all types of problems that affect mission-critical distributed systems. This chapter explained how Ethernet problems, such as the Ethernet Capture Effect, may be masked by slowing down the host and not witnessing some of the race conditions that can appear in target systems because of queuing of timer firings and other related issues. Time dilation also currently has a higher memory footprint than some network simulators that are able to reduce memory requirements via shared data structures, operating system emulation, etc. While these network simulators might be useful to determine if the target system works if provided with certain operating systems or configurations and modeling of environments, implementations of time dilation such as DieCast allow developers to test their business logic for distributed or networked applications on actual operating systems with excellent approximations of time source progression, disk queuing, and network throughput and latency.

REFERENCES

1. Basu, A., M. Bozga, and J. Sifakis. 2006. "Modeling Heterogeneous Real-Time Components in BIP." In *Fourth IEEE International Conference on Software Engineering and Formal Methods*, Pune, India 3–12.
2. Mogul, J. C. 2006. "Emergent (Mis)behavior vs. Complex Software Systems." 2006. In *1st ACM SIGOPS/EuroSys European Conference on Computer Systems*, 293–304. Belgium: Leuven.
3. Ramakrishnan, K. R. and H. Yang. 1994. "The Ethernet Capture." In *Proceedings of IEEE 19th Local Computer Networks Conference*. Minneapolis, MN.
4. Gupta, D., K. Yokum, M. McNett, A. C. Snoeren, G. M. Voelker, and A. Vahdat. 2006. "To Infinity and Beyond: Time-Warped Network Emulation." In *Proceedings of the 3rd USENIX Symposium on Networked Systems Design and Implementation*, San Jose, CA.
5. Einstein, A. 1905. "Zur Elektrodynamik bewegter Körper." *Annalen der Physik* 17: 891.
6. Gupta, Diwaker, Kashi V. Vishwanath, and Amin Vahdat. 2008. "DieCast: Testing Distributed Systems with an Accurate Scale Model." In *5th USENIX Symposium on Networked System Design and Implementation*, 407–22. San Francisco, CA.
7. Cheng, B. H. C., L. A. Campbell, and E. Y. Wang. 2000. "Enabling Automated Analysis through the Formalization of Object-Oriented Modeling Diagrams." In *International Conference on Dependable Systems and Networks*, New York, 305–14.
8. Batory, Don, Jacob Neal Sarvela, and Axel Rauschmayer. 2004. "Scaling Step-Wise Refinement." *IEEE Transactions on Software Engineering*. Vol. 30:6, Piscataway, NJ.
9. Bliudze, S. and J. Sifakis. 2008. "Causal Semantics for the Algebra of Connectors." In *Formal Methods for Components and Objects*, Sophia Antipolis, France, 179–99.
10. Engels, G., J. M. Küster, R. Heckel, and L. Groenewegen. 2001. "A Methodology for Specifying and Analyzing Consistency of Object-Oriented Behavioral Models." *ACM SIGSOFT Software Engineering Notes* 26 (5): 186–95.
11. Henzinger, T. and Sifakis, J. 2006. "The Embedded Systems Design Challenge." In *Lecture Notes in Computer Science*, Springer, 1–15.
12. Clarke, E., A. Emerson, and J. Sifakis. 2009. "Model Checking: Algorithmic Verification and Debugging." In *Communications of the ACM*, New York, 74–84.
13. Schmidt, D. 2006. "Model-Driven Engineering." In *IEEE Computer Society*, Piscataway, NJ, 25–32.
14. Oxford English Dictionary. 2010. Oxford, UK: Oxford University Press.
15. Wainer, G. and Mosterman, P. 2010. *Discrete-Event Modeling and Simulation: Theory and Applications*. Boca Raton, FL: CRC Press.
16. Fujimoto, R. M., K. Perumalla, A. Park, H. Wu, M. H. Ammar, and G. F. Riley. 2003. "Large-Scale Network Simulation: How Big? How Fast?" In *11th IEEE International Symposium on Modeling, Analysis, and Simulation of Computer and Telecommunications Systems*, 116–23. Orlando, FL.
17. Varga, A. 2001. "The OMNeT++ Discrete Event Simulation System." In *Proceedings of the European Simulation Multiconference (ESM)*. Prague, Czech Republic.
18. Banks, J., J. Carson, B. Nelson, and D. Nicol. 2009. *Discrete-Event System Simulation*. 5th ed. Upper Saddle River, NJ: Prentice Hall.

19. Rao, D. M. and P. A. Wilsey. 2002. "An Ultra-Large-Scale Simulation Framework." *Journal of Parallel and Distributed Computing* 62 (11): 1670–93.
20. Riley, G. F. and Ammar, M. H. 2002. "Simulating Large Networks—How Big Is Big Enough?" In *Grand Challenges in Modeling and Simulation*.
21. White, B., Lepreau, J., Stoller, L., Ricci, R., Guruprasad, S., Newbold, M., Hibler, M., Barb, C., and Joglekar, A. 2002. "An Integrated Experimental Environment for Distributed Systems and Networks." In *5th Symposium on Operating Systems Design and Implementation (OSDI)*. Boston, MA.
22. Hill, J., J. Edmondson, A. Gokhale, and D. C. Schmidt. 2010. "Tools for Continuously Evaluating Distributed System Qualities." *IEEE Software*, Vol, 27:4, 65–71.
23. Barham, P., et al. 2003. "Xen and the Art of Virtualization." In *Proceedings of the 19th ACM Symposium on Operating System Principles*.
24. Vahdat, A., et al. 2002. "Scalability and Accuracy in a Large-Scale Network Emulator." In *Proceedings of the 5th Symposium on Operating Systems Design and Implementation*, 271–84.
25. Martin, D. E., McBrayer, T. J., and Wilsey, P. A. 1996. "WARPED: A Time Warp Simulation Kernel for Analysis and Application Development." In *29th Hawaii Internetional Conference on System Sciences*.
26. Rao, D. M. and P. A. Wilsey. 1999. "Simulation of Ultra-large Communication Networks." In *Seventh IEEE International Symposium on Modeling, Analysis, and Simulation of Computer and Telecommunications Systems*, 112–9.
27. Loyall, J., et al. 2009. "QoS Enabled Dissemination of Managed Information Objects in a Publish-Subscribe-Query Information Broker." In *SPIE Defense Transformation and Net-Centric Systems Conference*. Orlando.
28. Lee, S., Moon, B., Park, C., Kim, J., Kim, S. 2008. "A Case for Flash Memory SSD in Enterprise Database Applications." In *ACM SIGMOD International Conference on Management of Data*. Vancouver.
29. Caulfield, A. M., Grupp, L. M., Swanson, S. 2009. "Gordon: Using Flash Memory to Build Fast, Power-Efficient Clusters for Data-Intensive Applications." In *14th International Conference on Architectural Support for Programming Languages and Operating Systems*. Washington, D.C.
30. RUBiS. 2012. http://rubis.objectweb.org.

18 Simulation for Operator Training in Production Machinery

Gerhard Rath

CONTENTS

18.1 Introduction 440
    18.1.1 Ultra Flexible Reversing Mill 440
    18.1.2 Simulation for Operator Training 440
    18.1.3 Modeling 441
    18.1.4 Real-Time Simulation 443
    18.1.5 Game-Based Learning, Game Programming, and Simulation 443
18.2 Simulation of Plant and Machinery 444
    18.2.1 Languages for Simulation 445
    18.2.2 Recent Developments and Trends 446
18.3 Simulator for Rolling Mill Operator Training 446
    18.3.1 Requirement Analysis of the Training Simulator 447
    18.3.2 System Design of the Training Simulator 448
    18.3.3 Reduction of Complexity 451
        18.3.3.1 Hardware Expense 452
        18.3.3.2 Reduction of Possible Machine States 452
        18.3.3.3 Reduction of States by the Control Program 453
        18.3.3.4 Depth of Simulation 453
    18.3.4 Simulation Tool Selection and Program Coding 455
        18.3.4.1 Simulation Tool 455
        18.3.4.2 Example Details 456
18.4 Experiences With Simulation-Based Training 457
References 458

This chapter describes the development of and the experience with a training simulator for a steel rolling mill. The goal of the training is how to replace the rolls, which is normally controlled by the electronic system, but has to be done in certain exceptional situations by the operators or the maintenance personnel without automatic sequence control. The actual machine is not available for training, since the production is fully automated and is running around the clock. This operator training simulator (OTS) replaces the functionality of the machine in a loop with a replica of the electronic control system (hardware-in-the-loop, HIL). Simulation is applied to train the operators and the maintenance staff. To keep the staff well trained without the necessity to interrupt the production routine of the machine was the motivation to develop the simulator.

The introduction describes the basic considerations that were necessary for a successful development project. In the second section, the status of simulation in industrial production plants is discussed, where it is noted that simulation is of fast growing significance in this field. The third section presents details of the simulator design and some interesting particularities that were found during the project. Finally, the experiences acquired over several months are given.

18.1   INTRODUCTION

After a short description of the actual simulation problem, a brief overview of past and present trends in simulation-based operator training is given. The third subsection shows some important aspects of modeling from the view of software engineering. For simulation of fast physical systems such as machines, considerations of real-time aspects are important. Finally, a look at the impact of modern computer games to the development of simulation software is useful before starting the modeling work.

18.1.1   ULTRA FLEXIBLE REVERSING MILL

Voestalpine Schienen GmbH is a leading manufacturer of rails and is exploiting a new ultra flexible reversing mill consisting of three identical mill stands [1]. The automation system of the mill provides all necessary control for production, which includes the basic sequential and closed-loop control of the movements of the mechanical equipment and the technological process control such as minimum tension control.

In normal operation, the complex roll exchange is a process of maintenance and runs fully automated under supervision of the control system. With the help of hydraulic and electric systems, three systems of horizontal and vertical rolls are exchanged. The process goes through about one hundred steps and requires hundreds of actuators and an even larger amount of sensors. Sometimes a failure in the system occurs, and the automatic cylce stops. Then, after a repair, the steps must be carried out without automated sequencing. The danger of making a mistake and damaging some components is high. Such mistakes increase the downtime of the machine significantly. More and regular sessions with a systematic training program are necessary for faster performance. A long period of continuous production, which is of course a desirable advantage, turns out to be a problem for the maintenance personnel, since they have fewer opportunities to train and hone their skills. This was the reason for the development of the training simulator.

18.1.2   SIMULATION FOR OPERATOR TRAINING

The need for training simulation emerged when human mistakes during the operation of technical devices started causing harm to people. Beginning with mechanical simulators in the first decades of aviation, the first computer simulations appeared in the 1960s. Flight simulators for the training of operators of airplanes [2] and spacecraft were indispensable. In addition to modeling and simulation, pedagogical frameworks based on training modules were developed to lead the trainee step-by-step through the basics of operation.

The high cost of such equipment is why such simulators were introduced very late to industrial equipment. Nuclear power plants with their high risks were the first to see computer simulation employed [3]. The chemical industry and fossil power plants followed. Recent progress in computer technology, hardware, and software have reduced the cost and enabled the use of simulation also in industries with lower risk and impact, such as heavy machinery [4].

While conventional training requires access to resources dedicated to production (machine, plant, people), simulation on the computer reduces the danger caused by mistakes of the trainees, and also the cost of potential damage. Beside these points, making the original system available for training reduces the production capacity and is a more recent argument in favor of OTS. For accurate simulation, models for the physical or chemical processes are controlled by a replica of the deployed control system. Having the simulation in a loop with a piece of original hardware is referred to as HIL [5,6].

Other fields of training simulation emerge for highly complex systems. For example, the optimization of the market of electrical power supply [7] necessitates optimal operation of the power grid [8]. For medical appliances, training simulation is also well established [9]. A demographic problem, the retirement of aged experts that results in a lack of training facilities for the apprentices, is proposed to be solved with training in simulation [10].

Embedded simulation is a recent development in military appliances to bring training closer to high fidelity environments. When not in action, weapons can temporarily switch to a simulation mode for training. This allows training not only in training centers and classrooms but also in the actual place of operation. The environment is more realistic and training can occur more often during standby times. Civil equivalents of this technique are conceivable.

18.1.3 MODELING

Before a simulation of a system can be carried out, a model must be established. Different definitions for a model exist in several standards or books; a useful one can be found in Lieberman [11]:

> A model can be defined as the simplification of a complex system for the purpose of communicating specific details.
>
> **Definition of a model**
> *Lieberman*

The emphasis should be on the word "communication," telling us that the purpose of a successful modeling process is not to create a model, but to inform a hopefully well-defined group of people about some selected details of the system under study that is modeled. A model should not be built until the modeler understands the intended audience. The same author describes an important property of a model:

> A model operates much like a lens: It focuses attention on items while obscuring or omitting everything else!

> A mo **much like a lens: It focuses a** items
> while **omitting everything else!**

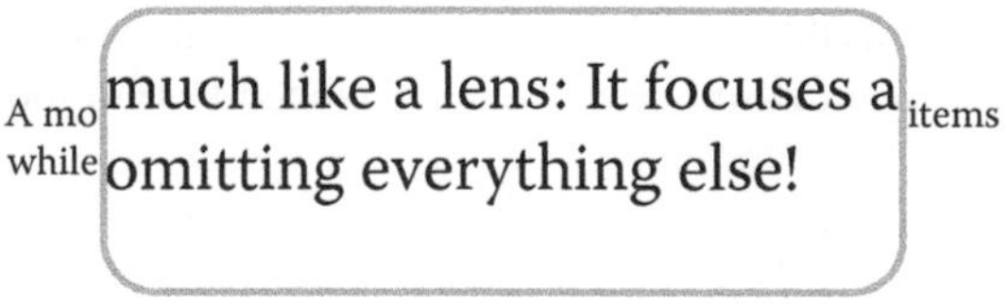

FIGURE 18.1 Focusing effect of a model. (Adopted from Lieberman, B. A. "The Art of Modeling, Part I: Constructing an Analytical Framework," 2003, accessed July 3, 2010, http://download.boulder.ibm.com/ibmdl/pub/software/dw/rationaledge/aug03/fmodelingbl.pdf.)

Figure 18.1 explains this property. We and our customer have to accept that a model cannot represent all details of the real system. Trying this, we would fail.

If a system should be modeled for different purposes, different modeling procedures, methods, and tools are applied [12]. In general, separate models for the same plant may be required for, for example, presentation of a machine to salespeople, investigating technical details of the plant behavior, optimizing production and logistic properties, and training of operators.

After the requirement analysis for the modeling development is completed, the design may follow one of the strategies:

1. Top-down
2. Bottom-up
3. Inside-out
4. Hardest first

These are the same strategies we know from the software development process. The application of the first three strategies is described in Lieberman [13]. For the steel rolling mill project described in this chapter, the "hardest first" approach had to be applied, since the behavior of some critical parts of the system were not known exactly and had to be explored first.

Before starting the modeling process, a classification is helpful to select the best-suited language or tool for the actual purpose. A useful classification may be according to Figure 18.2. For *continuous models*, time is a primary simulation variable. This is the way we describe dynamic systems starting from ordinary differential equations (ODE), applying Laplace transform and transfer functions [14]. *Time-based models* running on a computer are inherently of a discrete nature. This means that modeling of dynamic systems requires the z-transform to obtain digital transfer functions. Modern modeling languages such as Simulink® [15] are capable of masking this difference and even allow running discrete-time and continuous-time models concurrently within one system. This simplifies the modeling task and saves us from working with z-transform or similar techniques. *Event-based models* are used when the system primarily reacts to events happening at certain time points, and the state of the system does not change over time between two subsequent events. Finite state machines (FSM) are one example of a basic method to implement such a model. Again, it is state of the art to have continuous-time, discrete-time, and event-driven

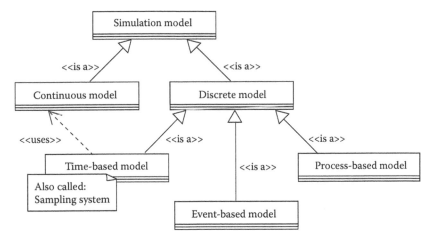

FIGURE 18.2   Classification of mathematical models for simulation.

models combined in a single computer model. Finally, *process-based models* were developed to map workflow processes of practical value into a simulation model, for example, Business Process Modeling Notation [16].

18.1.4 REAL-TIME SIMULATION

A simulator for operator training requires a real-time response. This means the time behavior of the simulation must match with the physical system. The definition "real time" demands that the computations that are necessary in a given time interval can be completed in this specified time interval [17]. Not only the calculation of the models, but also data communication must abide by this restriction [18]. Usually, the real-time property is obtained by the utilization of a dedicated operating system. In computers for human–machine interface (HMI), often a standard operating system with some real-time extension is used. Even if the "hard" real-time property cannot be guaranteed, such systems are proven to work properly as simulators. The chosen environment for the actual steel rolling mill project, Winmod®, is one of this type [19].

18.1.5 GAME-BASED LEARNING, GAME PROGRAMMING, AND SIMULATION

Modern computer games are an important pacemaker for developments in the computer industry. So it is worth it to take a look at game programming when investigating the facilities of modern simulation.

Game-based learning (GBL) is an integrated part of eLearning [20]. Even if no GBL is intended, the target audience, the trainees on the simulator, may have grown up with computer games and acquired a different cognitive character ("gamer generation" [21]). Games are applied in learning environments to increase the motivation. Fantasy, rules/goals, stimulus, challenge, mystery, control, and feedback are the characteristics of a game that augment the learning effect [22].

Exploring the possibilities of new media causes enthusiasm, but one must be cautious about the eLearning hype. A game is only a game in a narrower sense, if there is no purpose behind it [23], and people who often play games are more critical about serious games [24].

For the designer of a training simulator, the following aspects may be important.

Playing the Simulation Game. A simulation itself is to an extent a game that models parts of the actual world [25]. The topic of simulation for training is given by the actual task, but for presentation, visualization, etc., we can gain inspiration from eLearning and GBL on how to design an attractive environment for the simulation.

Trainer or GBL Environment. Most OTS for industrial production systems require a trainer to guide through a session. This trainer usually observes and rates the session. If the reaction to a fault needs to be trained, the trainer provokes a failure at a certain instant. Substituting this trainer with software would be a step towards GBL.

Game Programming. For the developer of simulation software, game programming is a rich source of methods and techniques. All required design patterns on how to create successfully virtual worlds for a trainee are developed and published. For example, work has been done on how to design the architecture for simulation in real-time and visualize it in 3D [26]. To model the dynamic behavior of game characters or machines, work that employs FSM, parallel or nondeterministic automata can be found [27]. Scripting techniques that simplify rich but complex programming languages as those employed in design patterns are ideally suited for industrial programmers.

18.2   SIMULATION OF PLANT AND MACHINERY

Because of high costs, the application of simulation has been limited to industry and organizations with a high level of risk. Today simulation is also becoming prevalent in the production industry. Shorter product cycles, flexible production, risk reduction, and more competition give a good return on investment to simulation techniques.

This highly dynamic process is still ongoing, so the vocabulary of simulation is not standardized yet for the production industry. In contrast, for flight simulators, the requirements were defined much earlier [28]. The following classification and keywords are based on some definitions of the German chemical industry [29], on the specifications of a supplier for simulation systems [19], and on personal discussions [30].

Control System Tester. Such simulators are used to test the function of the control systems; this is a verification of the control software.

Process Trainer. Process trainer simulations are not used to verify the function of the automation system, but focus on the simulation of process dynamics. The trainees learn to understand a complex, physical, or chemical process.

Virtual Startup. The control system can be tested and commissioned while connected to virtual machinery. This is designated a virtual startup. Since the control system is the same that will later be working with the physical plant, it should be mentioned that, from the viewpoint of the control system, it is a real startup [30]!

Factory Acceptance Test (FAT). A simulation model of a complex production plant and its control system can be subjected to a FAT. Logistic chains, dynamic and logical properties, capacities etc. can be verified with the participation of the customer before delivery and assembly of the actual system. This is progress in the sense of "Simultaneous Engineering" [31]. Mistakes in the requirement specifications, planning mistakes, and critical states of the machinery are detected without risk in an early phase and save time during the startup of the plant.

Functional Validation. With the customer involved, the software is checked against the requirements. It is not the goal here to find programming errors, which should be already removed by verification, but misunderstandings between the supplier and the customer [30]. Without simulation, usually this validation is carried out during commissioning of the plant. With the help of simulation, it is possible to correct such errors much earlier.

Design and Planning. In an early project phase, simulation allows us to find mistakes in planning caused by incompatibilities between different engineering departments. For example, the variables used for controller programming may not correspond to the inputs and outputs as defined by the electric engineers. Without simulation, such errors can be detected only very late during the assembly or the startup phase.

Instruction and Training. OTS work with a replica of the automation system. Training and instruction can proceed without the physical plant. Usually this is used to shorten the startup of the actual plant. OTS are working in the loop with the actual process control system (HIL). It is desired to include defined scenarios of plant failures ("what if" scenarios [29]).

Maintenance and Optimization. After the startup of the real plant, the simulation helps optimize parameters and find causes of problems. The simulation can run parallel to the physical plant and avoids the necessity of interaction with and perhaps disturbance of production.

Presentation. A virtual presentation of the plant can support the salespeople and brings advantages over the competitors. Of course, the quality of the presentation is more important than the technical accuracy.

It is desired that one system should provide all these features, which is the goal of developers of modern simulators [19].

In production machinery, it is not possible to obtain a fully detailed physical fidelity of simulation since the reaction times are much shorter than in chemical processes. This is shown in an example in Section 18.3.3 and is one of the reasons why a certain component is modeled for different purposes with different tools [12].

18.2.1 LANGUAGES FOR SIMULATION

The problem to choose the most suitable language for a simulation of an industrial equipment is closely related to the selection of languages for the electronic control of the system; consequently, at first a look to these languages is required.

The control of machinery and plants in the production industry has been done with electronic systems since the 1980s. Before this, relay control was the state of the art. Free programmable systems (PLC, programmable logic control) emerged in the 1990s [32,33]. Those systems provided a restricted command set for programming, on the one hand to minimize the danger of mistakes and on the other hand to guarantee that persons different from the programmer could read, understand, and modify the code. From this point of view, programmers for industrial systems must be specialists in the specific process, more so than in ultimate software techniques.

Standardization of languages led to Continuous Function Chart and the languages defined in the IEC 1131 standard becoming the most important representatives [34].

The languages are based on graphical concepts such as electric circuit diagrams and FSM and on textual concepts such as assembler, Basic, and Pascal. Object orientation is implemented to a limited extent. New objects may be created "by composition," which means putting existing objects together to create new ones. Multiple inheritance and dynamic binding are not used.

The basic idea for all these restrictions and simplifications is to construct control programs that can be grasped by other process experts independent from their writers.

Now the same also applies for simulation programs. Machine and process experts should be enabled to develop models and their behavior, and the visualization and real-time monitoring of the simulated process. For example, MathWorks recently published an extenstion for Simulink to create IEC-1131-compatible programs [35].

The choice of the language for simulation is a trade-off between the versatility of a modern computer language and the maintainability of the software by people who are more process experts than software experts. For the actual project of the rolling mill simulator, the selected system [19] provides a graphical programming suite similar to Function Block Diagram, which is one of the languages defined in the IEC 1131 standard.

18.2.2 RECENT DEVELOPMENTS AND TRENDS

All components of control systems for production machinery must provide interfaces to supplemental products of other manufacturers, for example, standard fieldbus systems or OPC communication (OLE for process control). Products for simulation must follow this trend since for different purposes different simulation environments are established.

Suppliers generate many documents to design a system. The ideal case is to generate a simulation completely automated from these documents with a minimum of further programming effort.

In regard to operator training, a simulation must provide possibilities to train exceptional situations. For example, the nuclear power industry has well-defined training scenarios for exceptions. Many industrial simulation systems work with a trainer who provokes an error during the session or defines the errors immediately before the start. Script-based or FSM techniques should make it easier to define training scenarios.

18.3 SIMULATOR FOR ROLLING MILL OPERATOR TRAINING

This section describes development work for the actual simulator in a time-ordered manner. First, the requirements are specified, then the design of the software components follows. Taking into consideration how to reduce the enormous complexity to a reasonable extent is important before selecting a tool and doing the coding work.

18.3.1   REQUIREMENT ANALYSIS OF THE TRAINING SIMULATOR

From the basic considerations described in Section 18.2, some conclusions are drawn: The language for developing training simulators should be one of the typical industrial languages so that machine experts can understand and change the program code of the simulator. C++ or other object-oriented languages, which offer more freedom to the programmer, but may result in unreadable code, are not favored. A training session should not have a game character and no success indicators of a training session should be recorded, since this is demanded by the trade union. The behavior of some critical machine parts are typically not known in sufficient detail so as to initiate their modeling without further preliminary analysis. Consequently, the hardest-first approach to design is to be applied.

The application of a use-case analysis turned out to be a successful part in the phase of requirement specifications. A simplified use-case diagram is shown in Figure 18.3. The machine operator commands the automated roll exchange as well as the manual exchange. In case of problems, a second actor, the maintenance expert, works on the "manual roll exchange." For training purposes, the simulator should substitute the physical machine for a use case "manual roll exchange."

During the automated production, the roll exchange also runs autonomously, and the operator supervises the procedure. In the case of a failure in the system, the roll exchange must be done by the operator, thought not manually, on a low level of automation. Several hundred drives must be coordinated in about one hundred steps. After a repair, the machine must be brought into a condition where the automatic

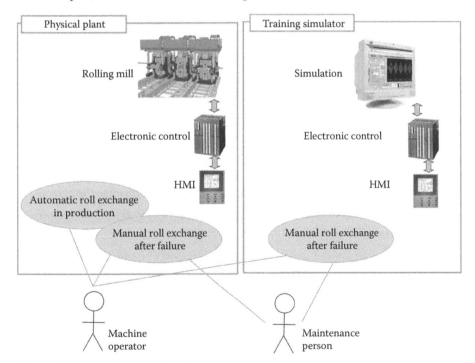

FIGURE 18.3   Use cases of the roll exchange task.

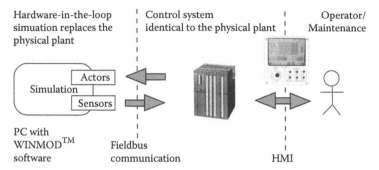

FIGURE 18.4   Hardware configuration of the training simulator.

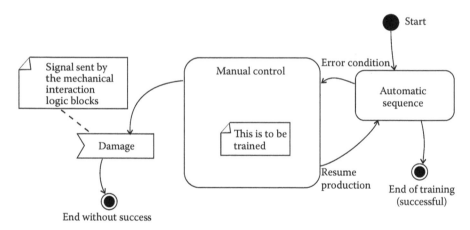

FIGURE 18.5   State diagram of the training session.

sequence may continue. Figure 18.4 shows the configuration of the actual system and the data flows among the components. The operator works with his HMI and a replica of the control system. Several hundred inputs and outputs must be communicated in real time between the control system and the simulator. To avoid the cost of copper wires, serial communication over an industrial fieldbus was the choice.

18.3.2 System Design of the Training Simulator

As a first design step, based on the requirement analysis and on the introductory considerations, the course of the training session was defined. It can be documented as a state diagram (see Figure 18.5), according to UML notation [36]. Two exits are designed. One is the successful end, when the machine can resume its normal production again. When the trainee causes damage to the system after a mistake, the session is terminated ("game-over" situation).

An important aspect can be found in Figure 18.5: the automatic sequence (which is not a topic for the training) is a milestone for testing the system. All components of the physical simulation are actuated by the control system and the system is running in a continuous mode.

The next design step is to find a structure for the simulation software. The simulator should replace the physical machine and provide its behavior to the replica of the electronic control system (see Figure 18.4). The artifacts of the simulation software are shown in Figure 18.6. Apart from the software component simulating the machine, there exists a visualization part. This is not for the trainee, who works with the HMI of the control system, but is required for administration and adjustments of the simulator. Since the operator of the physical machine cannot observe any process and must rely entirely on her/his operating panel, a three-dimensional visualization was not a requirement of the simulation system.

The exception generator is the component of the software in Figure 18.6 for defining and varying the training scenario. It means that the automatic cycle (see Figure 18.5) breaks and the actual training can start.

Another component of the system in Figure 18.6 is the input/output system. Via communication over a fieldbus, this part has to provide the state of about one thousand data points in real time.

The main component of the training simulator is the actual modeling software. A machine or plant consists of an incredible amount of things moving around, delivering signals or interacting with some other parts. To classify this, the machine

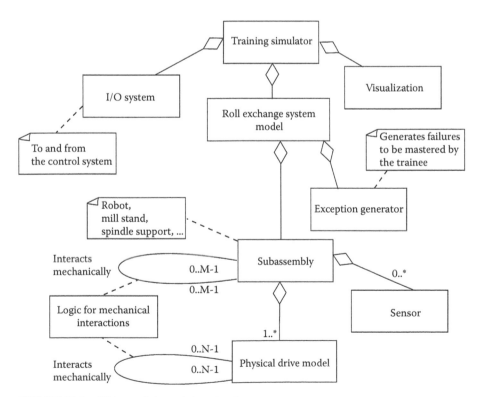

FIGURE 18.6   Objects of the training simulator.

was considered to consist of components called subassemblies in Figure 18.6. A subassembly is a constructive mechanical part of the machine. It can be a single lever, a carriage for the rolls, or a complex handling system such as a robot. Artifacts called sensors are mounted on these subassemblies to provide information about positions, pressures, etc., while physics and kinematics of these assemblies are to be modeled. One assembly can carry drive units, which are hydraulic or pneumatic cylinders and motors, or electrical drives. Their task is to convert the output signals of the controller into a mechanical force and to apply this to the mechanic subassembly. The motion as a reaction can be found by solving the ODE. Since unrestricted motion is not possible for all components, additionally collisions must be detected. Such interactions can occur among subassemblies or among drives (in a few cases). An interaction can cause damage, which then is a severe mistake of the trainee. Interactions without damage are usually blocking situations, since the machine consists of rigid bodies. Those situations may be resolved by a reversal motion of the drives and are not an illegal condition of the system.

The states of a drive during its simulation lifecycle are shown in Figure 18.7. A normal, successful training end does not terminate the simulation, since the design of the deployed production plant also is made for perpetual operation.

Two points have important implications regarding the design personnel, which are documented in the data-flow diagram in Figure 18.8. The simulator should model the behavior of the machine and its components; consequently, the designers are experts in the mechanical system. The input and output data of the simulation match exactly the sensors and actuators of the deployed system. But this exact mapping concept must be violated. Supplemental data flows must be added (see Figure 18.8):

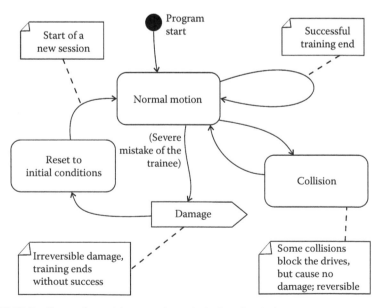

FIGURE 18.7   States that a drive runs through during simulation.

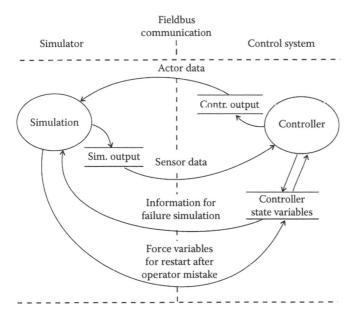

FIGURE 18.8 Data flow between simulator and controller.

1. The injection of a system fault that the trainee should master is not done by a trainer but automatically in a random manner. This makes it necessary to read the status data out of the control system to determine the instant when a component should fail.

2. If the trainee causes damage, the entire system must be restarted. It would be too time-consuming to restart the simulator and the control system again. Unfortunately, the machine components cannot fast rewind to their initial positions and states, since the control system does not accept this and would trigger an alarm condition. The solution is to overrule such error conditions by writing internal state variables of the controller.

Both tasks require knowledge of the controller internals. Consequently, it does not suffice to have experts for the physics modeling only, but also control experts who are familiar with the actual controller must be on the team throughout the project.

Furthermore, any intervention in the control system bears the danger that its behavior is changed for the subsequent training session. A mismatch with the physical system would be the consequence.

Now the basic design of the simulator is done. Before a decision for a certain software tool can be made, the complexity of the system should be investigated. Computing time limits the model number of mechanical components and the large number of possible interactions can increase the coding work excessively.

18.3.3 REDUCTION OF COMPLEXITY

The most important issue throughout the development process of a simulator is to reduce the complexity to a reasonable extent.

18.3.3.1 Hardware Expense

The simulator replaces the machine or plant and communicates with the clone of the control system. An obvious technique is to connect both systems with wires, as in the physical plant. The communication channel in Figure 18.4 would consist of wires for every input and output. Looking at the cost, this is only affordable for small systems. Consequently, the usual way to establish this communication is to use a fieldbus system, which enables the transfer of thousands of data points with one single cable.

As a further reasonable reduction step, some use a simulation inside the controller running in parallel. Since the exchange of data is done via variables, no additional hardware is required. This minimal configuration is applied for two purposes:

1. To test some critical parts of the control software
2. To replace and simulate parts of the physical plant during service operations to establish a normal production cycle

Perhaps this can be called *embedded simulation, on-board simulation,* or *distributed simulation*. It should be mentioned that for safety-related systems, this configuration is not acceptable for FAT.

18.3.3.2 Reduction of Possible Machine States

The components of a typical production machine have interactions that are to be programmed for simulation. The amount of these artifacts should be estimated on the example of the actual project. In the actual project, we call these machine components subassemblies and drives (see Figure 18.6). If each unit can interact with every other, the maximum possible number is a combination by two:

$$L_{max} = \begin{pmatrix} N \\ 2 \end{pmatrix} + \begin{pmatrix} M \\ 2 \end{pmatrix} \tag{18.1}$$

with

N Number of drive units
M Number of mechanical subassemblies
L Number of possble interactions

With the example of $N = 40$ and $M = 8$ for one part of the actual rolling mill model, the maximum number of possible interactions will be $L = 808$. It would not be affordable to program all these interactions. Fortunately, this is only the worst case. The nature of heavy machinery yields a reduction of the number of possible interactions. Manipulators usually are designed to form a linear chain, where each drive can have an interaction with another at the end of its path. For example, a piece of load is handed over to the next at the end position of the first manipulator. Consequently, the amount of interactions is reduced to

$$L = (N - 1) + (M - 1) = N + M - 2 \tag{18.2}$$

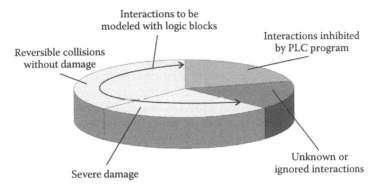

Interactions to be
modeled with logic blocks

Interactions inhibited
by PLC program

Reversible collisions
without damage

Unknown or
ignored interactions

Severe damage

FIGURE 18.9 Interactions of machine parts (collisions).

For the actual example, the number of interactions is reduced to $L = 46$, which is now realizable.

Additionally, many drives are clamping or locking devices. They have interactions that are possible only at one end of their stroke, which is a further reduction.

18.3.3.3 Reduction of States by the Control Program

The number of states that must be programmed for the simulation of the machine can be reduced further when we look at Figure 18.9.

Any control logic has programmed manifold *interlocks* that protect the plant from dangerous conditions and from the most probable operator mistakes. It is possible to exclude these situations to reduce the coding effort, since the PLC will not allow motions leading to them. This saves a lot of development time and also avoids overloading the computing capacity of the simulator. This technique will not be allowed if the purpose of the simulation is to test or commission the control software. But for operator training, this advantage can be taken.

Some collisions are reversible. They block the motion but do not lead to damage. The simulation must provide the opportunity for the operator to take back the move and to continue the training. It is remarkable that the programming of these collisions only requires influencing the motions of parts, but not changing the state of the system. No state logic is necessary to model this kind of interaction.

The number of collisions causing severe damage that must be simulated is now much lower than expected. They must be modeled with state logic that determines the course of the training session.

Furthermore, there always exist interactions with extremely improbable occurence. Provided that the system is not safety critical, we can trust the trainees not to act like monkeys on a typewriter [37] and do not model these situations.

Finally, in every automated system there are unknown states that may cause unexpected incidents even years after commission. But it is not the purpose of an OTS to find such failures.

18.3.3.4 Depth of Simulation

Dynamic systems in nature are preferably modeled with systems of ODEs, which can be solved numerically with solvers. Applying, for example, Newton's law yields a model for every mass in a mechanical system. If further we had ideal springs and viscous friction for the interaction among the masses, the resulting model would be a linear time invariant system, which is preferred for the analysis of dynamic systems [14]. This raises the hope to find a model for every thing in the mechanical system, since it consists of discrete, nearly ideal rigid bodies, but this expectation is misleading. Nonlinear springs, Coulomb friction, thermodynamical gas process, and complex kinematics lead to nonlinear equations of motion. Modern simulation environments can cover these effects, but at the expense of increased computing time. Another setback is the occurrence of time responses spread over several powers of ten, which leads to so-called stiff systems in a numerical sense. They are solvable with dedicated numerical solvers that require a higher sampling rate of data and even can cause instabilities in the result [38]. Hydraulic drives are an archetype of a stiff system [39]. Figure 18.10a shows an example from the actual project that is documented in this chapter. A mechanical load model on the right side of Figure 18.10a calculates speed and position from the force input with the help of Newton's equation of motion. Force is coming from the cylinder model, accepting incoming oil flows to its chambers. The oil pressures are fed back to the proportional valve that determines the oil flow. Tubes are represented by a first-order capacity. To prevent negative cylinder pressures, a counterbalance valve throttles the outgoing oil stream. Finally, a load-sensing pump reduces the supply pressure to avoid loss of power, when high pressure is not required. On a standard PC, this system is not computable in real time. The computation of the load position in the system of Figure 18.10a motion needs ten times more than the counterpart in reality. Since simulation for operator training necessitates the real-time property, we cannot handle even one such drive. Thinking of the modeling discipline given in Section 18.1.3, the system must be simplified by omitting every detail not necessarily required, which is difficult, cumbersome, and

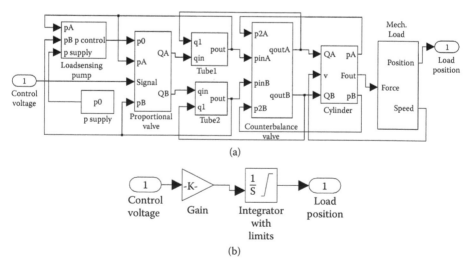

FIGURE 18.10 Models of hydraulic cylinder drives. (a) Extended functionality, including load sensing pump, proportional valve, tube dynamics, counterbalance valves, cylinder, load; (b) ultimate simplification of a drive.

time consuming yet critical work during the requirement analysis. For many drives in the actual project, a minimal version of a simple integrator with upper and lower limits as shown in Figure 18.10b was sufficient.

Doing this reduction allowed hundreds of models to run on the computer in real time together with other tasks such as communication and visualization. In conclusion, a thorough simplification during the requirement analysis is recommended to obtain a well-performing simulator.

18.3.4   SIMULATION TOOL SELECTION AND PROGRAM CODING

The requirement analysis in the previous sections was a prerequisite for definitively choosing a simulation tool. The training sequence was defined, the model components were classified, and the basic behavior of moving machine parts was identified. All steps were carried out with the aim to minimize the development expense.

Most simulation problems in automation are designed as linear processes. For example, a flight landing simulation requires the operator to follow an exact sequence. Any deviation from the sequence causes the disruption of the training. Or, a chemical process has definite sequence steps that must be fulfilled; otherwise, the product would be lost.

Quite contrary to these, the maintenance operator in the rolling mill has many degrees of freedom without being punished for an unnecessary, incomplete, or even an incorrect action. In most cases, the person would waste time, but would not destroy machine components. Such a system has a huge amount of possible states, and the preferable modeling method to enable determining which state sequences are irrelevant is the discrete state space approach. This supports a state space search (such as effectively employed in, for example, computer chess) to whittle down all possible scenarios. As a result of the thorough analysis and design described in the previous sections, it turned out that all requirements can be fulfilled with a standard tool. Specific methods of artificial intelligence are not necessary.

18.3.4.1   Simulation Tool

As the result of the requirement analysis, it was possible to choose a simulation tool that is standard for modeling plant and machinery. The product of choice was Winmod [19].

18.3.4.1.1   Discrete Time Simulation

This simulation principle is the standard for modeling of physical or chemical processes. The actual simulator for the rolling mill is running in the loop with the logic control (HIL, Figure 18.4). The time increments have the same order of magnitude as the control program, which is about 0.05 s. Usually, a real-time operating system is required for such short cycle times. Winmod runs under Microsoft Windows with components embedded in some system routines with high priority. Of course, this cannot provide hard real-time behavior [17], but it is sufficient for simulation of systems of low dynamics. Since communication with the control device must fulfil the hard real-time condition, delegated fieldbus hardware (Profibus) in a slot of the simulation computer is used.

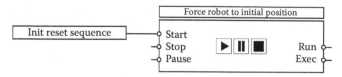

FIGURE 18.11   Block for generating signal patterns over time.

18.3.4.1.2   Rudimentary Support of Events

To describe dynamic systems that are driven by events, automation engineers use graphical tools based on the FSM formalism. Defined in the IEC 1131 standard, a language called sequential function chart allows programming of preferably linear chains of discrete steps. In addition to this, Winmod provides a tool to generate signal sequences over time (pattern generation, Figure 18.11). The sequence is defined with a simple text code and the graphical representation has an input to start the sequence. We used this feature to overrule the behavior of the machine model for initialization. This step beyond the physical reality is difficult, since the virtual machine is under the supervision of the actual controller, and error conditions must be avoided. The third possibility to handle events is a library of basic logic elements such as flip flops. All these state-oriented components run within the discrete time system and there is no separate discrete event control.

18.3.4.1.3   Graphical Programming

Since engineers prefer graphical descriptions, a representation similar to electric or electronic circuitry is provided. In UML 2.0, this corresponds to the communication diagram.

18.3.4.1.4   Object Creation by Composition

Object orientation is an essential concept for all kinds of software systems. Here it is preferred that new objects can be constructed only by combining objects, all derived from classes in a framework library. Some advanced concepts, such as multiple inheritance and object creation during runtime, are not allowed.

18.3.4.1.5   Import of Variables

The most powerful feature of a tool designated for machinery simulation is the capability to import variables, inputs, and outputs of the control system directly from its source code. This saves the work to define more than five hundred items for the actual project.

18.3.4.2   Example Details

The program code is a file of graphical drawings, consisting of blocks connected with signal lines (objects and message exchange in software terms). Here two important artifacts from Figure 18.6 are explained.

18.3.4.2.1   Integrator as Drive Design Pattern

All drives and subassemblies are modeled with first-order behavior (see Figure 18.12). This makes the basic numerical integrator to a design pattern for these artifacts (left).

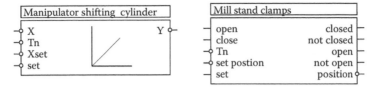

FIGURE 18.12   Simple drive modeled with integrator (left side); a more complex drive (right).

More complex drives are composed of several blocks but always have a similar representation (see Figure 18.12, right).

18.3.4.2.2   Exception Generator

During a training session, the trainee should resolve exceptional situations. For this reason, faults should be generated, for example, by blocking a cylinder or stalling a signal due to a broken sensor. The trainer may enter a number to select a predefined fault situation. The fault is triggered when the controller (and together with it, the entire plant) is in a certain state (see Figure 18.8). If the trainee is working alone on the simulator without a trainer, one fault out of a predefined list may be triggered at random. Of course, training without exeption is also possible.

18.4   EXPERIENCES WITH SIMULATION-BASED TRAINING

A training simulator was developed for the roll replacement procedure in a steel rolling mill. During a test period of several months, the effect of the training was observed. The most important criterion for the success of a training simulator is the acceptance by the personnel, which was excellent in the actual project. Periodic training sessions were carried out using the equipment since its introduction. After a period of some months, the following benefits became apparent and showed that the simulator meets expectations:

1. More training can take place than on the physical machine that is dedicated to production.
2. Training sessions can take place in a planned manner at scheduled times and not at arbitrary times during a production stop.
3. Training sessions on the simulator reduce the risk of damage to the physical machine.
4. The training program is standardized and contains all essential exercises. In contrast, on the deployed machine, seldom are all desired functions operable at one time.
5. Operators act faster and more accurate after being trained on the simulator. Resuming production after a fault is faster than before the introduction of the simulator.
6. Sometimes the cause of a failure in the complex system remains unclear. Simulation allows experimental repetitions to find the cause of such errors.

7. New components of the control software can be tested on the simulator before their implementation on the controller of the deployed machine.
8. The simulator helps in understanding the functions of the control system. It can also be applied for training of control engineers.
9. Software bugs in the actual control program were found, even severe ones.
10. Minor mistakes in the HMI could be detected, for example, wrong labels or swapped signals.

REFERENCES

1. Pfeiler, H., N. Köck, J. Schröder, and L. Maestrutti. 2003. "The New Rail Mill of Voestalpine Schienen at Donawitz." *MPT International* 26 (6): 40–4.
2. Page, R. L. 2000. "Brief History of Fligth Simulation." In *SimTecT 2000 Proceedings,* pp. 11–7. Sydney.
3. Perkins, T. 1985. "Simulation Technology in Operator Training. Full-Scope Plant-Specific Simulators Are Part of the New Reality." *IAEA Bulletin* 27 (3): 18–24.
4. Freedman, P. 2000. "New Drill Jumbo Operator Training Simulator." *IEEE Canadian Review* 34: 16–8.
5. Isermann, R., J. Schaffnit, and S. Sinsel. 1999. "Hardware-in-the:Loop Simulation for the Design and Testing of Engine-Control Systems." *Control Engineering Practice* 7 (5): 643–53.
6. Resmerita, S., P. Derler, W. Pree, and K. Butts. 2012. "The Validator Tool Suite: Filling the Gap Between Conventional Software-in-the-Loop and Hardware-in-the-Loop Simulation Environments." In *Real-Time Simulation Technologies: Principles, Methodologies, and Applications,* edited by K. Popovici, and P. J. Mosterman. Boca Raton, FL: CRC Press.
7. Shahidehpour, M., H. Yamin, and Z. Li. 2002. *Market Operations in Electric Power Systems: Forecasting, Scheduling, and Risk Management.* New York: John Wiley & Sons.
8. Haq, E., M. Rothleder, B. Moukaddem, S. Chowdhury, K. Abdul-Rahman, J. G. Frame, A. Mansingh, T. Teredesai, and N. Wang. 2009. "Use of a Grid Operator Training Simulator in Testing New Real-Time Market of California ISO." In *IEEE PES Power & Energy Society General Meeting, Calgery, Alberta,* pp. 1–8.
9. Issenberg, S. B., W. C. McGaghie, E. R. Petrusa, L. D. Gordon, and R. J. Scalese. 2005. "Features and Uses of High-Fidelity Medical Simulations That Lead to Effective Learning: A BEME Systematic Review." *Medical Teacher* 27: 10–28.
10. Hodgkinson, G. P., N. Daley, and R. Payne. 1995. "Knowledge of, and Attitudes towards, the Demographic Time Bomb." *International Journal of Manpower. An Interdisciplinary Journal on Human Resources, Management & Labour Economics* 16 (8): 59–76.
11. Lieberman, B. A. 2003. "The Art of Modeling, Part I: Constructing an Analytical Framework." Accessed July 3rd 2010. http://download.boulder.ibm.com/ibmdl/pub/software/ dw/rationaledge/aug03/fmodelingbl.pdf.
12. Seidl, E., C. Spielmann, and G. Rath. June 2010. Personal Discussion. Andritz AG.
13. Lieberman, B. A. 2007. *The Art of Software Modeling,* pp. 17. Boca Raton: Auerbach Publications.
14. Rowell, D. and D. N. Wormley. 1997. *System Dynamics: An Introduction.* Upper Saddle River, NJ: Prentice-Hall.
15. Colgren, R. 2006. *Basic MATLAB, Simulink And Stateflow.* Reston, VA: AIAA American Institute of Aeronautics & Ast.
16. White, S. A., D. Miers, and L. Fischer. 2008. *BPMN Modeling and Reference Guide— Understanding and Using BPMN.* Lighthouse Pt, FL: Future Strategies.

17. Kopetz, H. 1997. *Real-Time Systems: Design Principles for Distributed Embedded Applications.* Netherlands: Kluwer Academic Publishers.
18. Kweon, S. K., K. G. Shin, and Q. Zheng. 1999. "Statistical Real-Time Communication over Ethernet for Manufacturing Automation Systems." In *Proceedings of the 5th IEEE Real-Time Technology and Applications Symposium,* pp. 192–202. Vancouver, Canada.
19. Winmod. "Real Time Simulation Center for Automation." Accessed July 10th 2010. http://www.winmod.de.
20. Rosen, A. 2009. *E-learning 2.0: Proven Practices and Emerging Technologies to Achieve Results.* New York: Amacom.
21. Sauvé, L., L. Renaud, D. Kaufman, and J. S. Marquis. 2007. "Distinguishing between Games and Simulations: A Systematic Review." *Journal of Educational Technology & Society* 10 (3): 247–56.
22. Garris, R., R. Ahlers, and J. E. Driskell. "Games, Motivation and Learning: A Research and Practice Model." *Simulation and Gaming* 33 (4): 441–67.
23. Kickmeier-Rust, M. D. 2009. "Talking Digital Educational Games." In *Proceedings of the 1st International Open Workshop on Intelligent Personalization and Adaptation in Digital Educational Games,* pp. 55–66. Graz, Austria.
24. Pannese, L. and M. Carlesi. 2007. "Games and Learning Come Together to Maximise Effectiveness: The Challenge of Bridging the Gap." *British Journal of Educational Technology* 38 (3): 438–54.
25. Schiffler, A. 2006. "A Heuristic Taxonomy of Computer Games." Accessed June 10th 2010. http://www.ferzkopp.net/joomla/content/view/77/15/.
26. McShaffry, M. 2009. *Game Coding Complete.* Boston: Course Technology.
27. Dalmau, D. S.-C. 2003. *Core Techniques and Algorithms in Game Programming.* Indianapolis, IN: New Riders Publishing.
28. Rehmann, A. J. and M. C. Reynolds. 1995. *A Handbook of Flight Simulation Fidelity Requirements for Human Factors Research.* Technical report, CSERIAC. Springfield, Virginia: National Technical Information Service.
29. Klatt, K.-U. 2009. Trainingssimulation. *ATP* 51 (1): 66–71.
30. Mewes, J. and G. Rath. July 2010. Personal Discussion, CEO of Mewes & Partner, Berlin.
31. Bullinger, H.-J. and J. Warschat. 1996. *Concurrent Simultaneous Engineering Systems.* London: Springer.
32. Parr, E. A. 2003. *Programmable Controllers: An Engineer's Guide.* 3rd ed. Oxford: Newnes.
33. Bolton, W. 2009. *Programmable Controllers.* 5th ed. Oxford: Newnes.
34. Lewis, R. W. 1996. *Programming Industrial Control Systems Using IEC 1131-3.* The Institution of Electrical Engineers.
35. MathWorks. 2010. "Simulink® PLC Coder™." Accessed July 3rd 2010. http://www.mathworks.com/products/sl-plc-coder/
36. Pilone, D. and N. Pitman. 2005. *UML 2.0 in a Nutshell.* Cambridge, MA: OReilly & Associates.
37. Borel, É. 1913. "Statistical Mechanics and Irreversibility (Mécanique Statistique et Irréversibilité)." Journal of Physics (*Journal of Physique.*) 3 (5): 189–96.
38. Crosbie, R. 2012. "Real-Time Simulation Using Hybrid Models." In *Real-Time Simulation Technologies: Principles, Methodologies, and Applications,* edited by K. Popovici, and P. J. Mosterman. Boca Raton, FL: CRC Press.
39. Hodgson, J., R. Hyde, and S. Sharma. 2012. "Systematic Derivation of Hybrid System Models for Hydraulic Systems." In *Real-Time Simulation Technologies: Principles, Methodologies, and Applications,* edited by K. Popovici, and P. J. Mosterman. Boca Raton, FL: CRC Press.

19 Real-Time Simulation Platform for Controller Design, Test, and Redesign

Savaş Şahin, Yalçın İşler, and Cüneyt Güzeliş

CONTENTS

19.1 Introduction 462
19.2 Structure and Functions of the CDTRP 466
19.3 Taxonomy of Real-Time Simulation Modes Realized by CDTRP 467
19.4 Categorization of CDTRP Modes Based on Their Suitability for
Design, Test, and Redesign Stages 468
19.5 Implementation of the Plant Emulator Card with PIC Microcontroller 470
19.6 Experimental Setup of the Developed CDTRP 471
19.7 Verification and Validation of the CDTRP Based on Benchmark
Plants.. 475
    19.7.1 Benchmark Plants Implemented in CDTRP for Verification of
    Operating Modes 475
    19.7.2 Controller Design, Test, and Redesign by the CDTRP on a
    Physical Plant: The DC Motor Case 475
        19.7.2.1 Implementation of the Physical Plant Together with Its
        Physical Actuator and Sensor Units 476
        19.7.2.2 Identification of DC Motor to Obtain a Model to Be
        Simulated and Emulated 476
        19.7.2.3 Recreating the Disturbance and Parameter
        Perturbation Effects 478
        19.7.2.4 Controller Design–Test–Redesign Process 478
        19.7.2.5 The Simulation, Emulation, and Physical Measurement
        Results Obtained along the Entire Design Process
        Implemented by CDTRP 479
        19.7.2.6 Recreating the Parameter Perturbations in the
        S-E-R Mode 480
        19.7.2.7 Recreating Noise Disturbance in the S-E-R Mode 481
        19.7.2.8 Response to Single Short-in-Time Large-in-Amplitude
        Pulse Disturbance in the S-E-R and S-R-R Modes 482

19.7.3 Investigation of Reliable Operating Frequency of Mixed Modes
of CDTRP: Coupled Oscillators as Benchmarks 482
19.7.3.1 Analog Hardware Implementation of Synchronized
Lorenz Chaotic Systems 484
19.7.3.2 Synchronization of (Lorenz Transmitter) Emulator
with a Physical Analog (Lorenz Receiver) Plant 484
19.7.3.3 Synchronization of the (Lorenz Receiver) Emulator
with Physical Analog (Lorenz Transmitter) Hardware 486
19.7.3.4 Synchronization of (Lorenz Transmitter) Simulator
with (Lorenz Receiver) Emulator 488
19.7.3.5 Effect of Feedback on Chaotic Synchronization 488
19.7.3.6 Synchronization of (Linear Undamped Pendulum
Receiver) Simulator/Emulator with (Signal Generator
Transmitter) Simulator 489
19.7.3.7 Synchronization of (Linear Undamped Pendulum)
Receiver Emulator with (Signal Generator) Transmitter 490
19.7.3.8 Synchronization of a Physical Analog (Lorenz
Receiver) Hardware with Transmitter Emulator 493
19.8 Conclusions 493
References 496

19.1 INTRODUCTION

Control is a problem of finding a suitable (control) input driving a given system to a desired behavior. In general, there is extensive literature introducing the field of control and providing background material on solving control problems [1–3]. Having a system displaying a desired behavior is usually achieved by designing a controller that produces the necessary control in terms of the fed back actual (plant) output and a reference signal representing the desired (plant) output. The design of a controller working well for a given real plant in an actual environment requires the consideration of the behavior of the actual plant under the actual operating conditions [4–11]. In one extreme, such a controller design problem can be attempted to be solved by examining the simulated controller on a simulated plant under simulated operating conditions [2,12], while in another extreme, the approach relies on testing and tuning the controller hardware on the physical plant in the actual environment [13,14]. Testing the proposed controllers' performance on the simulated or physical plant is followed by a redesigning or parameter tuning process performed offline or online. Both approaches have their own advantages/disadvantages and also difficulties for experimentation. For most of the cases, examining the controller candidates directly on the physical plant may not be possible in the laboratory environment or may be dangerous because of possible damages that may result [15,16]. In between these extremes lies an approach that mimics the physical plant in real time and in actual environmental conditions, especially in combination with the analog/digital interface units. However, such an approach is not only complicated in software simulations but it also yields, with great probability, unreliable simulators that are highly sensitive to the unavoidable modeling errors that each simulated unit embodies [17,18].

In spite of these drawbacks, in general, simulation of control systems is preferred (1) for understanding the behavior of the plant together with its actuator and sensory devices based on their identified models obtained before an expensive and belabored implementation (i.e., for analysis) and (2) for testing the designed controllers whether the design specification are met (i.e., for synthesis). Analysis and/or synthesis platforms developed for general and also for certain specific purposes are documented in other works [16–27]. In the controller synthesis case, which is the main concern in this chapter, testing the controllers is followed by a redesign and/or tuning procedure. The developed controller design–test–redesign platform (CDTRP) consists of a simulator together with a software manager on a personal computer (PC) that is implemented with a graphical user interface (GUI), a microcontroller based emulator, and a hardware peripheral unit. CDTRP is intended to have the advantages of using simulated or emulated plants in the controller design and also of testing the candidate (simulated, emulated, or actual) controllers on the almost actual operating conditions.

Depending on the implementation of the controller, plant, and peripheral unit as simulation, emulation, or physical hardware, the developed CDTRP can be operated in 24 different real-time operation modes (Tables 19.1 through 19.3). These operation modes are also referred to as *real-time simulation modes* [19–27], meaning the samething throughout this chapter. The simulator and emulator in all the 24 modes of CDTRP are designed for performing real-time simulations; however, they can be run faster or slower for different purposes. For example, a fast running plant emulator or simulator can be used for model reference adaptive control. So, the simulation modes that can be realized in the platform are not restricted to the mentioned 24 real-time modes. Some of the real-time simulation modes correspond to the well-known "hardware-in-the-loop simulation" [19–27], where the controller is realized as hardware and

TABLE 19.1

Taxonomy of Real-Time Simulation Modes of CDTRP. The Peripheral Unit Is Implemented in the Simulator (PC)

Plant Controller	Simulator (in PC)	Emulator	Physical (Hardware)
Simulator (in PC)	S-S-S	S-E-S	S-P-S
Emulator	E-S-S	E-E-S	E-P-S
Physical (Hardware)	P-S-S	P-E-S	P-P-S

TABLE 19.2

Taxonomy of Real-Time Simulation Modes of CDRTP. The Peripheral Unit Is Implemented in the Emulator

Plant Controller	Simulator (in PC)	Emulator	Physical (Hardware)
Simulator (in PC)	S-S-E	S-E-E	S-R-E
Emulator	E-S-E	E-E-E	E-P-E
Physical (Hardware)	P-S-E	P-E-E	P-P-E

TABLE 19.3

Taxonomy of Real-Time Simulation Modes of CDRTP. The Peripheral Unit Is Implemented as Physical Hardware

Physical (Hardware)	Simulator (in PC)	Emulator	Physical (Hardware)
Simulator (in PC)	–	S-E-P	S-P-P
Emulator	–	E-E-P	E-P-P
Physical (Hardware)	–	P-E-P	P-P-P

Note: The first column modes are realized in CDTRP without using an additional analog interface extenstion for the PC.

the plant is implemented on the PC or in the emulator (Tables 19.1 through 19.3). Some of the other modes correspond to the well-known "control prototyping" and "software-in-the-loop simulation" [19, 20]. Both simulate the controller on the PC but differ from each other in the plant part. The first one is the physical and the second is implemented on the PC (Tables 19.1 through 19.3). The simulation modes of the developed CDTRP include not only the abovementioned "hardware-in-the-loop simulation," "control prototyping," and "software-in-the-loop simulation," but also, up to the knowledge of the authors, all the other simulation modes in the literature.

The CDTRP platform is developed according to a controller design–test–redesign methodology, which can be stated as follows:

- To provide a high level of flexibility of choosing and testing controller and plant models from a wide variety by the simulator unit in the early stages of the controller design.
- To create environmental conditions as close as possible to the real world by the emulator and peripheral units in the final stage of the controller design process.

So, the CDTRP can be used for the following:

- The design–test–redesign of the controllers under the framework of a chosen specific controller design method (such as an adaptive or robust method)
- Comparison of the performances of the different controller design methods, techniques, and algorithms on the simulated, emulated, or physical plants with the emphasis of the controller design in real time and the actual environment, and so supporting the selection of the best controller for a specific applications, on the other hand serving as a test-bed for researchers in examining their immature controller design method in its development phase
- Verification and validation of a plant model [28–31] based on the simulated and emulated plants with emphasis on running in real time and in the actual environment

- Controller design requiring a parameter training procedure based on the measurements and also calculations on an emulated (identified) plant model as in done artificial neural networks–based controller design methods [24,32–35]
- Low cost real-time implementation of control systems based on the benchmark plants that are of educational value but difficult or impossible to realize in an educational laboratory [36]

Similar real-time simulation platforms have been realized in the literature [6,10, 11,21,22,37,38]; however,

- They are not dedicated to being a general purpose design–test–redesign controller platform.
- They are restricted either to a specific control application (e.g., robot, specific electrical motors, pantograph, or dynamometer) or to a certain type of simulation mode such as software-in-the-loop or hardware-in-the-loop.
- None of them possesses a (physical) hardware peripheral unit that comprises all of the (actual) analog and digital actuators, sensory devices, the external disturbance, parameter perturbation signal derivers, and analog/digital controller hardware components, and so they have the ability to recreate an actual environment in a more restricted sense than the CDTRP.
- The frequency limits for the real-time simulation modes realized by these platforms have not been reported in contrast to the CDTRP.
- They do not have a GUI unit capable of monitoring and controlling the overall simulation platform such as the one that the CDTRP includes.

This chapter is organized as follows. In Section 19.2, the developed platform CDTRP is briefly described presenting its subunits (i.e., the emulator card, the software simulator on the PC, the hardware peripheral unit, and the GUI managing the overall platform). Structural properties and functions of the emulator, simulator, and peripheral unit; the facilities supplied by the GUI; and a block diagram for the emulator card that is made up of the PIC microcontroller and its input–output interfaces are given in this section. The real-time operating modes provided by the CDTRP are described in Section 19.3 by the introduced taxonomy together with a soft categorization based on their suitability to the design, test, and redesign stages in a controller design process in Section 19.4. The implementation of a plant emulator card and experimental setup of the developed CDTRP are given in Sections 19.5 and 19.6. In Section 19.7.1, the implementation of the operating modes of the CDTRP are demonstrated on several benchmark plants for, in a sense, verification of the simulation platform. The experimental results on the implementation of design–test–redesign procedure of a physical plant (i.e., a micro DC motor) are given in Section 19.7.2. In Section 19.7.3, the coupled Lorenz chaotic systems and pendulums are used as tools for investigating the frequency range for the plant and controller dynamics to be simulated and/or emulated reliably in different real-time operating modes of the CDTRP. The conclusions of the developed CDTRP are given in Section 19.8.

19.2 STRUCTURE AND FUNCTIONS OF THE CDTRP

The CDTRP comprises four main units, which are a real-time simulator running on a PC, a real-time plant emulator realized in a plant emulator card, a hardware peripheral unit for recreating a physical environment for the plant emulator, and a GUI on the PC that manages the entire platform. The structure and the interconnection of the subunits of CDTRP are depicted in Figure 19.1. The first unit is the software simulator implemented on the PC. It aims to simulate the controller, the plant, and/or the peripheral unit components depending on the operating modes of the platform (i.e., it simulates, for instance, the plant if it is active in the chosen mode of operation). The simulation program consists of MATLAB® 7.04 code and it requires the controller algorithm and plant model as MATLAB code. The simulator is designed to run essentially in real time; however, it can run in any time step faster or slower than real time, which may be preferred depending on the application.

The second unit is the plant emulator card whose core is the PIC microcontroller 18F452. The microcontroller is devoted to emulate the controller, the plant, and/or the peripheral unit components depending on the operating modes of the platform. The plant emulator card also possesses digital and analog interfaces for the communication of the plant emulator with the other units of CDTRP (i.e., the GUI and the hardware peripheral unit card). The PIC is programmed by Custom Computer Services C Program Compiler Version 4.084 software run on the PC before installing it on the plant emulator card, which allows the PIC software to be managed by the GUI and to communicate with the PC and the hardware peripheral unit card. Note that the details on the hardware realization of the emulator card are given in Section 19.5.

The third unit of the CDTRP, that is, the hardware peripheral unit card, is the most flexible part of the platform: depending on the application, it contains a (analog and/or digital) hardware controller, actuators, sensory devices, and signal drivers corresponding to the external disturbance and parameter perturbations to recreate the physical environmental conditions for the plant emulator.

The fourth unit of the CDTRP, the GUI, provides the management of the entire platform. The GUI is implemented with more than 2000 lines of MATLAB code in the MATLAB GUI designer tool. The GUI serves as a monitoring and controlling unit for the platform. The GUI manages all operating modes listed in Tables 19.1

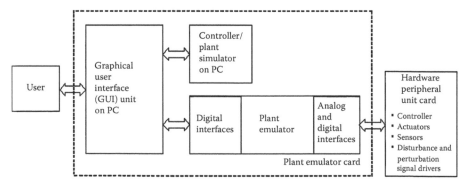

FIGURE 19.1   Structure of controller design–test–redesign platform.

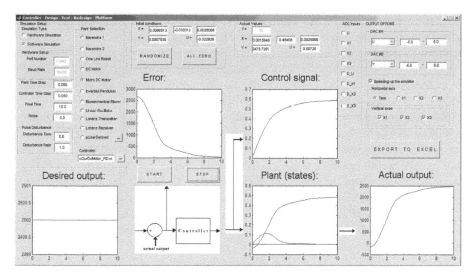

FIGURE 19.2 Front panel of the graphical user interface for the platform.

through 19.3 and the environmental conditions provided by the hardware peripheral card. A view of the front panel of the GUI is given in Figure 19.2 where the monitoring and controlling tools supplied to the users by the GUI are visible. The main features and management (i.e., monitoring and controlling) facilities of the GUI are presented in more detail in other work [39].

19.3 TAXONOMY OF REAL-TIME SIMULATION MODES REALIZED BY CDTRP

With the features of the GUI mentioned in Section 19.2, the developed simulation platform CDRTP becomes self-contained and it can implement 24 different real-time operating modes given in Tables 19.1 through 19.3.

The operating modes given in Tables 19.1 through 19.3 are obtained depending on the implementation of controller, plant, and peripheral unit as the simulator (PC), emulator, or physical (analog/digital) hardware. Other than S-S-S, E-E-E, and R-R-R modes, all modes will be called as mixed modes since at least two of the controller, plant, and peripheral unit are implemented in different units of CDTRP (i.e., the simulator (PC), the emulator, and the physical hardware). Note that the controllers are applied to the plants within the unity feedback gain configuration in all modes. However, the modes are by no means restricted to this particular (yet quite general) feedback configuration, which means the other possible configurations can be created by making some modifications to the introduced GUI. The simulator and emulator of CDRTP can be run faster than real time or run without hard time limits. However, these cases (mentioned in Isermann [19]) are beyond consideration in the presented work, since the focus is on test and design of controllers under the criterion of working well for real-time operations.

The first column modes in Table 19.3 cannot be realized in CDTRP without using an analog interface feeding the output of the physical peripheral unit to the PC where the plant is simulated.* The actuator, sensors, disturbance, and parameter perturbation effects that are created in the peripheral unit using analog/digital hardware can be embedded in the emulator by means of the emulator interface. So, the S-E-R, E-E-R, and R-E-R modes of CDTRP constitute a contribution to the real-time simulation literature, since there is no such interface possibility of the simulators running on PCs and the available emulators in the literature.

R-R-R operating mode corresponds to an actual control system and not a simulation mode. S-S-S mode where all parts of the control system are simulated on a PC is the most commonly used mode in the literature so that many software tools are available. All the real-time S-S-S, S-E-S, E-S-S, E-E-S, S-S-E, S-E-E, E-S-E, and E-E-E operating modes where the controller, plant, and peripheral unit are all implemented in the simulator (PC), or possibly in the emulator, would be considered the so-called "software-in-the-loop" simulation in the literature [19,20]. In a similar way, the modes S-R-S, E-R-S, R-R-S, S-R-E, E-R-E, S-R-R, and E-R-R where the plant under test is physical would be considered the so-called "prototyping modes" [19] and the modes R-S-S, R-E-S, R-S-E, R-E-E, and R-R-E where the controller is implemented as physical hardware while the plant is simulated would be considered the so-called "hardware-in-the-loop" simulation [19–27]. It should be noted that giving a single name for different modes is confusing, and so it may be better to use the three-letter based index provided in Tables 19.1 through 19.3 to distinguish different modes. An alternative may be the separation of the modes into subclasses such as simulated-controller, simulated-plant, simulated-peripheral, emulated-controller, emulated-plant, emulated-peripheral, physical-controller, physical-plant, or physical-peripheral classes of modes. Each of these modes then refers to a set of modes whose common property is specified by the class name, for example, the real-plant class consists of all modes where the plant is real but the controller and peripheral may be simulated, emulated, or physical. Throughout the chapter, three-letter indices will be used for the individual modes and the lastly mentioned classes will be used for the associated set of modes whenever appropriate.

In a further refinement, the set of operating modes can be extended by assigning extra letters for each element of peripheral subunit, that is, actuator, sensor, disturbance, and perturbation, depending on their implementation in simulator or emulator or as physical hardware. A hardware realization for controllers and components of the peripheral unit is application specific. In other words, depending on the application, the hardware may be a fully analog system (as in the chaotic synchronization application presented in Section 19.7.3) or a microcontroller or digital signal processor card equipped with analog interface to communicate with the emulated or physical plant.

19.4 CATEGORIZATION OF CDTRP MODES BASED ON THEIR SUITABILITY FOR DESIGN, TEST, AND REDESIGN STAGES

According to the main issue addressed, which is to design controllers that achieve high performance at the physical plant under actual environmental conditions, a controller design procedure following a three stage path is proposed.

---

* Of course, such an extension is possible but at the expense of additional cost.

(Initial) Design Stage: Use a simulation mode of CDTRP that enables implementing any kind of controller, plant, and peripheral components with a relatively much higher memory and speed provided by CPU as compared to the PIC processor and with a high flexibility of changing the models and parameters of controller, plant, and peripheral components in an efficient way. Of course, the most suitable mode for the initial design stage is S-S-S. Depending on the control application, on the chosen (initial) set of candidate control methods, and also on the experience and knowledge of the control system designer, the modes listed in the first rows and columns of Tables 19.1 through 19.3 are typically preferred in the initial design stage.

Test Stage: Use an operating mode of CDTRP that examines the control methods in terms of their performances under conditions as close as possible to the physical world and so provides a tool for eliminating the controller candidates that give poor performances on the physical plant in the actual environment. In other words, the modes suitable for the test stage are the ones that provide physical performance features, providing sufficient information to make a correct decision on their usability in deployed applications. All these modes provide a flexible implementation efficiency with the ability to create and change the models and their parameters for controller, plant, and peripheral components, though, of course, to a lesser extent as compared to the initial design stage. In this sense, the modes listed in the last rows and columns of Tables 19.1 through 19.3 would be preferred in the test stage. However, even the S-S-S mode can be used for test purposes to eliminate some controller candidates in the early phases of the design process.

Before starting the design process, one of the first issues that should be clarified is the final operating mode for the control system under consideration. To understand beforehand how the physical control system will behave under actual conditions, the final test mode should be chosen close to the actual operating conditions. Depending on the control application, hardware realization possibilities, plant dynamics, and research/educational needs, the final test mode may be chosen as the R-R-R mode or any other mode reflecting the physics sufficiently yet implementable in laboratory conditions. For instance, R-R-R appears to be the most appropriate choice for final test of analog chaotic control systems [40], since their simulation and emulation do not reflect physics because of their large bandwidth dynamics.* As another example, the S-R-R mode of CDTRP is chosen for the final test in the control application where the PC is used for hardware implementation of the controller (see [5,9,10,22,24,27] and the controller design process for the micro DC motor in Section 19.7.2).

Redesign Stage: Redesign may be defined as going back to the design stage for updating models and/or parameters of controller, plant, and/or peripheral units and then testing new controller candidates in the same or in new environments. In one extreme case, redesign requires enlarging the set of control methods to be applied. In the other extreme case, it requires tuning the controller parameters

---

* See also the coupled Lorenz systems example in Section 19.7.3 studied for another purpose.

only. In the former, redesign may end after many loops of design–test–redesign. In this setting, the operating modes that can be used in the redesign stage would be the ones suitable to the initial design and test stages. However, it can be stated that the operating modes listed in the last rows and columns of Tables 19.1 through 19.3 would more likely be preferred in the redesign stage since a small set of models and parameters remain to be tested. So, less implementation flexibility is necessary and less distance to the physical plant after early phases of the design process.

The above design procedure defines a soft category specifying which operating modes are suitable to the initial design, test, and redesign stages. In fact, the suitability of an operating mode to a considered stage of a controller design process is highly dependent on the control application, that is, on the plant, actuator, sensors, and environmental conditions. So, it appears to be impossible to provide a crisp category of modes based on their suitability to the design, test, and redesign stages. For a specific control application, it may be possible to determine the best suitable operating mode to the design, test, and redesign stages by successive implementation of some operating modes. Herein, the iteration process conducts assessment and evaluation of the obtained controller's performance for each implemented mode to observe which mode yields the controller having the best deployed performance.

19.5 IMPLEMENTATION OF THE PLANT EMULATOR CARD WITH PIC MICROCONTROLLER

Microcontrollers that are single-chip computers with limited computer features [41] are widely used for control applications, usually to implement controllers together with digital and analog input interfaces. The plant emulator card of CDTRP has been realized with a PIC18F452 microcontroller. The main reasons for this choice are as follows: (1) It is a low cost and easily programmed device, and so it can be reproduced easily by instructors and students for educational purposes, by engineers for industrial applications, and also by researchers for testing their controllers on emulated plants. (2) Its capability suffices to implement many benchmark plants and even synchronized chaotic systems (Section 19.7.3) and also analog and digital interfaces necessary to communicate with peripheral units for recreating an actual environment in the developed CDTRP.

The schematic diagram of the implemented hardware of the plant emulator card is provided in Figure 19.3. The PIC18F452 microcontroller-based plant emulator card has (1) 32 KB of internal flash Program Memory, (2) 1536-byte RAM area, (3) 256-byte internal EEPROM, (4) eight channel analog inputs via 10-bit A/D converter, (5) four channel digital input/outputs, and (6) two analog output ports via D/A converter. An LCD character display 4 four lines and 20 columns is also connected to the microcontroller to indicate the outputs and the control inputs of the emulated plant with respect to current time. The PIC18F452 microcontroller is programmed with 5672 lines of C code, which use 95% of the ROM and 27% of the RAM of the microcontroller.

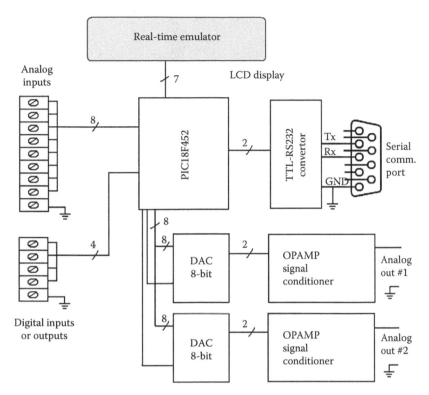

FIGURE 19.3   Schematic diagram of the hardware of the plant emulator card.

An image of the hardware realization of the plant emulator card is provided in Figure 19.4, where the physical locations of the blocks depicted in the schematic diagram of the plant emulator card are annotated with descriptive tags.

19.6   EXPERIMENTAL SETUP OF THE DEVELOPED CDTRP

The experimental setup of the developed CDTRP platform is provided in Figure 19.5. It has been used for the design–test–redesigning controllers of the benchmark plants in Table 19.4 and other plants by means of operating modes given in Tables 19.1 through 19.3. The CDTRP platform setup is composed of a plant emulator card, a PC, a signal generator, power supplies, and an oscilloscope. The specifications of the plant emulator card have already been provided in Section 19.2. The PC is chosen as having a Centrino processor and a 1 GB memory, which constitute a minimum configuration necessary for running Microsoft Windows XP and MATLAB 7.04 software. The signal generator as part of the hardware peripheral unit card is included to provide external analog signal noise (D_U). Power sources constituting another part of the hardware peripheral unit card are used while DC variacs (i.e., variable power supply) are changed manually to generate the perturbations of the plant variables (D_X1, D_X2, and D_X3).

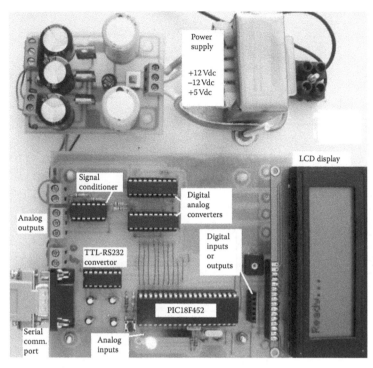

FIGURE 19.4 The hardware of the plant emulator card.

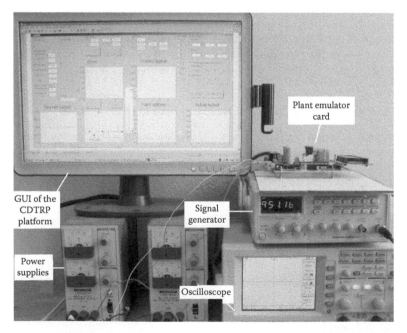

FIGURE 19.5 The experimental setup for the entire controller design–test–redesign platform.

TABLE 19.4
Benchmark Plants of Control Design–Test–Redesign Platform

Benchmark Plant (BP)	State Model	Descriptions and Associated References	Operating Modes
BP1	$x_1(k+1) = \sin(x_1(k)) + u(k)(5 + \cos(x_1(k)u(k)))$ $y(k) = x_1(k)$	A first-order plant: $u(k)$ control input, $y(k)$ output [34]	S-S-S, S-E-S, S-E-E, S-E-R
BP2	$x_1(k+1) = 0.1x_1(k) + 2\dfrac{u(k)+x_2(k)}{1+(u(k)+x_2(k))^2}$ $x_2(k+1) = 0.1x_2(k) + u(k)\left(2 + \dfrac{u^2(k)}{1+x_1^2(k)+x_2^2(k)}\right)$ $y(k) = x_1(k) + x_2(k)$	A second-order plant: control input $u(k)$, output $y(k)$, states of the plant $x_1(k)$ and $x_2(k)$ [34,42]	S-S-S, S-E-S, S-E-E, S-E-R
BP3	$\ddot{q}_1 = -\dfrac{MgL}{I}\sin q_1 - \dfrac{k}{I}(q_1 - q_2)$ $\ddot{q}_2 = \dfrac{k}{J}(q_1 - q_2) + \dfrac{1}{J}u$ $y = q_1$	One-link robot (flexible joint mechanism): control torque input $u(k)$, the output $y(k)$, and the angular positions of the plant $q_1(k)$ and $q_2(k)$. Parameters: moments of inertia $I = 0.031$ kg m² and $J = 0.004$ kg m², mass-gravity–distance $MgL = 0.8$ N m, spring constant $k = 31$ N/rad [33,43–47]	S-S-S, S-E-S, S-E-E, S-E-R
BP4	$\dot{x}_1 = -\dfrac{R}{L}x_1 - \dfrac{k_m}{L}x_2 + \dfrac{1}{L}u$ $\dot{x}_2 = \dfrac{k_m}{J}x_1 - \dfrac{b_m}{J}x_2$ $y = x_2$	A DC motor: $u(k)$ control input, $y(k)$ output, $x_1(k)$ and $x_2(k)$ states of the plant. Parameters: $R = 5\,\Omega$, $L = 0.5$ H, inertia $J = 0.1$ kg m², damping constant $b_m = 0.2$ N m/rad/s, and constant $k_m = 1$ V s [48]	S-S-S, S-E-S, S-E-E, S-E-R
BP5	$\dot{x}_1 = -2x_1 + 8448u$ $y = x_1$	Realized physical micro DC motor: control input $u(k)$, output $y(k)$, states of the plant $x_1(k)$	S-S-S, S-E-S, S-E-E, S-E-R, S-R-S, S-R-R

(continued)

TABLE 19.4

Benchmark Plants of Control Design–Test–Redesign Platform (*Continued*)

Benchmark Plant (BP)	State Model	Descriptions and Associated References	Operating Modes
BP6	$\dot{x}_1 = x_2$ $\dot{x}_2 = \dfrac{M+m}{Ml}x_1 - \dfrac{1}{Ml}u$ $y = x_1$	An inverted pendulum: control input $u(k)$, output $y(k)$, states of the plant $x_1(k)$ and $x_2(k)$. Parameters: $M = 2$ kg, mass of the ball $m = 0.1$ kg, and $l = 0.5$ m [14]	S-S-S, S-E-S, S-E-E, S-E-R
BP7	$\dot{x}_1 = x_2$ $\dot{x}_2 = -\dfrac{B}{I}x_1 - \dfrac{r}{I}u$ $y = x_1$	A biomechanical elbow: control input $u(k)$ is difference between flexion muscle F_f and extensor muscle F_e forces, output $y(k)$ states of the plant $x_1(k)$ and $x_2(k)$. Parameters: the inertia of the link $I = 0.25$ kg m^2, the moment arms $r = 0.04$ m, and damping constant $B = 0.2$ N m s/rad [49]	S-S-S, S-E-S, S-E-E, S-E-R
BP8	$\dot{x}_{r1} = -wx_{r2}$ $\dot{x}_{r2} = wx_{r1}$	An oscillator system (undamped linear pendulum): states of the plant $x_{r1}(k)$ and $x_{r2}(k)$. If the oscillator system's state $x_{r1}(k)$ is chosen as $x_{r1}(k)$, which could be one of the states of another oscillator, this system is used for synchronization of coupled oscillators. Parameter: $w > 0$	S-S-S, S-E-S, S-E-E, S-E-R, R-E-S, R-E-E, R-E-R
BP9	$\dot{x}_{r1} = \sigma(x_{r2} - x_{r1})$ $\dot{x}_{r2} = rx_{r1} - x_{r2} - 20x_{r1}x_{r3}$ $\dot{x}_{r3} = 5x_{r1}x_{r2} - bx_{r3}$	Realized Lorenz chaotic transmitter: states of the plant $x_1(k)$, $x_2(k)$, and $x_3(k)$. The equations are chosen as in Cuomo and Oppenheim [50] but with different parameters $\sigma = 2.12$, $r = 120.5$, and $b = 10.6$ to reduce into the main harmonics around 1.5 Hz	S-S-S, S-E-S, S-E-E, S-E-R, E-R-S, E-R-E, E-R-R
BP10	$\dot{x}_{r1} = \sigma(x_{r2} - x_{r1})$ $\dot{x}_{r2} = rx_{r1} - x_{r2} - 20x_{r1}x_{r3}$ $\dot{x}_{r3} = 5x_{r1}x_{r2} - bx_{r3}$	Realized Lorenz chaotic receiver: states of the plant $x_1(k)$, $x_2(k)$, and $x_3(k)$. Parameters: $\sigma = 2.12$, $r = 120.5$, and $b = 10.6$.	S-S-S, S-E-S, S-E-E, S-E-R, R-E-S, R-E-E, R-E-R

19.7 VERIFICATION AND VALIDATION OF THE CDTRP BASED ON BENCHMARK PLANTS

This subsection presents experimentation results realized for verification and validation of the operating modes of the CDTRP on three physical plants and the results obtained by the investigations on reliable operating frequency of mixed modes of the CDTRP are presented in this subsection.

19.7.1 BENCHMARK PLANTS IMPLEMENTED IN CDTRP FOR VERIFICATION OF OPERATING MODES

In this section, the benchmark plants that are implemented as simulation, emulation, or physical hardware for verification of the operating modes of the CDTRP are listed. A set of benchmark plants that can be chosen by the users via the plant selection button in the front panel of the GUI are included in the menu of the CDTRP to help the users expediently analyze their control methods on these selected benchmark plants. This menu is open to be extended by the users to cover a larger class of benchmarks or any other plants of interest. The plants must be defined by at most three-dimensional state equations.*

The benchmark plants available in the menu are listed in Table 19.4 and these are also used for verification of the operating modes of the developed CDTRP. The benchmark plants in Table 19.4 are chosen from benchmarks extensively used in the literature (which implies that they are difficult to identify and/or difficult to control) to compare the system identification methods and/or to compare the controller design methods or they are of practical importance in some other sense.

19.7.2 CONTROLLER DESIGN, TEST, AND REDESIGN BY THE CDTRP ON A PHYSICAL PLANT: THE DC MOTOR CASE

This section is devoted to demonstrating how to implement the controller design–test–redesign procedure introduced in the previous section to reach a controller working well for a physical plant operated under the actual environmental conditions. A micro DC motor is chosen as the plant and the final operating mode is determined as S-R-R, where the PC is used as the controller, the physical plant is the micro DC motor, the sensor, and actuator, that is, the encoder and driver are realized in a PIC microcontroller driver card (see Figure 19.6). In the initial design stage of the controller design–test–redesign procedure, the S-S-S mode was found to be the most suitable mode for one because of the simulation efficiency of the simple yet realistic DC motor models but also because of its implementation flexibility as observed in a set of experiments conducted in the S-S-S, S-E-S, and S-E-E modes.

In the test stage, the S-E-E mode was first implemented for testing the considered controller design methods on the emulated plant under emulated environmental conditions. Then, the controller candidates were analyzed in the S-E-R mode. It was found that the S-E-R mode is the most suitable mode for testing the considered control

---

* This restriction is because of the capacity limit of the chosen emulator hardware (i.e., PIC18F452).

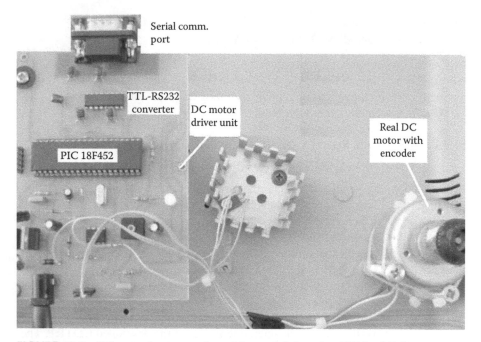

FIGURE 19.6   DC motor hardware whose driver unit is based on PIC for S-R-R mode.

methods, since it enables analysis of the controller candidates under quite realistic conditions, while still providing flexibility in simulating different controller candidates efficiently. In the redesign stage, the S-R-S and S-R-E modes were implemented at the beginning and then the S-R-R mode was implemented as the final mode where the parameters of the controller that were found to be the best in the test stage were tuned.

The details of the above explained steps of the implementation of the controller design–test–redesign procedure for the DC motor example are presented next.

19.7.2.1   Implementation of the Physical Plant Together with Its Physical Actuator and Sensor Units

A micro DC motor was considered. The DC motor driver card (see Figure 19.6) was realized with a PIC18F452 microcontroller driver unit that has a serial interface to communicate with the PC. The speed of the motor was considered as the output and was measured by an encoder to provide feedback to the system. The microcontroller-based driver unit drives the DC motor speed revolutions per minute (rpm) via pulse width modulation (PWM) changing between 0% and 100%. That is to say, in the implemented S-R-S, S-R-E, and S-R-R operating modes, the control signal from CDTRP is converted into a PWM signal to control physical DC motor speed rpm.

19.7.2.2   Identification of DC Motor to Obtain a Model to Be Simulated and Emulated

To simulate and also emulate the physical DC motor in the simulated-plant and emulated-plant modes that are to be used for the design and test stages, a system identification procedure was applied to the physical DC motor and then the simplest yet realistic model for the considered micro DC motor was obtained.*

A step function, which is obtained by changing PWM sharply, was fed to the physical DC motor plant. The input–output data pairs of the plant necessary for the identification were measured via the hyper-terminal of the PC. After the process of data gathering, the transfer function of the plant was found under the assumption of a single-input single-output linear dynamic system for the plant. It is known [44,51] that the physical DC motor plant can be well modeled as a first-order delayed dynamic system according to the response of the plant due to the step input. The step response of the first-order system defined with three parameters is provided in the Laplace and time domain, respectively, as follows:

$$H(s) = \frac{K}{Ts+1} e^{-Ls} \tag{19.1}$$

$$h(t) = K(1 - e^{-(t-L)/T}) \cong K(1 - e^{-t/T}) \tag{19.2}$$

where T is the time constant, L is the dead-time, and K is the gain [52]. If the dead-time L is sufficiently small compared to the time constant T of the plant, the step response of the system can be approximated as shown in Equation 19.2. Thus, a first-order micro DC motor model (BP5 in Table 19.4) was obtained with the measured maximum motor speed 4224 rpm, the time constant 0.5 s, and the dead-time 0.011 s. This model was used for the S-S-S, S-E-S, S-E-E, S-E-E, and S-E-R operating modes of the CDRTP. Although the first-order system is sufficient to model the physical micro DC motor for testing the candidate controllers, a second-order model is also identified and implemented for validating the emulator of the platform by analyzing the emulation performance of the platform with a more realistic model. The considered second-order delay dynamic system model is defined in the Laplace domain as follows:

$$H(s) = \frac{Kw_n^2}{s^2 + 2\zeta w_n s + w_n^2} e^{-Ls} \tag{19.3}$$

$$P_o = e^{-\frac{\zeta\pi}{\sqrt{1-\zeta^2}}} \tag{19.4}$$

$$T_s = 3\tau = \frac{3}{\zeta w_n} \tag{19.5}$$

---

* It should be noted that although more complicated models better suited to the real data measured from the DC motor can be derived, it was preferred to work with this simple model for the plant. Not only does the simple model allow taking advantage of efficient simulation and emulation of the DC motor, but it also allows more focus on testing the controllers' performances on the considered DC motor and on its simulated/emulated model in a comparative way, rather than on simulating/emulating the DC motor more realistically by more complicated models. In fact, it was observed that choosing a simple first-order dynamic model for the DC motor that was identified by using a step response method provides sufficiently close responses to the ones measured from the physical DC motor and also enables seeing the differences among the performances of the different controllers implemented.

where K is the gain, L is the dead-time, ζ is the damping ratio, and w_n is the natural frequency. The damping ratio and natural frequency were determined from the P_0 overshoot percentage in Equation 19.4 and T_s settling time in Equation 19.5 [52]. The second-order micro DC motor model was obtained based on the measured P_0 of 10%, T_s of 4.46, and the dead-time of 0.011 s.

19.7.2.3 Recreating the Disturbance and Parameter Perturbation Effects

The disturbance and parameter perturbation effects that can be implemented in the emulator by means of the signal generator and power supply components of the hardware peripheral unit were produced in the implemented S-E-R and S-R-R modes to recreate an actual environment. Since there is no such interface possibility of the simulators running on the PCs and emulators in the simulation/emulation platforms known in the literature, this is a unique feature of the developed CDTRP platform.

19.7.2.4 Controller Design–Test–Redesign Process

The DC motor speed tracking problem was chosen as a case study. The following four different types of controllers were designed using mainly the S-S-S, S-E-S, and S-E-E modes, then tested using the S-E-E and S-E-R modes on the identified model BP5, and finally using the S-R-S, S-R-E, and S-R-R modes on the realized micro DC motor: (1) a proportional-integral-derivative (PID) controller designed by the Ziegler–Nichols (ZN) method [13,14,52], (2) a PID controller designed by the Chien, Hrones, and Reswick (CHR) method [53], (3) a robust controller designed by the partitioned robust control (PRC) method [54], and (4) a direct adaptive controller designed by the Model Reference Adaptive Control (MRAC) method [54].

A PID controller that defines the control signal u in terms of the error is given as follows:

$$u = K_p e + K_i \int e \, dt + K_d \dot{e} \tag{19.6}$$

where e is the error between the desired and actual output of the plant, $\dot{e}$ is the derivative with respect to time of the error, and K_p, K_i, and K_d are the proportional, integral, and derivative gain parameters, respectively. In the first method, the parameters of the PID controller were calculated based on the BP5 plant model as $K_p = 54.54$, $K_i = 2479.3$, and $K_d = 0.3$ by using the ZN step response method. In the second method (CHR), which was preferred for yielding minimum overshooting [12], the PID controller parameters were calculated* as $K_p = 43.18$, $K_i = 1635.6$, and $K_d = 0.1995$. The PRC is designed as having two separate parts: (1) proportional-derivative (PD) control u_e and (2) auxiliary control u_y [54,55]. In this method, given a plant model $\dot{x} = -ax + bu$ (e.g., BP5 model), the control signal was calculated as follows:

$$u = u_e + u_y = K_v \dot{e} + K_p e - ax - \Delta ax \tag{19.7}$$

where K_v and K_p gains related to the control input u_e were chosen greater than zero, and for the auxiliary controller $u_y \Delta a = 1.2a$ was chosen for removing the parameter

---

* The considered PID controller is actually redesigned at this step within the terminology introduced in this chapter.

perturbations and plant uncertainties. In the fourth method applied, the MRAC controller is composed of a first-order reference model $\dot{x}_m$, two adaptive controller parameters $\dot{u}_y^k$, $\dot{u}_r^k$, and a control signal u as follows:

$$
\begin{aligned}
\dot{x}_m &= -a_m x_m + b_m r \\
\dot{u}_y &= -\lambda (x_m - x) y a_m (e^{-a_m}) \\
\dot{u}_r &= -\lambda (x_m - x) r a_m (e^{-a_m}) \\
u &= u_r + u_y
\end{aligned}
\tag{19.8}
$$

where y stands for the actual output, r stands for the reference signal, the designed parameter was chosen as $\lambda = 0.5$, and the values of the reference model parameters a_m and b_m are two times the values of the BP5 plant parameters. The equations in Equation 19.8 generate the control signal.

19.7.2.5 The Simulation, Emulation, and Physical Measurement Results Obtained along the Entire Design Process Implemented by CDTRP

The step responses of the physical DC motor and the BP5 implemented in the emulator (both controlled by the same controllers designed with the above-mentioned methods) for the desired output of 2500 rpm are given in Figure 19.7a through d,

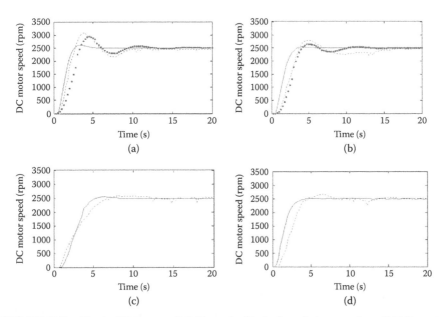

FIGURE 19.7 Physical DC motor S-R-R results (dashed), real-time emulator S-E-S results for first-order model (solid), and real-time emulator S-E-S results for second-order model (dotted) with (a) a proportional-integral-derivative (PID) controller whose parameters are designed by the Ziegler–Nichols method, (b) a PID controller whose parameters are designed by the Chien, Hrones, and Reswick method, (c) a partitioned robust controller, and (d) a Model Reference Adaptive controller.

respectively. It can be seen from the responses depicted in Figure 19.7 that one may prefer the first two controllers, that is, the ones designed by the ZN and CHR methods, since the step responses of the emulator have relatively short rise times and are close to the reference signal, that is, the step function. However, the responses of the physical DC motor controlled by these two controllers for the same step input are not close to the emulated first-order model responses in the sense that the physical DC motor demonstrates second-order dynamic behavior rather than first-order.* On the contrary, the responses to the step input of the physical DC motor and the emulated first-order model are very close to each other for the MRAC case.

The following can then be concluded: (1) it can be expected that the real plant will behave similar to the emulator for the PRC and MRAC even if the plant model poorly reflects the behavior of the physical plant and (2) it can be expected that the physical plant will behave similar to the emulator for the PID controllers designed by the ZN and CHR methods only when the plant model is realistic so as to be capable of reflecting the behavior of the real plant well. So, the analysis results obtained from the developed CDTRP in the emulated-plant and also the simulated-plant modes are reliable on the realistic plant models for any kind of controller design methods and further reliable even on poor plant models for the PRC and MRAC design methods.

19.7.2.6 Recreating the Parameter Perturbations in the S-E-R Mode

Parameter perturbations were created as a multiplicative effect for the model parameters. The perturbation signals were provided by DC power source in the 0–5 V (DC) range and were fed to the hardware peripheral unit card via the D_X1 port that was activated by the ADC options in the GUI, where 1 V for D_X1 corresponds to the nominal plant parameters case. The observed responses of the emulated BP5 controlled by the PRC for four different parameter values are given in Figure 19.8a. The responses observed for the MRAC case are given in Figure 19.8b. In addition, the responses obtained for the PRC and MRAC are compared to each other in terms of their performances for two parameter perturbation values in Figure 19.9. No results are depicted for the PID controller case, since it behaves very poorly in the face of parameter variations. It can be said that the responses under parameter variations are close to each other for the PRC and MRAC. As observed in the above analysis part, (1) choosing more realistic models yields better emulation results to be obtained by the CDTRP platform and (2) for the MRAC and to a lesser extent for the PRC cases, the responses of the emulator are very close to the physical plant. In light of these facts, it may be concluded that the performances of the controllers on the physical plants under parameter perturbations can be examined using S-E-R mode in the manner that the CDTRP implements.

19.7.2.7 Recreating Noise Disturbance in the S-E-R Mode

Noise disturbances were created as additive to the control signal (i.e., input of the plant model). The noise signal was provided by the signal generator whose range is given in Table 19.5. It was fed to the hardware peripheral unit card via D_U port

---

* As seen in Figures 19.7a and b, the responses of the emulated second-order system model are much closer to the real DC motor responses.

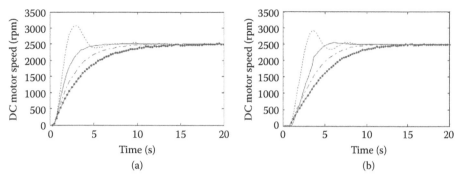

FIGURE 19.8 Responses of BP5 to the step input for parameter variation coefficients of 0.5 (dashed, --), 1 (solid, -), 1.5 (dotted-solid, -·), and 2 (dotted on solid, ---). Note that the coefficient values are the factors multiplying the nominal parameter value to obtain the perturbed parameter. The responses in (a) and (b) are for the partitioned robust controller and the Model Reference Adaptive controller, respectively.

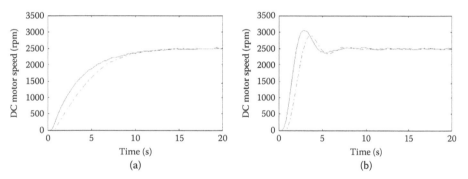

FIGURE 19.9 Responses of BP5 to the step input for the parameter variation coefficients of 2 in (a) and 0.5 in (b). The responses (solid) and (dotted-solid, -·) were obtained for the partitioned robust controller and the Model Reference Adaptive controller, respectively.

TABLE 19.5
Noise Disturbances Range Scaling

Volt (AC) from Signal Generator	Volt (DC) for ADC
+1 V	5 V
0 V	2.5 V
−1 V	0 V

that is activated by the ADC options in the GUI. The noise signal amplitude can be scaled via the front panel of the GUI for CDTRP applications. For the noise signal $du = 0.1 \sin(2\pi f)$ with $f = 1 \text{Hz}$, which was created by the noise scaling factor of 0.1 set by using the front panel of the GUI, the corresponding responses obtained for PID, PRC, and MRAC are given in Figure 19.10. As expected, the CDTRP confirms

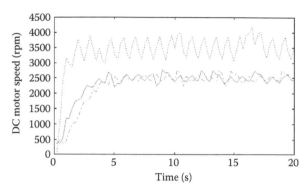

FIGURE 19.10 Responses of the emulator controlled by proportional-integral-derivative controller (dashed, --), partitioned robust controller (solid, -), and Model Reference Adaptive controller (dotted-solid, -·) under the noise du = 0.1 sin(2π*l*t).

that the PID has quite poor performance under the applied additive noise, whereas the PRC and MRAC perform well. It should be noted that the MRAC shows the best steady state performance.

19.7.2.8 Response to Single Short-in-Time Large-in-Amplitude Pulse Disturbance in the S-E-R and S-R-R Modes

Recreating the effects of relatively small amplitude noise disturbances in the emulation have been presented above. Now, another kind of disturbance effect, that is, a single short-in-time and large-in-amplitude pulse disturbance effect was created and applied to both the emulated BP5 model and the physical DC motor. This enables a better understanding of the validity of the developed platform in mimicking the behavior of the physical plants under actual disturbances. The disturbance was created as a multiplicative effect to the output of the model. The pulse time and amplitude were chosen in the front panel of the GUI. In the tests, the pulses were created after the transient regime (e.g., at 10 s as shown in Figure 19.11) to mimic the disturbances appearing in the steady state working conditions for the plant. The measured responses of the physical DC motor and the emulator controlled by the PID, robust, and MRAC controllers are given in Figure 19.11.

19.7.3 INVESTIGATION OF RELIABLE OPERATING FREQUENCY OF MIXED MODES OF CDTRP: COUPLED OSCILLATORS AS BENCHMARKS

All mixed operating modes of CDTRP require at least two of the controller, plant, and peripheral components to be implemented in different units of CDTRP, that is, in two of the simulator (PC), the emulator, and the hardware peripheral unit. So, the mixed modes necessitate real-time compatibility of the PC, emulator, and hardware peripheral unit. In other words, these units have to communicate in a (real-time) synchronized fashion. To examine this ability of the CDTRP, Lorenz systems–based chaotic synchronization is chosen as the benchmark application. As will be seen below, chaotic synchronization is a good example to understand the operating (frequency) limits of the CDTRP.

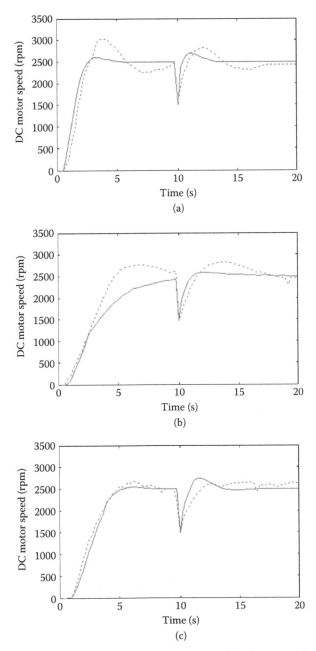

FIGURE 19.11 Responses of the physical DC motor and the emulator controlled by the proportional-integral-derivative (whose parameters are calculated with the Ziegler-Nichols method), robust, and Model Reference Adaptive controllers are given, respectively, in (a), (b), and (c). The pulse amplitude is chosen as $k = 0.6$ and the time when the pulse is applied as 10 s. Note that (dashed, --) is for the physical plant and (solid, -) is for the emulator of CDTRP.

The following were observed* in a set of experiments conducted:

1. The emulator and the simulator cannot be synchronized to each other and to external hardware because of the high frequency dynamics intrinsic to the chaotic (Lorenz) system.
2. As a consequence of the maximum achievable sampling frequency that can be realized in the MATLAB environment used for the PC simulator and in the emulator implemented by using the PIC microcontroller, the simulator and the emulator of the developed CDTRP can be synchronized to each other when implementing the dynamics up to 25 Hz while the simulator and the emulator can implement the dynamics up to 300 Hz and 25 Hz, respectively, if they are operated as uncoupled.
3. The frequency range of the simulated/emulated system dynamics for which mutual synchronization among the emulator, the simulator, and the external hardware are achievable can be enlarged by using a suitable feedback control.

In the sequel, the experimental results obtained for a set of different kinds of implementations of synchronized coupled Lorenz systems are given first and then the coupled linear undamped pendulums are examined to find the limit for the frequency that enables synchronous operation for the implemented dynamics.

19.7.3.1   Analog Hardware Implementation of Synchronized Lorenz Chaotic Systems

Before examining the CDTRP in the E-R-R and R-E-R operating modes for the synchronized system of coupled Lorenz systems, the systems were first implemented as analog hardware (i.e., in the R-R-R mode). The master–slave configuration proposed in Cuomo et al. [56] was used for the implementation. The transmitter and receiver circuits of the Lorenz chaotic systems, whose scaled state equations are given as BP9 and BP10, respectively, in Table 19.4, were realized with the circuit configuration in Figure 19.12 using the analog multiplier AD633, opamp LF353, and passive circuit elements (R1 = R2 = R6 = R7 = 100 kΩ, R3 = R5 = R8 = R10 = 10 kΩ, R4 = R9 = 1 MΩ, RV1 = RV3 = 100 kΩ, RV2 = RV4 = 220 kΩ, and C1 = ... = C6 = 100 nF).† The synchronization result for the analog hardware realization of the master–slave synchronization of the coupled systems is shown in Figure 19.13.

19.7.3.2   Synchronization of (Lorenz Ttransmitter) Emulator with a Physical Analog (Lorenz Receiver) Plant

As a second implementation of master–slave synchronization of coupled Lorenz systems, the CDTRP was operated in the E-R-R mode such that the transmitter BP9

---

* It should be noted that the last (interesting) observation can be interpreted as follows: The interfaces among the simulator, emulator, and external hardware have delays due to the interrupt routines, which can be modeled by (complex frequency) poles determining an upper cutoff frequency. And, where the synchronization is achieved, these poles that limit the frequency range can be shifted to a higher frequency point by using feedback control.

† Note that the circuit configuration was chosen as given in Paul [57] but with the BP9-BP10 coefficients as σ = 10, r = 56.6, and b = 5.02. For the coefficients, the Lorenz systems then produce chaotic oscillations with the main harmonics around 8.5 Hz.

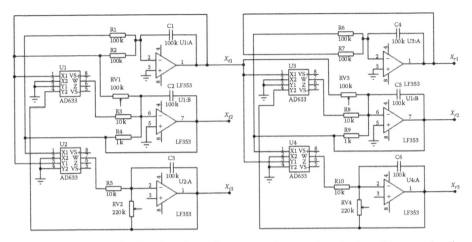

FIGURE 19.12 Realized analog circuit for master–slave synchronization of Lorenz chaotic systems.

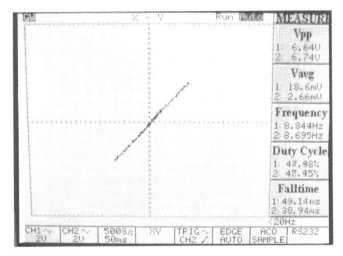

FIGURE 19.13 A snapshot in the *X-Y* mode of the oscilloscope where the signals in the *X* and *Y* channels are the first state variables of the master and slave Lorenz circuits, respectively.

was implemented in the emulator of the CDTRP and the receiver implemented as analog hardware shown in Figure 19.12 was used as the plant. The X1 state was observed from the hardware peripheral unit card via the DAC port that was activated by the DAC options in the GUI. The implementation of the system in E-R-R mode is given in Figure 19.14. As seen in Figure 19.15, when operating in the E-R-R mode, the emulator of the CDTRP can roughly mimic the chaotic behavior of the Lorenz system. In this E-R-R mode, the master Lorenz system was realized in the emulator and the slave Lorenz system was realized in the analog hardware. During the experiments, such master–slave coupled Lorenz systems were never observed as synchronized.

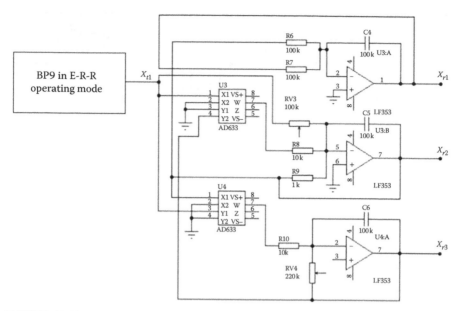

FIGURE 19.14 Implementation of the E-R-R mode of the controller design–test–redesign platform for the synchronized Lorenz systems.

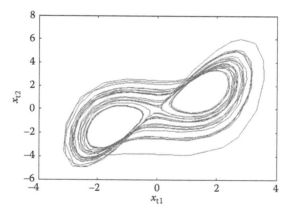

FIGURE 19.15 Phase portrait for the x_{t1} and x_{t2} states of the emulated BP9 Lorenz chaotic system as observed via the graphical user interface when the controller design–test–redesign platform is operated in the E-R-R mode.

19.7.3.3 Synchronization of the (Lorenz Receiver) Emulator with Physical Analog (Lorenz Transmitter) Hardware

As a third implementation of master–slave synchronization of coupled Lorenz systems, the CDTRP was operated in the R-E-R mode such that the receiver BP10 was implemented in the emulator of the CDTRP and the transmitter implemented as analog hardware shown in Figure 19.12 was used as the master. The X1 state was applied to the emulator of the CDTRP via the hardware peripheral unit card by the

X1 port that was activated by the ADC options in the GUI. The implementation of the R-E-R mode is given in Figure 19.16.

As seen in Figure 19.17, when operating in the R-E-R mode, the emulator of the CDTRP fails to mimic the chaotic behavior of the Lorenz system. In this R-E-R

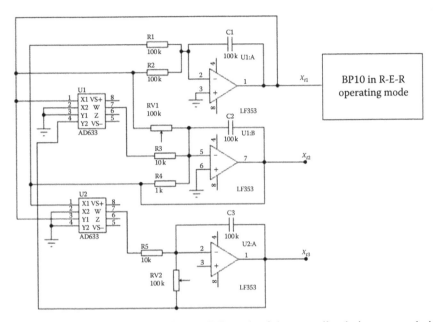

FIGURE 19.16 Implementation of the R-E-R mode of the controller design–test–redesign platform for the synchronized Lorenz systems.

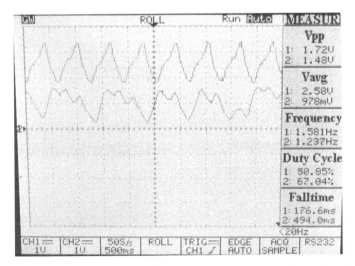

FIGURE 19.17 First state variables of the analog transmitter and the emulated receiver in the R-E-R mode for synchronized Lorenz systems. The receiver state is the top trace and the transmitter state is the bottom trace.

mode of operation, the master Lorenz system was implemented in the analog hardware card and the slave Lorenz system was implemented in the emulator. During the experiments conducted for this realization of the master–slave Lorenz system, synchronization was never observed.

19.7.3.4 Synchronization of (Lorenz Transmitter) Simulator with (Lorenz Receiver) Eemulator

As a fourth implementation of master–slave synchronization of coupled Lorenz systems, the CDTRP was operated in the S-E-E mode such that the receiver BP10 was implemented in the emulator of the CDTRP and the transmitter was implemented in the simulator of the CDTRP. As seen in Figure 19.18, although both the simulator and emulator can roughly mimic the chaotic behavior of the Lorenz system, they fail to be synchronized to each other in the master–slave configuration.

19.7.3.5 Effect of Feedback on Chaotic Synchronization

The above implemented master–slave configuration for chaotic synchronization is, indeed, an open loop control system where the receiver is the plant and the transmitter output is the reference signal to be tracked by this plant. One can argue that not only the lack of implementing the high frequency components of the chaotic signals is the source of failure to achieve the chaotic synchronization, but also the master–slave configuration is another source of dissynchronization as this open loop control configuration is sensitive to internal/external disturbances and delays. To clarify this point, in a unity feedback closed loop configuration, a PID controller with the parameters $K_p = 1$, $K_i = 100$, and $K_d = 0.01$ is used to provide a suitable control input to the receiver Lorenz system for deriving its output to track the reference chaotic signal produced by the transmitter Lorenz system. As seen in Figure 19.19, the PID controller with unity feedback closed-loop configuration provides the desired synchronization for the Lorenz receiver system whose output

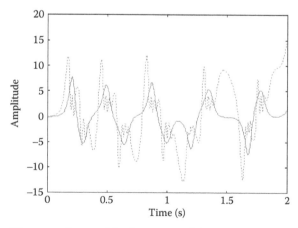

FIGURE 19.18 Time waveforms of the first states of the Lorenz transmitter and receiver, which are implemented in the simulator and the emulator, respectively.

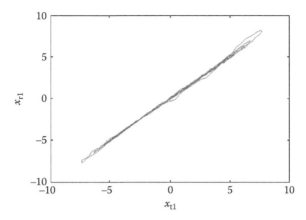

FIGURE 19.19 First state variable of the receiver Lorenz system as compared to the reference signal, which is the first state variable of the transmitter Lorenz system in the S-E-E mode.

tracks the reference chaotic signal at least for the main harmonics corresponding to low frequency components.

The above-mentioned limits of the CDTRP in implementing the synchronized Lorenz systems are natural consequences of the chaotic dynamics of the Lorenz systems that intrinsically possess high frequency components. Moreover, the implementation is limited by the maximum achievable sampling frequencies that can be realized in the MATLAB environment used for the PC simulator and in the emulator implemented by the PIC microcontroller. As seen in the last closed-loop implementation of coupled Lorenz systems based on a simple PID controller, the synchronization can be achieved by using a suitable feedback at least for the main (low frequency) harmonics of the chaotic signals. The exact frequency range for the control system dynamics that allows synchronous operations of the units of the CDTRP platform in implementing these dynamics was investigated by considering what may be the simplest yet challenging example, namely the linear undamped pendulums. The corresponding results are presented in the next section.

19.7.3.6 Synchronization of (Linear Undamped Pendulum Receiver) Simulator/Emulator with (Signal Generator Transmitter) Simulator

To determine the reliable operating frequency range for the simulator and emulator when communicating with each other, first, a signal generator was used as a transmitter in a master–slave configuration for deriving the emulator or simulator where the BP8 pendulum model was implemented (Figure 19.20a). Then, a PID controller with parameters $K_p = 1$, $K_i = 100$, and $K_d = 0.001$ in the unity feedback closed-loop configuration was used to control the receiver pendulum to track the output of the signal generator. For both the configurations, the CDTRP was operated in the S-E-E mode and also in the S-S-S mode (Figure 19.20b.). It was observed that both the

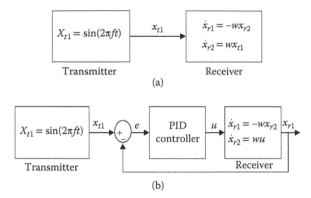

FIGURE 19.20   (a) Receiver emulator or simulator was derived by a simulated signal generator (x_{t1}) in open-loop master–slave configuration for S-S-S or S-E-E modes of the controller design–test–redesign platform. (b) Receiver emulator or simulator was derived by a simulated signal generator (x_{t1}) in closed-loop for S-S-S or S-E-E modes of the controller design–test–redesign platform.

implemented modes yield almost identical results up to $f = 25$ Hz, as shown by the results obtained for the S-S-S mode given in Figure 19.21.

On the one hand, in the open-loop configuration, the synchronization that was achieved for $f = 1$ Hz was observed to fail beyond $f = 10$ Hz. On the other hand, in the closed-loop configuration, it was observed that the synchronizations sustained up to $f = 100$ Hz and $f = 25$ Hz in the S-S-S mode and the S-E-E mode, respectively. These observations actually determine the limit of the operating frequency for the control systems dynamics whose real-time implementations in the S-E-E and S-S-S modes of the CDTRP are reliable in the sense that they can be considered as valid real-time implementations.

19.7.3.7   Synchronization of (Linear Undamped Pendulum) Receiver Emulator with (Signal Generator) Transmitter

To determine the reliable frequency range for the emulator in the R-E-E and R-E-R modes of CDTRP where it receives a signal from external analog hardware, the receiver BP8 was implemented in the emulator and analog signal generator test equipment was used for the transmitter as shown in Figure 19.22. The x_t analog signal was applied to the emulator via the hardware peripheral unit card by the U port that was activated by the ADC options in the GUI. The difference between the experiments done in the S-E-E mode shown in Figure 19.20 and in the R-E-E and R-E-R modes shown in Figure 19.22 is in the transmitter part such that in the former one the transmitter is realized in the simulator and in the other ones the transmitter is realized in analog hardware. The analog signal received by the emulator and also the output of the pendulum created in the emulator are transferred from the emulator to the GUI. Therefore, the unique additional source of limiting the frequency range of the emulated dynamics in this experiment is the usage of the input U port of the emulator operated by the ADC. As shown in Figure 19.23, it was observed that in the open-loop

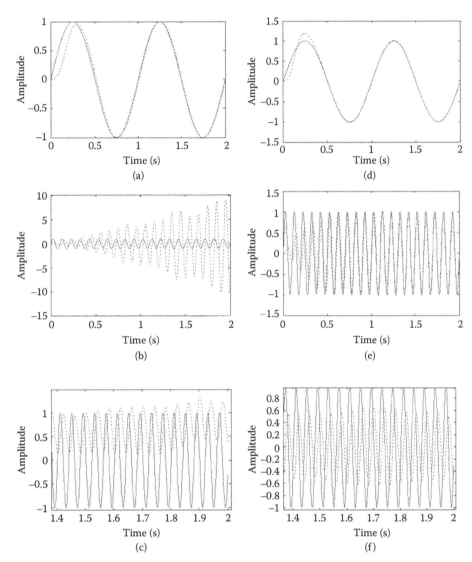

FIGURE 19.21 (a) Master–slave synchronization at $f = 1$ Hz. (b) Master–slave synchronization failure at $f = 10$ Hz. (c) Master–slave synchronization failure at $f = 100$ Hz. (d) Proportional-integral-derivative (PID)-based closed-loop synchronization at $f = 1$ Hz. (e) PID-based closed-loop synchronization at $f = 10$ Hz. (f) PID-based closed-loop synchronization at $f = 100$ Hz. Note that (dashed, --) is for the receiver (pendulum) signal and (solid, -) for the transmitter (generator) signal in S-S-S mode of the controller design–test–redesign platform.

configuration of Figure 19.22, the synchronization was achieved up to $f = 1.52$ Hz. It should be noted that the relatively narrower frequency range than the one obtained for S-E-E mode was due to the ADC interface of the PIC microcontroller.

Figure 19.22b shows that the above given frequency limit can be extended up to $f = 2.92$ Hz (Figure 19.24) by using a PID controller with the parameters $K_p = 1$,

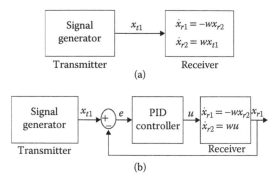

FIGURE 19.22   (a) Receiver (pendulum) emulator was derived by analog signal generator. (b) Receiver (pendulum) emulator controlled by a proportional-integral-derivative controller tracks transmitter signal in R-E-E and R-E-R modes.

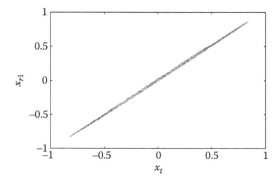

FIGURE 19.23   Master–slave synchronization between analog generator signal and first state of pendulum realized in the emulator as observed from data transferred to graphical user interface.

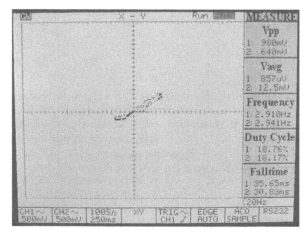

FIGURE 19.24   A snapshot on the X-Y mode of the oscilloscope where the signals in the X and Y channels are the signal generator signal and the pendulum emulator output, respectively, in the R-E-E and R-E-R modes.

$K_i = 100$, and $K_d = 0.001$ in the unity feedback closed-loop configuration for controlling the receiver pendulum to track the output of the analog signal generator. The PID controller was implemented as analog hardware (Leybold LH 734 06 PID-Controller Lab Equipment), so the CDTRP was operated in the (closed-loop) R-E-E and R-E-R modes. Note that the frequency limit is lower than $f = 25$ Hz, which is the one observed for the closed-loop (signal generator–pendulum) synchronization in the S-E-E mode since the analog output of the DAC interface of the plant emulator card was also used in addition to the analog input of the ADC interface of the emulator.* The above results show that the real-time implementations of the control systems dynamics realized in the R-E-E and R-E-R modes are reliable up to the $f = 2.92$ Hz frequency.

19.7.3.8 Synchronization of a Physical Analog (Lorenz Receiver) Hardware with Transmitter Emulator

As the last implementation, the emulator implementing transmitter was used for deriving an analog receiver, that is, the Lorenz system. Then, the reliable frequency range for the emulator in the E-R-E and E-R-R modes of CDTRP was examined. The pure sinusoidal signal generated in the emulator was applied via the DAC interface of the emulator in an additive manner to the first state of the Lorenz system implemented in the hardware peripheral unit card (see Figure 19.25). Note that the Lorenz system was realized by the capacitances for reducing the main harmonics of the chaotic signal around the maximum reliable real-time operation frequency of the emulator. The source of limiting the frequency range of the implemented dynamics in this experiment is the usage of analog output of the DAC interface of the plant emulator card. As shown in Figures 19.26 and 19.28, it was observed that the synchronization is achieved up to $f = 4.05$ Hz in the open-loop configuration of Figure 19.28 and achieved up to $f = 9.09$ Hz for the closed-loop configuration with a PID controller with the parameters $K_p = 1$, $K_i = 100$, and $K_d = 0.001$ in the unity feedback in Figure 19.27 (where Leybold LH 734 06 PID-Controller Lab Equipment was used again).

It should be noted that the Lorenz system derived by the pure sinusoidal transmitter signal did not exhibit a chaotic Lorenz signal for the large amplitude values of transmitter signal anymore and that the synchronization was indeed achieved for the main harmonic of the disturbed Lorenz signal. So the phase synchronizations seen in Figures 19.26 and 19.27 were not exact because of the subharmonics. The subharmonics appear as a consequence of the considered operating mode, which requires the implementation of a part of the Lorenz system in the analog hardware and the other part in the emulator so that their interface is a source of nonlinearity causing subharmonics. The results show that the real-time implementations of the control systems dynamics realized in the E-R-E and E-R-R modes are reliable up to the $f = 9.09$ Hz frequency.

---

* Observe from Figure 19.22 that the analog output of the emulator was fed to the oscilloscope and also to the analog PID hardware.

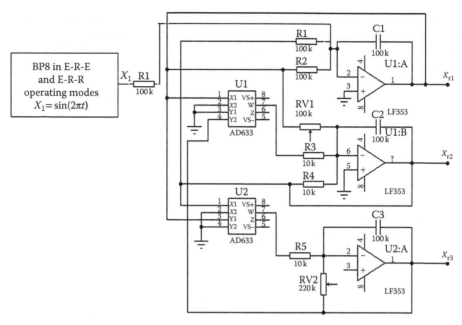

FIGURE 19.25 Transmitter emulator derives analog Lorenz system receiver in E-R-E and E-R-R modes of the controller design–test–redesign platform (a chaotic state modulation system where a transmitter signal is injected into the chaotic system as additive to the first state).

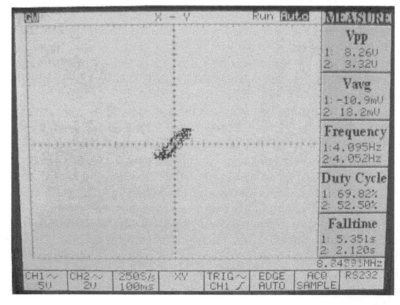

FIGURE 19.26 A snapshot on the *X-Y* mode of the oscilloscope where the signals in the *X* and *Y* channels are the transmitter signal and the first-state variable of the Lorenz system, respectively.

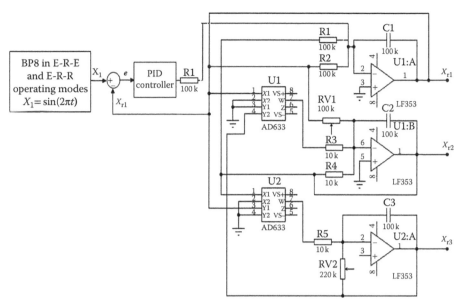

FIGURE 19.27 A snapshot in the *X-Y* mode of the oscilloscope where the signals in the *X* and *Y* channels are the transmitter signal and the first-state variable of the Lorenz system, respectively, in proportional-integral-derivative-based closed-loop control.

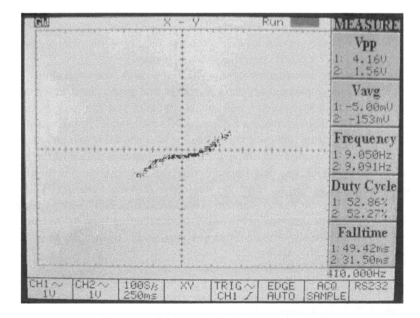

FIGURE 19.28 Proportional-integral-derivative-based closed-loop control for Lorenz system receiver to track output of the transmitter emulator in E-R-E and E-R-R modes of the controller design–test–redesign platform.

19.8 CONCLUSIONS

A real-time simulation platform called CDTRP has been developed, verified, and validated. The developed platform consists of a simulator together with a software manager in PC (i.e., GUI), microcontroller-based emulator, and hardware peripheral unit. This platform is devoted to controller design, test, and redesign with an emphasis on recreating actual operating conditions for examining and analyzing the actual performance of designed controllers while still having the opportunity to modify and tune the controller candidates based on their performances in a flexible way and without causing any damage to the physical plant or creating dangerous situations. CDTRP enables the embedding, via its hardware peripheral unit, of actual disturbances and any part or accessories of a physical plant. For example, actuators and sensors can be implemented in laboratory conditions while the other parts of a plant are implemented in the simulator or emulator of the platform. This configurability enables users to approximate the physical plants and their actual environments as closely as desired. So, the developed CDTRP can be used for the design–test–redesign of controllers, comparison of the performances of the different controller algorithms, and controller parameter training procedures.

The real-time simulation modes (e.g., "hardware-in-the-loop simulation," "control prototyping," and "software-in-the-loop simulation") are realized with the developed CDTRP because it can be operated in 24 different real-time operating modes where the controller, plant, and peripheral unit are implemented either in the simulator, in the emulator, or realized by external analog or digital hardware depending on the application and requirements on, for example, memory and time complexity. A subset of the operating modes that are introduced in this chapter contribute to the real-time simulation literature while the remaining ones correspond to the well-known simulation modes in the literature. The operating modes are described in a novel taxonomy and categorized based on their suitability to the design, test, and redesign stages of the proposed controller design process.

As observed by the investigation conducted in this work, the capabilities of the CDTRP platform in implementing the controllers and plants are limited by the hardware and software realizations used for its simulator and emulator units. This includes the MATLAB environment used for the PC simulator, a PIC microcontroller to implement the emulator, and the communication between these two and the external hardware units. The frequency range for a reliable implementation of the operating modes of the CDTRP was determined by measuring the phase synchronization between the coupled oscillators, which are realized in different units of CDTRP.

The proposed design, test, and redesign procedure can be used in any simulation platform that provides a hierarchy of operating modes ranging from the flexible ones to the ones close to physics, and the developed platform is open to be improved in some focused control applications.

REFERENCES

1. Doyle, J. C., B. A. Francis, and A. R. Tannenbaum. 1992. *Feedback Control Theory.* New York: Macmillan.
2. Goodwin, G. C., S. F. Graebe, and M. E. Salgado. 2001. *Control System Design.* Engelwood Cliffs, NJ: Prentice-Hall.

3. Mandal, A. K. 2006. *Introduction to Control Engineering Modeling, Analysis and Design*. New Delhi, India: New Age International Limited Publishers.
4. Keel, L., J. Rego, and S. Bhattacharyya. 2003. "A New Approach to Digital PID Controller Design." *IEEE Transactions on Automatic Control* 48 (4): 687–92.
5. Yamamoto, T., K. Takao, and T. Yamada. 2009. "Design of a Data-Driven PID Controller." *IEEE Transactions on Control Systems Technology* 17 (1): 29–39.
6. Mehta, S. and J. Chiasson. 1998. "Nonlinear Control of a Series DC Motor: Theory and Experiment." *IEEE Transactions on Industrial Electronics* 45 (1): 134–41.
7. Pellegrinetti, G. and J. Bentsman. 1996. "Nonlinear Control Oriented Boiler Modeling—A Benchmark Problem for Controller Design." *IEEE Transactions on Control Systems Technology* 4 (1): 57–63.
8. Güvenç, B. A. and L. Güvenç. 2002. "Robust Two Degree-of-Freedom Add-on Controller Design for Automatic Steering." *IEEE Transactions on Control Systems Technology* 10: 137–48.
9. Lin, F. J. 1997. "Real-Time IP Position Controller Design with Torque Feedforward Control for PM Synchronous Motor." *IEEE Transactions on Industrial Electronics* 44: 398–407.
10. Rodriguez, F. and A. Emadi. 2007. "A Novel Digital Control Technique for Brushless DC Motor Drives." *IEEE Transactions on Industrial Electronics* 54 (5): 2365–73.
11. Betin, F., A. Sivert, A. Yazidi, and G.-A. Capolino. 2007. "Determination of Scaling Factors for Fuzzy Logic Control Using the Sliding-Mode Approach: Application to Control of a DC Machine Drive." *IEEE Transactions on Industrial Electronics* 54: 296.
12. Boyd, S. P. and C. H. Barratt. 1991. *Linear Controller Design: Limits of Performance*. Englewood Cliffs, NJ: Prentice-Hall.
13. Ziegler, J. G. and N. B. Nichols. 1942. "Optimum Setting for Automatic Controllers." *ASME Transactions* 64: 759–68.
14. Ogata, K. 1997. *Modern Control Engineering*. 3rd ed. Englewood Cliffs, NJ: Prentice-Hall.
15. Bishop, R. H. 2008. *The Mechatronics Handbook*. 2nd ed. Boca Raton, FL: CRC Press.
16. Zeigler, B. P. and J. Kim. 1993. "Extending the DEVS-Scheme Knowledge-Based Simulation Environment for Real-Time Event-Based Control." *IEEE Transactions on Robotics and Automation* 9 (3): 351–6.
17. Maclay, D. 1997. "Simulation Gets into the Loop." *Institute of Electrical and Electronics Engineers* 43 (3): 109–12.
18. Bacic, M. 2005. "On Hardware-in-the-Loop Simulation." *Proceedings of the 44th IEEE Conference on Decision and Control, and the European Control Conference 2005*, Seville, Spain, December 12–15, 2005, pp. 3194–98.
19. Isermann, R., J. Schaffnit, and S. Sinsel. 1999. "Hardware-in-the-Loop Simulation for the Design and Testing of Engine-Control Systems." *Control Engineering Practice* 7(5): 643–53.
20. Scharpf, J., R. Hoepler, and J. Hillyard. 2012. "Real Time Simulation in the Automotive Industry." In *Real-Time Simulation Technologies: Principles, Methodologies, and Applications*, edited by K. Popovici and P. J. Mosterman. Boca Raton, FL: CRC Press. ISBN 9781439846650.
21. Xu, X. and E. Azarnasab. 2012. "Progressive Simulation-Based Design for Networked Real-Time Embedded Systems". In *Real-Time Simulation Technologies: Principles, Methodologies, and Applications*, edited by K. Popovici and P. J. Mosterman. Boca Raton, FL: CRC Press. ISBN 9781439846650.
22. Facchinetti, A. and M. Mauri. 2009. "Hardware-in-the-Loop Overhead Line Emulator for Active Pantograph Testing." *IEEE Transactions on Industrial Electronics* 56 (10): 4071–78.
23. Steurer, M., C. Edrington, M. Sloderbeck, W. Ren, and J. Langston. 2009. "A Megawatt-Scale Power Hardware-in-the-Loop Simulation Setup for Motor Drives." *IEEE Transactions on Industrial Electronics*, PP 1-1. doi: 10.1109/TIE.2009.2036639.

24. Li, H., M. Steurer, K. Shi, S. Woodruff, and D. Zhang. 2006."Development of a Unified Design, Test, and Research Platform for Wind Energy Systems Based on Hardware-in-the-Loop Real-Time Simulation." *IEEE Transactions on Industrial Electronics* 53 (4): 1144–51.
25. Dufour, C., T. Ishikawa, S. Abourida, and J. Belanger. 2007. "Modern Hardware-In-the-Loop Simulation Technology for Fuel Cell Hybrid Electric Vehicles." *Proceeding of the 2007 IEEE Vehicle Power and Propulsion Conference (VPPC-07)*, Arlington, Texas, September 9–12, 2007, pp. 432–39.
26. Hanselmann, H. 1996. "Hardware-in-the-Loop Simulation Testing and its Integration into a CACSD Toolset." *The IEEE International Symposium on Computer Aided Control System Design*, Dearborn, MI, 152–6.
27. Lu, B., X. Wu, H. Figueroa, and A. Monti. 2007. "A Low-Cost Real-Time Hardware-in-the-Loop Testing Approach of Power Electronics Controls." *IEEE Transactions on Industrial Electronics* 54: 919.
28. Smith, R. S. and J. C. Doyle. 1992. "Model Validation: A Connection Between Robust Control and Identification." *IEEE Transactions on Automatic Control* 37: 942–52.
29. Balcı, O. 2003. "Verification, Validation, and Certification of Modeling and Simulation Applications." In *Proceedings of the 2003 Winter Simulation Conference*, ed. S. Chick, P. J. Sanchez, E. Ferrin, and D. J. Morrice, Piscataway, New Jersey: IEEE. Vol. 1, pp. 150–8.
30. Sargent, R. G. 2004. "Validation and Verification of Simulation Models." In *Proceedings of the 2004 winter simulation conference*. Washington, D. C., pp. 17–28.
31. Özer, M., Y. İşler, and H. Özer. 2004. "A Computer Software for Simulating Single-Compartmental Model of Neurons." *Computer Methods and Programs in Biomedicine* 75 (1): 51–7.
32. Suykens, J. A. K., J. Vandewalle, and B. De Moor. 1996. *Artificial Neural Networks for Modeling and Control of Non-Linear Systems*. Boston, MA: Kluwer.
33. Spooner, J. T., M. Maggiore, R. Ordonez, and K. M. Passino. 2002. *Stable Adaptive Control and Estimation for Nonlinear Systems: Neural and Fuzzy Approximator Techniques*. New York: Wiley.
34. Narendra, K. S. 1996. "Neural Networks for Control Theory and Practice." *Proceedings of the IEEE* 84 (10): 1385–406.
35. Fukuda, T. and T. Shibata. 1992. "Theory and Applications of Neural Networks for Industrial Control Systems." *IEEE Transactions on Industrial Electronics* 39: 472–91.
36. Şahin, S., M. Ölmez, and Y. İşler. 2010. "Microcontroller-Based Experimental Setup and Experiments for SCADA Education." *IEEE Transactions on Education* 53 (3): 437–44.
37. Tarte, Y., Y. Q. Chen, W. Ren, and K. L. Moore. 2006. "Fractional Horsepower Dynamometer: A General Purpose Hardware-in-the-Loop Real-Time Simulation Platform for Nonlinear Control Research and Education." *Proceedings of the 45th IEEE Conference on Decision & Control*, Manchester Grand Hyatt Hotel San Diego, CA, USA, December 13–15, 2006, pp. 3912–17.
38. Wang, L. F., K. C. Tan, and V. Pralzlad. 2000. "Developing Khepera Robot Applications in a Webots Environment." *Proceedings of the International Symposium on Human Micromechatronics and Human Science*, October 22–25, Nagoya, Japan, pp. 71–6.
39. Şahin, S., Y. İşler, and C. Güzeliş. 2010. "A Microcontroller Based Test Platform for Controller Design." *IEEE International Symposium on Industrial Electronics 2010: ISIE'2010*, July 4–7, 2010, 36–41. Bari/Italy.
40. Şahin, S. and C. Güzeliş. 2010. "Dynamical Feedback Chaotification Method with Application on DC Motor." *IEEE Antennas and Propagation Magazine* 52 (6): 222–33.
41. Ibrahim, D. 2006. *Microcontroller Based Applied Digital Control*. West Sussex: John Wiley & Sons, Ltd.

42. Narendra, K. S. and S. Mukhopadhyay. 1997. "Adaptive Control Using Neural Networks and Approximate Models." *IEEE Transactions on Neural Networks* 8 (3): 475–85.
43. Slotine, J. J. E. and W. Li. 1991. *Applied Nonlinear Control.* Upper Saddle River, NJ: Prentice Hall.
44. Khalil, H. 1996. *Nonlinear Systems.* 2nd ed. Upper Saddle River, NJ: Prentice Hall.
45. Spong, M., K. Khorasani, and P. V. Kokotovic. 1987. "An Integral Manifold Approach to the Feedback Control of Flexible Joint Robots." *IEEE Journal Robotic and Automatics* 291–300.
46. Abdollahi, F., H. A. Talebi, and R. V. Patel. 2006. "A Stable Neural Network-Based Observer with Application to Flexible-Joint Manipulators." *IEEE Transactions on Neural Networks* 17: 118.
47. Ghorbel, F., J. Y. Hung, and M. W. Spong. 1989. "Adaptive Control of Flexible Joint Manipulators." *IEEE Control Systems Magazine* 9: 9–13.
48. Lewis, F. L., D. M. Dawson, and C. T. Abdallah. 2004. *Robot Manipulator Control: Theory and Practice.* New York, NY: Marcel Dekker.
49. Micera, S., A. M. Sabatini, and P. Dario. "Adaptive Fuzzy Control of Electrically Stimulated Muscles for Arms Movements." *Medical and Biological Engineering Computing* 37: 680–85.
50. Cuomo, K. M. and A. V. Oppenheim. 1993. "Circuit Implementation of Synchronized Chaos with Applications to Communications." *Physics Review Letter* 71 (1): 65–8.
51. Wang, L. 2009. *Model Predictive Control System Design and Implementation Using MATLAB.* London: Springer.
52. Astrom, K. J. and T. Hagglund. 1995. *PID Controllers: Theory, Design, and Tuning.* 2nd ed. Research Triangle Park, NC: Instrument Society of America.
53. Chien, K. L., J. A. Hrones, and J. B. Reswick. 1952. "On the Automatic Control of Generalized Passive Systems." *Transactions ASME* 74: 175–85.
54. Craig, J. J. 1986. *Introduction to Robotics: Mechanics and Control.* Reading, MA: Addison-Wesley.
55. Hsia, T. C. S., "A New Technique for Robust Control of Servo Systems." *IEEE Transactions On Industrial. Electronics* 36: 1–7.
56. Cuomo, K. M., A. V. Oppenheim, and S. H. Strogatz. "Synchronization of Lorenz-Based Chaotic Circuits with Applications to Communications." *IEEE Transactions on Circuits and Systems II* 40: 626–33.
57. Paul, H. 2009. "Build a Lorenz Attractor." (2009, August 10). [Online]. Available: http://frank.harvard.edu/~paulh/misc/lorenz.htm.

20 Automotive Real-Time Simulation

Modeling and Applications

Johannes Scharpf, Robert Höpler, and Jeffrey Hillyard

CONTENTS

20.1 Introduction 501
20.2 RT Techniques for Automotive Systems 503
20.3 Achieving RT in Engine Simulation.. 506
    20.3.1 Choosing the Right Type of Model 506
    20.3.2 Data Preparation 508
    20.3.3 Model Verification and Validation.. 510
    20.3.4 From the Model to a RT Executable.. 511
20.4 RT Implementations in the Automotive Industry 512
    20.4.1 Diagnosing a Faulty EGR Valve... 513
    20.4.2 In-Cylinder Pressure Feedback Control 518
20.5 Future Developments... 519
References.. 519

20.1 INTRODUCTION

The automotive industry is continuously demanding shorter development cycles in order to maintain competitiveness. For this reason, car manufacturers have always been adopting state-of-the-art strategies to shorten the time to market such as simultaneous engineering, rapid product development, and front loading of the development process. These are examples where simulation techniques are key enablers. The use of simulation in the development process leads to an early increase of knowledge. Consequently, subsequent processes can start sooner, which amounts to time and cost savings. Nowadays, simulation is used throughout all phases of product development, in the conceptual design and requirements phase, during design of hardware and software, and in computer-aided production planning. However, it still is most efficient in early stages of the development process. Particularly, the modeling and simulation of dynamic systems accelerates development in early phases, for example, in feasibility studies by evaluating the potential of new topologies using virtual prototypes or in the automated and model-based design of control strategies.

Simulation allows technical prototypes to mature quicker and at the same time reduce their number. Moreover, fewer experiments on test stands are necessary.

The behavior of technical systems can be described and investigated by dynamic system modeling and simulation. Real-time (RT) techniques serve as a natural supplement to modeling and simulation and allow for the running simulation to "connect" with real-world systems in various ways. Hardware-in-the-loop (HIL) methods are an explicit incarnation of this coupling of virtual and technical systems in that one or more virtual components of the simulation model are replaced with physical ones. The physical parts of the technical system interact with the numerical model that simulates the realistic physical behavior of the remaining virtual components as sketched in Figure 20.1.

The HIL system has two main requirements that must be met: Firstly, the interface has to provide real physical entities such as force, torque, current, etc. This is done by actuators, sensors, and their controls. If the system to be interfaced is a computer such as an embedded system, coupling is somewhat simpler because it takes place at an electrical signal level. This method is beneficial especially for the development and testing of electronic control units of computer-based systems, for example, engine control units (ECUs) in Schuette and Ploeger (2007). Secondly, physical behavior of the simulation model must execute synchronously with the physical system parts, or in *real time*. A closer look reveals that it is sufficient to have the relevant physical information, or *states*, from the simulation available only at every point in time when virtual and physical parts, which may also be human operators, must interact through the interface.

This chapter presents some aspects of how automotive development benefits from RT and HIL methods. Section 20.2 provides an overview of challenges and restrictions of modeling automotive systems in a RT context. Simulating multidomain systems such as passenger cars requires special care when choosing modeling approaches, approximations, and mathematical formalisms. Section 20.3 focuses on the RT simulation and control of a combustion engine—itself a classic multidomain system—as a prime example. Section 20.4 highlights several applications of the RT methods and models but also addresses the limitations of these approaches. Finally, Section 20.5 presents some concluding thoughts.

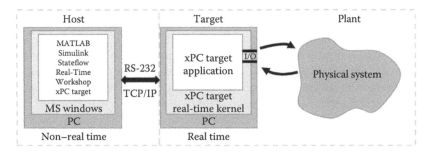

FIGURE 20.1 Example of a HIL system layout (From Mosterman, P. J. et al., *Handbook of Networked and Embedded Control Systems*, edited by Dimitrios Hristu-Varsakelis and William S. Levine, 419–46, Birkhäuser, Boston, 2005. With permission.)

20.2   RT TECHNIQUES FOR AUTOMOTIVE SYSTEMS

Numerical RT simulation of the behavior of any dynamic system is based on specific formulations of the dynamics. These models are mathematical relationships represented either as differential equations, difference equations, or hybrid approaches (Mosterman and Biswas 2002). Particularly, the modeling of automotive systems imposes several challenges:

- A vehicle comprises a plurality of physical domains.
- The time scales of the dynamics cover several orders of magnitude.
- A combinatorial variety of vehicle types and configurations exists.
- A vital need for disclosure of intellectual property by the original equipment manufacturers (OEM) is evident.

Although today's vehicles rely increasingly on computers, they are still classical multidomain systems. Table 20.1 lists the predominant domains involved and common ways to model these domains in the field of RT simulation.

According to the continuous nature of almost all technical parts, their physics is described by partial differential equations (PDEs) and hence numerically best through the finite element method (FEM) and related methods. As shown in Table 20.1, this abstraction of physical reality is not often considered in the field of automotive RT simulation. With this in mind, which abstractions are considered and for what reasons?

Many types of formalisms, such as FEM or computational fluid dynamics (CFD) with their fine-grained discretization in space and time, can be ruled out for RT

TABLE 20.1
Modeling Domains in Automotive RT Simulation

Domain	Technical aspect	Typical modeling formalism
Rigid and elastic body mechanics	Drivetrain, engine crank, chassis, axles, tires	Multibody dynamics, machine dynamics
Electrical	Electrical system, ancillaries, drives	Analog/digital circuit simulation
Chemistry	Combustion, battery, exhaust after-treatment	Rate equations of net reactions
Thermodynamics	Combustion, air/exhaust, battery	Equilibrium thermodynamics, filling and emptying, one-dimensional flows
Hydraulics	Fuel system, brakes, steering	Filling and emptying, one-dimensional flows
Thermal	Cooling system	Phenomenological linear thermodiffusion
Control	ECU	Linear control, neural networks (Meder et al. 2007)
Computer science	ECU, smart components	Automata, statecharts, RT-UML (Douglass 2004)

applications. Complex models that must observe an immense number of variables over an extended period of time, such as those for simulating gas flow, challenge current computing hardware. If that is the case, then how can one accelerate the computational ability of a technical system model? Computational demand is proportional to the number of simulation state variables $x(t)$. Reducing to a lower number of simulation state variables is beneficial for two reasons: First, the smaller system can be formulated by ordinary nonlinear differential equations (ODEs) or differential algebraic equations (DAE), where

$$0 = f(x, \dot{x}, u, p, t)$$

governs the time evolution of $x(t)$ (Vidyasagar 1993). The variable u represents the control inputs, and the variable p represents all parameters that do not appear as time derivatives. This approach often delivers satisfying quality of simulation results. The vast amount of simulation variables present in an FEM formulation is condensed to several physically meaningful states and allows for control laws to be designed much easier. The restrictions of this process are obvious: one must be careful when fundamental assumptions for the reduced model will not hold or when interesting phenomena will be completely neglected. In the field of internal combustion engine (ICE) simulation, these are certainly regions where one approaches the limits of thermodynamic equilibrium. The most prominent ones are the near-supersonic processes that take place in a turbocharger or the processes of combustion and gas exchange in a cylinder.

Secondly, an often overseen issue is that the reduction to a few simulation states affects the necessity for obtaining suitable data. Efforts to parameterize the model may reduce considerably because the governing differential equations take lumped values. For instance, a total mass is required instead of a mass distribution, or a scalar value or characteristic map of a heat transfer coefficient is used instead of a thermodiffusion process. Particularly in the automotive field, a valid and consistent set of data leading to parameters p for the parameterization of simulation models is difficult to obtain. Intellectual property issues are one reason for this. The main reason, however, lies in the early, or conceptual, phase of technical design where employing simulation tools is most advantageous: necessary data at this point in time are either scarce, unconsolidated, or simply not available for use in the simulation. This situation is aggravated by the fact that good data are essential for reliable simulation results. The reader is referred to Section 20.3.2 for more details.

Apart from the mathematical formulation of the causal behavior of an automotive system, there is a second distinctly important aspect to RT modeling and simulation: the notion of *time*. The most obvious way in which the treatment of time affects RT simulation is through the time discretization within the numerical simulation. Numerical integration schemes make time discrete by using fixed or variable step sizes and project the approximation of the exact ODE/DAE solution onto a sampling grid (Brenan et al. 1987). This grid must satisfy several requirements in order to lead to a meaningful solution:

- The computational load induced by the evaluation of the simulation model at each grid point must match the computing power of the RT platform to avoid task overruns. Depending on the type of RT system, such task overruns either can be tolerable or must be avoided at all costs (Schuette and Ploeger 2007). Automotive applications such as vehicle dynamics test rigs traditionally use sampling rates of 1 kHz. This restricts the integration scheme to being a fixed-step method where the number of grid points is proportional to computational load. The Forward Euler method (Brenan et al. 1987) is the most prominent among them in that it guarantees a finite execution time.
- The eigenfrequencies of the model impose an upper boundary on the used step size. Especially for engine applications, this leads to sampling rates of at least 100 kHz as mentioned in Section 20.3.1 for calculating the in-cylinder pressure.
- Automotive systems often contain highly reactive subsystems. As a consequence, it is, for example, for a HIL application, not sufficient (although necessary) to have the simulation run in synchrony with a physical process. Synchronicity with the physical process is superseded by the more stringent requirement of reactivity. Well-defined reaction times to external stimulation are crucial in embedded ECU applications. Particularly, the reaction of the cylinder and combustion model to the timing-critical fuel injection signals must take place within a time frame of tens of microseconds (see Section 20.4.2).

Amenability to RT execution is also affected by the way time itself is modeled, which tends to be less obvious as it is more implicitly incorporated in the modeling formalism. The states in a continuous-time model, for example, formulated as an ODE/DAE system, may change smoothly over time. In contrast with this type of model, there are discrete-event models where the state is an element belonging to a discrete set and will change abruptly over time, which is well suited to efficiently model abstract-switching behavior. This discontinuous behavior is often modeled as a finite-state machine, statecharts, or a state-transition diagram. In RT applications where computational efficiency is paramount, RT simulation models are usually combinations of continuous and discrete behavior, also known as hybrid dynamic systems. Although well-suited to RT applications, these introduce additional complexity in terms of numerics and modeling; see Mosterman and Biswas (2002) for more details.

There is a large variety of tools for RT modeling and simulation. Some approaches provide the possibility to model any hybrid dynamic system, without restrictions to the physical domain, in a modular and block-oriented way. Prominent examples are native Simulink® (Simulink 2004) and Modelica (Modelica 2010). While the first employs causal modeling, the latter is an acausal modeling formalism, which has certain advantages, for example, for model inversion. Multidomain tools with a special focus on automotive systems are, for example, ASCET-MD (ASCET-MD 2009), for the development of embedded automotive control systems, GT-Suite (GT-Suite 2009), and AMESim (AMESim 2009) for zero and one-dimensional modeling of physical systems with an option to generate RT capable code.

20.3   ACHIEVING RT IN ENGINE SIMULATION

This section shows the tasks and challenges in practical RT modeling of automotive systems. Like a passenger vehicle, the combustion engine is a complex multiphysics system in its own right. Without neglecting generality, this section focuses on the modeling of combustion engines, particularly their crucial intake/exhaust gas and fuel subsystems.

20.3.1   CHOOSING THE RIGHT TYPE OF MODEL

There are a considerable number of approaches to modeling combustion engines, but just a few are amenable to RT applications. A physical formulation of the dynamics will be stressed because of its superior properties in terms of its ability

1. To extrapolate into regions where no measurements are available
2. To reuse models and parameters
3. To reduce the effort required for parameterization
4. To immediately understand the modeled phenomena

In this section, black-box approaches such as neural networks or model order reduction are considered a complement to PDE- or ODE-based methods (Meder et al. 2007). The number of domains involved is considerable, leading to a heterogeneous physical description and rich dynamics of the complete system. That is why most resulting models are either restricted to being a single domain model or a hybrid of both black-box and physical formulation, often resulting in gray-box models.

Among the most complicated, and also most interesting, dynamics in an automotive system may be the gas-exchange phenomena during a working cycle in the cylinder and the intake air and exhaust paths. Turbulent flows of multiphase media within complex geometries, heat transfer, and intricate chemical reactions, for example, during combustion, call for CFD models because of their resolution capabilities in time and space (Chung 2003). These models, however, are still far from being RT capable on standard hardware, for example, FIRE (AVL FIRE 2006), mainly because of the discretization in two or three dimensions. Classical methods of order reduction or the usage of neural networks with a focus on control theory are sometimes applied to reduce complexity. Papadimitriou, for example, shows a method to transform complex one-dimensional flow models into an RT neural network model of a combustion engine (Papadimitriou et al. 2005).

A more physical way of reducing the complexity of PDE solutions is the use of modal and lumped formulations of the dynamics in which symmetries are exploited or assumed. The engine cycle calculation (ECC) partitions gas volumes into ideally mixed finite volumes and ends up with a set of coupled ODEs that are solved while satisfying the conservation laws, as shown in Merker et al. (2006). The technical analogy is a set of plenums or coolers coupled by nozzles. The very complex and nonequilibrium aspects such as heat release rate and heat transfer are augmented as parametric or semiphysical models (see, e.g., Vibe [1970]) that allow for the adaptation of the model to varying operating points. An investigation of the eigenfrequencies of these models reveals that typical sampling rates must lie in the order of magnitude of 100 kHz to obtain realistic results for, for example, the in-cylinder pressure of a high-rev gasoline engine. Especially, the heat transfer to the cylinder walls, the piston, and the cylinder head is treated usually by empirical models as this process is much too complex to model from first principles. A more realistic modeling of a direct injection combustion process or the formation of engine emissions can be approximated within the ECC. The cylinder volume then has to be partitioned into two to several hundred coupled zones (Heywood 1989). The ECC is a powerful method offering a wide range of application. Common applications are development of control strategies, sensitivity analysis, combustion process development, and ECU testing. The computational complexity is at the limits of current processor power providing cause to the emergence of commercial RT models (DYNA4Engine 2009; ASM InCylinder 2006).

Mean value engine models (MVEMs) are a classical way of modeling the processes inside the combustion engine. Similar to ECC, these are based on a DAE/ODE formulation, but the physical states are considered within a more macroscopic time scale. States are considered mean values with respect to a time range of three to five engine revolutions (Jensen et al. 1991). This is sufficient to model fast changing processes such as the opening of the exhaust gas recirculation (EGR) valve. Desired states are engine speed, engine torque, and manifold pressure to name just a few. In terms of complexity and computational effort, they rank below the ECC (Hendricks and Sorenson 1990). For the description of engine phenomena that are too complex for modeling the physics from first principles, dependent and independent variables are separated to be able to employ simple empirical models or mapped data from measurements (Jensen et al. 1991). The strength of MVEMs lies in their ability to be easily adapted to other engines. Their compact mathematical formulation requires only a comparably small number of parameters and can be calibrated to a particular engine with a small set of measurements (Hendricks and Sorenson 1990). MVEMs are often utilized in the development of control strategies and in model-based diagnostics. See Section 20.4 for some HIL applications.

State-free approaches such as engine torque maps represent a simple way to model engine behavior. These are very popular, especially in ECU applications, because they are computationally robust and efficient while offering fairly high accuracy if using data obtained by (often relatively few) engine measurements. On the other hand, maps are almost ruled out for modeling transient behavior accurately (Gheorghiu 1996), and they can potentially introduce undesired nonlinearities into the dynamics of a controlled system.

In the end, the purpose of RT engine simulation is to deliver a system's response with respect to time that is physically correct and free of contradictions. Therefore, the choice of an appropriate model for a specific investigation or task is an intricate trade-off between available computing power, required sample rates and response times, and validity of the modeled dynamics. Figure 20.2 relates the computational complexity to time scales of characteristic phenomena in engine modeling. The gray area depicts the region of applications currently possible on standard processing hardware.

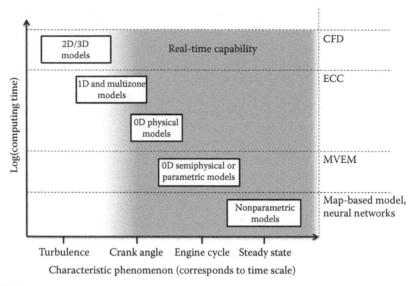

FIGURE 20.2 Relationship between engine model complexity and modeled phenomena.

20.3.2 Data Preparation

As pointed out in Section 20.2, a model of the dynamics of a technical system is composed of (1) a description of the physical behavior, (2) model parameters p, and (3) initial conditions. In order for a model to be able to sufficiently replace the physical component of a system, it is imperative that the data used are sufficiently precise and correct. Insufficient care while preparing model data or modeling will almost certainly lead to erroneous simulation results. Furthermore, data preparation takes a significant amount of time as part of the process of modeling and simulation. For these reasons, the steps of data preparation as well as model verification and validation are investigated in more depth here.

Data is almost never directly suitable to be used "as is" with the simulation model. Because of this, data preparation, sometimes known as preprocessing, is necessary to extract and transform raw data into a form suitable for the model. Data preparation is, in essence, a model of the parameters for the simulation model. For RT models, two practicalities stand out. Firstly, any calculation that can be done prior to simulating decreases the computational load. Secondly, the data must be adapted regarding structure and type to suit the model, if only because RT models often approximate complicated physical relationships, as shown in Section 20.3.1, by using lookup tables.

Data preparation requires prior knowledge of the simulation model to know which data are required as both data and model are intimately linked. Designing the data preparation process can take up a substantial amount of time and end up being an iterative process. Data preparation can roughly be divided into (1) data extraction and (2) data transformation for our purposes here. The process is schematically depicted in Figure 20.3.

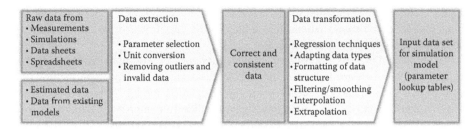

FIGURE 20.3   Flow diagram of data preparation steps.

The sources of data are manifold. They could be measurements, simulation results, data sheets, or spread sheets provided from a supplier. For the case that not all input data is available, missing data must be estimated until the remaining data are provided. This may be sufficient to get the model up and running and to even start with testing. The following three steps will lead to correct and consistent data regarding units and values:

- Parameter selection is the process of identifying the correct parameters necessary to parameterize the model. There may often be many more parameters measured or provided (e.g., in a data sheet) than is necessary.
- Often the provided data is not specific in terms of physical units, coordinate systems, and measurement conditions. A typical problem that arises in engine applications is the absence of reference conditions for a table of turbocharger operating points, which renders the data practically unusable (Moraal and Kolmanovsky 1999). Some modeling languages such as Modelica or Simscape™ have variable classifications that support using correct units or even the type of variable such as force or mass.
- Removing outliers and invalid data will be necessary as soon as measurements are intended to be used as model inputs or for the extraction of model parameters. Problems during experiments can be sensor failures or incorrect settings that can lead to invalid data.

Data transformation will process the data according to the model requirements regarding quality, type, and structure of the input data. Data quality refers to the smoothness of data and the availability of data as a usable format. Filtering and smoothing remove discontinuities in the input data. This can help prevent the simulation of models based on such data from aborting and increases computational efficiency even for cases where a simulation could run but must take small time steps over a discontinuity to satisfy tolerances. Because data were measured at scattered points, it is often necessary to carry out interpolation to map the values onto an equidistant grid. Extrapolation will extend input data to cover regions that lie beyond the range of the measurements. Regression techniques (Izenman 2008; Ljung 1999) are used to identify model input parameters from measured data. The type of parameter depends on the number of input variables and can be, for example, scalars or n-dimensional lookup tables. Finally, all input data must conform to the requirements of the simulation model with respect to file format, memory specifications, and similar formal matters. It is a part of the data transformation process to ensure such requirements are fulfilled.

An automated process can be very valuable for data preparation because, if implemented, the user is rewarded with a reliable, time-saving, and convenient method of creating the input data set necessary for the simulation model. This is especially advantageous when the input data have been modified or improved. Directly after data preparation, it is helpful to visualize the results. Simply visualizing data in the form of plots is not only a way of identifying errors in the data, it is also a way of understanding the simulation model and its implementation. One beneficial side effect of preparing data is revealed when the user of the simulation model also carries out the data preparation but was not involved in the modeling: The user is then forced to view the data that indicate how the model was implemented. In this manner, the user becomes more aware of the model while working with the data.

20.3.3 MODEL VERIFICATION AND VALIDATION

Along with building a simulation model and preparing data to be used in the model, the process of verifying and validating (V&V) the simulation model is an essential step in striving toward accurate and reliable simulation results. Performing V&V is important for all types of simulation models but even more so for RT simulation applications where coupling to real-world subsystems requires quantitative correctness of the model behavior.

Model verification can be defined as ensuring that the computer program of the computerized model and its implementation are correct (Sargent 2007). This could be the transformation of a flowchart or a mathematical model consisting of equations into a computer program that can be executed. Model validation may be defined as substantiation that a computerized model within its domain of applicability possesses a satisfactory range of accuracy consistent with the intended application of the model (Schlesinger et al. 1979). The relationships between the computerized model, the conceptual model, and the technical system that is being modeled (the "real system") are illustrated as the Sargent Circle shown in Figure 20.4. The cross-dependencies stand out in the sense that the entire process of V&V is iterative. Finding a mistake late in the model validation phase may even require reevaluating the conceptual model and modifying it. Such a change can have an extensive influence on the model behavior, leading to an entirely new testing process that must be started from the very beginning. This can have far-reaching consequences as there may be other models or components being designed that depend on having an accurate model and the design of which now must be postponed. It is for this reason that modeling errors should be caught as soon as possible.

There are many principles that should be accounted for to ensure successful V&V. Some of the most prominent ones are

- The simulation model is only valid for the tested conditions.
- A valid model does not imply credible and accurate results.
- Successful subsystem testing does not guarantee model credibility in its entirety.

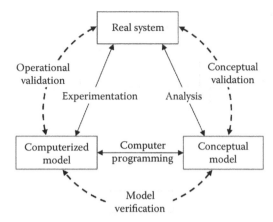

FIGURE 20.4 Simplistic overview of the process of model verification and validation. (From Sargent, R. G., *Proceedings of the 2007 Winter Simulation Conference*, Washington, DC, 2007. With permission.)

Carrying out V&V requires extensive testing for dynamic system models. One general method of doing this is presented by Lehmann (2003). Different types of tests are examined in Sargent (2007), for example:

- Compare simulation results to real measured values from the technical system.
- Compare simulation results to results from other valid models with at least comparable complexity.
- Test extreme inputs or initial conditions.
- Obtain expert opinion on the input–output relationship of model.
- Perform steady-state and transient testing.

As suggested above, there can never be 100% validity. Ideally, the best chances for success derive from testing thoroughly and often by starting at the function level, continuing up through the subsystem and component level, and finally ending at the top system level. Each simulation model is unique, though, and so benefits from customized and detailed testing procedures. Moreover, testing thoroughly and efficiently is further complicated because the model implementation blends complex behavior on mathematical, numerical, and execution levels as described in the following section.

20.3.4 FROM THE MODEL TO A RT EXECUTABLE

Simulation in this context mainly denotes numerical time integration. Given the model of a dynamic system $0 = f(x, \dot{x}, u, p, t)$ with parameters p, control inputs u, and the initial states $x(t = t_{start})$, one can solve the initial-value problem by means of numerical integration schemes. As pointed out in Section 20.2, under RT conditions one is mostly restricted to using schemes with a guaranteed execution time and, therefore, mainly to explicit and fixed-step-size methods. Unfortunately, these methods in particular provide limited accuracy and stability when solving stiff, discontinuous, or constrained systems that often arise in the field of engineering (Eich-Soellner and Führer 1998). In order to reconcile accuracy and efficiency, integration schemes can be carefully chosen for each domain and RT model or submodel. For example, higher order single-step methods, or approximations of implicit methods where iterative solutions are truncated (Eich-Soellner and Führer 1998), can be implemented in order to meet RT requirements. Some high-performance RT solutions are even able to switch the equations *and* integration schemes depending on the current engine model state (DYNA4Engine 2009).

Code generation is the final step influencing the performance of an executable RT simulator. Though tedious, some dynamics model descriptions, controller, and integration schemes are still implemented by manual coding, for example, for reasons of limited resources in embedded systems. However, nowadays this approach is restricted to specific automotive applications and very expensive because of the combinatorial complexity in terms of vehicle types and configurations. When using modeling formalisms such as Simulink, the high-level, possibly graphical, model description can be automatically transferred to C code or binaries for a specific RT platform or embedded system. Prevalent tools are Real-Time Workshop® or TargetLink for generating code from Simulink models. Dymola (2009) compiles textual or graphical models in Modelica. Exploiting a symbolic representation of the model offers interesting options to automatically enhance computational performance. Mixed-mode integration, and particularly the method of *inline-integration* (Elmqvist et al. 1995), leads to potentially large but computationally efficient code, especially attractive in RT applications.

20.4 RT IMPLEMENTATIONS IN THE AUTOMOTIVE INDUSTRY

RT methods couple virtual and physical technical subsystems. The goal is to test physical and virtual components under practical conditions without requiring the complete system. Actual target hardware ECUs are tested using signals generated from an engine simulation on HIL test stands. Earlier stages in a model-based design process might rely on more virtual configurations: in the software-in-the-loop (SIL) configuration generated code of a control unit model is connected with a plant model. The system might neglect, for example, sensor effects and does not even run in RT. The same holds for processor-in-the-loop (PIL), but in this case, controller code runs on the target platform (Mosterman et al. 2004). Methods like HIL are attractive options in model-based development of control schemes for dynamic systems. Within the automotive industry these methods are not restricted to controller design and parameterization but are also applied (1) to quality assurance and approval of functions for testing the reaction to short circuits, cable failure, electromagnetic compatibility, and fail-safe behavior, (2) to compatibility within the control unit network in terms of interfaces and communication, and (3) to the testing of diagnostic functions (Schuette and Ploeger 2007).

There are a number of advantages of HIL methods in the field of combustion engines:

- Reproducibility: A simulation environment eliminates unwanted influences, such as those resulting from varying ambient pressure or temperature, and there is no limit to the amount of testing since there is no wear on virtual components.
- Cost effectiveness: Fuel and preconditioning of media are not necessary, which leads to less energy consumption and therefore significantly reduced operating costs.
- Development time: HIL component testing allows for fewer engine test bed or roller test bench experiments.
- Automation: Tests can be run in an automated fashion, without requiring human interaction, which enables many configurations to be used simply by loading a different engine model.
- Safety: Testing extreme modes of operation does not carry the risk of damaging expensive prototypes or injuring a test operator.

The additional effort required in setting up a HIL experiment consists of the implementation of an appropriate RT simulation model that is sufficiently accurate and validated with experimental data. In some cases (e.g., when no prototype of a new engine is available), it will also be necessary to create a detailed, predictive model that can be used to validate the RT model.

The following applications outline the power of RT methods notably in the field of ECUs.

20.4.1 Diagnosing a Faulty EGR Valve

Component diagnostics have become a crucial issue in the field of combustion engines. Legislation and the necessity to support repair technicians in error searching are the driving forces in the ongoing expansion of On-Board Diagnostics (OBD) requirements. It allows the control of engine emissions not by direct measurement of the exhaust gas but by monitoring the components that either directly or indirectly influence the quality of the emissions. A further goal of OBD is the automatic identification of the faulty component or part being responsible for poor emissions, which is stored as an error code in memory. This enables a technician to read the error memory and receive advice as to which parts require replacement in order to remedy the problem. This can avoid time-consuming error searching and saves costs as a result. In this section, diagnostic functions for an EGR valve using HIL methods are investigated.

The goal of EGR is to reduce NO_x emissions by adding inert gas to the cylinder charge. The exhaust gas is channeled from the exhaust manifold, cooled with engine coolant, and fed into the fresh intake air flowing toward the cylinders. The amount of recirculated exhaust gas is controlled by the EGR valve. The system layout is shown in Figure 20.5.

The EGR valve is actuated by a DC motor or by a vacuum actuator as shown in Figure 20.6. The necessary vacuum is provided by the vacuum pump of the engine and controlled by a pressure transducer in order to achieve the desired EGR valve lift also shown in Figure 20.6.

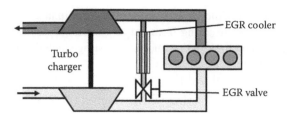

FIGURE 20.5   Engine layout of a high-pressure EGR system.

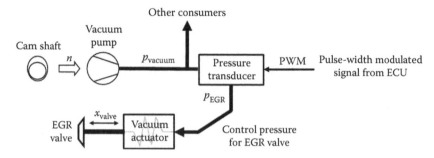

FIGURE 20.6   Method of actuation of the EGR valve.

The EGR system is subject to very high temperatures and exposed to deposits from the soot in the exhaust gas. This makes the system susceptible to failure. As EGR is used to lower NO_x emissions, any deviation from the prescribed behavior will have a substantial impact on emission quality. A variety of fault possibilities is shown in Table 20.2.

It can be seen that different faults can lead to the same effect. In order to uniquely identify the cause for a fault, it is therefore not sufficient to solely detect it but also to determine particular characteristics that can be traced back to a failed component. There are many causalities in the engine behavior, which can be used for fault detection and identification, and often it is not clear even to the experienced engineer as to which ones are most relevant. For this reason, the HIL method is a valuable tool for the development of diagnostic functions as it reveals the causalities in a reproducible and cost-effective manner. A prerequisite is the usage of an appropriate RT engine model.

The model equations are derived by applying Newton's Second Law to the EGR valve plate and valve stem. Besides the inertial force due to the mass of the moving valve, acting loads occur from the friction force, the return spring force, and the force as a result of the vacuum (see Figure 20.7). There are additional loads resulting from the pressure difference between inlet and exhaust port. These loads are neglected because they are small and can hardly be quantified once the valve has left its closed position.

Newton's Second Law as applied with the forces acting on the valve

$$0 = F_v(p_v) + F_s(x) + F_{Fr}(\dot{x}) - m\ddot{x}$$

is used to determine the valve position x. The force from the vacuum actuator F_v depends on the pressure difference because of the vacuum p_v, the spring force F_s

TABLE 20.2

Various Faults of an EGR System

Fault	Potential causes	Effect
Rupture in pipe	Vibration, thermal expansion	Leakage of exhaust gas
Blockage in heat exchanger	Soot deposits	No EGR, high NO_x emissions
Fouling of heat exchanger	Soot deposits	Reduced cooling, reduced amount of EGR
EGR valve stuck open	Broken return spring, friction due to soot deposits	Too much EGR, high soot emissions
EGR valve stuck closed	Detached vacuum hose, friction due to deposits, failure of vacuum pump, pressure transducer failure	No EGR, high NO_x emissions
Slow dynamic response of EGR valve	Vacuum hose leakage, faulty vacuum pump, friction due to deposits	Reduced emission quality

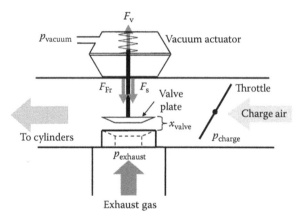

FIGURE 20.7 EGR valve system overview.

depends on the valve position, the friction force F_{Fr}, implemented as a Coulomb–Stribeck function, depends on the valve velocity, and the mass m represents the sum of the moving masses.

The EGR mass flow is calculated using the equation for an orifice

$$\dot{m} = 2A(x)\mu\sqrt{p_0\rho_0}\sqrt{\frac{\kappa}{\kappa-1}\left(\pi^{\frac{2}{\kappa}} - \pi^{\frac{\kappa-1}{\kappa}}\right)} \quad \text{with} \quad \pi = \frac{p_1}{p_0}$$

and depends on the gas states before and after the orifice, which are represented by density of the exhaust gas ρ, the pressure p, and the adiabatic exponent κ (Merker et al. 2006). The index 0 represents the upstream state of the orifice, and the index 1 represents the downstream state. The parameter $A(x)$ is the flow area through the orifice dependent on valve lift. With the known values of EGR mass flow $\dot{m}_{\mathrm{EGR}}$, intake air mass flow $\dot{m}_{\mathrm{air}}$, and their respective heat capacities c, the cylinder mass flow $\dot{m}_{\mathrm{cyl}}$, and temperature T_{cyl} can be calculated:

$$T_{\mathrm{cyl}} = \frac{c_{\mathrm{EGR}}\dot{m}_{\mathrm{EGR}}T_{\mathrm{EGR}} + c_{\mathrm{air}}\dot{m}_{\mathrm{air}}T_{\mathrm{air}}}{c_{\mathrm{EGR}}\dot{m}_{\mathrm{EGR}} + c_{\mathrm{air}}\dot{m}_{\mathrm{air}}}$$

The model is embedded in a mean-value model of a diesel engine as presented, for example, in Jensen et al. (1991) and in Guzzella and Amstutz (1998). The main advantage of the model is that faults can be easily simulated, which renders it very convenient for the testing of diagnostic functions on the HIL test bed. For simplicity, only the last three fault cases from Table 20.2 are considered further. These faults are introduced into the engine model by setting the valve lift to its minimum or maximum level, which represents the errors of the EGR valve stuck in a closed or open position, respectively. The increased friction is simulated by scaling the friction force by a factor of about 1.6, resulting in slow dynamic response of the valve.

Figure 20.8 shows a section of the cycle that was used for the OBD investigation. The engine speed and load were set independently as is possible on an engine test

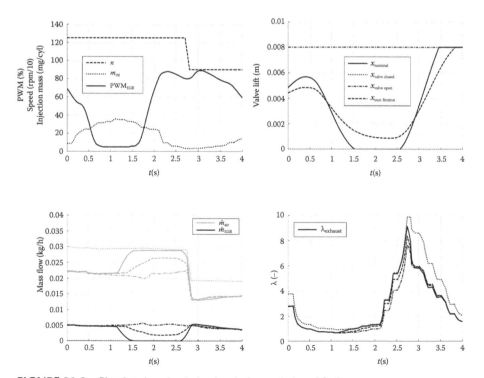

FIGURE 20.8   Simulated engine behavior during an induced fault occurrence.

bed. The corresponding actuator signal for the EGR valve (PWM_{EGR}) is shown in the top-left plot of Figure 20.8. The top-right plot of this figure shows the EGR valve lift for the fault cases. The legend of this figure is representative for the subsequent figures where it is not shown completely for the sake of clarity. The valve positions for the valve stuck in open and closed position, respectively, are visible. One can also clearly see the reduced dynamic behavior of the valve with increased friction of the valve shaft.

The bottom two plots show the influence of the EGR valve faults on engine parameters. Their behavior matches the expectations well. For diagnostic functions, the parameters air mass flow and λ are of special interest as there are sensors for them in modern diesel engines that can be used for fault detection. The combustion-air ratio λ is defined as the actual air–fuel ratio divided by the stoichiometric air–fuel ratio. Model-based diagnostics (Isermann 2005) use the deviation of a sensor value from a model value, also referred to as a residual. The sensor value includes the effects of the fault while the model represents the nominal state as it is based on actuator signals and signals not affected by the fault. For diagnostics purposes, the goal is to gather as much information as possible about the fault from the residuals in order to identify the fault size, location, and time of occurrence. The residuals of the air mass flow and λ based on the simulation results given in Figure 20.8 are shown in Figure 20.9.

In creating diagnostic functions, fault patterns have to be defined and implemented as code. As this process is based on heuristic knowledge about the faults, the implementation of diagnostic functions for control units is an iterative process, which is ideally performed on a HIL test stand for reasons of cost and time savings. The diagnostic functions can be directly implemented on a prototype ECU and tested for correct operation. The residuals not only contain the information about the fault but also include model and measurement deviations and deviations due to component tolerances. It is the task of the diagnostic function to take these factors into account. In spite of careful design of the diagnostic function, final adjustments and validation with a real engine are still necessary, though.

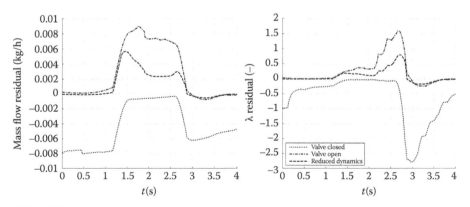

FIGURE 20.9   Residuals of mass flow and λ during the induced fault occurrence.

20.4.2 IN-CYLINDER PRESSURE FEEDBACK CONTROL

Changing legislation is forcing car manufacturers worldwide to reduce emissions and fuel consumption. Some manufacturers and suppliers are turning to in-cylinder pressure feedback control as a possibility to meet the upcoming emissions regulations, such as the EURO 6 and Tier II standards. Especially in transient operating conditions, engine emissions have much potential to be reduced by sophisticated strategies that precisely control the fuel injection and combustion process (Kuberczyk et al. 2010). Current developments in pressure sensors that can better withstand the harsh environment inside an engine are enabling new control strategies even without costly air mass flow sensors and, if otherwise present, NO_x sensors (Klein 2009). In-cylinder pressure closed-loop control allows for an instantaneous modification of the fuel injection pattern based on the current operating conditions and the in-cylinder gas state for each cylinder. This allows for the compensation of unwanted disturbances and asymmetries between cylinders due to engine wear and variability of injectors and valves, hence optimizing the combustion of each cylinder individually (Nieuwstadt and Kolmanovsky 1999).

Employing HIL methods is a typical manner to develop these types of model-based control strategies. A HIL system requires a detailed RT simulation model providing the cylinder pressure depending on at least the current injection pattern and timing as well as the cylinder gas state, that is, temperature, pressure, and gas composition. The ECC approach presented in Section 20.3.1 is able to compute these values that describe the gas state in RT on standard computing hardware as shown for one operating point in Figure 20.10.

Commercial packages based on ECC, such as DYNA4Engine, ASM InCylinder, and GT Power RT, sample the pressure signal at a rate below 10 kHz and are able to capture all the prominent events such as ignition delay and start of ignition as well as the correct amount of energy released at the necessary time. The simulation step size must be significantly smaller than the commonly used 1 ms as evident when considering that 1 ms at an engine speed of 5000 rpm corresponds to 30 crank angle degrees. This requires a step size that is at least one order of magnitude smaller to be able to simulate the fuel injection and combustion sufficiently accurate for an ECU.

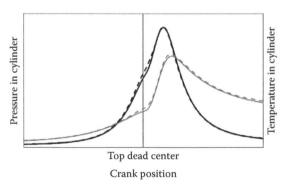

FIGURE 20.10 Normalized in-cylinder pressure and temperature around top dead center—measured (dashed) and simulated (solid).

In order to obtain a realistic time-resolved simulation of the in-cylinder pressure and temperature as shown, some practical issues arise:

- Time delays: The sampling of signals in the I/O of the test rig delays signals an amount which corresponds to the characteristic time scales of the combustion process.
- Sensors: Because of the extreme conditions inside the cylinder, the sensor degrades and its output drifts so that a model of the of the sensor behavior is mandatory.
- Calibration: Some control schemes evaluate absolute values and not just the shape of the pressure curve, requiring correct calibration of the model.

However, the parts of the simulation model outside of the combustion process (e.g., the air path) normally only require a 1 ms step size, a fact exploited by means of a multirate simulation. Sampling rates are adapted to the eigenfrequencies of each subsystem. The task with the high sample rate would support the combustion process, the injection signals, and the cylinder pressure signal, whereas the slower task would apply to the rest of the model (Schuette and Ploeger 2007). RT simulation of heavy duty engines with a multitude of cylinders, for example, in marine applications, can be implemented by parallelization of the multirate model across multiple processors (DYNA4Engine 2009).

20.5 FUTURE DEVELOPMENTS

RT and HIL methods will certainly benefit from parallel computing, especially through the introduction of multicore processors. Particularly in the field of combustion engines, parallelization may pave the way toward RT simulation of one-dimensional gas dynamics, multizone cylinder models, and even CFD models that are beyond the current capabilities of standard RT hardware.

This will go well with the drastically increasing complexity of combustion engines due to emission relevant components such as selective catalytic reduction (SCR) catalytic converters or low pressure EGR, the usage of biofuels, and the demand for further improvement of fuel economy. The new era of electric and hybrid vehicles challenges the field of combustion engine control through a multitude of innovative topologies and use cases unknown to traditional automotive design. Simulation and particularly RT techniques will be able to close the gap in knowledge and experience while investigating the complex interactions between electronic, computational, and mechanical subsystems that will emerge from the integrated development of reliable and economical automotive systems.

REFERENCES

AMESim. 2009. *AMESim User Manual*. Leuven, Belgium: LMS International.
ASCET-MD. 2009. *ASCET V5.1 User Guide*. Stuttgart, Germany: ETAS.
ASM InCylinder. 2006. *ASM InCylinder Simulation Package*. Paderborn, Germany: dSpace GmbH.

AVL FIRE. 2006. *FIRE v8.5 Manual*. Graz, Austria: AVL List GmbH.

Brenan, K. E., S. L. Campbell, and L. R. Petzold. 1989. *Numerical Solution of Initial-Value Problems in Differential-Algebraic Equations*. Amsterdam, The Netherlands: Elsevier.

Chung, T. J. 2003. *Computational Fluid Dynamics*. Cambridge: Cambridge University Press.

Douglass, B. P. 2004. *Real Time UML: Advances in the UML for Real-Time Systems*. 3rd ed. Boston, MA: Addison-Wesley.

Dymola. 2009. *Dymola 7.4 User Guide*. Paris, France: Dassault Systèmes.

DYNA4Engine. 2009. *DYNA4Engine THEMOS User Guide*. Munich, Germany: TESIS DYNAware GmbH.

Eich-Soellner, E., and C. Führer. 1998. *Numerical Methods in Multibody Dynamics*. Stuttgart, Germany: Teubner.

Elmqvist, H., M. Otter, and F. E. Cellier. 1995. "Inline Integration: A New Mixed Symbolic/ Numeric Approach for Solving Differential-Algebraic Equation Systems." In *Proceedings of the ESM'95, SCS European Simulation Multi-Conference*, Prague, Czech Republic.

Gheorghiu, V. 1996. *Modelle für die Echtzeitsimulation von Ottomotoren*. Essen, Germany: Haus der Technik.

GT-Suite. 2009. *GT-Suite V7.0*. Westmont, IL: Gamma Technologies.

Guzzella, L., and A. Amstutz. 1998. "Control of Diesel Engines." *IEEE Control Systems Magazine* 18(5): 53–71.

Hendricks, E., and S. Sorenson. 1990. "Mean Value Modelling of Spark Ignition Engines." SAE Technical Paper 900616.

Heywood, J. B. 1989. *Internal Combustion Engine Fundamentals*. New York: McGraw-Hill Higher Education.

Isermann, R. 2005. "Model-Based Fault-Detection and Diagnosis—Status and Applications." *Annual Reviews in Control* 29 (1): 71–85.

Izenman, A. J. 2008. *Modern Multivariate Statistical Techniques: Regression, Classification, and Manifold Learning*. New York: Springer.

Jensen, J.-P., A. F. Kristensen, S. C. Sorenson, N. Houbak, and E. Hendricks. 1991. "Mean Value Modeling of a Small Turbocharged Diesel Engine." SAE Technical Paper 910070.

Klein, P. 2009. *Zylinderdruckbasierte Füllungserfassung für Verbrennungsmotoren*. PhD diss., Universität Siegen, Germany.

Kuberczyk, R., S. Dobler, B. Vahlensieck, and M. Mohr. 2010. "Reducing NO_x and Particulate Emissions in Electrified Drivelines." In *Getriebe in Fahrzeugen*. Friedrichshafen, Germany, June 22–23.

Lehmann, E. 2003. *Time Partition Testing*. PhD diss., Technische Universität Berlin, Germany.

Ljung, L. 1999. *System Identification: Theory for the User*. 2nd ed. Upper Saddle River, NJ: Prentice Hall.

Meder, G., A. Mitterer, H. Konrad, G. Krämer, and N. Siegl. 2007. "Development and Calibration of Model-Based Controller Functions by the Example of a Six Cylinder Engine with Full-Variable Valvetrain." *Automatisierungstechnik* 55 (7): 339–45.

Merker, G., C. Schwarz, and R. Teichmann. 2011. *Combustion Engines Development: Mixture Formation, Combustion, Emissions and Simulation*. Berlin, Germany: Springer.

Modelica. 2010. *Modelica Language Specification, V3.2*. Linköping, Sweden: Modelica Association.

Moraal, P., and I. Kolmanovsky. 1999. "Turbocharger Modeling for Automotive Control Applications." SAE Technical Paper 1999-01-0908.

Mosterman, P. J., and G. Biswas. 2002. "A Hybrid Modeling and Simulation Methodology for Dynamic Physical Systems." *Simulation: Transactions of the Society for Modeling and Simulation International* 178 (1): 5–17.

Mosterman, P. J., S. Prabhu, A. Dowd, J. Glass, T. Erkkinen, J. Kluza, and R. Shenoy. 2005. "Embedded Real-Time Control via MATLAB, Simulink, and xPC Target." In *Handbook of Networked and Embedded Control Systems*, edited by Dimitrios Hristu-Varsakelis and William S. Levine, 419–46. Boston, MA: Birkhäuser.

Mosterman, P. J., S. Prabhu, and T. Erkkinen. 2004. "An Industrial Embedded Control System Design Process." In *Proceedings of the Inaugural CDEN Design Conference (CDEN'04)*, 02B6-1 through 02B6-11, Montreal, Quebec, Canada, July 29–30, 2004.

Nieuwstadt, M. J., and I. Kolmanovsky. 1999. "Cylinder Balancing of Direct Injection Engines." In *Proceedings for the American Control Conference*, San Diego, CA.

Papadimitriou, I., M. Warner, J. Silvestri, J. Lennblad, and S. Tabar. 2005. "Neural Network Based Fast-Running Engine Models for Control-Oriented Applications." SAE Technical Paper 2005-01-0072.

Sargent, R. G. 2007. "Verification and Validation of Simulation Models." In *Proceedings of the 2007 Winter Simulation Conference*, Washington, DC.

Schlesinger, S. et al. 1979. "Terminology for Model Credibility." *Simulation* 34 3:101–5.

Schuette, H., and M. Ploeger. 2007. "Hardware-in-the-Loop Testing of Engine Control Units—A Technical Survey." SAE Technical Paper 2007-01-0500.

Simulink. 2004. *Using Simulink*. Natick, MA: The MathWorks.

Vibe, I. I. 1970. *Brennverlauf und Kreisprozess von Verbrennungsmotoren* [German translation from Russian]. Berlin, Germany: VEB-Verlag Technik.

Vidyasagar, M. 1993. *Nonlinear Systems Analysis*. 2nd ed. Upper Saddle River, NJ: Prentice Hall.

21 Specification and Simulation of Automotive Functionality Using AUTOSAR

Marco Di Natale

CONTENTS

21.1 Introduction 523
21.2 Introduction to AUTOSAR...526
    21.2.1 VFB Level 527
    21.2.2 Behavioral Level..528
    21.2.3 RTE Level..529
    21.2.4 AUTOSAR Process 530
    21.2.5 AUTOSAR Timing Model 532
    21.2.6 AUTOSAR Tools and the Role of Tools...536
21.3 Use of AUTOSAR Models for Simulation And Analysis 537
    21.3.1 Use of AUTOSAR in the Development Process 537
    21.3.2 Simulation at the VFB (and Behavior) Level..................................538
    21.3.3 Simulation at the RTE Level 538
    21.3.4 Simulation at the BSW Level: From ECU-Level Simulation to
            Full Architecture Modeling..539
21.4 Model-To-Model Integration and Translation: From Simulink® to
    AUTOSAR and From AUTOSAR to Simulink® 540
21.5 Simulation, Timing Analysis, and Code Generation: Consistency Issues
    and Model Alignment..545
21.6 Conclusions...546
Acknowledgments..546
References...546

21.1 INTRODUCTION

The Automotive Open System Architecture (AUTOSAR) development partnership, which includes several carmakers or original equipment manufacturers (OEMs), car electronics (Tier 1) suppliers, and tool and software vendors, has been created to develop an open industry standard for automotive software architectures. The

AUTOSAR standardization effort spans across all software levels, from device drivers to the operating system, the communication abstraction layers, the network stacks, and also the specification of application-level components.

The current version of the standard includes several specifications, which can be roughly classified into

- A standard specification for the definition of application software components (SWCs), their interfaces, and an environment for their cooperation at the functional level.
- A standard for the definition of the cooperation of *component behaviors*, including the definition of the events that are relevant for executing the *runnables* of components, which are defined as procedures executed in reaction to events.
- A reference software *platform architecture*, encapsulating the basic software (BSW) (operating system and device drivers included) and providing a middleware layer for the execution of components and their interactions (including communication and exchange of signals and events).
- A standard format for the (coarse grain) description of distributed *hardware architectures*, the description of the mapping of runnables to tasks, and the mapping of those onto the execution architecture.
- A *metamodel* formalizing all the above.

In addition, AUTOSAR defines a development model (or, more appropriately, major steps and cornerstones in the development process) with automated support by tools and a common interchange format, based on XML, for model information at all stages in the process.

The AUTOSAR project has focused on the concepts of location independence, interface standardization, and code portability. The initial purpose was to enable the transition from Federated Architectures to Integrated Architectures. In a Federated Architecture, suppliers provide a node or electronic control unit (ECU) to be integrated in a network for each major functionality, which caused a proliferation of nodes and complex dependencies on network messages. In an Integrated Architecture, OEMs provide the specification of SWCs to suppliers, who are responsible for their development according to a standardized interface. OEMs can then integrate SWCs on an architecture platform of their choice. This type of portability and modularity requires the definition of a standard for SWCs, which includes the definition of their interfaces and a standard application program interface (API) for interacting with the software abstracting the physical platform. Hence, the AUTOSAR model has been strongly oriented to the representation of typical interactions among software packages and its semantics is strongly oriented to the modeling of cooperating procedures, rather than the representation of a formal model of computation (MoC) with a mathematical underpinning.

Later, with the evolution of the standard, the goal has become more ambitious and AUTOSAR is becoming a candidate for full-fledged system-level modeling, including capabilities for simulation and (worst-case) timing analysis or possibly even other types of analysis (such as reliability) that are required on system-level models.

The need for better control over the emergent behavior, the opportunity for analysis (by simulation or other formal method), and the need for controlling integration of heterogeneous models (for example Simulink® subsystems [1]) led the consortium to the specification of the AUTOSAR metamodel, which however, is still far from being based on a formal MoC.

To give an example, the notion of time and timed events was only introduced in the latest (4.0) release of the standard [2], with the goal of enabling worst-case timing analysis of response times at the level of specifications. Unfortunately, it appears that the timed event model is at best cumbersome if the purpose is, for example, to build a common discrete-time base for a system-wide AUTOSAR model integrating synchronous reactive (SR) models as subsystems. In addition, currently there is no tool support for the use of timed events.

With respect to the goal of simulation, AUTOSAR is clearly oriented toward the following: the simulation of functionality considering the structure of the SWCs and their actual code implementation; the implications of the mapping of the runnables to the elements of the execution architecture; and of scheduling issues and other platform-related delays, such as communication latency and operating system or device-driver delays. As such, AUTOSAR models and tools can be used for both hardware-in-the-loop (HIL) or software-in-the loop simulations (see also the classification in the chapter at Ref. [3]) and can be performed in real time or in logical time.

AUTOSAR is not suited to the representation of continuous-time systems. Therefore, its use in a model-in-the-loop framework will require integration with other tools (such as Simulink). These tools either provide the plant model and cooperate in a cosimulation environment or produce the code implementations for the models of the missing (continuous-time) systems. AUTOSAR models are a better fit to simulate the controller part. However, the lack of support for a detailed behavioral model makes their use unlikely in the early stages. AUTOSAR runnables are intended to be simple entry points for a code procedure. Even when the runnable behavior is available as a dataflow or as a state machine (or other type of model), either a cosimulation environment linking other tools (such as Simulink) to an AUTOSAR simulation engine or (more likely) the actual code implementation is required.

In conclusion, with respect to the framework outlined in the previous chapter, an AUTOSAR model can be simulated either right before or right after the availability of a real-time executable. The underlying model is a discrete event model based on the available framework of the run-time environment (RTE) events (described in the following sections). An AUTOSAR model can be constructed for the simulation of discrete-time systems by leveraging timed events (although creating a global-time framework is not necessarily an easy task, given that periodic timed events are local to each component).

Also, with respect to the task of model verification, the lack of a clear semantics with respect to time makes any type of formal verification of correctness in model-to-model transformations difficult or even impossible altogether.

We provide first a short introduction to AUTOSAR and its main concepts. Next, we discuss issues related to the possibility of using an AUTOSAR description of the system for simulating its behavior at increasing levels of accuracy with respect to the model of the execution platform. Finally, issues related to the necessity of maintaining a consistent view of the models that are used for simulation, timing analysis, and code generation are discussed.

21.2 INTRODUCTION TO AUTOSAR

AUTOSAR was created to develop an open industry standard with a common software infrastructure based on standardized interfaces for SWC specification and integration at different layers. In agreement with the move toward an integrated architecture, the AUTOSAR goal is to enable the definition of SWCs that can execute independently from their placement in a complex, distributed platform.

The AUTOSAR architecture has three major domains. In the first, at the level of the *Virtual Functional Bus* (VFB), the system is described as a collection of application SWCs, cooperating over ports in a purely functional architecture (top part of Figure 21.1, which outlines the tool-assisted match between the AUTOSAR functional model and the execution platform). The VFB consists of the set of logical connections linking the interfaces of the cooperating SWCs. AUTOSAR separates communication from computation and synchronization. The component interface is defined as a collection of ports, and the execution of the procedures of a component is represented by a set of *runnables*, activated in response to events and defined in a *behavior level*.

At runtime, the set of SWCs is executed on a standard software abstraction layer. This layer is automatically generated according to the specification of the (possibly distributed) execution platform supporting the computations and according to the specifications of the VFB layer and the behaviors of the components. The software

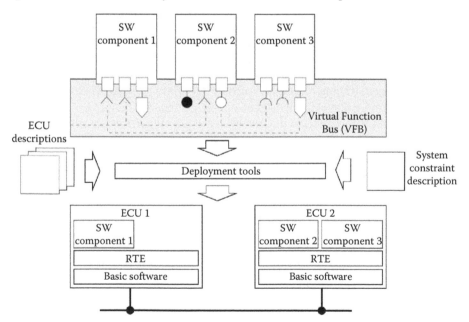

FIGURE 21.1 The AUTOSAR functional model and a deployment onto a distributed architecture.

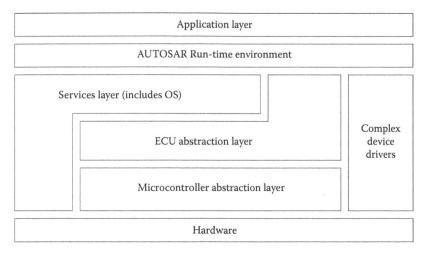

FIGURE 21.2   The AUTOSAR standard software platform.

layer between the platform and the realization of the SWCs using the standard API is called RTE (run-time environment). The RTE is the runtime implementation of the VFB and has the responsibility of implementing the virtual communication occurring over the component ports redirecting it to a concrete communication protocol. The concrete implementation differs according to the task mapping and the ECU allocation of the communicating runnables. In addition, the RTE generates events resulting from the component interactions and is responsible for the generation of periodic events.

The RTE, in turn, runs on a standardized architecture of additional abstraction layers, device drivers, and operating system functions, which is collectively referred to as *BSW* (an outline of the standard SW platform is in Figure 21.2).

The third element of the specification is the AUTOSAR process itself, or better the definition of a basic workflow that connects the main development stages, with the expected results and the definition of the tools to be used in each stage.

21.2.1   VFB Level

The first level for the definition of an AUTOSAR system is the Functional Model, where SWCs encapsulating sets of system-level functions are connected and cooperate over a VFB (Figure 21.1). Later, they are bound to an execution platform consisting of ECUs connected by communication buses. AUTOSAR components may be defined hierarchically. At the bottom of the hierarchy, components are "atomic," meaning that each instance is assigned to one ECU (cannot be distributed).

In AUTOSAR, a component defines its required and provided interface through ports, which are the only means to interact with other components. Ports may provide or require access to data or services. A port defining a provided point of access is defined as a P-Port, and a required port is called R-Port. An Interface, attached to each port, may be of type client/server, defining the signatures of the operations that are invoked or provided, or of type sender/receiver, defining a model of the data structures

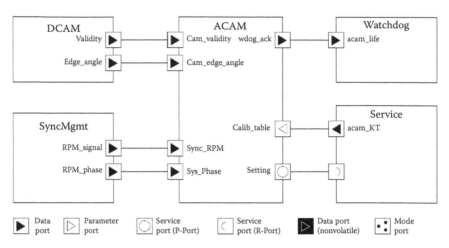

FIGURE 21.3 AUTOSAR components and ports.

that are written or read. The formal description of the interface is in the *software component template*, an XML file, which includes the formal specification of the data types and the description of the communication behavior, as specified by attributes.

Those attributes define (among others) the length of the queues for the data ports and the behavior of receivers (blocking, nonblocking, etc.) and senders (send cyclic, etc.). Figure 21.3 shows an example of AUTOSAR components interacting over sender/receiver ports and client/server ports.

The port syntax identifies several types of data ports: regular data type, parameter ports, and nonvolatile ports. Furthermore, the language syntax identifies special ports for the definition of mode switches and also provides for a distinction between application ports and ports that connect application components to AUTOSAR services.

21.2.2 BEHAVIORAL LEVEL

In AUTOSAR, the behavior of atomic SWCs is represented by a set of *runnable entities* (*runnables* for short) that are in essence entry points of procedures or, as stated in the standardization documents, atomic schedulable units (sequential segments of code, to be executed under the control of a scheduler), communicating with each other over the component *ports*. AUTOSAR provides several mechanisms for runnables to access the data items in the ports for *sender/receiver* communication and the services of *client/server* communication ports. In particular, access to ports of type Send/Receive can be *implicit* or *explicit*. In the case of an explicit access, it is the job of the developer to insert calls to the standard RTE API functions for reading from and writing into ports inside the runnables. If a port is accessed in an implicit way by a runnable, then the port contents are copied from the port automatically by the RTE before starting the runnable code (for receive ports) or written at the end of the runnable execution (for sender ports). The activation model, the communication among runnables, and the synchronization in the execution of runnables (local and remote) is specified in a middle-level RTE layer.

21.2.3   RTE LEVEL

The run-time environment (RTE) is the runtime implementation of the VFB. The RTE is local to each ECU. It is a complex layer that must provide location independence in the request of services and data communication among components. The RTE also takes care of the management and forwarding of events to runnables (for triggering their execution or releasing them from wait points). In practice, in the case of local communication, service requests are simply remapped to local function calls. In the case of remote services, the RTE forwards the call to a stub that performs marshalling of parameters in a message, selects the appropriate communication network and the corresponding device driver, transmits the message, and waits for the reply message. When the reply message arrives, the RTE extracts the returned value and closes the remote service request by returning to the caller and generating (if requested) the appropriate event.

The available RTE events are the following:

- *Timing Event* triggering periodical execution of runnables.
- *DataReceivedEvent* upon reception of a sender/receiver communication.
- *OperationInvokedEvent* for the invocation of client/server service.
- *DataSendCompleteEvent* upon sending a sender/receiver communication.

Each runnable must be associated with at least one RTE event, defining its activation. The scope of each RTE event is limited to the component. For example, two periodic events with the same period activating runnables on different components are not guaranteed to be synchronous. Actually, their relative phase is unspecified unless explicitly modeled by using synchronization components (typically at the BSW level). In case a runnable is activated by more than one event, it must be reentrant or the developer must take care of protecting its internal state variables. A *WaitPoint* can be used to block a runnable while waiting for an event. Runnables that may block during their execution are called type 2, as opposed to the others, defined as type 1 (this definition impacts the selection of the operating system scheduler and of the scheduling policies). For instance, for an operation defined as part of a component interface, the behavior specifies which runnable is activated as a consequence of the operation invocation event.

The only possible definition of an AUTOSAR MoC emerges from the composition of the behavior options that are attached to each interface and runnable specifications, with no guarantee of consistency. The concept of time and time-related events, which is necessary for the formal definition of an MoC, has only appeared in release 4.0 of the standard. The time model is briefly addressed and discussed in Section 21.2.5.

Finally, there is the description of components/runnables at the implementation level. The language for the description of the internal behavior of runnables is not part of AUTOSAR. AUTOSAR does not require a model-based development flow: a component may be handwritten or generated from a model. For the definition of the models of the runnables, it relies on external tools such as Simulink and ASCET, which brings the issue of the composition of heterogeneous models. The lowest (most concrete) level of description specifies a reference to the code files implementing the runnables of the components and the resource consumption of SWCs. This resource requirements model includes the worst-case execution time of runnables, for which a special section is reserved in the implementation description of the runnables.

AUTOSAR defines the methodology and tool support to build a concrete system of ECUs. This includes the configuration and generation of the runtime environment (RTE) and the BSW (including the real-time operating system) on each ECU. The generation of the RTE and BSW layer is the responsibility of the *deployment tools*, which define the mapping of the function to the architecture and the generation of code based on the model of the execution platform, defined (in XML format) by the ECU *description files*.

Finally, the definition and realization of components is not limited to the application level. Designers can define runnables and components at the level of the BSW as well, with less restrictive rules. These runnables and components are included in the code automatically generated by the deployment tools and can be directly activated by events generated at the operating system level (interrupt handlers, counters, and alarms in AUTOSAR).

21.2.4 AUTOSAR PROCESS

AUTOSAR requires a common technical approach for selected steps in the development process, from the system-level configuration to the generation of the executables for each ECU. The AUTOSAR Methodology (a graphical representation is in Figure 21.4) is the description of these activities, with their inputs and outputs. The AUTOSAR Methodology is formally not a true process description since it does not prescribe the order in which the activities should be performed, neither the actors involved.

The first step is the creation of the *System Configuration Input*, which defines the SWCs, the hardware execution platform, and the system constraints that apply to the

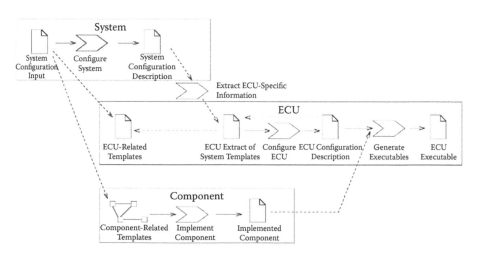

FIGURE 21.4 The AUTOSAR process.

mapping of the components onto the ECUs. The format for the formal description is defined in a standard XML schema and includes the following templates.

- *Software Components*: Each SWC is described in terms of its ports and interfaces, including the data types and the port attributes.
- *ECU Resources*: Each ECU requires specifications regarding the processor unit(s), the available memory, and the available peripheral devices, including sensors and actuators.
- *System Constraints*: Regarding the mapping of the signals into bus messages as well as the mapping of the component runnables into tasks and the allocation of tasks into the ECUs.

The System Configuration Input includes or references various constraints. These constraints are very simple and mostly consist of forced or forbidden mapping configurations. In reality, the template allows for the definition of estimates on the availability of resources on ECUs, thereby providing less intuitive constraints to mapping configurations.

The *Configure System* activity has the responsibility to map all the SWCs to ECUs and also to define the system communication matrix. This matrix describes the message frames exchanged over the system networks, that is, the mapping of the signal information (the data exchanged over ports) into the frames. Deriving such a mapping is extremely complex and consists of a system optimization problem, in which system metrics of interest, including performance and possibly reliability, extensibility, and composability should be maximized considering constraints on resources and timing requirements [4,5]. In the mind of the AUTOSAR developers, this synthesis and optimization process should be assisted by a set of tools, defined as *System Configuration*. The output of this activity is the *System Configuration Description* including system information (e.g., bus mapping and topology) and the mapping of SWCs to ECUs.

Further steps must be performed for each ECU in the system. The activity denoted as *Extract ECU-Specific Information* extracts the information from the System Configuration Description for a specific ECU, thereby generating the *ECU Extract of System Configuration*. This step is rather simple since it consists of a projection of the elements of the System Configuration Description that are allocated to a specific ECU.

At the ECU level, the system configuration stage is mirrored by the *Configure ECU* activity, where the RTE and the BSW modules are configured for each ECU. This includes the definition of the task model (the assignment of runnables to tasks) and the configuration of the scheduler and of the BSW components (driver included). The result of the activity is included in the *ECU Configuration Description*. The configuration is based on the information extracted from the *ECU Extract of System Configuration*, the definition of the *Available SWC Implementations*, and the *BSW Module Description*. The latter contains the vendor-specific information for the ECU configuration.

This activity defines the detailed scheduling information, the configuration of the communication module, the operating system, and other AUTOSAR services. Moreover, this is the time when an implementation must be provided for each atomic SWC. In contrast to the extraction of ECU-specific information, the configuration activity is actually a complex design synthesis and optimization step.

In *Build Executable* (last step), an executable is generated based on the configuration of the ECU. This step typically involves generating code (e.g., for the RTE and the BSW), compiling code (compiling generated code or compiling SWCs available as source code), and linking everything together into an executable.

Parallel to these steps are several steps performed for every application SWC (to be integrated later into the system), including the generation of the component's API and the implementation of the component's functionality.

The initial work in this context starts with providing the necessary parts of the SWC description. That means at least the *Component Internal Behavior Description* as part of the SWC-related templates has to be filled in. The internal behavior describes the scheduling-relevant aspects of a component, that is, the runnable entities and the events they respond to. Furthermore, the behavior specifies how a component (or more precisely which runnable) responds to events like receiving data elements. However, it does not describe the detailed functional behavior of the component, which is only filled in later, when the *Component Implementation* (typically the C sources) is provided, together with additional implementation-specific information and information about the further build process steps (e.g., compiler settings, optimizations, etc.).

21.2.5 AUTOSAR TIMING MODEL

Starting from version 4.0, AUTOSAR includes a *Specification of Timing Extensions* [6]: a language overlaid on top of the existing AUTOSAR concepts with the primary purpose of enabling the definition of timing contracts and the (worst-case) schedulability analysis of the timing behavior of a system. Unfortunately, the keywords "simulation," "model of computation," or "model-to-model transformation" never appear among the list of the requirements for the timing model, which is indeed lacking with respect to the needs of a formal semantics for simulation and verification.

With respect to the AUTOSAR process, the timing specification extends the AUTOSAR metamodel, without necessitating a separate template. Also, the extension is introduced without the indication of any additional process steps. Hence, the description of the previous subsection remains fundamentally unchanged.

In the specification of timing extensions for AUTOSAR, the event is the basic entity. It is used to refer to an observable behavior within a system (e.g., the activation of a *RunnableEntity* and the transmission of a frame) at a point in time. The AUTOSAR timed event is formally derived from a generic abstract concept defined as *TimedDescriptionEvent*. Unfortunately, timed events are not specializations of RTE events, but define a partly independent structure of timing annotations and constraints. As a consequence, it becomes problematic to maintain and enforce the consistency between the RTE events defining the behavior semantics (i.e., the activation and the synchronization among runnables) and the timed events providing time attributes and constraints that apply to the behaviors. The description of the timing elements follows the organization of the AUTOSAR-level views (Figure 21.5).

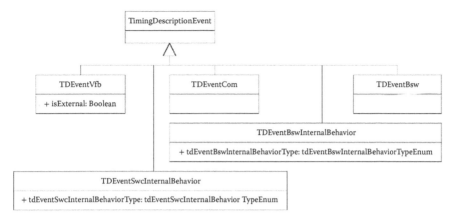

FIGURE 21.5   Types of events by modeling level.

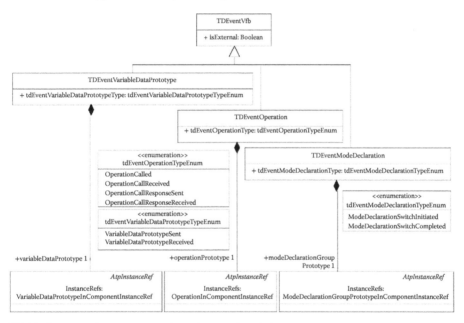

FIGURE 21.6   Timed events defined for the virtual functional bus level.

Timing events and constraints can be applied to a *VFB-level view*. Functional end-to-end timing constraints (including possibly the physical sensors and actuators) can be captured in this view. Of course, the VFB view is independent of the platform design and of the allocation of the execution resources. Therefore, the analysis does not include any timing issues from resource scheduling, nor the description of platform overheads. Also, the internal behavior of components is not considered. In conclusion, the constraints that can be defined and analyzed are execution intervals, as opposed to deadlines, defined between events on component ports. As shown in Figure 21.6, the timed events allowed at this level mirror the corresponding RTE events on ports (such as, for example, the time at which data is sent or received, an operation called, or an operation call received). When adding the *internal behavior description*, timed events can refer to the activation, start, and termination (see Figure 21.7) of the execution of RunnableEntities. This specification still mostly applies to the definition, and possibly verification, of end-to-end deadline constraints.

After the BSW and ECU generation stage, more information is available on the resources of the execution platform, and the allocation of runnables to tasks and of tasks to ECUs. Also, the model now includes the specification of schedulers and the composition of the message frames. Therefore, the events that are defined at these levels also apply to tasks and message frames. This is the level where schedulability analysis can be performed and also the level where timed events could be leveraged to simulate the impact of platform-dependent delays.

In general, at all levels, the occurrence of a timing event can be described by an *Event Triggering Constraint*. AUTOSAR offers four basic types of event triggering constraints as described in Figure 21.8: *periodic events*, *sporadic events*, which are characterized by a minimum interarrival time, and *bursty events*, in which the specification defines periodic bursts occurring with a given period, and the composition of each burst in terms of number of events and minimum interarrival time. Finally, streams determined by the *position in time of each event and fully arbitrary patterns* complete the available categories.

In addition, the extension defines constraints and properties that apply to pairs or sets of events (their relations). One notable example of a relation among events is the event chain, with which the concept of end-to-end timing constraint is associated (Figure 21.9).

For the purposes of simulation, an important requirement is the possibility of defining causality dependencies, that is, to enforce a partial order of execution among the system actions and the possibility of enforcing synchronization among the actions. Before the timing extensions, expressing a causal dependency and synchronizing the events triggering the execution of runnables was extremely difficult if even possible at all (an example of this difficulty is provided in the following section dealing with model-to-model transformations).

Unfortunately, the new timing extensions only provide a partial solution. Causality constraints can indeed be expressed by an *Execution Order Constraint*, which is used to express the functional dependency between ExecutableEntities (runnables) and to restrict their order of execution at runtime. The AUTOSAR

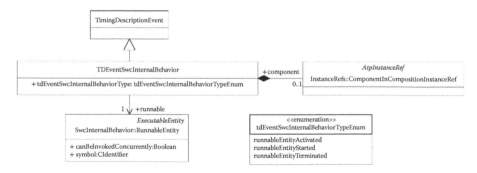

FIGURE 21.7   Timed events that apply to the behavior description.

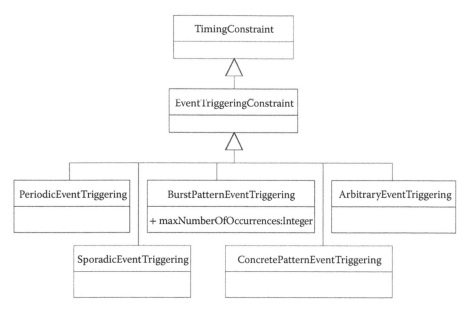

FIGURE 21.8   Metamodel of timing events in AUTOSAR 4.0.

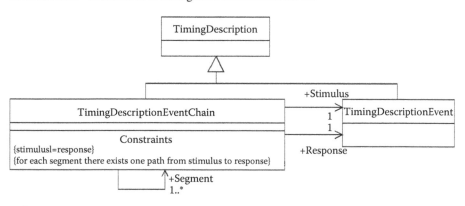

FIGURE 21.9   Metamodel of timing event chains in AUTOSAR 4.0.

Synchronization Timing Constraint, however, is not intended to be used to define a consistent time base for timed events and, in turn, for the system behavior. The definition of a set of system-wide synchronized timed events could have allowed the AUTOSAR modeling of discrete-time systems and the translation of synchronous models in AUTOSAR. However, the original purpose of the specification is different and synchronization constraints do not apply to sets of individual events but to event chains, with the possibility of synchronizing the initial stimulus or the final response times. The motivation for the current definition of timed events is to reduce the pessimism in the worst-case analysis of end-to-end response times. Indeed, the synchronization constraint is intended to be defined together with a tolerance, indicating a true specification constraint on the runtime behavior of complex chains of reactions rather than (for example) a method for enforcing a discrete-time framework.

Another set of constraints can be used for the purpose of synchronization: the offset constraints. These constraints are somewhat more suited to the synchronization of events since they explicitly target individual events rather than chains. However, an offset always refers to a single pair of events (a source and a target), which once again makes its use cumbersome for a system-wide synchronization of reactions.

In conclusion, it is worth noting that version 4.0 of the standard is very recent and still not supported by commercial tools. The hope is that future evolve and become more suited to the goals of simulation and allow for better integration (meaning on formal grounds) of heterogeneous models.

21.2.6 AUTOSAR Tools and the Role of Tools

From the description of the AUTOSAR process, it is very clear how the use of tools is recommended or even mandatory in several activities, such as the code generation and configuration stages. Indeed, there are currently several AUTOSAR tools available on the market. In addition, the AUTOSAR consortium members realized that there is a common or core module that is required by all tool providers. This core module includes the functions for the management of the metamodel and the enforcement of its rules in the construction of a model, the representation of the model in memory, the generation and parsing of the input and output files at each stage of the AUTOSAR process in the standard ARXML format, and basic capabilities for model editing.

The consortium members launched a common initiative to create an open project (Artop, an open source project, although open for contribution and use by the consortium members only [7]) that leverages the metamodeling and modeling features of the Eclipse platform [8] to develop a common foundation that could be exploited by all AUTOSAR tools developers, vendors, and users. The code of the Artop project is available to the members of the consortium that can contribute to its development and use it in the form they see fit, either directly (in the case of OEMs and Tier 1 suppliers) or, more likely, as a common starting point for the development of commercial tools in the case of tool vendors. Artop does not have simulation and analysis capabilities and does not include code generators or the mapping synthesis capability for matching a functional model to an execution platform. These components are intended to be developed as custom extensions by tool vendors.

Currently, tool vendors tend to specialize in the different process activities that are requested by the standard. Several companies offer tools for the design of AUTOSAR models. Examples of products in this area are SystemDesk by dSPACE [9], DaVinci by Vector [10], and AUTOSAR Builder by Geensoft [11]. These tools include RTE generators, but not the configuration of the BSW. Also, they offer (somewhat limited) simulation capabilities (mostly at the VFB and behavior level as discussed in more detail in the following section).

Other companies provide products, such as Tresos from Elektrobit [12], specialized in the stages of BSW generation and configuration, including the operating system.

In general, the mapping of the functional model into the execution platform is quite far from being a synthesis process supported by an optimization engine. Currently, tools typically provide a default configuration of the task model and require a complete user specification with respect to the mapping of the tasks onto the ECUs.

Limited support for AUTOSAR is also offered by other modeling tools, including, for example, Simulink. The support does not include full model-to-model transformation capabilities (not surprisingly, given the sematics distance between the two models), but focuses on these areas:

- Integration of code generated from a Simulink model with the RTE generated by AUTOSAR tools
- Generation of a component or runnable specification by wrapping a (virtual) Simulink subsystem to allow its inclusion in an AUTOSAR model
- Importing an AUTOSAR component specification into Simulink by automatically generating an equivalent subsystem interface model

21.3 USE OF AUTOSAR MODELS FOR SIMULATION AND ANALYSIS

Before going into the technical details on the use of AUTOSAR models for simulation and the discussion of the issues related to model-to-model transformations, it is important to recall what information is necessary for building an AUTOSAR model and at what stage in the development process an AUTOSAR description of the system should be available. In other words, is the AUTOSAR description the first and authoritative description of the system functions or is it produced later in the development based on some other models or outputs? Also, we should ask what could be the contribution of simulation performed on an AUTOSAR model to the understanding of the system dynamics. This question is relative to the possible advantages and outputs of a system-level simulation (or analysis of other types, including timing and reliability) performed on AUTOSAR models. That is, what can an AUTOSAR model tell us that cannot be obtained in another way?

An initial short answer could be the following: given that AUTOSAR is fundamentally an empty container for functional behavior descriptions and given its limitations in the modeling of continuous-time systems, it is probably (as of now) not a realistic starting point for system-level modeling and analysis, but rather a valuable intermediate formalism that allows a better evaluation of the impact of the software design and coding and of the software–hardware execution platform on the performance of the controls and the other application functions.

21.3.1 USE OF AUTOSAR IN THE DEVELOPMENT PROCESS

A possible use of the AUTOSAR formalism and tools in a development process is therefore the following:

- The system functionality, including the system controls, are modeled in a suitable language (e.g., Simulink), together with the plant model, and validated by simulation with logical execution time assumptions. In case the controller (re)uses AUTOSAR SWCs, possibly imported in Simulink.
- The execution platform for the system is selected and modeled in an AUTOSAR environment.

- The functional model is translated into a corresponding AUTOSAR set of components, some functionality is possibly realized by reusing existing component models, and the virtual AUTOSAR model resulting from the composition is mapped onto the execution architecture according to the AUTOSAR process.
- The result of the mapping stage, performed by the AUTOSAR tools, is the definition of the communication matrix, the task model, and the configuration of the RTE and BSW levels. This definition now allows the simulation of the system functions considering platform-related issues, including the simulation of not only the actual code implementing the application functions but also the code for the communication services and the basic services, with the operating system and the device drivers and the actual network message set.

Although other options are possible, this is a flow that exploits AUTOSAR strengths and attempts to overcome its limitations.

21.3.2 SIMULATION AT THE VFB (AND BEHAVIOR) LEVEL

This is the first option, the one commonly provided by commercial tools and possibly the one without clear advantages with respect to other competing methods since the peculiarity of an AUTOSAR model should be the availability of architecture-specific information, including the tasking and message model and the scheduling information. VFB-level simulation focuses on the higher level of the functional behavior of applications within the context of three layers of abstraction: the interface definition, the behavior definition, and the implementation definition.

For the interface definition layer, the ports and the associated interfaces of a SWC communicating with other components and underlying BSW services are validated and verified. For the behavioral layer, runnables are validated and verified in terms of their execution semantics with respect to their triggering and activation sources. Since there is no model for the RTE and BSW, the events that are relevant for simulation are the set of (simulated) RTE periodic events and the Call and Write events on, respectively, Sender and Receiver ports.

The implementation layer addresses the validation and verification of resource consumption. In particular, if this layer is fully defined, the time when events are generated can be inferred using the worst-case execution time attribute of each runnable. If this information is not available, the runnable execution is simulated to occur in zero logical time [13]. In a VFB-level simulation, the RTE is emulated on the selected simulation platform with its operating system, omitting details about the actual RTE implementation, the task model, and the execution platform.

21.3.3 SIMULATION AT THE RTE LEVEL

The next level is the simulation of the system including the set of RTE calls and the RTE overheads. This requires availability of an RTE generator that provides the corresponding code. The simulation at the RTE level adds little to the previous step since it allows to verify the system functions with the *actual code* that performs the data communication over ports and the generation of events.

21.3.4   SIMULATION AT THE BSW LEVEL: FROM ECU-LEVEL SIMULATION TO FULL ARCHITECTURE MODELING

This final simulation level includes the possibility of modeling components at the BSW level, including the tasking model, the operating system, and the device drivers. Simulation at the BSW level can be performed not only at the ECU level but also at the system level, assuming availability of an architecture model, which allows the emulation of the ECUs and the network. Also, availability of the entire software stack at this point fully enables HIL simulation. This is indeed supported by the dSPACE development toolchain allowing ECU-level HIL simulation on the (micro) autobox platforms.

For the timing evaluation at the ECU level with platform models, there is little support from commercial tools.

An example of performance evaluation of AUTOSAR-based systems applied to a power window case study is described in Ref. [14], where an AUTOSAR application description is translated into a Discrete Event System Specification (DEVS) and then simulated using a DEVS engine. DEVS is a formalism for modeling discrete event systems in a hierarchical and modular way [15,16]. In Ref. [14], a DEVS model of the AUTOSAR BSW layers was used to simulate the real-time characteristics of the power window system. The execution times of all relevant actions were estimated by measurements on a MPC560xP microcontroller running at 64 MHz with an AUTOSAR BSW generated using Tresos AutoCore. The source code was compiled using the MULTI compiler of Green Hills Software. The measures included the execution time of all the runnables for any possible state (without the calls to the RTE); the execution time required for activating or suspending tasks and the context switch times; and the execution time of the writing and receiving of messages in every part of the communication stack including the RTE. The semantics of DEVS is indeed suited for matching the event model of AUTOSAR, and the simulation results were found to be a good match of the observed execution on the actual hardware target. The model included not only the AUTOSAR BSW but also a model of the controller area network (CAN) bus.

Another example of ECU-level modeling effort coupling the description of AUTOSAR SWCs with the BSW generation and an architecture model defined in SystemC is provided in Ref. [17]. As in the previously cited work, the objective is to evaluate the timing behavior of AUTOSAR components taking into account the underlying hardware and communication paths.

In Ref. [17], a formal correspondence was established between AUTOSAR modeling concepts and the corresponding SystemC design concepts [18] at different levels of abstraction (e.g., AUTOSAR runnables map onto SystemC processes). When considering the RTE and the BSW levels, the AUTOSAR design can be translated into a SystemC Programmer View where timing information is provided by timing annotations (in terms of execution cycles). Therefore, if the implementation software exists (e.g., after generation by the RTE generators and BSW configurators), the timing behavior of the software can be evaluated by cycle-level simulation on SystemC models of the hardware platform using the SystemC simulation engine [19].

Additional research works [18,19] discuss simulation options for distributed platforms. In this case, of course, a model of the network adapter and the communication network must be provided. The case of CAN communication is discussed in the models in Ref. [14], and a FlexRay network model is part of the use case showing the applicability of the method proposed in Ref. [17].

21.4 MODEL-TO-MODEL INTEGRATION AND TRANSLATION: FROM SIMULINK® TO AUTOSAR AND FROM AUTOSAR TO SIMULINK®

Because of the AUTOSAR limitations, including the difficulty in the definition and integration of plant models, it is quite unlikely that an AUTOSAR model is used for the early stages of control validation. Therefore, it is important to allow designers to develop controls first in a suitable environment, such as, for example Simulink. Later, the information can be translated into an AUTOSAR model that can provide the additional definition of the execution architecture and the software architecture, including the task model, with the delays and overheads that are characteristic of the execution platform, such as scheduling latencies and communication delays.

In this second step, however, there are several difficulties involved. Simulink is based on a synchronous/reactive MoC, even if without a formal definition of the model semantics. In a Simulink model in which the continuous part of the model is solved with a fixed-step solver (which is a practical requirement when the final target is generating a computer-executable implementation), all the system blocks are modeled to be executed according to a system-wide synchronous discrete-time framework. Activation events occur instantanously and synchronously across the system. Computations and communications are performed in zero logical time.

In contrast, AUTOSAR is not based on a formal MoC. Completeness of behavior modeling is not mandatory since runnables are only required to be entry points to a sequential program. RTE events are local, and their scope (including periodic events) is at the component level. Therefore, the system-level behavior is defined as emerging from the cooperation of components. Several issues arise because of this choice.

- Communication semantics is partly specified using port attributes at the VFB-functional level and partly specified using the RTE events that define the activation of runnables following communication-related events at the RTE level. This would require semantics rules and tool support for guaranteeing the consistency of the two definitions. Similarly, timed events add yet another superstructure of specifications, which must be kept in agreement with the others.
- Activation and synchronization semantics among runnables is specified using RTE events that are local to each ECU. This makes the realization of a system-level semantics difficult. Consider, for example, the

synchronous model of Figure 21.10 (in the middle) in which a block b13, activated at 20 ms and part of a subsystem communicates with two other blocks on a different subsystem: b21, activated at 10 ms (oversampling the output stream of b13), and b22, activated at 20 ms and communicating with a unit delay. According to the synchronous semantics, the activation and communication pattern for the example is represented at the bottom of the figure. The corresponding AUTOSAR system is depicted at the top of Figure 21.10. A component corresponds to each subsystem and a runnable to each block with matching index (ρ13 is the implementation of b13). The activation and communication events are represented in the figure in a nonstandard notation (in practice, when using commercial tools, they would be part of an XML description). Also, the connections between runnables and ports and the state and communication variables are expressed in a nonstandard graphical description. Assume also that

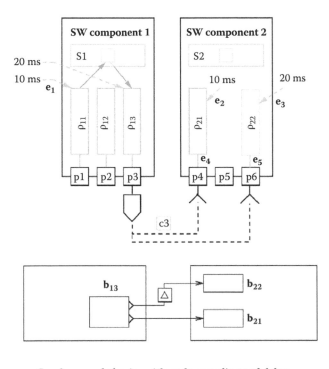

Synchronous behavior with undersampling and delay

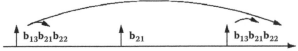

FIGURE 21.10 Semantics issues when mapping a typical synchronous behavior in AUTOSAR.

the components are implemented on different ECUs. The synchronous semantics illustrated in the bottom part of the figure cannot be easily defined using the available RTE events:

- The activation events of the blocks are synchronized, whereas RTE periodic events are local to each ECU and, in principle, with an unspecified relative phase.
- The causal dependency between the inputs and outputs of blocks b21 and b22 defines a partial order (a set of precedence constraints) in the execution of blocks. Expressing this partial order using the RTE timer events or the events on data ports is not trivial. In our example, $\rho 21$ must be activated after the production of data from $\rho 13$ but only once every two activations of $\rho 13$.
- Finally, communication with a unit delay is difficult to express without adding a dedicated runnable.

In reality, even if a translator from a Simulink to an AUTOSAR model would be highly appropriate for matching the needs of a model-based development flow, it is the inverse transformation that has received most of the attention and support from both sides (the AUTOSAR consortium and MathWorks®).

As part of the AUTOSAR specification, the document "Applying Simulink to AUTOSAR" [20] defines the rules to translate an AUTOSAR component (or a set of components) into a corresponding set of Simulink subsystems. Although one of the purposes of the document is to allow simulation of AUTOSAR components in a Simulink environment, a large part of the document is dedicated to the definition of translation methods that allow seamless integration of the code generated using the tools by MathWorks tools with the code generated by AUTOSAR tools for the RTE layer.

Also, the Real-Time Workshop® and Simulink Coder® [21] code generator products for the Simulink simulation environment today include a large section that is dedicated to the production of AUTOSAR components specifications and AUTOSAR-compatible code from sections of a Simulink model. However, a careful read of these user manuals reveals that in essence what is provided is not a general-purpose translator from Simulink to AUTOSAR, but a set of modeling guidelines and configurations of the code generators to produce an implementation of Simulink subsystems that is compliant with the AUTOSAR interfacing conventions and the RTE event semantics, including the automatic generation of AUTOSAR (XML) component specifications.

In essence, the model-to-model translation is straightforward (the code generation issues are more involved). AUTOSAR runnables are made to correspond to Simulink subsystems. When the AUTOSAR component has a single runnable, the subsystem models both the component and its runnable. When multiple runnables are part of the AUTOSAR component, each runnable corresponds to an atomic subsystem and the component to a wrapping (virtual) subsystem.

Sender/receiver ports map onto Simulink ports. A required port or R-Port is a Simulink input and a P-Port is a Simulink output port, with the corresponding definition of interfaces mapped onto Simulink BusObjects. Some more complexities are necessary to represent the client/server (method-oriented) interfaces of AUTOSAR. The AUTOSAR standard declares the client/server communication among application components out of scope and limits itself to modeling the client ports of application components that are requesting (standardized) services from the BSW or the ECU abstraction. Also, in the case of ports modeling BSW services, the correspondence consists of a block (Stateflow® or S-function) obtaining as input the set of required arguments and calling internally the desired function with its standardized signature. Of course, capturing the semantics for the time at which the service is invoked is entirely another story, given that this semantics is not explicit in the AUTOSAR model.

With runnables, however, the situation is clear: runnables are executed in response to RTE events, which can be periodic (time driven) or generated by operations on ports. Correspondingly, in Ref. [20], a canonical pattern is defined where runnables are function-call subsystems in Simulink and for each SWC the corresponding Simulink subsystem defines an additional port, dedicated to the set of RTE events that trigger its runnables. The RTE events are collectively joined in a bus that is input to this special port. A bus selector is then used to route the appropriate events as triggers to the runnable subsystem. An example is shown in Figure 21.11 showing two cooperating components with their RTE event ports and the internals of the Component 1 subsystem in Figure 21.12 showing the bus selector with the two outgoing function call signals for the runnables.

Inter-runnable communication is handled by inter-runnable ports, which correspond to additional input and output ports on the runnables subsystems. The implicit or explicit communication model affects the code generation mode but not the simulation model, where access to the ports always occurs implicitly according to the Simulink semantics.

When it comes to the composition of the AUTOSAR components and the system-level simulation, the recommended canonical pattern is the one represented in Figure 21.13. In this pattern, the RTE events are generated by a single subsystem

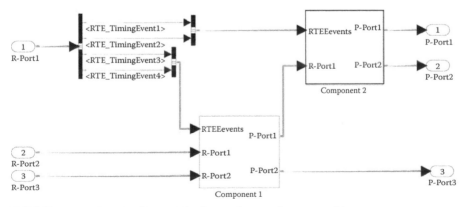

FIGURE 21.11   Simulink® model for forwarding runtime executable events to components and realizing data communication.

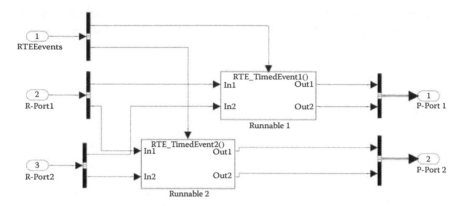

FIGURE 21.12   Simulink® model for an AUTOSAR component with two runnables.

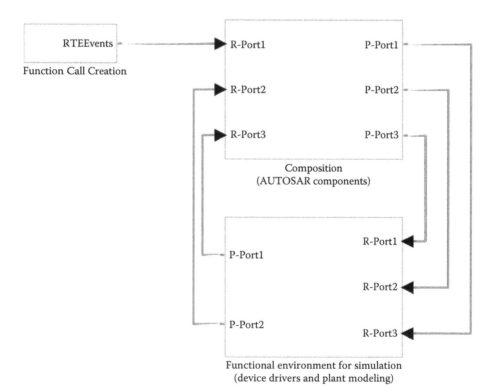

FIGURE 21.13   Canonical pattern for the Simulink® representation of an AUTOSAR composite.

labeled as *Function Call Creation*. The rationale for such a centralized generation is to collect in a single block the logic controlling the execution order of runnables. The internals of this subsystem are not specified in Ref. [20] as it states that "it is up to the user to implement this in an appropriate way."

21.5   SIMULATION, TIMING ANALYSIS, AND CODE GENERATION: CONSISTENCY ISSUES AND MODEL ALIGNMENT

An AUTOSAR model has (at least) three different objectives: it can be used to simulate the functional behavior, it can be used to perform worst-case timing analysis using suitable tools and, of course, it is used to generate the code for the implementation of the model on a target platform. The same model should be used for the three purposes or at least very limited changes should be required to obtain the three different objectives of validation by timing analysis, simulation, and finally automatic implementation by code generation.

To illustrate some of the issues, let us consider a fuel injection system. In this control application, several functions must be executed in response to events that correspond to a given rotation angle of the engine shaft. In practice, a crankshaft position sensor provides a reference hardware signal that is sporadic in nature and depends on the rotation speed of the shaft. This signal triggers a set of functions, typically mapped for execution onto a single task.

In AUTOSAR, a sporadic signal of this type cannot be directly defined as an RTE event. However, inside the application model, the runnables that are executed in response to the crankshaft position signal must be activated in response to an RTE event, given that this is the only legal way for a runnable to be activated in AUTOSAR.

One possible solution is the following. A BSW runnable represents the interface between the hardware signal coming from the crankshaft position sensor and the application tasks. This BSW runnable writes into a sender/receiver port to forward to the application the sensor position signal. The SWCs containing application runnables that must be activated in response to this signal will define a matching port. Inside them, the runnables are then defined to be activated on the event of data received on the port.

The BSW component is the main variation point with respect to the three model versions. For simulation purposes, the BSW runnable is activated periodically by a counter/alarm pair. Inside, it contains the simulation stub that generates the stream of the crankshaft sensor signals. These signals are represented by writes into the sender/receiver port on which all the application-level runnables that must perform actions in correspondence to these events are waiting.

For code generation purposes, the BSW component is replaced by a device driver that will explicitly call the RTE API function for writing into the data port at the end of its execution.

Finally, for the purpose of worst-case timing analysis, it is necessary to have information about the worst-case arrival rate of the activation events for the sporadic task onto which all the runnables reacting to the crankshaft position sensor events are mapped. In AUTOSAR 4.0, this indication could come directly from a timed event triggering information (Figure 21.8). In past releases of the standard, a timing analysis tool would have to deduce the rate from the rate of the writes into the corresponding data port, which would require navigating the communication model graph to the BSW runnable and the corresponding alarm (which, however, can only be periodic since no AUTOSAR versions up to 3.2 allow sporadic activation events).

21.6 CONCLUSIONS

This chapter provides an introduction to AUTOSAR as a language for system-level modeling and SWC modeling. Focus is on the issues related to the use of AUTOSAR models for simulation purposes and on the integration of AUTOSAR models with heterogeneous models (such as Simulink) for behavioral modeling. Also, the new timing model of AUTOSAR is analyzed with respect to its capability of defining a discrete-time base for simulation purposes or, in general, a timing model that allows seamless migration or integration of SR models. In conclusion, AUTOSAR is a promising solution for filling some of the gaps of modern methodologies with respect to platform modeling and system-level modeling and analysis, but still lacks the support of a formal (based on mathematical rules) MoC and of timed events.

ACKNOWLEDGMENTS

The author wishes to thank Giacomo Gentile and Nicola Ariotti of Magneti Marelli SpA for their insightful discussions on the modeling of the fuel injection case study.

REFERENCES

1. MathWorks. The MathWorks Simulink and Stateflow User's Manuals, available at http://www.mathworks.it/help/toolbox/simulink/ (04/04/12).
2. The AUTOSAR Consortium. The AUTOSAR Specification v 4.0, available at www.autosar.org (04/04/12).
3. Scharpf, Johannes, Robert Hoepler, and Jeffrey Hillyard. 2012. "Real-Time Simulation in the Automotive Industry." In *Real-Time Simulation Technologies: Principles, Methodologies, and Applications*, edited by Katalin Popovici, and Pieter J. Mosterman. Boca Raton, FL: CRC Press.
4. Wei, Zheng, Qi Zhu, Marco Di Natale, and Alberto Sangiovanni-Vincentelli. 2007. "Definition of Task Allocation and Priority Assignment in Hard Real-Time Distributed System." In *Real-Time Systems Symposium*, December 3–6, Tucson, AZ, pp. 161–70.
5. Davare, Abhijit, Qi Zhu, Marco Di Natale, Claudio Pinello, Sri Kanajan, and Alberto Sangiovanni-Vincentelli. 2007. "Period Optimization for Hard Real-Time Distributed Automotive Systems." In *Proceedings of the Design Automation Conference*, June 4–8, San Diego, CA, pp. 278–83.
6. The AUTOSAR Consortium. The AUTOSAR Timing Extensions, available at http://www.autosar.org/download/R4.0/AUTOSAR_TPS_TimingExtensions.pdf (04/04/12).
7. The AUTOSAR Consortium Artop Group. The AUTOSAR Tool Platform User Group, available at http://www.artop.org/ (04/04/12).
8. The Eclipse Foundation. The Eclipse Foundation open source community website, www.eclipse.org (04/04/12).
9. dSPACE. The dSPACE SystemDesk, product documentation available at http://www.dspace.com/en/inc/home/products/sw/system_architecture_software/systemdesk.cfm (04/04/12).
10. Vector. Vector AUTOSAR Solutions, product information available at http://www.vector.com/vi_autosar_solutions_en.html (04/04/12).
11. Geensoft. The AUTOSAR Builder, product information available at http://www.geensoft.com/ja/article/autosarbuilder/ (04/04/12).

12. Elektrobit. EB tresos, product information available at http://www.automotive.elektrobit .com/home/ecu-software/eb-tresos-product-line.html (04/04/12).

13. Templ, Josef, Andreas Naderlinger, Patricia Derler, Peter Hintenaus, Wolfgang Pree, and Stefan Resmerita. 2012. "Modeling and Simulation of Timing Behavior with the Timing Definition Language (TDL)." In *Real-Time Simulation Technologies: Principles, Methodologies, and Applications*, edited by Katalin Popovici, and Pieter J. Mosterman. Boca Raton, FL: CRC Press.

14. Reindl, Florian. Performance evaluation of AUTOSAR based systems, available at http://www.eb-tresos-blog.com/2010/12/performance-evaluation-of-autosar-based-systems/ (04/04/12).

15. Bernard, Zeigler, Tag Gon Kim, and Herbert Praehofer. 2000. *Theory of Modeling and Simulation*, 2nd ed. New York: Academic Press.

16. Gabriel, Wainer, and Pieter J. Mosterman. 2010. *Discrete-Event Modeling and Simulation: Theory and Applications. Computational Analysis, Synthesis, and Design of Dynamic Systems*. CRC Press, Boca Raton, FL, ISBN 9781420072334, December, 2010.

17. Krause, Matthias, Oliver Bringmann, Andre Hergenhan, Gokhan Tabanoglu, and Wolfgang Rosentiel. 2007. "Timing Simulation of Interconnected AUTOSAR Software-Components." In *Proceedings of the DATE Conference*, April 16–20, Nice, France, pp. 1–6.

18. Grotker, Thorsten, Stan Liao, Grant Martin, and Stuort Swan. 2002. *System Design with SystemC*. Springer. Kluwer Academic Publisher, New york (now Springer), available on line at http://www.springerlink.com/content/p16217/?p=d25f5bad8b794111b6797650 cabf04c9& pi=0 (04/04/12).

19. OSCI. The Open SystemC Initiative, web site http://www.accellera.org/home/ (04/04/12).

20. The AUTOSAR Consortium. The AUTOSAR Simulink Style Guide, available at http:// www.autosar.org/download/AUTOSAR_SimulinkStyleguide.pdf (04/04/12).

21. Simulink Coder, product web page http://www.mathworks.it/products/simulink-coder/ index.html (04/04/12).

22 Modelica as a Platform for Real-Time Simulation

John J. Batteh, Michael M. Tiller,
and Dietmar Winkler

CONTENTS

22.1 Introduction 549
22.2 Modelica Overview...551
    22.2.1 History 551
    22.2.2 Modelica Example 552
22.3 Modelica Features..556
    22.3.1 Open Platform...556
    22.3.2 Acausal Modeling...557
    22.3.3 Symbolic Manipulation 561
    22.3.4 Inverse Models..566
    22.3.5 Model Configuration...569
    22.3.6 Real-Time Language Extensions and Interfaces 570
    22.3.7 Tool-Specific Features 571
22.4 Application Examples..572
22.5 Conclusions..576
References...576

22.1 INTRODUCTION

With the growing adoption of model-based system development to shorten product development cycles and reduce costs and hardware prototypes, modeling of physical and control systems has emerged as a critical element in modern systems engineering efforts. The systems engineering V is a standard way to describe the processes that encompass a system engineering effort. While the V model has been refined as applied in particular industries, the core process steps are largely similar across applications. One representation of the systems engineering V as applied to intelligent transportation systems is shown in Figure 22.1. Modeling and simulation supports the systems engineering V throughout the entire product development process. In the early stages of the V, upfront modeling supports initial concept assessment and target and requirements cascade moving down the left side of the V. Computer-aided engineering (CAE) methods such as finite element analysis (FEA) and computational fluid dynamics (CFD) are typically employed

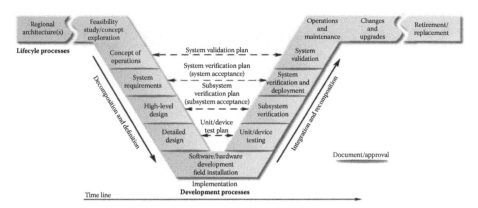

FIGURE 22.1 Systems engineering V. (From National ITS Architecture Team, *Systems Engineering for Intelligent Transportation Systems*, Report no. FHWA-HOP-07-069, United States Department of Transportation, 2007, http://www.ops.fhwa.dot.gov/publications/seitsguide/index.htm (accessed June 30, 2010). With permission.)

at the bottom of the V to support detailed design and development efforts. Moving back up the right side of the V, the integration and verification phases routinely involve verification and validation (V&V) using model-based representations of the engineered system. After system deployment, models can be used to help understand and diagnose issues in the field and can even provide simulation-based maintenance opportunities. Although the type of modeling naturally evolves with engineering tasks and data availability throughout the product development process, it is quite clear that models can and do play a critical role in complex systems engineering efforts.

While upstream modeling efforts may be wholly contained in the virtual realm, the necessity to integrate models with hardware increases during the later stages of the product development process as virtual prototypes are replaced by their physical representations. Computational efficiency becomes a key issue during this stage of the integration as models must simulate in real time to satisfy hardware interface requirements. Satisfying real-time simulation constraints is often one of the most challenging tasks for model developers. Making appropriate trade-offs between model complexity and model fidelity requires keen understanding of the underlying system dynamics and the spectrum of time scales inherent in the modeled system and input drivers.

This chapter introduces the modeling language Modelica [1] as a modeling platform to support real-time simulations. A brief introduction to the Modelica language and its fundamental language features will be given. The focus of this work is on the features of the Modelica language that make it desirable for modeling of physical systems for real-time targets. While the Modelica language itself is open and can be interpreted by a wide variety of free and commercial tools [54] for potential generation of real-time executables, the intent is not to focus on any particular tool or application but rather the language itself.

22.2 MODELICA OVERVIEW

Modelica is a nonproprietary, object-oriented modeling language for multidomain physical systems. Models in Modelica are described by differential, algebraic, and discrete equations. Modelica has been used to model a wide variety of complex physical systems in the mechanical, electrical, thermal, hydraulic, pneumatic, and fluid domains with significant industrial application in the aerospace, automotive, and process industries. An extensive list of publications, including the full proceedings from all Modelica conferences, is available on the publication page of the Modelica website [2]. As a complement to modeling of acausal physical behavior, Modelica also supports modeling of systems with prescribed input/output (I/O) relationships (i.e., causal systems), such as control systems and hierarchical state machines. This section gives a brief overview of the Modelica language and its fundamental language features.

22.2.1 History

The Modelica language has evolved as a collaboration between computer scientists and engineers to provide a modeling language to describe the behavior of multidomain physical systems in an open format not tied to a particular tool. The roots of the Modelica language can be traced back to Hilding Elmqvist's PhD thesis [3] in which he designed and implemented the Dymola modeling language. The Dymola language used an object-oriented, equation-based approach to formulate models of physical systems. Rather than forcing modelers to pose the problem using ordinary differential equations (ODEs) for which many solvers exist, the Dymola language allowed the problem to be posed as differential-algebraic equations (DAEs), which is generally considered a more natural way to describe physical problems [4]. Sophisticated symbolic algorithms were used in the language implementation to transform the mathematical model into a form capable of solution by existing numerical solvers. As symbolic algorithms advanced, in particular, the Pantelides algorithm [5] for DAE index reduction, it became possible to solve an even larger class of problems using the new approach pioneered by the Dymola language. Within the context of the ESPRIT project "Simulation in Europe Basic Research Working Group (SiE-WG)," Hilding Elmqvist in 1996 sought to bring together a group of object-oriented modeling language developers and engineering simulation experts in an attempt to develop a new modeling language to describe physical systems over a wide range of engineering domains. After a series of 19 meetings over the course of the next 3 years, version 1.3 of the Modelica language specification was released in December 1999.

The Modelica language specification is maintained by the Modelica Association, a nonprofit, nongovernmental organization established in 2000. The Modelica Association consists of individual and organizational members who actively participate in the design of the Modelica language. Since the initial release of the language specification, several major updates to the language have been released that incorporate additional language elements. The current version of the Modelica specification is version 3.2 [1]. The current and the previous versions of the Modelica specification are available on the Modelica website [6]. In addition to the Modelica language specification, the Modelica Association also develops and maintains the Modelica Standard Library (MSL), a large, free, multidomain library of Modelica models. The MSL serves as a common link between free and commercial library developers to promote compatibility between implementations. A listing of free and commercial libraries is available on the libraries page of the Modelica website [7]. For more history on the Modelica language, the interested reader is referred to the two published books on the language [8,9]. The most current information on the Modelica language can be found on the Modelica website at www.modelica.org [10].

22.2.2 MODELICA EXAMPLE

To illustrate a few key features of the Modelica language, consider the example shown in Figure 22.2. This multiphysics model consists of a battery connected to a motor driving a load. The battery model includes the heat-generation effects of the dissipative resistance and the resulting degradation in voltage due to thermal effects. The thermal network model includes a lumped model of the battery thermal mass along with simple flow-based cooling through convection. A controller controls the battery cooling by monitoring the battery temperature and actuating the flow rate in the cooling system to achieve the desired battery temperature. This multidomain physical system model contains elements from the mechanical (rotational), electrical, and thermal domains. The model was constructed both from component models in the MSL and from models implemented by the authors.

The Modelica source code for the example model is shown in Figure 22.3. The Modelica source code completely describes the dynamics of the system, but a

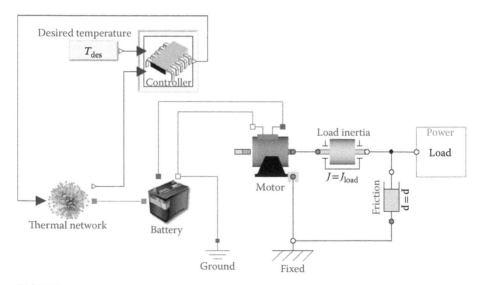

FIGURE 22.2 Modelica example model.

```
model MotorWithThermalBattery
  parameter Modelica.SIunits.Inertis Jload=0.15 "load inereia";
  parameter Modelica.SIunits.RotationalDampingConstant d=0 "damping constant";
  parameter Modelica.SIunits.Temperature Tdes=315 "desired temperature";
  Implementations.BatteryThermal battery a;
  Implementations.PermanentMagnetMotor motor a;
  Modelica.Electrical.Analog.Basic.Ground ground a;
  a;
  Modelica.Mechanics.Rotational.Components.Fixed fixed a;
  Modelica.Mechanics.Rotational.Components.Inertia load_inertia (J=Jload) a;
  Implementations.BatteryThermalNetwork thermal_network a;
  replaceable Implementations.TemperatureControlPID controller constrainedby
    Interfaces.TemperatureControl a;
  Modelica_Blocks_Sources_RealExpression desired_temperature(y=Tdes) a;
  Implementations.LoadPower load a;
  Modelica.Mechanics.Rotational.Components.Damper friction(d=d) a;

equation
  connect(fixed.flange, motor.support)a;
  connect(battery.p, motor.pin_ap) a;
  connect(battery.n, motot.pin_an) a;
  connect(motor.flange, load_inertia.flange_a) a;
  connect(battery.n, ground.p) a;
  connect(thermal_network.battery, battery.thermal_port) a;
  connect(desired_temperature.y, controller.desired_temperature) a;
  connect(controller.flow_command, thermal_network.flow_command) a;
  connect(thermal_nctwork.battery_temperature, controller.temperature) a;
  connect(load.flange, load_inertia.flange_b) a;
  connect(friction.flange_b, load_inertia.flange_b) a;
  connect(friction.flange_a, fixed.flange) a;
end MotorWithThermalBattery;
```

FIGURE 22.3 Modelica code for example model.

compiler is necessary to actually simulate the example. As will be described later in this chapter in more detail, the compiler flattens the equations in the Modelica source code and combines them into a causal set from which computer code can be generated for integration with existing numerical solvers.

To provide further understanding of the dynamics of the model, Figure 22.4 shows simulation results as generated by the commercial tool Dymola* [11]. The load applied to the system is shown along with the resulting battery terminal voltage, battery temperature, and the controlled flow command for battery cooling. Note that the cooling system has a fixed cooling capacity that is insufficient to maintain the battery temperature under high-load conditions. Since the battery voltage is a function of the temperature, the desired load power is not met under all conditions. This example model will be used in Sections 22.3.4 and 22.3.5 to illustrate key concepts related to real-time modeling capability and model configuration.

A key language feature in Modelica is the connector concept. Connectors are used to define the interfaces of the models. A unique connector is defined once for each physical domain. A connector definition primarily consists of two kinds of variables: a potential variable and a flow variable. These variables can also be described as "across and through" or "effort and flow" variables. Special semantics apply to

---

* Dymola was used to generate all simulation results shown throughout this work though nearly any Modelica-compliant tool could have been used.

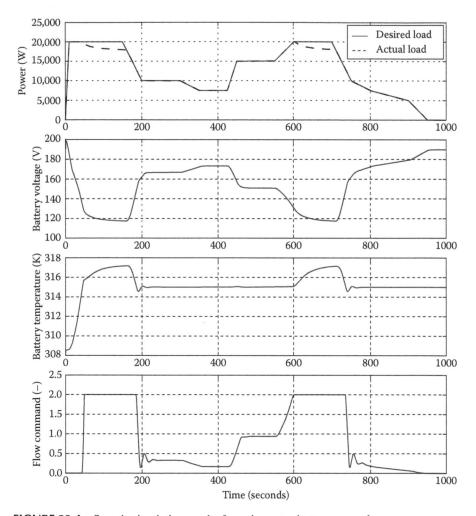

FIGURE 22.4   Sample simulation results from the motor battery example.

connections made between connectors, and such connections result in equations between variables in the connectors. In a connection set, all matching potential variables are equated, and all matching flow variables are summed to zero at the connection point. Referring to the example model in Figure 22.2, the rotational connectors are depicted as gray circles, the electrical connectors as blue squares, and the thermal connectors as red squares connected between the battery and the thermal network components. The definitions of the connectors from the MSL are shown in Figure 22.5.

The connector concept is extremely powerful as it facilitates natural, acausal modeling of physical systems where connections between components in the Modelica model mimic connections seen in the physical world. As will be discussed in more detail, acausal models include no a priori assumptions about

```
         Connector Pin "Pin of an electrical component"
            SI.Voltage v "Potential at the pin"
            flow SI.Current i "Current flowing into the pin";
            a;
         end Pin;
```

```
      connector Flange_a
        "1-dim.rotational flange of a shaft (filled square icon)"
        SI.Angle phi "Absolute rotation angle of flange";
        flow SI.Torque tau "Cut torque in the flange";
        a;
      end Flang_a;
```

```
partial connector HeatPort "Thermal port for 1-dim-heat transfer"
   Modelica.SIunits.Temperature T "Port temperature";
   flow Modelica. SIunits.HeatFlowRate Q_flow
     "Heat flow rate (positive if flowing from outside into the component)";
   a;
end Heatport;
```

FIGURE 22.5 Connector definitions.

causality and allow quick and effortless model construction and reconfiguration. For example, a connection between rotational connectors is equivalent to a rigid connection (i.e., a kinematic constraint or an inertialess shaft) between physical elements. In the electrical domain, a connection between components is equivalent to a lossless wire connection (i.e., a perfect short). Just as physical systems are constructed through connections between hardware components, Modelica models are constructed through connections between equivalent virtual components. In this way, the virtual representation is consistent with system design schematics used by engineers.

Another key feature of the connector-based approach is the ability to seamlessly integrate models from various sources. The use of the standard connectors defined in the MSL ensures model compatibility between the MSL, user-implemented, free, and commercial libraries. Thus, Modelica with its open language specification is also an ideal model exchange format. As mentioned previously, the example model in Figure 22.2 consists primarily of models from the MSL. At the top level of the model, the motor, damper, and inertia components are from the MSL. The thermal network, battery, load, and controller were primarily implemented as subsystems constructed from lower-level components of the MSL. The thermal network subsystem model is shown in Figure 22.6. The only author-implemented model in the thermal network is the component that calculates the heat transfer coefficient for the convection element based on the cooling flow rate. With established multidomain connectors, model implementations using those connectors in the MSL, and other core language features to promote model reuse and configurability, the Modelica language provides a powerful platform for model-based systems engineering.

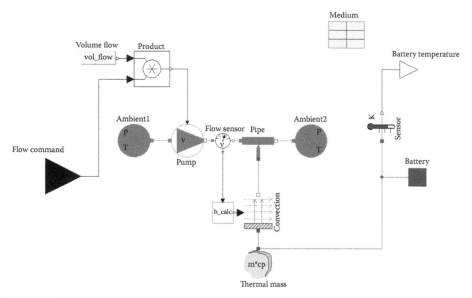

FIGURE 22.6   Thermal network subsystem from example model.

22.3   MODELICA FEATURES

With that brief introduction to the Modelica language, this section elaborates more specifically on Modelica language features and associated attributes that make Modelica a desirable platform for real-time simulation. The focus will be primarily on Modelica although a few tool-specific implementations relative to real-time simulation capability are briefly discussed.

22.3.1   OPEN PLATFORM

Modelica is a nonproprietary language and thus offers an open platform for model and application development. The nonproprietary nature of the language allows intellectual property captured in models to be maintained separate from the tools thereby allowing model developers to move freely among Modelica-capable tools based on their relative merits. This situation is in stark contrast to other simulation tools where the models are integrated with the tools in a proprietary way and thus are not portable to other tools. While conformance to the same language syntax allows tool-independent model formulation, it should be noted that simulating the same model identically between tools also requires that the execution engine is formalized and used as reference semantics for each of the tools.

Another opportunity that arises from the open platform is that of accessibility. Since the Modelica specification is open and freely available, any interested party is welcome to create a parser and even a compiler to interpret Modelica source code. With source code for many solvers already available, it is certainly feasible to create custom tools for Modelica model development and simulation. For example, the Modelica Software Development Kit (SDK) [12] provides an API and Modelica compiler to allow Modelica code to be embedded into existing software and tools.

The open platform also supports innovation from small companies and universities. Several open-source, Modelica-based projects have been initiated. OpenModelica [13] is an open-source Modelica modeling and simulation environment. JModelica.org [14] is an open-source Modelica-based platform for simulation and optimization. Scicos [15] developed at INRIA is a modeling and simulation environment that includes partial support for Modelica. While these offerings may not be as comprehensive in their support of the Modelica language as existing commercial tools, they certainly illustrate potential for innovative offerings based on the open Modelica platform.

Besides accessibility to the Modelica language for custom tool development, another benefit of the open platform is the large quantity of high-quality models that are available in the MSL and free Modelica libraries. Rather than working on simplified, academic problems, the existing Modelica model base provides high-quality, relevant models in multiple engineering domains that can be used to fully support tool and algorithm development. These models allow tool and algorithm innovators to work on real technical problems with models that are already available thus reducing the burden of creating complex examples to showcase new algorithm or tool capabilities.

22.3.2 ACAUSAL MODELING

Modelica supports two of the most common modeling formalisms for continuous systems: block diagram and acausal modeling [8]. The block diagram approach involves constructing systems of component blocks to calculate unknown quantities from known quantities. The block diagram, or causal, approach is very well suited to modeling of controllers where the signal flow concept is quite natural. The sensor signals are known inputs, and the controller model is responsible for calculating actuation signals based on the sensed signals. The block diagram approach is not as well suited to physical system modeling in which the relevant mathematical description for the behavioral dynamics consists of conservation equations. For example, what would be the "input" to a resistor?

Acausal modeling involves modeling components from a free body or first principles sense without a priori assumptions about I/O causality. Acausal models specify relationships in which potentials across components drive flow of conserved quantities. Physical components are naturally acausal and are naturally described by the Modelica connector-based approach. Figure 22.7 shows an example of a mechanical system modeled with both acausal and block diagram formalisms in Modelica. For those familiar with the physical system schematics, the acausal representation is a natural virtual representation. Furthermore, the acausal component models can be reused regardless of the causality imposed by the system while the implementation of the block diagram model changes drastically because of fundamental changes in the inputs and thus the model structure.

Another benefit of the acausal modeling formalism as implemented in Modelica is that components can generally be connected in a physical way without restriction. This feature has a profound impact both on model management and configuration and on model computational efficiency, a critical factor in real-time simulation. One of the key challenges in real-time simulations is striking a balance between model fidelity and computational expense. To achieve this balance, idealizations to improve performance are often required.

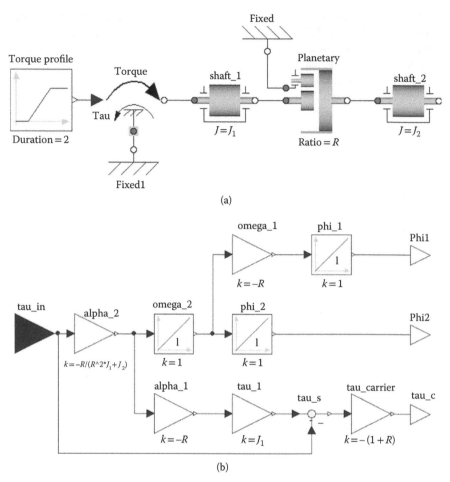

FIGURE 22.7   Sample model in (a) acausal and (b) block diagram formalisms.

To illustrate this point, a vehicle modeling example is considered. There are many compliant elements in a vehicle drivetrain, and these elements can lead to significant noise, vibration, and harshness (NVH) issues in the driveline as excited under various driving conditions. Capturing the NVH effects in the driveline necessitates the modeling of the key compliances in the system and results in higher-order dynamics in the vehicle drivetrain. These higher-order dynamics are, in fact, the focus of the NVH modeling effort. In addition to the NVH model, another class of vehicle models is focused on the simulation of vehicle performance and fuel economy. Performance and fuel economy models are typically simulated over long time scales, including drive cycles that can be thousands of seconds in length. On this time scale, the higher-frequency NVH effects do not typically impact the vehicle-level results significantly. Thus, the idealization of the drivetrain as rigid is a perfectly reasonable assumption for a performance and fuel economy model.

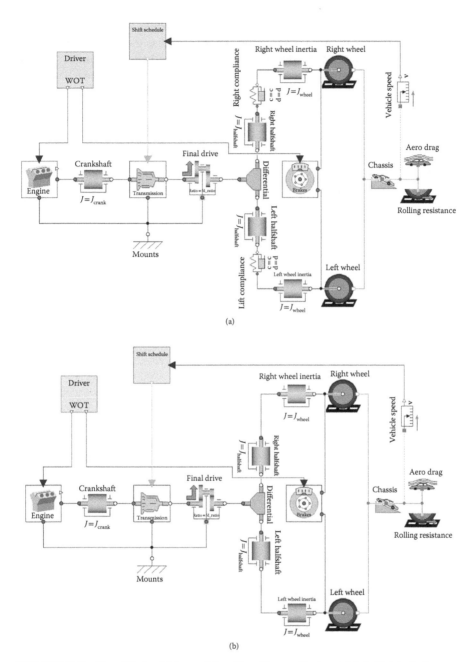

FIGURE 22.8 Simplified vehicle models with a (a) compliant and (b) rigid drivetrain.

To illustrate the two classes of vehicle models, consider Figure 22.8, which shows a simplified vehicle model with a compliant and rigid drivetrain. The two models are identical save for the spring-damper element that represents the compliance of the half shafts. Note that the rigid model actually has two inertia elements connected together representing the wheel and half shaft inertias. While many physical modeling tools have difficulties with such configurations (because they lead to high-index DAEs), this is not the case in Modelica, which was designed with high-index systems in mind. Connecting two inertias together might seem odd, especially in a flat model such as the ones shown in Figure 22.8. This topic will be addressed shortly as part of the configuration management discussion.

Now consider the situation where the compliant model is used for performance and fuel economy simulations, perhaps because of limitations in the modeling tool to handle rigid, kinematic connections. In an attempt to stiffen the compliant model to mimic the behavior of a truly rigid model, the stiffness of the compliant element is increased. Figure 22.9 shows simulation results from a wide open throttle (WOT) simulation of the rigid and compliant vehicle model. The stiffness of the compliant vehicle model has been set to roughly twice the typical half shaft stiffness to approach the behavior of the rigid model. The vehicle-level behavior of the two models is nearly identical as can be seen from the vehicle speed and the engine speed comparisons. However, the computational expense is nearly twice as great for the compliant model as illustrated by the CPU time required for the simulation. In particular, note

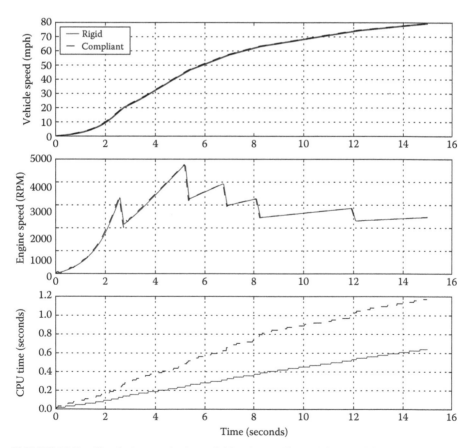

FIGURE 22.9 Simulation results from rigid and compliant vehicle model.

the jumps in CPU time corresponding to each shift event in the compliant model. While this evaluation was performed with an open loop WOT test to try and isolate the compliant effects, it should be noted also that the WOT simulation is not nearly as dynamic as a drive cycle where the driver inputs could be changing much more rapidly thus introducing even more shift events and likely resulting in even larger differences in simulation time between the compliant and the rigid models.

While the simple vehicle model shown in Figure 22.8 modeled the drivetrain in a flat fashion at the top level of the model without introducing another level of hierarchy, this modeling approach was primarily used to make it easier to illustrate the differences between the rigid and the compliant models. In fact, even this simple model includes enough components and connections that the top level of the model is becoming crowded and would benefit in terms of readability from the grouping of additional components into logical subsystems. As mentioned previously, it may seem that the potential need for the compliant element in the vehicle model is an artifact of the modeling of the half shaft and wheel as two separate inertias instead of a single effective inertia. While it is easy to see the two inertias connected together in the flat model, it would not be as obvious if the two inertias were part of separate subsystems.

Another benefit of grouping components into subsystems is to take advantage of an architecture-based modeling and model configuration approach. This topic will be discussed in more detail later in this chapter. An architecture-based modeling approach hinges on the ability to arbitrarily connect components. If the modeling tool cannot handle arbitrarily connected components, one remedy for the underlying conflict between the configuration management and the model mathematics is to insert unphysical interfaces, such as stiff springs in a mechanical system to ensure that the two inertias are not connected together, at the expense of computational efficiency. With no restrictions on component connections in Modelica, Modelica can support architecture-based model configuration while maintaining the natural physical interfaces between components, even if those interfaces include rigid connections.

22.3.3 SYMBOLIC MANIPULATION

While acausal models are clearly preferable to block diagram implementations for physical systems modeling, acausal models do introduce simulation challenges. While ODEs are convenient to solve, most physical problems in science and engineering are naturally described by DAEs. Acausal models require the solution of DAEs that include both ODEs and algebraic equations representing the system constraints. The efficient solution of DAEs, particularly high-index DAEs that result from structurally singular systems that have constraints between states, requires a combination of symbolic and numerical approaches and is an active research topic. The Modelica language has been designed to specifically protect for these approaches and optimizations to support the generation of efficient code from DAEs.

While Modelica provides the language to allow the expression of mathematical models, it does not prescribe the method for solving the resulting DAEs. The solution method and solver integration falls into the realm of the Modelica compiler, be it commercial or open source as described previously. This section briefly describes some of the symbolic and numerical techniques used in the solution of DAEs. The intent is not to focus on the algorithms or techniques but is instead to focus on the Modelica language elements that support them. The interested reader is referred to the works by Cellier et al. [4], Cellier and Kofman [16], Celier [53], and Anderson [17] for more details on the algorithms and numerical techniques briefly introduced here.

The mathematical description of models in Modelica consists of Boolean, discrete, and DAEs. On the basis of a Modelica model, the resulting set of Boolean equations, discrete equations, and DAEs can be obtained. However, there are no general-purpose solvers for these sets of equations. Direct numerical DAE solvers typically result in slow simulations. The standard approach in Modelica-based tools is to translate the DAE into an ODE form for solution. Since a Modelica model preserves the equations that describe the relationships between variables in symbolic form, it is possible to perform symbolic analysis and manipulation to aid in the solution of the resulting set of equations. Symbolic analysis helps determine an efficient way to develop a causal set of equations that can be solved numerically or even symbolically if possible.

To generate a causal set of equations, the first task is to understand the structure of the problem. One way to understand the problem structure is through a structure incidence matrix [16,17]. The rows of the matrix are indexed by the equations and the columns by the variables or unknowns. If an equation contains a given variable, the number one is placed in the corresponding entry in the matrix. Using the information provided in the matrix, a rule-based approach can be used to determine which variable should be solved from each equation. Another method to generate a causal set of equations is an algorithm based on graph theory and first proposed by Tarjan [18]. In this method, a structure digraph shows the equations and unknowns as nodes in vertical columns with lines drawn between nodes if an unknown appears in a given equation. On the basis of the information in the structured digraph, the algorithm defines an approach that can sort the equations into a causal, executable sequence. The resulting equivalent structure incidence matrix is lower triangular, and therefore, there is an equation to compute each of the unknowns from variables that have already been computed. The Pantelides algorithm [5] is a popular causalization algorithm because of its compact recursive implementation [16].

While equation sorting is a key component in the solving of DAEs, it is not entirely sufficient for full triangularization, in particular, in case of cyclic dependencies between variables, or so-called *algebraic loops*. Instead of a true lower triangular incidence matrix, the sorting algorithms result in block lower triangular form with the blocks containing the equations that are part of the algebraic loops. Depending on the nature of the problem, the loop equations could be either linear or nonlinear, and there are many techniques to solve the resulting equations. Blocks containing linear equations can be solved efficiently (and symbolically if they are small enough), but nonlinear equations will generally require Newton iterations.

An approach that has a profound impact on computational efficiency for algebraic loops is tearing. A tearing algorithm seeks to break apart a system of equations by assuming values for variables and then solving the resulting set of equations with residual equations for the torn variables based on the system of equations. By analyzing the structure of the underlying system, it is possible to identify tearing variables that can have a drastic impact on computational efficiency [16]. Tearing often leads to a reduction in the number of iteration variables for solving the nonlinear equations. It may also result in significant decoupling of simultaneous systems of equations with the potential for more efficient solution of the decoupled equations based on the known value of the tearing variable.

A key element to computationally efficient modeling is the handling of events where logical expressions change value [16]. Event handling is especially important since discontinuous equations or changes in model behavior must be handled as discrete events for numerical integration schemes. Since the integration is typically interrupted to resolve each event, unnecessary events can severely hamper computational speed. Modelica includes language semantics to control the handling of events. The semantics of the *smooth* operator indicate to the Modelica compiler that an expression is continuously differentiable up to the order provided by the user [1]. Since the expression on which the *smooth* operator acts can involve complex branching constructs such as *if* statements, the *smooth* operator aids the Modelica compiler in identifying the structure of the problem regarding continuity of the variables and partial derivatives of the variables in the expression and potentially avoiding unnecessary events at branching conditions since the modeler has guaranteed continuity up to a particular order. The *noEvent* operator in Modelica is also used to avoid the generation of events by controlling the generation of crossing functions from complex expressions with branching conditions based on variables of type real [1].

As noted previously, minimizing the number and size of nonlinear equations that must be solved can also have a significant impact on computational efficiency. The *semiLinear* operator in Modelica is used to help the Modelica compiler identify situations where an expression is linear with respect to a given variable but with two different slopes when the variable is positive and negative [1]. This operator can help the compiler to generate sets of linear equations rather than nonlinear ones. The *semiLinear* operator is symbolic in nature and gives the underlying tool a greater understanding of the modeler's intent, which can help resolve some ambiguities in certain classes of models. One example of such *semiLinear* operator usage is in the handling of reversing flow in fluid systems where the flow enthalpy is either the upstream enthalpy or the downstream enthalpy based on the direction of mass flow [19].

Another example where symbolic information is very useful is in state selection. For acausal models, state selection is not trivial since variables introduced in completely different subsystems can end up being kinematically coupled through the connection graph. Therefore, it is necessary to symbolically analyze the entire model to identify such cases and resolve a unique set of states. Furthermore, the selection of appropriate states for integration can drastically affect computational efficiency. For example, the modeler might want to control the states that are selected as part of the symbolic manipulation process based on knowledge of the problem or details of the formulation. In thermodynamic problems, the conservation equations are written in terms of mass and energy, but it is typically preferable to select the intensive variables pressure and temperature as states rather than the extensive variables that depend on the size of the system and can vary greatly over a given simulation. Furthermore, consider that property relationships, such as enthalpy and internal energy, are often explicitly written as nonlinear functions of pressure and temperature, and any other state selection would require the solution of a nonlinear equation to determine the intensive variables. The Modelica language includes the *stateSelect* language element to help the modeler control the selection of states [1]. The *stateSelect* element also allows the modeler to exert varying degrees of control over the state selection. Using the *stateSelect* attribute, it is possible to specify that a given variable is always selected as a state, that it should be preferred as a state, that it should not have any preference with regard to state selection, that it should be avoided as a state, or, finally, that it should never be chosen as a state. The Modelica language even allows initial equations to be specified for variables that are not states. For example, in a thermodynamic system with mass and energy as states, pressure and temperature can be specified for initialization. Another example of the impact of state selection on computational efficiency is in three phase electrical systems where the dq0 or Park transformation [20] can be applied to calculate alternative currents that exhibit more DC-like behavior. The choice of the alternative currents as states can drastically improve computational speed and can be easily achieved in Modelica using the *stateSelect* construct [21].

To illustrate the impact of equation sorting, state selection, and causality on computational efficiency, consider the following electrical example reproduced from the excellent tutorial by Bernhard Bachmann [21] on the mathematical aspects of object-oriented modeling. The model shown in Figure 22.10 is a simple electrical circuit consisting of sinusoidal voltage source, inductor, and a nonlinear resistor implementation. Two equivalent models are formulated for the nonlinear resistor with $i = f(v)$ causality as shown in Equation 22.1 and $v = f(i)$ causality as shown in Equation 22.2:

$$i = a \times \tanh (b \times v) \tag{22.1}$$

$$v = \frac{1}{2b} \log \left(\frac{1 + \dfrac{i}{a}}{1 - \dfrac{i}{a}} \right) \tag{22.2}$$

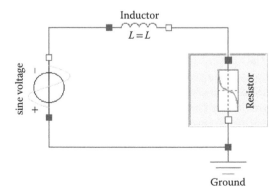

FIGURE 22.10   Electrical example model with nonlinear resistor.

Figure 22.11 shows the simulation results from the circuit with the two resistor implementations. The voltage and the current in the resistor are identical as expected, but the computational expense is significantly higher for the $i = f(v)$ formulation. The explanation for this effect is a result of the interaction between the state selection and the causality of the nonlinear equation. For this model, the typical state selection where differentiated variables are automatically chosen as states results in the current as a state. With the $i = f(v)$ causality, a nonlinear equation must be solved to

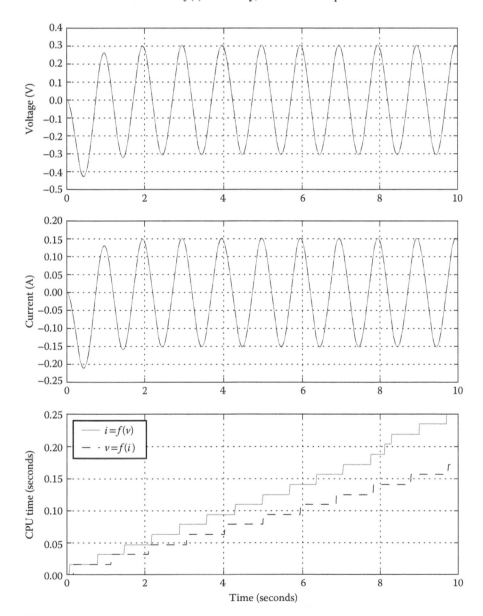

FIGURE 22.11   Simulation results from electrical example with nonlinear resistor.

determine the voltage drop across the resistor. With the $v = f(i)$ causality, a solution of nonlinear equations is not required as the nonlinear equation can simply be evaluated for the voltage drop. Bachmann [21] notes that the causality that results in the most efficient solution depends on the model topology. If the inductor and the resistor were placed in parallel rather than in series as in Figure 22.10, then the $i = f(v)$ implementation would be the more efficient formulation. This example also highlights why it is so important to have symbolic model representations where the formulation details are visible such that these issues can be diagnosed and understood.

22.3.4 INVERSE MODELS

Another benefit of acausal modeling is the support for inverse models. An inverse model is formulated by specifying model outputs rather than model inputs. The resulting executable model must then calculate the model inputs such that the model outputs are as prescribed. Causality is reversed in an inverse model since outputs are known and inputs are unknown thus requiring an acausal model formulation. Inverse models are sometimes denoted as backward models as opposed to forward models that are conventionally driven with known or calculated inputs. Inverse models can be used in place of controller logic to support rapid, upfront design concept assessment and feasibility without requiring a detailed controller design and tuning effort that would otherwise be required to reasonably assess the concept design [22]. Inverse models are also extremely useful for providing insight into control system design. A few examples of inverse modeling with Modelica involve the formulation of driver models that perfectly follow a prescribed vehicle speed trace for powertrain simulations [23], energy consumption studies in aircraft equipment systems in response to prescribed load traces [24], and model-embedded control applications [22].

Inverse modeling approaches can also be used to impact computational efficiency. Inverse models can be used to implement perfect control in lieu of proportional–integral–derivative (PID) controller implementations with high gains. While high gains can improve the ability of the system to follow a desired trace, significant tuning is often required to achieve desired control response. A side effect of high gains is the introduction of high-frequency inputs that can excite high-frequency response in the model thus affecting computational efficiency. The situation is especially acute if the high-frequency response is beyond the bandwidth of interest for the simulation. Ultimately, however, the high-gain implementation is simply trying to mimic the perfect control response.

Inverse models can also be used to implement localized perfect control in the context of a complex control architecture where a particular control feature is replaced by a perfect control implementation. One can imagine this model inversion acting as a sort of "perfect control," which, assuming the system is sufficiently invertible, ensures that the desired response is exactly achieved. By implementing perfect control, one or more states in the model can typically be removed thus reducing the total number of states in the system as well as potentially eliminating eigenmodes that slow down the integration as well. Reduction in the number of states is especially useful for model embedded control applications where the computational expense grows nonlinearly with the number of states in the system [22]. It should be noted that, depending on the linearity of the model being controlled, perfect control can result in nonlinear systems that were not present in the forward facing model. The ability of a Modelica compiler to tear nonlinear systems is the key to minimizing the impact of nonlinear systems generated by a perfect control approach. If the nonlinear systems are too extensive or prove difficult to solve, it is certainly possible that the computational efficiency of the perfect control approach could be degraded compared to the forward facing implementation. The trade-off between nonlinear systems and more states with potentially linear systems must be evaluated on a per problem basis.

As an illustration of perfect control in Modelica, consider again the model shown in Figure 22.2. Figure 22.12 shows the original controller implementation with a PID controller and also a perfect control implementation. Note that the PID controller has been replaced by a block from the MSL to handle inverse problems. Simulations were conducted by simply replacing the original controller with the perfect controller using the Modelica language features for model configuration. Figure 22.13 shows results from running the simulation with the perfect control. The flow command limits have

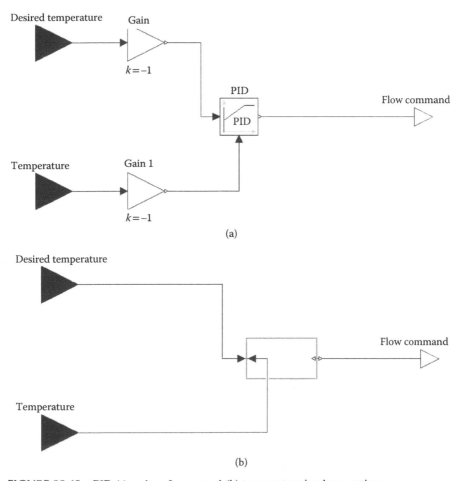

FIGURE 22.12   PID (a) and perfect control (b) temperature implementations.

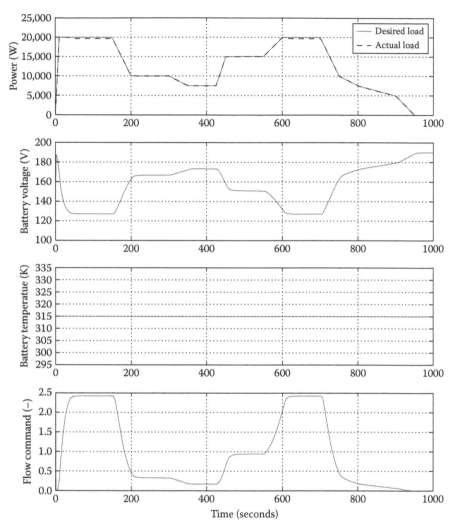

FIGURE 22.13 Sample simulation results from the motor battery example with perfect controller.

been removed to illustrate the commands issued by the perfect controller to achieve the desired battery temperature. With perfect temperature control, the system can also meet the desired load command. It should be noted that the unconstrained perfect control simulation yields the dynamic range necessary by the controller to achieve the desired output response. Thus, the results are not identical to those in Figure 22.4 using the PID controller for which the flow command was limited. Note, however, that there is evidence of some oscillatory behavior in the flow command signal from the PID controller implementation in Figure 22.4. For this sample problem, no effort was expended to tune the controller to potentially reduce or eliminate the oscillatory commands for the PID implementation; obviously, no tuning is required for the perfect control implementation to identically achieve the desired target.

22.3.5 Model Configuration

With a distributed, acausal model development approach, system models are constructed through connections between component or subsystem models. While it is possible to assess model fidelity with testing at the component and subsystem level, the impact of individual model implementations on computational efficiency is often not revealed until the system model is assembled and tested. As models transition from desktop simulations to real-time platforms, variants of individual models may be required to support real-time execution.

Model configuration and variant management are key components of effective model management for any model-based systems development effort. The Modelica language includes several language features that directly support model configuration. To support plug-and-play model configuration, the Modelica language allows configuration management through the replaceable language construct. The replaceable semantics allow model variants to be created by replacing individual models with other compatible models. With a strong typing system, compatible models can be easily identified. Returning to the motor battery example, consider the construction of the model with the perfect controller. The perfect controller model variant is constructed by replacing the original controller with the perfect controller (note the replaceable qualifier before the instantiation of the controller component in Figure 22.3). The Modelica code to create the model variant is shown in Figure 22.14. Since Modelica supports inheritance, the model variant is created by inheriting from the original model shown in Figure 22.3 and simply replacing the original controller implementation with the PID implementation. While this example illustrates configuration of a top-level model component, model configuration through the replaceable construct can occur at any level of the model hierarchy. The ability to create model variants without code duplication increases modeling efficiency and aids in effective model management. The replaceable construct allows plug-and-play configuration of different versions of the same model with selected model variants to support real-time applications.

As shown in the motor battery example, Modelica models can be easily constructed by connecting component models, and model variants can be created using the replaceable construct. The next level in model configuration management uses prewired model architectures of configurable components for plug-and-play construction of model variants. Figure 22.15 shows a sample model architecture to support vehicle system modeling [23]. Model architectures offer an even more powerful platform for managing model variants as many different applications can be supported with the same architecture. The same architecture can support models of varying levels of detail, a key feature for real-time modeling.

```
model MotorwithThermalBatteryPerfect
  extends motorWithThermalBattery (redeclare
      Implementations.TemperatureControlPerfect controller);
    a;
  end MotorwithThermalBatteryPerfect;
```

FIGURE 22.14 Modelica code for example model with perfect controller.

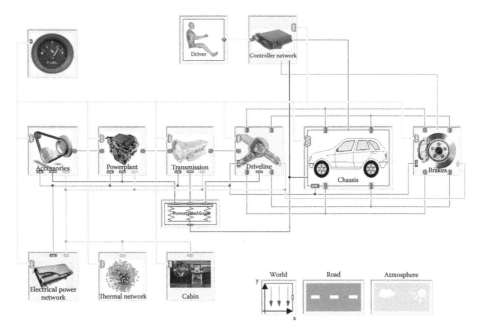

FIGURE 22.15 Vehicle model architecture. (From Batteh, J., and M. Tiller, *Proceedings of the 7th International Modelica Conference*, Como, Italy, pp. 823–32, Modelica Association, 2009. With permission.)

22.3.6 REAL-TIME LANGUAGE EXTENSIONS AND INTERFACES

As a result of increased interest in real-time computing and model-based control system development, new language elements have been introduced in Modelica for use in embedded systems. Stemming from the ITEA2 EUROSYSLIB and ITEA2 MODELISAR projects, Modelica language extensions have been designed and included in the Modelica 3.2 Specification [1] that allow models of embedded systems to be partitioned into various elements to support model-in-the-loop, software-in-the-loop (SiL), hardware-in-the-loop (HiL) applications, rapid prototyping, [25,26] and code generation [27]. The language elements facilitate the mapping between the logical system architecture where the functional and logical behavior of the control system are specified and the technical system architecture where the concrete implementation of the logic is defined. The intent is to allow the Modelica tool to automatically translate between the logical and the technical systems given mapping information formally defined in Modelica semantics. A key feature of the implementation allows the mapping to the technical system to be defined without modification of the logical system so that a single logical model can be configured for different embedded system use cases.

The new free Modelica_EmbeddedSystems library [27] provides access to the new language elements. The library provides interfaces and examples to define the partitioning of a distributed model into tasks and subtasks, the appropriate configuration between software and hardware elements, and the communication and I/O interfaces between various elements. The configuration definition includes specifications for the various tasks, subtasks, sampling rates, target processors, device drivers, integration methods, synchronization, initialization, and so on. These features allow a model of an embedded system to integrate both model-based and hardware elements in a state-of-the-art control system development environment. In conjunction with the Modelica_EmbeddedSystems library, the new Modelica_StateGraph2 library [28] offers improved hierarchical state machines for simulation of hybrid systems and real-time applications.

A related development from the ITEA2 MODELISAR project is the development of the Functional Mockup Interface (FMI), a tool-independent standard for model exchange and cosimulation [29]. The FMI is a result of collaboration between vendors of Modelica and non-Modelica tools to facilitate model exchange between simulation tools. The goal of this effort is to gain acceptance as an open standard in the CAE community to improve model exchange between suppliers and original equipment manufacturers (OEMs). The FMI standard consists of FMI for Model Exchange and FMI for Co-Simulation. The FMI looks to be a very promising development for real-time applications, including planned support for AUTOSAR, the upcoming standard for embedded system software in vehicles.

22.3.7 Tool-Specific Features

While the focus of this work is on the Modelica language, a Modelica compiler is required to generate executable code and is thus an integral part of the real-time platform. As such, there are a number of tool-specific features that have been developed to support real-time applications. These features, in conjunction with the Modelica language, are of paramount importance and merit the following overview.

The majority of the published results for real-time simulation of Modelica models were obtained using Dymola [11]. A good discussion of the various symbolic and numerical techniques employed in Dymola to solve DAEs and generate efficient code for real-time simulation is provided in the work by Elmqvist et al. [30]. Dymola has implemented a number of advanced symbolic/numeric techniques to support real-time simulation. A key challenge for multiphysics models is the disparate time scales in a model. Resolving the fastest time scales in stiff models with explicit fixed-step integration schemes requires small step sizes, which negatively impact computational efficiency. Implicit schemes allow larger time steps but require the solution of nonlinear systems of equations. Inline integration attempts to resolve this issue by combining the discretization formulas of the integration method with the model equations [31]. The resulting set of difference equations are then subject to symbolic manipulation in an attempt to generate efficient code for real-time simulation. Mixed-mode integration involves the use of implicit schemes for fast states and explicit schemes for slow states [32]. Event handling for fixed time steps is a challenge as iteration to find zero-crossings can cause computational overruns. Thus, Dymola has implemented several techniques for event handling such as predicting the occurrence of events and step size modification to detect events and synchronize with real time following event detection. These techniques can have a drastic impact on the ability to satisfy real-time constraints for complex physical systems in Modelica [33].

A common way of addressing real-time computational constraints is the development of model variants with varying complexity and potentially varying structure. Now consider the situation where a given simulation requires a more complex model during some phases of the simulation but a reduced order model can suffice otherwise. This modeling approach results in the potential for variable structure systems with structural changes at runtime, which are difficult to handle because of the semantics of the Modelica language. A derivative language of Modelica called Sol has been developed to provide a development platform for the investigation of technical solutions for variable structure systems [34]. Sol relaxes some of the semantics of the Modelica language regarding the creation and deletion of equations at runtime to more naturally handle variable structure systems. While not a commercial tool, this impressive, open research effort includes a formal definition of the language, a parser, and even a solution platform for numerical simulation. Published results demonstrate the application of the Sol framework to a rotational mechanical system driven by an engine model [34] and on a multibody trebuchet model [35]. The work is aimed at providing guidance and suggestions for potential improvements to the Modelica language. This effort is another example of the power of an open platform to support innovation.

22.4   APPLICATION EXAMPLES

This section provides some examples of real-time applications from the Modelica literature. Since Section 22.3 mostly discussed features of the Modelica language to support real-time work, the intent of this section is to provide a flavor for selected published real-time applications of Modelica. These applications encompass many different model-based system development activities, including HiL, SiL, and controller V&V. It should be noted that many of these examples illustrate a workflow whereby the Modelica tool leverages third-party products such as MathWorks Real-Time Workshop® [36] to generate code and compile an executable for the target real-time platform.

Otter et al. [37] demonstrated a very early application of Modelica and Dymola for real-time simulation of a detailed automatic gearbox. A model for the 4-speed automatic gearbox with a planetary and Ravigneaux gear set is developed in Modelica including the torque converter and the associated clutches to enable the various gears. The gearbox model is implemented with a simple drivetrain model to create a vehicle model. The paper described in detail several of the key Modelica models, including the implementation of the clutch model. Real-time simulation was performed using the code generated by Dymola, which was compiled using Real-Time Workshop and downloaded to a dSPACE platform for simulation. This early work demonstrated real-time capability from a Modelica model using Dymola's symbolic manipulation for DAEs.

As a demonstration of the mixed-mode integration and inline integration schemes in Dymola for real-time simulation, Schiela and Olsson [32] tested the schemes on a

6-degree-of-freedom (DOF) robot example and a vehicle drivetrain example with a diesel engine and associated air path and loads. Benchmarking was performed using standard explicit and implicit integration schemes in addition to the mixed-mode and inline schemes in Dymola. Significant reduction in computational time was reported with the newly implemented techniques.

In one of the earliest industrial applications of Modelica for real-time simulation, Toyota [33] evaluated the newly introduced capabilities in Dymola for inline and mixed-mode integration. This work was part of a larger effort to use Modelica models to support HiL controller verification efforts. The authors evaluated inline and mixed-mode integration schemes for two industry application problems that included stiff behavior. The first problem was an engine example consisting of a mean value engine model with a complete air flow path with orifice-based flow calculations including a variable geometry turbo and an exhaust gas recirculation system. The second problem was a hydraulic actuator system with valves and long pipes. The problems were simulated on different HiL platforms with the different integration schemes for benchmarking. The authors note that the advanced integration techniques result in improved real-time performance for the stiff problems including up to two orders of magnitude reduction in computational time when compared with explicit methods.

Elmqvist et al. [30,38] performed a detailed study of the modeling and real-time simulation issues associated with detailed vehicle powertrain and chassis models in Modelica. Common issues in automotive real-time simulations are described in the context of a gearbox example. The authors provide detailed information on the challenges, modeling approaches, and resulting real-time capability of a detailed automotive powertrain simulation. The authors demonstrated impressive real-time execution of a detailed vehicle dynamics model with 72 DOF on the Opal-RT platform using Dymola's elaborate symbolic processing capabilities. Elmqvist et al. [30] provide insight into the symbolic and numerical processing capabilities of Dymola. Several different integration schemes are benchmarked and the results compared for both computational accuracy and efficiency. This work demonstrated that detailed automotive models in Modelica could be simulated in real time.

Backman and Edvall [39] illustrated the use of a real-time dynamic process simulator for a pulp paper mill to support operator training for new equipment installs. Modelica models for the paper machines and processes were developed by integration of standard models of general equipment (valves, pumps, pipes, tanks, heat exchangers, etc.) with custom models of machinery specific to the paper plant. These reusable models were integrated with the actual control system and the actual operator displays to provide a simulation-based training environment that is identical to that of the actual physical plant. The models were compiled with Dymola into a Microsoft Windows application that is used as a real-time DDE server. This training environment helps to familiarize operators with new process dynamics, displays, and interlocking logic. Furthermore, it offers a safe environment for the simulation of a wide range of scenarios and disturbances, some potentially dangerous if performed in a physical sense, that affect the process conditions. In addition to training, the models can support control strategy development and verification before hardware deployment, resulting in significant reduction in test time.

Ferretti et al. [40] developed and implemented a concept to obtain real-time simulation code directly from Modelica models in an attempt to facilitate the use of existing Modelica models in real-time applications. The implementation involved the development of a special Modelica block that is used to define the I/O variables, communication with external tasks and hardware, and execution scheduling for the simulation. This module links to the Linux RTAI operating system and is capable of creating a soft real-time binary. Testing was performed on a 7-DOF multibody model for a robotic arm. While a promising approach, the resulting model was not real-time capable, and the authors recommended model refinement to simplify the computationally intensive dynamics, particularly nonlinear low-speed friction.

Kellner et al. [41] developed techniques for the parameterization of Modelica models in HiL environments without file I/O operations. On the basis of a strong desire to separate models from parameters, The ZF Group stores model parameters in external ASCII files that are linked at initialization. This approach poses a challenge for systems such as dSPACE that do not allow file I/O operations. The authors developed a static and dynamic parameterization approach to link parameter files into the execution code. The approach was demonstrated with Dymola through a passenger car vehicle model that was simulated on a dSPACE platform.

Morawietz et al. [42] developed a Modelica model library consisting of powertrain and electrical system models in Modelica to support the development of control strategies for intelligent vehicle energy management and fuel consumption minimization. The paper describes in detail the thermodynamic model of the engine, the alternator model, and the implementation of a neural network library. A dynamic engine thermal model and associated thermal network including the oil and coolant is presented to model the impact of warm-up on engine thermodynamics and fuel consumption. A dynamic HiL environment based on dSPACE hardware was built to measure component parameters and test energy management strategies. The HiL setup allowed the interaction of the modeled engine and electrical system with alternator and battery hardware. Sample HiL results are shown for the battery voltage and various currents in the system in response to load changes.

Winkler and Gühmann [43] integrated a vehicle model in Modelica with a dynamic engine test bench to support model-based calibration of engine and transmission control units. Since the engine is represented in hardware, the vehicle model concentrates on the dynamics of the rest of the powertrain and vehicle. The Modelica models were created with Dymola and ultimately used with RT-LAB real-time software running on a standard PC. The authors also outline a detailed synchronization procedure to ensure appropriate communication interfaces and coupling between the dynamic test bench and the HiL vehicle simulation. The resulting integrated simulations allow measurement of fuel consumption and emissions from the actual engine hardware, values that are often difficult to predict analytically, over different drive cycles for a given vehicle model. The virtual vehicle model can easily be modified to represent different powertrain configurations, including both conventional and hybrid vehicles.

Ebner et al. [44] implemented an HiL solution for the development and testing of hybrid electric vehicle (HEV) components. The objective of this work is to allow testing and validation of virtual systems in conjunction with actual hardware components. The real-time system is a Suse Linux Enterprise Real-Time (SLERT) operating system. The Modelica Real-Time Interface, developed by Arsenal Research, provides the connection between Dymola for model simulation and the I/O interfaces for interaction with the hardware systems for real-time synchronization. It also permits execution of the HiL simulation directly from Dymola. Test cases for a HEV and an electric two-wheeler are presented. The HiL system included Modelica vehicle simulation models integrated with hardware battery and electrical systems as well as a dynamic test bench.

Gäfvert et al. [45] implemented a Modelica-based solution for HiL simulation of dynamic liquid food processes. Modelica models for the liquid food processes were compiled using Dymola into a real-time process simulator running on a standard Windows PC and integrated with the production control strategy. This approach results in soft-real-time capability that is sufficient for the given application. The goal of this effort was to replace costly and time-consuming predelivery hardware testing using liquid surrogates with HiL testing using the process model. HiL simulation also offers opportunities for development, testing, and V&V of the production control strategy. A dairy pasteurization process was used to evaluate the real-time HiL platform. Testing with the HiL simulations over a range of fault conditions indicated that the model-based approach discovered nearly all problems identified with the hardware testing in addition to several others that were not discovered in hardware.

Yuan and Xie [46] describe the modeling of a hydraulic steering system including the dynamic models of key components such as the priority valve and steering control unit. A Modelica model of the hydraulic steering system is then implemented in Dymola. The resulting model is used in a real-time simulation in conjunction with a vehicle model from Carsim to capture the interaction between the steering system and the vehicle dynamics through human-in-the-loop testing. Opal-RT RT-LAB is used as the real-time target. The resulting HiL setup allows various driving maneuvers to be simulated to verify the performance of the hydraulic steering system.

Blochwitz and Beutlich [47] describe various steps for creating a real-time capable model from Modelica models in SimulationX [48]. Starting from a description of the key model, compiler, and solver optimizations required for real-time simulation, the authors describe the SimulationX Code Export Wizard that guides the user through the workflow to create real-time capable models in target-independent C code. The tool can interface with targets based on Simulink® [49] and Real-Time Workshop, dSPACE, SCALE-RT, and NI Veristand. Various model reduction and analysis techniques are described to aid the modeler in the nontrivial task of model complexity reduction for computational efficiency.

Bonvini et al. [50] demonstrated the use of Modelica for HiL simulation of a robot for control design. Although the application is simplified and largely academic in nature, the work is noteworthy for the open framework that it employed. Modelica is used to create a simplified model of a robot with two rotational DOF and also for creating the control algorithm. Through the open framework documented in the paper, a HiL system is shown with an Arduino microcontroller running the control algorithm from Modelica and a PC generating the desired trajectory, simulating the Modelica model of the robot, and visualizing the 3D movement. The open framework involves the use of OpenModelica for Modelica compilation and DAE generation through XML export, Java for generating executable code, and Python for visualization. This application, while academic in terms of scope and complexity, clearly showcases the unique possibilities with the open Modelica language in conjunction with other open-source tools.

Pitchaikani et al. [51] demonstrated a closed loop real-time simulation of a vehicle with a climate control system in Modelica. The climate control system model included the air and refrigerant loops with representations for the cabin, evaporator, expansion valve, condenser, compressor, blower, heater, and air inlet door. Real-time modeling of climate control systems is a challenge because of the complex, nonlinear physical phenomena inherent in the air and refrigerant loops and computationally expensive property calculations of the two media. The vehicle model was combined with a controller model in Simulink and simulated in an Opal-RT real-time environment. Drive cycle simulations with the integrated vehicle and controller model on the HiL platform were demonstrated in real time as part of a controller verification effort.

22.5   CONCLUSIONS

Multidomain physical system modeling with Modelica has proven to be a key technology for model-based systems development. This chapter provides an overview of the Modelica language and the key features that make it a desirable modeling platform to support real-time simulation. Several different sample models are provided to provide concrete examples of the features discussed. Selected simulation results highlight the computational impact of model formulation and symbolic manipulation. Published work is reviewed to convey a sense for the variety of real-world applications from many different domains that are supported by real-time simulation with Modelica.

With continued development by the Modelica Association, enhancements to the language and standard library are ongoing. The Modelica user base continues to grow as do the tools that support the Modelica language. This growth will continue to bring cutting edge modeling and simulation technology to the forefront to support both current and future real-time model-based system development applications.

REFERENCES

1. Modelica Association. 2010. *Modelica: A Unified Object-Oriented Language for Physical Systems Modeling—Language Specification*. Version 3.2. http://www.modelica .org/documents/ModelicaSpec32.pdf (accessed June 30, 2010).
2. Modelica Association. "Publications—Modelica Portal." Modelica and the Modelica Association—Modelica Portal. http://www.modelica.org/publications (accessed June 30, 2010).
3. Elmqvist, H. 1978. *A Structured Model Language for Large Continuous Systems*. Thesis, Department of Automatic Control, Lund University, Sweden. http://www.control.lth.se/ database/publications/article.pike?artkey=elm78dis (accessed June 30, 2010).
4. Cellier, F. E., H. Elmqvist, and M. Otter. 1995. "Modeling from Physical Principles." In *The Control Handbook*, edited by W. S. Levine, 99–108. Boca Raton, MA: CRC Press.

5. Pantelides, C. 1988. "The Consistent Initialization of Differential-Algebraic Systems." *SIAM Journal on Scientific and Statistical Computation* 9 (213): 213–31.
6. Modelica Association. "Documents—Modelica Portal." Modelica and the Modelica Association—Modelica Portal. http://www.modelica.org/documents (accessed June 30, 2010).
7. Modelica Association. "Modelica Libraries—Modelica Portal." Modelica and the Modelica Association—Modelica Portal. http://www.modelica.org/libraries (accessed June 30, 2010).
8. Tiller, M. 2001. *Introduction to Physical Modeling with Modelica*. Boston, MA: Kluwer Academic Publishers.
9. Fritzson, P. A. 2004. *Principles of Object-Oriented Modeling and Simulation with Modelica 2.1*. Piscataway, NJ: IEEE Press.
10. Modelica Association. "Home—Modelica Portal." Modelica and the Modelica Association—Modelica Portal. http://www.modelica.org (accessed June 30, 2010).
11. Dassault Systemes. 2008. *Dymola. Computer Software. Version 7.0*. Lund, Sweden: Dassault Systemes (Dynasim AB).
12. Deltatheta. Modelica SDK. http://www.deltatheta.com/products/modelicasdk/index.jsp (accessed June 30, 2010).
13. OpenModelica. Welcome to OpenModelica. http://www.openmodelica.org/ (accessed June 30, 2010).
14. Modelon AB. JModelica.org. http://www.jmodelica.org/ (accessed June 30, 2010).
15. Nikoukhah, R. "Scicos Homepage." Accueil—INRIA Rocquencourt. http://www-rocq.inria.fr/scicos/ (accessed June 30, 2010).
16. Cellier, F. E., and E. Kofman. 2006. *Continuous System Simulation*. New York: Springer.
17. Anderson, M. 1994. *Object-Oriented Modeling and Simulation of Hybrid Systems*. Thesis. Department of Automatic Control, Lund Institute of Technology, Sweden.
18. Tarjan, R. 1972. "Depth-First Search and Linear Graph Algorithms." *SIAM Journal of Computation* 1 (2): 146–60.
19. Casella, F., M. Otter, K. Proelss, C. Richter, and H. Tummescheit. 2006. "The Modelica Fluid and Media Library for Modeling of Incompressible and Compressible Thermo-Fluid Pipe Networks." In *Proceedings of the 5th International Modelica Conference*, Vienna, Austria, pp. 631–40. Modelica Association.
20. Park, R. H. 1929. "Two Reaction Theory of Synchronous Machines." *AIEE Transactions* 48: 716–30.
21. Bachmann, B. 2010. "Tutorial 2: Mathematical Aspects of Modeling and Simulation with Modelica." In *Proceedings of the 6th International Modelica Conference*, Bielefeld, Germany. http://www.modelica.org/events/modelica2006/Proceedings/tutorials/Tutorial2.pdf (accessed June 30, 2010).
22. Tate, E. D., M. Sasena, J. Gohl, and M. Tiller. 2008. "Model Embedded Control: A Method to Rapidly Synthesize Controllers in a Modeling Environment." In *Proceedings of the 6th International Modelica Conference*, Bielefeld, Germany, pp. 493–502. Modelica Association.
23. Batteh, J., and M. Tiller. 2009. "Implementation of an Extended Vehicle Model Architecture in Modelica for Hybrid Vehicle Modeling: Development and Applications." In *Proceedings of the 7th International Modelica Conference*, Como, Italy, pp. 823–32. Modelica Association.
24. Bals, J., G. Hofer, A. Pfeiffer, and C. Schallert. 2003. "Object-Oriented Inverse Modelling of Multi-Domain Aircraft Equipment Systems and Assessment with Modelica." In *Proceedings of the 3rd International Modelica Conference*, Linköping, Sweden, pp. 377–84. Modelica Association.

25. Xu, X., and E. Azarnasab. 2012. "Progressive Simulation-Based Design for Networked Real-Time Embedded Systems." In *Real-Time Simulation Technologies: Principles, Methodologies, and Applications*, edited by K. Popovici, P. J. Mosterman. Boca Raton, FL: CRC Press.

26. Scharpf, J., R. Hoepler, and J. Hillyard. 2012. "Real Time Simulation in the Automotive Industry." In *Real-Time Simulation Technologies: Principles, Methodologies, and Applications*, edited by K. Popovici, P. J. Mosterman. Boca Raton, FL: CRC Press.

27. Elmqvist, H., M. Otter, D. Henriksson, B. Thiele, and S. E. Mattsson. 2009. "Modelica for Embedded Systems." In *Proceedings of the 7th International Modelica Conference*, Como, Italy, pp. 354–63. Modelica Association.

28. Otter, M., M. Malmheden, H. Elmqvist, S. E. Mattsson, and C. Johnsson. 2009. "A New Formalism for Modeling of Reactive and Hybrid Systems." In *Proceedings of the 7th International Modelica Conference*, Como, Italy, pp. 364–77. Modelica Association.

29. Blochwitz, T., M. Otter, M. Arnold, C. Bausch, C. Clauss, H. Elmqvist, A. Junghanns, et al. 2011. "The Functional Mockup Interface for Tool Independent Exchange of Simulation Models." In *Proceedings of the 7th International Modelica Conference*, Dresden, Germany, pp. 823–32. Modelica Association.

30. Elmqvist, H., S. E. Mattsson, H. Olsson, J. Andreasson, M. Otter, C. Schweiger, and D. Brück. 2004. "Realtime Simulation of Detailed Vehicle and Powertrain Dynamics." In *Proceedings of the SAE 2004 World Congress & Exhibition*, Detroit, MI, Paper 2004-01-0768.

31. Elmqvist, H., S. E. Mattsson, and H. Olsson. "New Methods for Hardware-in-the-Loop Simulation of Stiff Models." In *Proceedings of the 2nd International Modelica Conference*, Oberpfaffenhofen, Germany, pp. 59–64. Modelica Association.

32. Schiela, A., and H. Olsson. 2000. "Mixed-Mode Integration for Real-Time Simulation." In *Modelica Workshop 2000 Proceedings*, Lund, Sweden, pp. 69–75. Modelica Association.

33. Soejima, S., and T. Matsuba. 2002. "Application of Mixed Mode Integration and New Implicit Inline Integration at Toyota." In *Proceedings of the 2nd International Modelica Conference*, Oberpfaffenhofen, Germany, pp. 65-1–65-6. Modelica Association.

34. Zimmer, D. 2008. "Introducing Sol: A General Methodology for Equation-Based Modeling of Variable-Structure Systems." In *Proceedings of the 6th International Modelica Conference*, Bielefeld, Germany, pp. 47–56. Modelica Association.

35. Zimmer, D. 2009. "An Application of Sol on Variable-Structure Systems with Higher Index." In *Proceedings of the 7th International Modelica Conference*, Como, Italy, pp. 225–32. Modelica Association.

36. MathWorks. Real-Time Workshop. http://www.mathworks.com/products/rtw (accessed June 30, 2010).

37. Otter, M., C. Schlegel, and H. Elmqvist. 1997. "Modeling and Realtime Simulation of an Automatic Gearbox Using Modelica." In *Proceedings of ESS'97 European Simulation Symposium*, Passau, Germany, pp. 115–21, International Society for Computer Simulation.

38. Elmqvist, H., S. E. Mattsson, J. Andreasson, M. Otter, C. Schweiger, and D. Brück. 2003. "Real-Time Simulation of Detailed Automotive Models." In *Proceedings of the 3rd International Modelica Conference*, Linköping, Sweden, pp. 29–38. Modelica Association.

39. Backman, J., and M. Edvall. 2005. "Using Modelica and Control Systems for Real-Time Simulations in the Pulp." In *Proceedings of the 4th International Modelica Conference*, Hamburg, Germany, pp. 579–83. Modelica Association.

40. Ferretti, G., M. Gritti, G. Magnani, G. Rizzi, and P. Rocco. 2005. "Real-Time Simulation of Modelica Models under Linux/RTAI." In *Proceedings of the 4th International Modelica Conference*, Hamburg, Germany, pp. 359–65. Modelica Association.

41. Kellner, M., M. Neumann, A. Banerjee, and P. Doshi. 2006. "Parametrization of Modelica Models on PC and Real Time Platforms." In *Proceedings of the 5th International Modelica Conference*, Vienna, Austria, pp. 267–73. Modelica Association.

42. Morawietz, L., S. Risse, H. Zellbeck, H. Reuss, and T. Christ. 2005. "Modeling an Automotive Power Train and Electrical Power Supply for HiL Applications using Modelica." In *Proceedings of the 4th International Modelica Conference*, Hamburg, Germany, pp. 301–07. Modelica Association.

43. Winkler, D., and C. Gühmann. 2006. "Synchronising a Modelica Real-Time Simulation Model with a Highly Dynamic Engine Test-Bench System." In *Proceedings of the 5th International Modelica Conference*, Vienna, Austria, pp. 275–81. Modelica Association.

44. Ebner, A., M. Ganchev, H. Oberguggenberger, and F. Pirker. 2008. "Real-Time Modelica Simulation on a Suse Linux Enterprise Real Time PC." In *Proceedings of the 6th International Modelica Conference*, Bielefeld, Germany, pp. 375–79. Modelica Association.

45. Gäfvert, M., T. Skoglund, H. Tummescheit, J. Windahl, H. Wikander, and P. Reuterswärd. 2008. "Real-Time HWIL Simulation of Liquid Food Process Lines." In *Proceedings of the 6th International Modelica Conference*, Bielefeld, Germany, pp. 709–15. Modelica Association.

46. Yuan, Q., and B. Xie. 2008. "Modeling and Simulation of a Hydraulic Steering System." *SAE International Journal of Commerical Vehicles* 1 (1): 488–94.

47. Blochwitz, T., and T. Beutlich. 2009. "Real-Time Simulations of Modelica-Based Models." In *Proceedings of the 7th International Modelica Conference*, Como, Italy, pp. 386–92. Modelica Association.

48. ITI GmbH. SimulationX. http://www.itisim.com/simulationx_505.html (accessed June 30, 2010).

49. MathWorks. Simulink. http://www.mathworks.com/products/simulink (accessed June 30, 2010).

50. Bonvini, M., F. Donida, and A. Leva. 2009. "Modelica as a Design Tool for Hardware-in-the-Loop Simulation." In *Proceedings of the 7th International Modelica Conference*, Como, Italy, pp. 378–85. Modelica Association.

51. Pitchaikani, A., K. Jebakumar, S. Venkataraman, and S. A. Sundaresan. 2009. "Real-Time Drive Cycle Simulation of Automotive Climate Control System." In *Proceedings of the 7th International Modelica Conference*, Como, Italy, pp. 839–46. Modelica Association.

52. National ITS Architecture Team. 2007. *Systems Engineering for Intelligent Transportation Systems*. Report no. FHWA-HOP-07-069. United States Department of Transportation. http://www.ops.fhwa.dot.gov/publications/seitsguide/index.htm (accessed June 30, 2010).

53. Cellier, F. E. 1991. *Continuous System Modeling*. New York: Springer-Verlag.

54. Modelica Association. 2010. *Modelica Tools—Modelica Portal*. Modelica and the Modelica Association—Modelica Portal. http://www.modelica.org/tools (accessed June 30, 2010).

23 Real-Time Simulation of Physical Systems Using Simscape™

Steve Miller and Jeff Wendlandt

CONTENTS

23.1 Introduction 581
    23.1.1 Examples of Real-Time Simulation..581
    23.1.2 Benefits of Real-Time Simulation...583
    23.1.3 Challenges of Real-Time Simulation..584
23.2 Moving from Desktop to Real-Time Simulation 585
    23.2.1 Procedure for Tuning Solver Settings...586
    23.2.2 Adjusting Models to Make Them Real-Time Capable 596
23.3 Results And Conclusions 597
References...597

23.1 INTRODUCTION

Real-time simulation of multidomain physical system models (mechanical, electrical, hydraulic, etc.) requires finding a combination of model complexity, solver choice, solver settings, and real-time target that permit execution in real time. A better understanding of the trade-offs involved in each of these areas makes it easier to achieve this goal and use Model-Based Design to reap the benefits of using virtual systems before building hardware prototypes. This chapter outlines the steps in moving from desktop to real-time simulation and illustrates this process using models built from MathWorks® products based on Simscape™ simulation technology. The steps described apply to real-time simulation regardless of which real-time hardware is used.

23.1.1 EXAMPLES OF REAL-TIME SIMULATION

Replacing physical devices, such as vehicles, robots, or planes, with virtual devices can drastically reduce the cost of testing control systems, software, and hardware. It can also improve the quality of the final product by enabling more complete testing of the entire system. Often it is necessary to run the computer simulation representing the virtual system in real time. This means that the inputs and outputs (I/O) in the virtual world of simulation must be read or updated synchronously with the

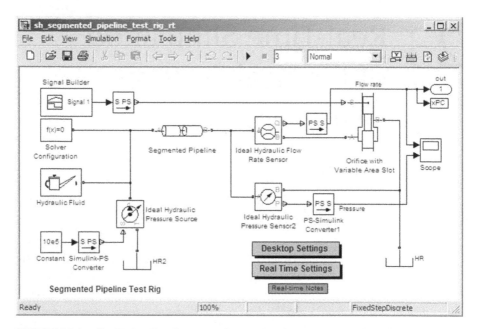

FIGURE 23.1   SimHydraulics demonstration model of water hammer in a pipeline.

physical world. When the simulation time reaches 5, 50, or 500 seconds, the exact same amount of time has passed in the physical world.

Configuring a model and the numerical integrator to simulate in this manner can be difficult. The simulation execution time per step must be consistent and sufficiently shorter than the time step of the simulation to permit any other tasks that the simulation environment must perform, such as reading sensor inputs or outputting transducer signals. This is a challenging prospect because the conditions vary during simulation. Switches open, valves close, and these occasional events can require more computations to achieve an accurate result. To be successful, the solver settings, simulation step size, and the level of model fidelity must be adjusted to find a combination that permits real-time simulation while delivering accurate results.

Advances in solver technology have made it easier to configure simulations to simulate in this fashion. Features added to simulation tools, such as fixed-cost algorithms (algorithms where the computational cost per time step is nearly constant) and local solvers added to Simscape from MathWorks, make it possible to simulate even complex models such as hydraulic pipelines in real time. As an example, we simulated the SimHydraulics® demonstration model sh_segmented_pipeline_test_ rig (Figure 23.1) on a desktop computer using solver ode15s, a maximum step size of 0.03 seconds, and a relative tolerance of 1e-3. After applying the process covered in this chapter to the model, accurate results were achieved when running the model in real time using xPC Target™ [1] from MathWorks on an Intel Core 2 Duo E6700 (2.66 GHz) with a simulation step size of 1 millisecond (Figure 23.2).

This is a particularly challenging model numerically for it includes the water hammer effect, which takes place when a valve is abruptly opened or closed creating

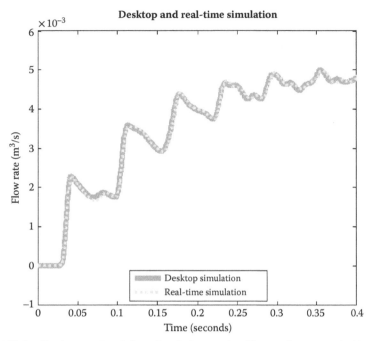

FIGURE 23.2   Desktop and real-time simulation results. The results are nearly identical.

rapid changes in flow rate. Even with the restrictions imposed by executing on a real-time target, the simulation results capture the oscillations in the hydraulic circuit. To develop this process, the steps described in this chapter for moving from desktop to real-time simulation have been applied to this and more than 30 other models in different physical domains built with MathWorks physical modeling products, all of which were able to run in real time.

23.1.2   Benefits of Real-Time Simulation

Real-time simulation is used in a number of steps in the development process and in some cases in the final product. In Model-Based Design, the plant model is used to develop and test the control and signal processing algorithms in desktop simulation. Once the designs are complete and the algorithms exist in production code, it is necessary to test that code as well as the production controller. Instead of connecting it directly to a hardware prototype, the plant model used in the design phase can be used to test the production code and processor if it is capable of running in real time. This is referred to as hardware-in-the-loop testing and offers many benefits, including the following:

1. Ability to test conditions that would damage equipment or personnel
2. Ability to test systems where no prototypes exist
3. Reduced costs in the later phases of development
4. Ability to test 24 hours a day, 7 days a week

In addition to the development process, real-time simulation is also used in the final product. Products that have a human in the simulation loop require real-time simulation. For example, flight simulators that are used to train pilots require real-time simulation of the plane, control system, weather conditions, and other aspects of their environment (see also Chapter 15).

23.1.3 CHALLENGES OF REAL-TIME SIMULATION

For a simulation to execute in real time, the amount of time spent calculating the solution for a given time step must be less than the duration of that time step. This requires that the execution time per simulation time step be bounded. Variable-step solvers, which are often used in desktop simulation, take smaller steps to accurately capture events that occur during the simulation. However, varying the step size is not an option for real-time simulation, and therefore, a fixed-step solver (implicit or explicit) must be used. This can make real-time simulation more challenging than desktop simulation. The model and fixed-step solvers must be configured so that system dynamics can be accurately captured without changing the step size. These requirements are constraints imposed by hard real-time systems, such as those used to test controller software and hardware. Soft real-time systems, such as video game physics, often have less stringent requirements.

A fixed-step solver that provides accurate results at a step size large enough to permit real-time simulation must be chosen. Most fixed-step solvers will produce equally accurate simulation results as a variable-step solver if a small enough step size is chosen. However, different fixed-step solver algorithms (implicit, explicit, lower/higher order, etc.) will require different step sizes to produce accurate results. They also require different amounts of computational effort per time step [2].

Once a solver is chosen, determining an appropriate step size is the next challenge. Increasing the time step to permit more time to calculate the result can lead to inaccurate results. Reducing the time step to improve the accuracy of the simulation results may make it impossible to execute in real time. Trial and error may be required to find the combination of settings that permit real-time simulation while producing accurate simulation results. The eigenvalues [3] of the system can give an indication of the time constants in the system and help determine what the required minimum step size is.

If this combination of settings cannot be found, it may be that the model contains effects that a fixed-step solver cannot handle at a step size that permits real-time simulation. These effects can be events in the simulation (hard stops, stick-slip friction, switches that open and close, etc.) or portions of the system that have a very small time constant (small masses attached to stiff springs, current or pressure oscillations, etc.). Identifying and modifying these elements is then required before searching for the combination of solver settings and step size that will permit real-time simulation.

Section 23.2 will cover the process of configuring a model used in desktop simulation for real-time simulation. The process involves determining if the model is built at an appropriate level of fidelity for real-time simulation, simplifying if necessary, and configuring solver settings to achieve accurate results.

23.2 MOVING FROM DESKTOP TO REAL-TIME SIMULATION

To move from desktop simulation to real-time simulation on the chosen real-time hardware, the following items must be adjusted until the simulation can execute in real time and deliver results sufficiently close to the results obtained from desktop simulation:

1. Solver choice
2. Number of solver iterations
3. Step size
4. Model size and fidelity

The procedure depicted in the flowchart in Figure 23.3 shows how to configure a model for real-time simulation.

This process has been applied to over 30 models containing hydraulic, electrical, mechanical, pneumatic, and thermal components that include a range of linear and

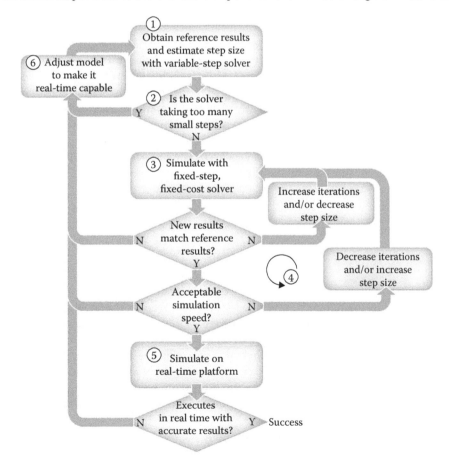

FIGURE 23.3 Flowchart depicting the process that helps engineers move from desktop simulation to real-time simulation.

nonlinear elements. In each case, real-time execution was achieved with very accurate results. The modeling concepts in these models included the following:

1. Combined multibody (3D-mechanical) and hydraulic or electrical systems
2. Dynamic, compressible fluid flow (hydraulic and pneumatic)
3. Physical phenomena (friction and clutch events)
4. Hybrid models (continuous-time plants plus discrete logic control systems)
5. Multirate systems (multiple sample rates within the model)

23.2.1 PROCEDURE FOR TUNING SOLVER SETTINGS

Step 1: Obtain a converged set of results with a variable-step solver.
To ensure that the results obtained with the fixed-step solver are accurate, a set of reference results are needed. These can be obtained by simulating the system with a variable-step solver and ensuring that the results are converged by tightening the error tolerances until the simulation results do not change. For Simscape models, the recommended variable-step solvers are ode15s and ode23t.

Step 2: Examine the step sizes during the simulation to determine if the model is likely to run with a large enough step size to permit real-time simulation.
A variable-step solver will vary the step size to stay within the error tolerances and to react to zero-crossing events [4]. If the solver abruptly reduces the step size to a small value (e.g., 1e-15s), this indicates that the solver is attempting to accurately identify a zero-crossing event. A fixed-step solver may have trouble capturing these events at a step size that is sufficiently large to permit real-time simulation.

The following MATLAB® commands can be used to generate a plot that shows how the time step varies during the simulation:

```
semilogy(tout(1:end-1),diff(tout),'-*');
title('Step Size vs. Simulation
Time','FontSize',14,'FontWeight','bold');
xlabel('Simulation Time (s)','FontSize',12);
ylabel('Step Size (s)','FontSize',12);
```

The plots in Figures 23.4 and 23.5 illustrate the concepts explained above. They are produced by executing the preceding MATLAB code on the simulation results from the SimHydraulics model shown in Figures 23.6 and 23.7, which contains a number of components that create zero-crossing events.

This analysis should provide a rough idea of a step size that can be used to run the simulation. Determining what effects are causing these events and modifying or eliminating them will make it easier to run the system with a fixed-step solver at a larger step size and produce results comparable to the variable-step simulation.

Step 3: Simulate the system with a fixed-step, fixed-cost solver and compare the results to the reference set of results obtained from the variable-step simulation.
As explained in Section 23.1.3, a fixed-step solver (implicit or explicit) must be used to run the simulation in real time. The chosen solver must provide robust performance and

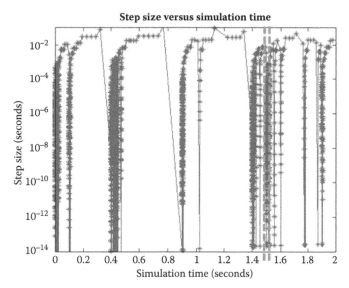

FIGURE 23.4 Plot of step size during variable-step simulation. Abrupt drops in step size indicate zero-crossing events. The amount of zero-crossing events and how easily the simulation recovers give a rough indication of how difficult it will be for a fixed-step solver to produce accurate results at the largest step size the variable-step solver uses.

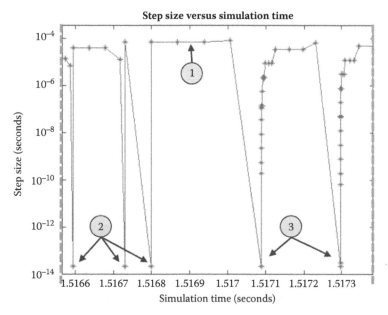

FIGURE 23.5 Plot of a smaller range of the step size during simulation. (1) indicates the step size that will meet the error tolerances for most of the simulation. (2) are examples of zero-crossing events where the solver recovered instantly and may not be difficult for the fixed-step solver. (3) are examples of zero-crossing events where the variable-step solver took longer to recover and will likely require a smaller step size for the fixed-step solver to deliver results with acceptable accuracy.

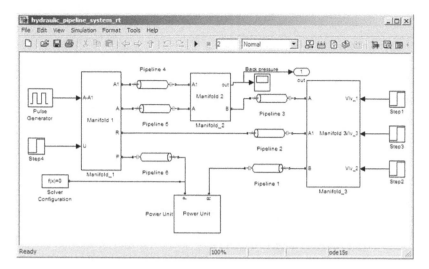

FIGURE 23.6 Hydraulic model of a pipeline system modeled in SimHydraulics containing valves, junctions, and a centrifugal pump.

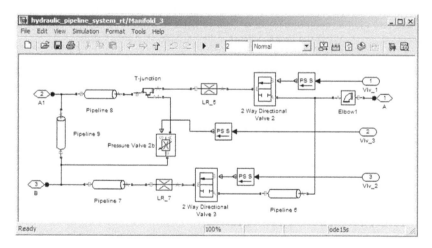

FIGURE 23.7 Portion of SimHydraulics model containing directional valves, flow control valves, and orifices.

deliver accurate results at a step size large enough to permit real-time simulation. The solver should be chosen to minimize the amount of computation required per time step while providing robust performance at the largest step size possible. To decide which type of fixed-step solver to use, it is necessary to determine if the model describes a stiff or a nonstiff problem. The problem is stiff if the solution the solver is seeking varies slowly, but there are other solutions within the error tolerances that vary rapidly [2].

Comparing the simulation results generated by a fixed-step implicit solver and a fixed-step explicit solver for the same model of a pneumatic system (Figure 23.8) shows a difference in accuracy that is dependent on step size (Figure 23.9) and model stiffness (Figure 23.10).

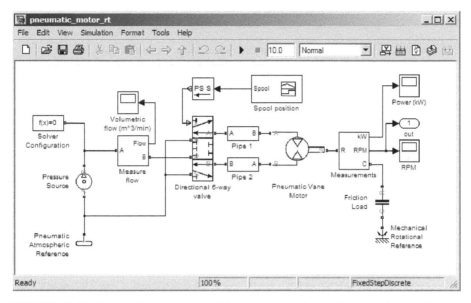

FIGURE 23.8 Pneumatic system containing a pump, valve, and motor, simulated using implicit and explicit fixed-step solvers.

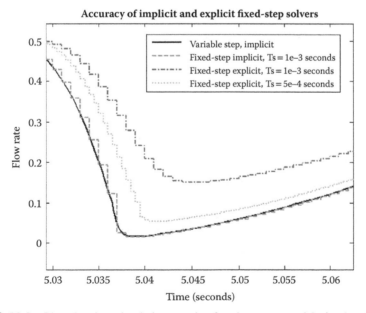

FIGURE 23.9 Plot showing simulation results for the same model simulated with a variable-step solver, fixed-step implicit solver, and fixed-step explicit solver. The explicit solver requires a smaller time step to achieve accuracy comparable to the implicit solver.

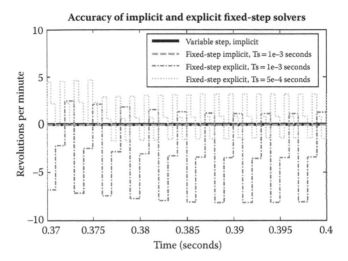

FIGURE 23.10 Plot showing simulation results for the same model simulated with a variable-step solver, fixed-step implicit solver, and fixed-step explicit solver. The oscillations in the fixed-step explicit solver simulation results suggest this is a stiff problem.

Explicit and implicit solvers use different numerical methods to solve the system of equations. An explicit algorithm samples the local gradient to find a solution, whereas an implicit algorithm uses matrix operations to solve a system of simultaneous equations that helps predict the evolution of the solution [2]. As a result, an implicit algorithm does more work per simulation step but can take larger steps. For stiff systems, implicit solvers should be used.

Both accuracy and computational effort must be taken into account when choosing a fixed-step solver. Simulating physical systems often involves multiple iterations per time step to converge on a solution. For a real-time simulation, the amount of computational effort per time step must be bounded. To have a bounded amount of execution time per simulation time step, it is necessary to limit the number of iterations per time step. This is known as a fixed-cost simulation. Fixed-cost simulation is used to prevent overruns, which occur when the execution time is longer than the sample time. Figure 23.11 shows how an overrun can occur if the number of iterations is not limited.

Iterations are necessary with implicit solvers. The iterations are handled automatically with variable-step solvers, but for the implicit fixed-step solver ode14x in Simulink®, the number of iterations per time step must be set. This is controlled by the parameter "Number Newton's iterations" in the Solver pane of the Configuration Parameters dialog box in Simulink.

Iterations are also often necessary for each Simscape physical network for both explicit and implicit solvers. The iterations in Simscape are limited by setting the checkbox, "Use fixed-cost runtime consistency iterations" and entering the number of nonlinear iterations in the Solver Configuration block (see Figure 23.12). If the local solver option is used, it is recommended to initially set the number of nonlinear iterations to two or three.

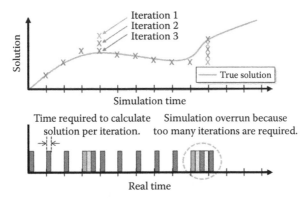

FIGURE 23.11   A fixed-step solver keeps the time step constant. Limiting any needed iterations per time step is necessary for fixed-cost simulation.

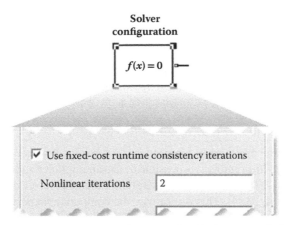

FIGURE 23.12   In Simscape, the Solver Configuration block permits limitation of the iterations per time step.

The amount of computational effort required by a solver varies with respect to a number of factors, including model complexity. To provide an indication of the relative cost for the fixed-step solvers available, a nonlinear model of a pneumatic actuation system (shown in Figure 23.8) containing a single Simscape physical network was simulated with each of the fixed-step solvers. These simulations were conducted at the same step size with similar settings for the total number of solver iterations. Figure 23.13 shows the normalized execution time.

From this plot, it is clear that for this example, most explicit fixed-step solvers require less computational effort than the implicit fixed-step solver ode14x. Although an explicit solver may require less computational effort, for stiff problems, an implicit solver is necessary for accurate results. For this example, the two local Simscape solvers (Backward Euler and Trapezoidal Rule) required the least computational effort. In most cases, they provide the best combination of speed and accuracy.

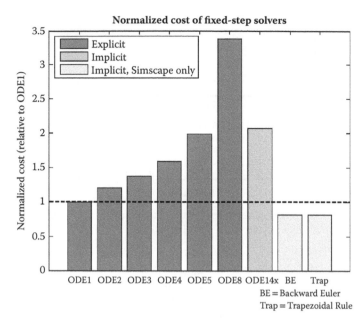

FIGURE 23.13 Plot of the normalized cost of all fixed-step solvers that can be used on Simscape models. The results were obtained by simulating a nonlinear model containing a single Simscape physical network with each solver at the same step size and similar settings for the total number of solver iterations.

A powerful option available in Simscape is to use a local solver on physical networks [5]. By using this option, it is possible to use an implicit fixed-step solver only on the stiff portions of the model and an explicit fixed-step solver on the remainder of the model (Figure 23.14). This minimizes the computations performed per time step, making it more likely the model will run in real time.

For Simscape models, the Backward Euler and the Trapezoidal Rule should always be tested and will most likely provide the best performance and the most flexibility because they can be configured per physical network. Figure 23.15 shows how to enable the local solver and the settings associated with it. The Backward Euler solver is designed to be robust and tends to damp out oscillations. The Trapezoidal Rule solver is designed to be more accurate and preserve oscillations. The Backward Euler algorithm tends to be less accurate than the Trapezoidal Rule but more numerically stable.

To summarize recommendations for setting up fixed-cost simulations:

1. If the system is nonstiff and is described by ordinary differential equations (ODEs), an explicit solver is usually the best choice.
2. If the system is stiff, an implicit solver (ode14x, Backward Euler, or Trapezoidal Rule) should be used and the number of iterations must be limited.
3. For Simscape models:
    a. Performing a fixed-cost simulation requires setting the number of iterations to prevent overruns. This is done by selecting the "Use fixed-cost

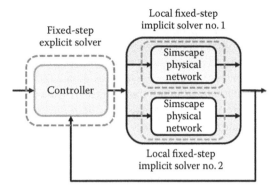

FIGURE 23.14 Using local solvers permits configuring implicit solvers on the stiff portions of the model and explicit solvers on the remainder of the model, minimizing execution time while maintaining accuracy.

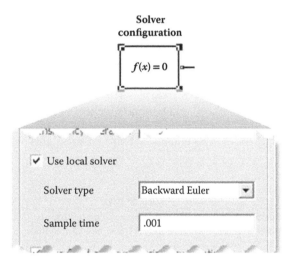

FIGURE 23.15 In Simscape, the Solver Configuration block permits configuration of local solvers on Simscape physical networks.

  runtime consistency iterations" setting in the Solver Configuration block attached to the Simscape physical network.

b. The local solvers in Simscape should always be tested. Using a local solver is often the best choice for fixed-cost simulations.

c. When performing fixed-cost simulation using the local solvers in Simscape, it is recommended to initially set the number of nonlinear iterations to two or three.

d. If you are using ode14x on a model with a Simscape physical network, to perform a fixed-cost simulation, it is necessary to enable fixed cost and set the number of nonlinear iterations in the Solver Configuration block.

Step 4: Find the combination of step size and number of nonlinear iterations where the step size is small enough to produce results that are sufficiently close to the set of reference results obtained from variable-step simulation and large enough so that there is enough safety margin to prevent an overrun.

During each time step, the real-time system must calculate the simulation results for the next time step (simulation execution) and read the inputs and write the outputs (processing I/O and other tasks). If this takes less than the specified time step, the processor remains idle during the remainder of the step. These quantities are illustrated in Figure 23.16.

The challenge is to find appropriate settings that provide accurate results while permitting real-time simulation. In each case, it is a trade-off of accuracy versus speed. Choosing a computationally intensive solver, increasing the number of nonlinear iterations, or reducing the step size both increases the accuracy and reduces the amount of idle time, raising the risk that the simulation will not run in real time. Adjusting these settings in the opposite direction will increase the amount of idle time but reduce accuracy.

It is necessary to leave sufficient safety margin to avoid an overrun when simulating in real time. If the amount of time spent processing inputs, outputs, and other tasks as well as the desired percentage of idle time are known, the amount of time available for simulation execution can be calculated as follows:

$$\text{Simulation Execution Budget} = \text{Step Size} - (\text{Processing Input/Output Time} + \text{Desired Percentage Idle Time} \times \text{Step Size})$$

Estimating the budget for the execution time helps ensure a feasible combination of settings is chosen.

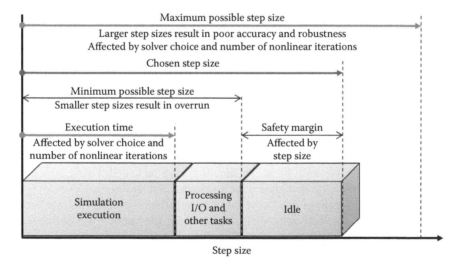

FIGURE 23.16 Trade-off involving solver choice, number of nonlinear iterations, and step size. For a given model, these must be chosen to deliver maximum accuracy and robustness with enough idle time to provide a sufficient safety margin.

The speed of simulation on the desktop can be used to estimate the execution time on the real-time target. There are many factors that affect the execution time on the real-time target, and therefore, simply comparing processor speed may not be sufficient. A better method is to measure the execution time during desktop simulation and then to determine the average execution time per time step on the real-time target for a given model. Knowing how these values relate for one model makes it possible to estimate execution time on the real-time target from the execution time during desktop simulation when testing other models.

Step 5: Using the selected solver, number of nonlinear iterations, and step size, simulate on the real-time platform and determine if the simulation can run in real time.
The tests run on the real-time platform should cover a representative set of tests to ensure that the worst-case scenario for the idle time is captured. This may include varying parameter values, inputs, and a range of conditions for the external hardware (reading/writing the I/O). The model needs to be robust enough to handle all situations that it may encounter.

Step 6: If the simulation does not run in real time on the selected real-time platform, it will be necessary to determine the cause and choose an appropriate solution.
If the simulation does not run in real time on the real-time platform, it may be due to the fact that the model is not real-time capable. The combination of effects captured in the model and the speed of the real-time platform may make it impossible to find solver settings that will permit it to run in real time (Figure 23.17).

If the simulation is not real-time capable, there are some options that can be explored:

1. Use a faster real-time computer.
2. Determine new settings that reduce the execution time (e.g., reducing the number of nonlinear iterations) or permit a larger step size.

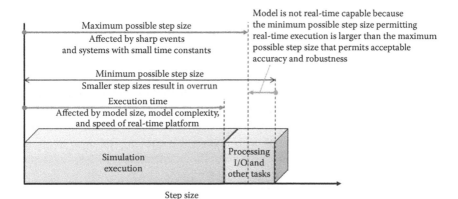

FIGURE 23.17 Diagram showing when a simulation is not real-time capable. The minimum possible step size permitting real-time execution is larger than the maximum possible step size that permits acceptable accuracy and robustness.

3. Eliminate effects that require significant computational effort or that require a small time step to accurately capture them.
4. If possible, configure the model and the real-time system to evaluate the physical networks in parallel. This can be done if for a given time step the networks are not dependent on one another. Experience with the generated code and the real-time target is required to use this option.

23.2.2 ADJUSTING MODELS TO MAKE THEM REAL-TIME CAPABLE

In the event that no settings can be found that permit the simulation to run in real time on the available real-time computer while delivering accurate results, it is necessary to modify or remove effects from the model that prevent real-time simulation. Here are two categories and some examples.

1. Elements that create events
   In this case, an event occurs so that the solution changes nearly instantaneously. The rapid change can be difficult for a fixed-step solver to step over and find the correct solution on the other side of the event. If it fails to find the solution, the solver may go unstable. Examples of elements that create these kinds of events include the following:
   a. Hard stops, backlash
   b. Stick-slip friction
   c. Switches or clutches
2. Elements with a small time constant
   In this case, an element or a group of elements has a very small time constant as compared to the desired simulation step size. These elements create fast dynamics that require a small step size so that a fixed-step solver can accurately capture the dynamics. Examples of systems that have a small time constant include the following:
   a. Small masses attached to stiff springs with minimal damping
   b. Electrical circuits with fast dynamics
   c. Hydraulic circuits with small compressible volumes

If a scripting environment that has commands permitting interrogation of the model, such as MATLAB, is available, identifying these components and parameters can be done very quickly, which narrows the search for the effects that must be modified. There are methods to automate these searches using tools such as the Simulink Model Advisor that makes it easier to apply these searches to other models.

Examining the eigenmodes of the system can indicate which states have the highest frequency, and mapping those states to the individual components may point to the source of the problem. For nonlinear models, this can only be done at an individual operating point and that operating point can be identified by looking for small step sizes during a variable-step simulation.

Once the effects have been identified, the next step is to modify or eliminate them. Methods that can be used to modify these effects include the following:

1. Replacing nonlinear component models with linearized versions of those models
2. Using lookup tables to simplify complex equations

3. Producing a simplified model by using system identification theory on the I/O data
4. Smoothing discontinuous functions (step changes) by using filters and other techniques.

Once the model is modified, the process described in Section 23.2 can be applied to identify the appropriate solver configuration and settings to enable real-time simulation.

23.3   RESULTS AND CONCLUSIONS

The procedure described in this chapter has been applied to over 30 physical models built using Simscape, SimHydraulics, and SimElectronics®, and all of them are able to run in real time. These models contain hydraulic, electrical, mechanical, pneumatic, and thermal elements and include applications such as hydromechanical servovalves, brushless DC motors, hydraulic pipelines with water hammer effects, and pneumatic actuation systems with stick-slip friction. Nearly all of the models are nonlinear. As an indication of the size of the models, after equation reduction, the smallest model had 4 states and the largest model had 117 states. The original number of states before equation reduction is typically much larger.

All simulations were performed on an Intel Core 2 Duo E6700 (2.66 GHz) that was running xPC Target from MathWorks. With each model, settings that permitted real-time simulation on the target and delivered accurate results with more than enough idle time to ensure robust simulation were found. The maximum percentage of a step spent in simulation execution was less than 18%, meaning that there was plenty of safety margin for processing I/O and other tasks. The average percentage spent in simulation execution was 3.9% and the minimum was 6e-4%.

This chapter covered the background of real-time simulation and described the steps in moving from desktop to real-time simulation. Balancing the trade-off of model fidelity and simulation speed is a core challenge of moving from desktop to real-time simulation. Simscape local solvers permit a lot of flexibility for adjusting the amount of computation done per time step by permitting different sample rates and setting the number of iterations per time step. The benefits of real-time simulation are significant, including reduced development costs and higher quality products. As a core element of Model-Based Design, it will continue to play an important role in product development processes.

REFERENCES

1. MathWorks. 2012. *xPC User's Guide*. Natick, MA: MathWorks.
2. Moler, C. 2003. *Stiff Differential Equations*. MATLAB News and Notes. http://www.mathworks.com/company/newsletters/news_notes/clevescorner/may03_cleve.html.
3. Strang, G. 1988. *Linear Algebra and Its Applications*, February. San Diego, CA: Harcourth Brace Jovanovich.
4. MathWorks. 2012. *Simulink User's Guide*. Natick, MA: MathWorks. http://www.mathworks.com/access/helpdesk/help/toolbox/simulink/ug/f7-8243.html#f7-9506.
5. MathWorks. 2012. *Simscape User's Guide*. Natick, MA: MathWorks.

24 Systematic Derivation of Hybrid System Models for Hydraulic Systems

Jeremy Hodgson, Rick Hyde, and Sanjiv Sharma

CONTENTS

24.1 Introduction 600
    24.1.1 Hydraulic Simulation Models...600
    24.1.2 Simulating Differential Algebraic Equations 600
    24.1.3 Parameter Abstraction of Small Masses...602
    24.1.4 Structure of Chapter 602
24.2 Network Equations for Incompressible Fluid Models 602
24.3 Simulation with Linear Velocity Constraints 604
    24.3.1 Reduced Order State Vector 604
    24.3.2 Minimum Order Momentum Equations..605
    24.3.3 Junction Pressures and Reaction Forces...605
24.4 Simulation with a Singular Mass Matrix...606
    24.4.1 Index of Differential Algebraic Equations 606
    24.4.2 Minimum Order Dynamic Equations..607
    24.4.3 Matrix Polynomial Equations for Algebraic States...........................608
    24.4.4 Solving Systems of Polynomial Equations 609
    24.4.5 Local Singularities in Algebraic Equations..610
24.5 Example: Pressure-Controlled Servo Valve 611
    24.5.1 Model Definition...611
    24.5.2 Model Dynamics...613
    24.5.3 Mode Transition (Finite State Machine)...614
    24.5.4 DAE Equations ("Positive" Mode) 615
    24.5.5 Simulation Results 617
24.6 Conclusions...620
24.7 Symbol List...620
Acknowledgment 620
References..621

24.1 INTRODUCTION

Simulation models of plant or actuator dynamics are utilized throughout the development cycle of new control systems. They can be used to validate the performance of a single controller in the presence of parameter variation or to investigate the behavior when the control loop is embedded within a larger system model. The models may be for desktop analysis or as part of a real-time test harness to validate the response of hardware components to simulated inputs. Whatever the application, efficient models are required whenever large numbers of simulations are to be carried out or where there are limits on the time available to update the model at each time step. To achieve this efficiency often requires that some simplifying assumptions must be made. This chapter aims to develop a systematic approach to developing models that maintain the inherent nonlinear behavior of a hydraulic circuit but that are based on a simplified representation of fluid dynamics.

24.1.1 HYDRAULIC SIMULATION MODELS

Individual hydraulic component models are traditionally built up from basic foundation blocks based on the dynamics of a volume of compressible fluid and the orifice equation. More complex system models can be developed either by connecting component models directly together or by using very small volumes of fluid as interface blocks at the junctions. These interface blocks may represent actual volumes of fluid or be artificial modeling artifacts used to connect self-contained subsystem models together. The benefits of this approach are that the model can be built up in a modular fashion and that the dynamics are described in terms of systems of ordinary differential equations (ODEs), for which there are numerous robust solvers available.

It is for this reason that many commercially available hydraulic simulation packages use this approach. The disadvantage is that small volumes of compressible fluid introduce very high-frequency dynamics into the model. These high-frequency modes require small time steps to be taken and hence results in longer simulation times. The problem can be partly alleviated by the use of "stiff" solvers if real-time code is not required. If real-time code is required, then one solution is to abstract the high-frequency dynamics by assuming that these small volumes of fluid are incompressible. It is shown in Ref. [1] that making this assumption leads to a model description in terms of differential algebraic equations (DAEs), that is, the model consists of a system of unconstrained dynamic equations, along with a number of algebraic constraints that the dynamic states must satisfy.

24.1.2 SIMULATING DIFFERENTIAL ALGEBRAIC EQUATIONS

Model descriptions in terms of DAEs arise in many applications, for example, in the simulation of electrical circuits, mechanical multibody dynamics, and chemical process models. An overview of current techniques used in the simulation of multibody dynamics is given in Ref. [2], which breaks down the approaches into three categories, namely Coupled-Force Balance, Lagrange, and Kane. The Coupled-Force Balance approach assumes that each body is free to move independently, and contact between bodies is maintained by applying external forces using stiff spring damper systems. This is analogous to connecting hydraulic components by small volumes of compressible fluid. The approach suffers from generating very stiff systems of ODEs, which, as previously discussed, can be difficult to simulate.

In the Lagrange approach, it is assumed that contact is defined by a system of algebraic constraints, which are adjoined to the unconstrained dynamic equations using what are termed "undetermined," or Lagrange, multipliers. A simplified form of the equations used to describe the dynamics of constrained rigid bodies is given in Equation 24.1, where $\mathbf{v}$ is the state vector of velocities, $\mathbf{M}$ is a positive-definite mass matrix, and $\mathbf{F}$ are the external forces acting on the system. The state constraints are described by the system of linear equality constraints on the velocities ($\mathbf{W}$), and the Lagrange multipliers λ then represent the internal constraint forces acting at the joints.

$$\mathbf{M\dot{v} + W\lambda = F(v)}$$
$$\mathbf{W^T v = 0}$$

$$(24.1)$$

The Lagrange approach to simulating this type of DAE is to augment the dynamic equations with the derivatives of the constraints, as shown in Equation 24.2. This results in a system of equations where the number of states equals the number of states in the original unconstrained system, plus the number of constraints. There are a number of problems with simulating this system of equations. Firstly, the initial state values must be consistent, that is, they must be chosen so that they satisfy the constraints. Secondly, because the constraints have been differentiated, any drift from the constraint manifold during simulation must be controlled. Finally, if the constraints are functions of the position states, then they must be differentiated twice before they can be augmented with the unconstrained dynamic equations. These acceleration constraints can be unstable and hence require additional stabilization to produce a stable system of equations.

$$\begin{bmatrix} \mathbf{M} & \mathbf{W} \\ \mathbf{W^T} & \mathbf{0} \end{bmatrix} \begin{bmatrix} \mathbf{\dot{v}} \\ \mathbf{\lambda} \end{bmatrix} = \begin{bmatrix} \mathbf{F(v)} \\ -\mathbf{\dot{W}^T v} \end{bmatrix}$$

$$(24.2)$$

The third approach to simulating multibody dynamic systems is based on the work of Kane [3], and a description of a practical simulator using this approach is given in Ref. [4]. The Kane method projects the unconstrained dynamic equations onto the constraint manifold, reducing the model to a minimal order system of ODEs, with a state vector equal in size to the number of unconstrained dynamic equations minus the number of constraints. The ability to generate a minimal order set of dynamic equations that can be simulated using standard ODE solvers makes the projection approach very appealing when simulating hydraulic models. There is one drawback, namely that a new set of minimal order ODEs must be derived whenever the number of constraints change. For example, the number of constraints will change in a hydraulic circuit model whenever a valve opens creating a new flow path or when a cylinder piston reaches its end-stop.

There are a number of approaches to handling systems with intermittent constraints. One method is to split the constraints into those that are always present and those that are likely to change. The projection approach can be used to reduce the number of equations based on the fixed number of constraints, and the remaining constraints can be implemented using the Lagrange approach as and when they become active.

This approach still suffers from the issues associated with the Lagrange approach. An alternative is to treat the system as a form of Hybrid System Model, whereby the model switches between systems of minimal order ODEs whenever the circuit topology changes. A discussion on the application of this approach to multibody system dynamics is given in Ref. [5], and an overview of a number of packages to support the simulation of this type of model is given in Ref. [6]. The decision that has to be made is whether to embed all possible sets of dynamic equations within the model before simulation or to derive new systems of equations online during the simulation. As the intent is to derive models that can run as efficiently as possible, a system by which all models are implemented before simulation is demonstrated in this chapter.

24.1.3 PARAMETER ABSTRACTION OF SMALL MASSES

The mass matrix $\mathbf{M}$ is always assumed to be of full rank when using the constraint embedding techniques described in Refs. [1,4]. This means that the mass matrix of the reduced order equations will also have full rank and hence be invertible during the simulation. To fully specify the dynamics of a hydraulic circuit may require that some very small masses are introduced to act as connecting elements between larger objects, and these small masses can introduce high-frequency dynamics, leading to slow simulations, and potential instabilities. This chapter extends the work by Pfeiffer and Borchsenius [1] by introducing an approach that can simulate the resulting minimum order dynamic equations when the mass matrix is singular due to making these small masses to zero.

24.1.4 STRUCTURE OF CHAPTER

The remainder of the chapter is organized as follows. In Section 24.2, the DAEs are derived that describe the dynamics of a hydraulic system when the fluid in the pipes, cylinders, or orifices is assumed to be incompressible. The approach by which these DAEs are reduced to a minimal order system of ODEs is then outlined in Section 24.3. In Section 24.4, the method of simulating the minimal order ODEs when the mass matrix is singular is described. The approach is demonstrated in Section 24.5 using, as an example, a circuit consisting of a servo valve controlling the fluid pressure in a closed pipe. Finally, conclusions are presented in Section 24.6.

24.2 NETWORK EQUATIONS FOR INCOMPRESSIBLE FLUID MODELS

The basis for a simulation model of connected hydraulic components containing incompressible fluid is a system of unconstrained momentum equations and a system of linear algebraic constraints. The unconstrained momentum equations are given in Equation 24.3, where $\mathbf{v}$ is the vector of fluid velocities augmented with states describing the motion of mechanical components such as cylinder pistons and valve spools. $\mathbf{M}$ is a diagonal matrix of fluid and component masses, and $\mathbf{F(v)}$ is the known forcing function given in Equation 24.4. For the type of hydraulic system under consideration, the forcing function consists of terms representing the external pressure sources at the interfaces to the system $(\mathbf{P_I})$, velocity-dependent friction terms $(\mathbf{W_V} \mathbf{v})$, and the pressure drop across open orifices $(\Delta \mathbf{P_O})$. The pressure drop across an orifice is assumed to obey the orifice equation in Equation 24.5. In this case, flow through an orifice is assumed to be turbulent, and the fluid density (ρ) is independent of pressure and temperature. This does not discount extending the equations to include changes in flow regime as described in Ref. [7], by making ρ and C_d functions of the current state values.

In addition to the known forcing functions, there are unknown internal constraint forces as illustrated in Figure 24.1. The constraint forces consist of pressure forces acting at fluid junctions $(\mathbf{P_J})$, reaction forces between components in contact with their end-stops $(\mathbf{F_R})$, and the pressure drop across closed orifices $(\Delta \mathbf{P_C})$.

$$\mathbf{M\dot{v}} + \mathbf{W_J P_J} + \mathbf{W_R F_R} + \mathbf{W_C \Delta P_C} = \mathbf{F(v)} \tag{24.3}$$

$$\mathbf{F(v)} = \mathbf{W_O \Delta P_O} + \mathbf{W_V v} + \mathbf{W_I P_I} \tag{24.4}$$

$$\Delta P_O = \left(\frac{\rho}{2}\right)\left(\frac{1}{C_d^2}\right) v|v| \tag{24.5}$$

The unknown constraint forces are related to a system of algebraic constraints. The velocity constraints include the requirement that the sum of the flows into a junction must be zero $(\mathbf{W_J})$, that flow through closed orifices $(\mathbf{W_C})$ must be zero, and that the motion of components that have reached their end-stops $(\mathbf{W_R})$ must be zero. The combined system of constraints can then be represented by $\mathbf{W}$ in Equation 24.6.

$$\begin{bmatrix} \mathbf{W_J^T} \\ \mathbf{W_R^T} \\ \mathbf{W_C^T} \end{bmatrix} \mathbf{v} = \mathbf{W^T v} = \mathbf{0} \tag{24.6}$$

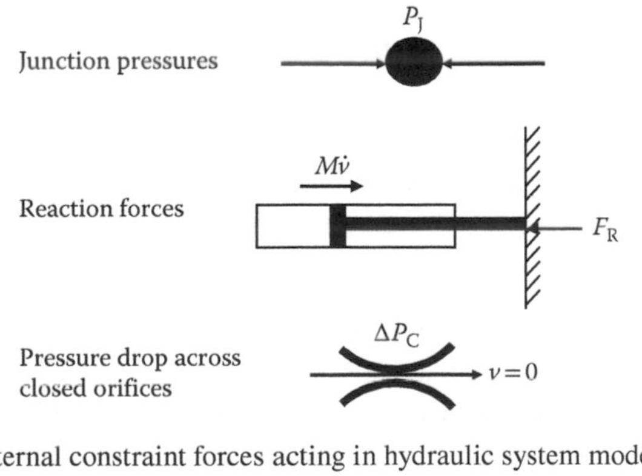

FIGURE 24.1   Internal constraint forces acting in hydraulic system models.

24.3 SIMULATION WITH LINEAR VELOCITY CONSTRAINTS

The aim in this section is to derive a minimal order system of ODEs, which satisfies the unconstrained dynamics in Equation 24.3, subject to the system of linear constraints on the velocities in Equation 24.6. The basis for the derivation is that a system of n dynamic equations subject to m constraints can be reduced to a system of equations with $v = n - m$ independent states.

24.3.1 REDUCED ORDER STATE VECTOR

The relationship between the full state vector ($\mathbf{v}$) and the reduced set of independent states ($\tilde{\mathbf{v}}$) is through the transformation matrix $\boldsymbol{\alpha}$ in Equation 24.7.

$$\mathbf{v} = \boldsymbol{\alpha}\,\tilde{\mathbf{v}} \tag{24.7}$$

The transformation matrix $\boldsymbol{\alpha}$ is not unique but has to be determined so that the full state vector will always satisfy the constraint equations. This can be achieved if the columns of $\boldsymbol{\alpha}$ form the basis vectors of the null space of the constraint equations. If $\boldsymbol{\alpha}$ satisfies that requirement, then the inner matrix product in Equation 24.8 will be satisfied for all values of the independent states.

$$\mathbf{W}^{\mathrm{T}}\mathbf{v} = \underbrace{(\mathbf{W}^{\mathrm{T}}\boldsymbol{\alpha})}_{=0}\tilde{\mathbf{v}} = 0 \tag{24.8}$$

One method for deriving $\boldsymbol{\alpha}$ is given in Ref. [8]. In this case, the independent states are selected to be a subset of the full state vector as shown in Equation 24.9.

$$\tilde{\mathbf{v}} = \begin{bmatrix} \mathbf{I} & \mathbf{0} \end{bmatrix}\begin{bmatrix} \tilde{\mathbf{v}} \\ \hat{\mathbf{v}} \end{bmatrix} \tag{24.9}$$

If the constraint matrix is partitioned as in Equation 24.10, then $\hat{\mathbf{v}}$ can be defined in terms of $\tilde{\mathbf{v}}$, and α is given by Equation 24.11.

$$\mathbf{W}^{\mathrm{T}}\mathbf{v} = \begin{bmatrix} \mathbf{W}_1^{\mathrm{T}} & \mathbf{W}_2^{\mathrm{T}} \end{bmatrix}\begin{bmatrix} \tilde{\mathbf{v}} \\ \hat{\mathbf{v}} \end{bmatrix} = 0 \tag{24.10}$$

$$\mathbf{v} = \begin{bmatrix} \tilde{\mathbf{v}} \\ \hat{\mathbf{v}} \end{bmatrix} = \boldsymbol{\alpha}\tilde{\mathbf{v}} = \begin{bmatrix} \mathbf{I}_v \\ -\mathbf{W}_2^{-\mathrm{T}}\mathbf{W}_1^{\mathrm{T}} \end{bmatrix}\tilde{\mathbf{v}} \tag{24.11}$$

This approach works well provided the selection of the independent states leads to a tractable system of linear equations (i.e., $\mathbf{W}_2$ is invertible). Making a suitable selection is not always straightforward either when there are a large number of states or when there are a large number of models to be analyzed. An alternative approach described in Ref. [9] uses the singular value decomposition (SVD) of the constraint matrix to derive the basis vectors of the null space at each time step during the simulation. This provides a robust and numerically stable approach, but can be computationally expensive and leads to the integration of states that are not directly related to the physical states of the system.

The approach taken is to determine which states can be selected as independent states before creating the simulation model. There are functions available in MathWorks Symbolic Math Toolbox™ [10] to derive the null space of the constraint matrix symbolically, and as the form of the resulting matrix is the same as in Equation 24.7, the independent states can be determined by a simple inspection of the rows of the matrix for a single entry of a unit gain. Within the simulation model, the null space matrix is regenerated using the indices of the selected states to be integrated and the formulation in Equation 24.11. The main benefit of doing this is that only a minimal amount of information must be stored with the model to define the independent state vector for each mode.

24.3.2 MINIMUM ORDER MOMENTUM EQUATIONS

The minimum order dynamic equations can be derived once the relationship between the full and the reduced order state vectors has been established. The time derivative of the full state vector is related to the reduced state vector by Equation 24.12.

$$\dot{\mathbf{v}} = \dot{\boldsymbol{\alpha}}\tilde{\mathbf{v}} + \boldsymbol{\alpha}\dot{\tilde{\mathbf{v}}} \tag{24.12}$$

Substituting Equation 24.12 into the unconstrained dynamic equations in Equations 24.3 and projecting the equations onto the constraint manifold by multiplying through by $\boldsymbol{\alpha}^T$ leads to a minimum order system of ODEs that have the form given by Equation 24.13. The forcing function $\tilde{\mathbf{F}}$ in Equation 24.13 is given by Equation 24.14, and the mass matrix $\tilde{\mathbf{M}}$ is given by Equation 24.15.

$$\tilde{\mathbf{M}}\dot{\tilde{\mathbf{v}}} = \tilde{\mathbf{F}}(\tilde{\mathbf{v}}) \tag{24.13}$$

$$\tilde{\mathbf{F}}(\tilde{\mathbf{v}}) = \boldsymbol{\alpha}^T \mathbf{F}(\tilde{\mathbf{v}}) - (\boldsymbol{\alpha}^T \mathbf{M}\dot{\boldsymbol{\alpha}})\tilde{\mathbf{v}} \tag{24.14}$$

$$\tilde{\mathbf{M}} = \boldsymbol{\alpha}^T \mathbf{M}\boldsymbol{\alpha} \tag{24.15}$$

24.3.3 JUNCTION PRESSURES AND REACTION FORCES

The reduction procedure described in Section 24.3.2 creates a system of ODEs that can be simulated using standard solvers. In the process, the constraint forces have been lost from the model description. The first stage to regenerate these forces is to derive an expression for the rate of change of the dependent states as a function of the independent states. If the constraint matrix is partitioned in Equation 24.10, then after differentiation, it can be rearranged to give the expression for $\dot{\hat{\mathbf{v}}}$ in Equation 24.16.

$$\dot{\hat{\mathbf{v}}} = -\mathbf{W}_2^{-T}\left(\dot{\mathbf{W}}^T\mathbf{v} + \mathbf{W}_1^T\dot{\tilde{\mathbf{v}}}\right) \tag{24.16}$$

The dynamic equations in Equation 24.3 can also be partitioned as shown in Equation 24.17. The unknown internal forces can be regenerated by substituting Equation 24.16 into Equation 24.17 and rearranging to give Equation 24.18.

$$
\begin{bmatrix} \mathbf{M}_1 & \mathbf{0} \\ \mathbf{0} & \mathbf{M}_2 \end{bmatrix} \begin{bmatrix} \dot{\tilde{\mathbf{v}}} \\ \dot{\hat{\mathbf{v}}} \end{bmatrix} + \begin{bmatrix} \mathbf{W}_1 \\ \mathbf{W}_2 \end{bmatrix} \begin{bmatrix} \mathbf{P}_J \\ \mathbf{F}_R \\ \Delta \mathbf{P}_C \end{bmatrix} = \begin{bmatrix} \mathbf{F}_1 \\ \mathbf{F}_2 \end{bmatrix} \tag{24.17}
$$

$$
\begin{bmatrix} \mathbf{P}_J \\ \mathbf{F}_R \\ \Delta \mathbf{P}_C \end{bmatrix} = \mathbf{W}_2^{-1} \left(\mathbf{F}_2 + \mathbf{M}_2 \mathbf{W}_2^{-T} \left(\dot{\mathbf{W}}^T \mathbf{v} + \mathbf{W}_1^T \dot{\tilde{\mathbf{v}}} \right) \right) \tag{24.18}
$$

24.4 SIMULATION WITH A SINGULAR MASS MATRIX

In Section 24.3, the mass matrix of the unconstrained dynamic equations must have full rank if the mass matrix of the reduced order equations ($\tilde{\mathbf{M}}$) is to be inverted during simulation. If some of the very small masses of fluid are made equal to zero to remove the high-frequency dynamics that they introduce, then this assumption will no longer hold. If $\tilde{\mathbf{M}}$ is singular, then the dynamic equations in Equation 24.13 are said to be quasi-linear DAEs, and an extensive description of the techniques available to solve this type of system is given in Ref. [11]. Sections 24.4.1 to 24.4.5 describe the method of determining the index of the resulting DAEs when the mass matrix is singular and also a method of simulating the equations.

24.4.1 INDEX OF DIFFERENTIAL ALGEBRAIC EQUATIONS

The index of the DAEs is found by extracting the "hidden" algebraic constraints from the original equations. The algebraic constraints $\mathbf{G}_0$ are determined by projecting the dynamic equations in Equation 24.13 onto the null space of the mass matrix using the projection $\boldsymbol{\beta}$ that satisfies Equation 24.19. The resulting constraint equations in Equation 24.20 define the manifold on which the states can travel.

The dynamics of the constrained system is then found by augmenting the original equations with motion at a local tangent to the constraint manifold, as shown in Equation 24.21. If the augmented mass matrix has full column rank, then the system has index 1. If the mass matrix does not have full column rank, then the above procedure is repeated. That is, the hidden constraints of the augmented system are determined and a new system of equations is derived by augmenting the dynamic equations with the derivative of the new constraints. The index of the DAEs is defined by the number of times the procedure must be repeated before the augmented mass matrix has full column rank.

$$
\boldsymbol{\beta}^T \tilde{\mathbf{M}} = \mathbf{0} \tag{24.19}
$$

$$
\mathbf{G}_0 = \boldsymbol{\beta}^T \tilde{\mathbf{F}}(\tilde{\mathbf{v}}) = 0 \tag{24.20}
$$

$$\begin{bmatrix} \tilde{\mathbf{M}} \\ \dfrac{\partial \mathbf{G}_0}{\partial \tilde{\mathbf{v}}} \end{bmatrix} \dot{\tilde{\mathbf{v}}} = \begin{bmatrix} \tilde{\mathbf{F}}(\tilde{\mathbf{v}}) \\ -\dfrac{\partial \mathbf{G}_0}{\partial t} \end{bmatrix} \tag{24.21}$$

The hydraulics models in Equation 24.13 have index 1, and it is feasible to simulate the system of overdetermined equations in Equation 24.21 directly. The problem is that the number of independent states is defined by the size of the state vector minus the number of hidden algebraic constraints. The initial values of the full state vector must therefore be consistent (i.e., lie on the constraint manifold), and any drift from the constraint manifold during the simulation must be controlled. The approach discussed in Ref. [11] overcomes these issues by selecting a minimum number of independent integrated states from the state vector. The integrated states can then be used to determine algebraically the remaining states so that the hidden constraints are satisfied. This is similar to the method described in Section 24.3, but now, the constraints are defined by nonlinear functions of the states, and they are implicit within the model description.

24.4.2 MINIMUM ORDER DYNAMIC EQUATIONS

A minimal order system of dynamic equations can be derived by partitioning the state vector $\tilde{\mathbf{v}}$ in Equation 24.22 into independent states $\tilde{\mathbf{v}}_i$ that are integrated in time and algebraic states $\tilde{\mathbf{v}}_a$ that are derived at each time step from a function of the integrated states. The reordering of the integrated and algebraic states into the full state vector is achieved through the permutation matrix $\mathbf{\Pi}$. The selection of which independent states are chosen from the full state vector is the same as in Section 24.3, namely by inspection of the symbolically derived null-space matrix $\boldsymbol{\beta}$.

$$\tilde{\mathbf{v}} = \underbrace{\begin{bmatrix} \mathbf{\Pi}_i & \mathbf{\Pi}_a \end{bmatrix}}_{\Pi} \begin{bmatrix} \tilde{\mathbf{v}}_i \\ \tilde{\mathbf{v}}_a \end{bmatrix} \tag{24.22}$$

The rate of change of the algebraic states is derived by differentiating the constraint equations in Equation 24.20 to give Equation 24.23, where the time dependency of the orifice areas in the constraint equations is accounted for by the third term in $\dot{t}$.

$$\frac{d\mathbf{G}_0}{dt} = 0 = \frac{\partial \mathbf{G}_0}{\partial \tilde{\mathbf{v}}_a} \dot{\tilde{\mathbf{v}}}_a + \frac{\partial \mathbf{G}_0}{\partial \tilde{\mathbf{v}}_i} \dot{\tilde{\mathbf{v}}}_i + \frac{\partial \mathbf{G}_0}{\partial t} \dot{t} \tag{24.23}$$

Rearranging Equation 24.23 into Equation 24.24, and substituting into the original equations in Equation 24.13, produces the minimum order dynamic equations in Equation 24.25, where the columns of $\mathbf{P}$ form the the basis vectors of the range space of the mass matrix. These equations represent a minimal order system of ODEs that satisfy $\mathbf{G}_0$ and where the resulting mass matrix is invertible. The problem is now to determine the relationship between the integrated and algebraic states when the constraint equations are nonlinear in nature.

$$\dot{\tilde{\mathbf{v}}}_a = -\left(\frac{\partial \mathbf{G}_0}{\partial \tilde{\mathbf{v}}_a}\right)^{-1}\left(\frac{\partial \mathbf{G}_0}{\partial \tilde{\mathbf{v}}_i}\dot{\tilde{\mathbf{v}}}_i + \frac{\partial \mathbf{G}_0}{\partial t}\right) \tag{24.24}$$

$$\mathbf{P}\tilde{\mathbf{M}}\left[\begin{array}{c} \mathbf{I} \\ -\left(\dfrac{\partial \mathbf{G}_0}{\partial \tilde{\mathbf{v}}_a}\right)^{-1}\dfrac{\partial \mathbf{G}_0}{\partial \tilde{\mathbf{v}}_i} \end{array}\right]\dot{\tilde{\mathbf{v}}}_i = \mathbf{P}\tilde{\mathbf{F}} + \mathbf{P}\tilde{\mathbf{M}}\left[\begin{array}{c} 0 \\ \left(\dfrac{\partial \mathbf{G}_0}{\partial \tilde{\mathbf{v}}_a}\right)^{-1}\dfrac{\partial \mathbf{G}_0}{\partial t} \end{array}\right] \tag{24.25}$$

24.4.3 MATRIX POLYNOMIAL EQUATIONS FOR ALGEBRAIC STATES

This section describes the derivation of the relationship between the integrated and the algebraic states in the reduced order model. On the basis of the original formulation of the forcing function in Equation 24.5, the algebraic constraints can be expressed in Equation 24.26 in terms of the partitioning of the state vector in Equation 24.7. The matrix representing the pressure drop across open orifices ($\mathbf{W}_0$) is partitioned in Equation 24.27, and the dependent states ($\hat{\mathbf{v}}$) are related to the independent states ($\tilde{\mathbf{v}}$) by Equation 24.28. The notation $\mathbf{v}\,|\mathbf{v}|$ in Equation 24.26 defines the signed square value of each element in the state vector.

$$\mathbf{G}_0 = (\boldsymbol{\alpha}\boldsymbol{\beta})^{\mathrm{T}}\left(\begin{array}{c} \begin{bmatrix} \mathbf{W}_{01} & \mathbf{W}_{02} \end{bmatrix}\begin{bmatrix} \tilde{\mathbf{v}} & |\tilde{\mathbf{v}}| \\ \hat{\mathbf{v}} & |\hat{\mathbf{v}}| \end{bmatrix} \\ +(\mathbf{W}_v\boldsymbol{\alpha} - \mathbf{M}\dot{\boldsymbol{\alpha}})\,\tilde{\mathbf{v}} \\ +\mathbf{W}_{\mathrm{I}}\mathbf{P}_{\mathrm{I}} \end{array}\right) \tag{24.26}$$

$$\begin{bmatrix} \mathbf{W}_{01} & \mathbf{W}_{02} \end{bmatrix} = \left(\frac{\rho}{2C_d^2}\right)\mathbf{W}_0 \tag{24.27}$$

$$\hat{\mathbf{v}} = \boldsymbol{\alpha}_1\tilde{\mathbf{v}} \tag{24.28}$$

The constraints in Equation 24.26 can be rearranged in Equation 24.29 for the unknowns $\tilde{\mathbf{v}}_a$, where the matrices $\mathbf{A}$, $\mathbf{B}$, $\mathbf{C}$, $\mathbf{D}$ are given by Equations 24.30 through 24.33.

$$\mathbf{G}_0 = \mathbf{A}\tilde{\mathbf{v}}_a\,|\tilde{\mathbf{v}}_a| + \mathbf{B}\tilde{\mathbf{v}}_a + \mathbf{C}\left(\tilde{\mathbf{v}}_i\right) + \mathbf{D}\hat{\mathbf{v}}|\hat{\mathbf{v}}| \tag{24.29}$$

$$\mathbf{A} = (\boldsymbol{\alpha}\boldsymbol{\beta})^{\mathrm{T}}\,\mathbf{W}_{01}\Pi_a \tag{24.30}$$

$$\mathbf{B} = (\boldsymbol{\alpha}\boldsymbol{\beta})^{\mathrm{T}}(\mathbf{W}_v\boldsymbol{\alpha} - \mathbf{M}\dot{\boldsymbol{\alpha}})\Pi_a \tag{24.31}$$

$$C(\tilde{\mathbf{v}}_i) = (\boldsymbol{\alpha\beta})^T \left(\mathbf{W}_{01}\Pi_i \tilde{\mathbf{v}}_i |\tilde{\mathbf{v}}_i| + (\mathbf{W}_v \boldsymbol{\alpha} - \mathbf{M}\dot{\boldsymbol{\alpha}})\Pi_i \tilde{\mathbf{v}}_i + \mathbf{W}_I \mathbf{P}_I \right) \tag{24.32}$$

$$\mathbf{D} = (\boldsymbol{\alpha\beta})^T \mathbf{W}_{02} \tag{24.33}$$

24.4.4 Solving Systems of Polynomial Equations

There are a number of ways to solve the system of polynomial equations in Equation 24.29 for the unknown states. The approach chosen is the iterative Newton (or Newton–Raphson) method given in Equation 24.34, where the Jacobian of the constraint matrix is given by Equation 24.35. This approach has proven to be robust, and standard ODE solvers can still be utilized as the iterative procedure is only invoked when the solver requires the derivatives of the independent integrated states. The iterative procedure is still considered an acceptable approach if the equations are to be implemented in a real-time simulation code, as the maximum number of iterations can be limited to guarantee that the solution will be found in the available time.

$$\tilde{\mathbf{v}}_a^{i+1} = \tilde{\mathbf{v}}_a^i - \left(\frac{\partial \mathbf{G}_0}{\partial \tilde{\mathbf{v}}_a} \right)^{-1} \mathbf{G}_0, \quad i = 1, 2, \dots, i_{max} \tag{24.34}$$

$$\frac{\partial \mathbf{G}_0}{\partial \tilde{\mathbf{v}}_a} = 2\mathbf{A}\,\mathrm{diag}\left(|\tilde{\mathbf{v}}_a|\right) + \mathbf{B} + 2\mathbf{D}\,\mathrm{diag}\left(|\hat{\mathbf{v}}|\right)\alpha_1 \Pi_a \tag{24.35}$$

Rather than inverting the Jacobian matrix, the velocity states are updated at each iteration by solving the system of linear equations in Equation 24.36 for the increment in the velocity estimates $\Delta\mathbf{v}$.

$$\frac{\partial \mathbf{G}_0}{\partial \tilde{\mathbf{v}}_a} \Delta\mathbf{v} = \mathbf{G}_0 \tag{24.36}$$

The equations in Equation 24.36 are solved using the $\mathbf{QR}$ decomposition of the Jacobian matrix in Equation 24.37, where $\mathbf{Q}$ is orthogonal, $\mathbf{R}$ is upper triangular, and $\mathbf{E}$ is a permutation matrix determined so that the values on the leading diagonal of $\mathbf{R}$ are in decreasing order. The $\mathbf{QR}$ decomposition is determined during the simulation using the FORTRAN $\mathbf{DGEQP3}$ function provided in the work by Anderson et al. [12] and described in the work by Quintana-Ort et al. [13]. The efficiency of these routines compared to other implementations is outlined in the work by Foster and Liu [14].

$$\frac{\partial \mathbf{G}_0}{\partial \tilde{\mathbf{v}}_a} = \mathbf{QRE}^T \tag{24.37}$$

Substituting the matrix $\mathbf{QR}$ decomposition into Equation 24.36 and multiplying through by $\mathbf{Q}^T$ leads to Equation 24.38. This equation can easily be solved for $\Delta\mathbf{z}$ as $\mathbf{R}$ is upper triangular. The calculated solution is reordered into the desired vector of increments using the permutation matrix in Equation 24.39.

$$\mathbf{R}\Delta\mathbf{z} = \mathbf{Q}^\mathrm{T}\mathbf{G}_0 \tag{24.38}$$

$$\Delta\mathbf{v} = \mathbf{E}\Delta\mathbf{z} \tag{24.39}$$

An additional benefit of decomposing the Jacobian in this way is that it provides an estimate for the rank of the matrix. The diagonal elements of $\mathbf{R}$ provide an approximation to the singular values, and any singular value near zero indicates that the matrix is rank deficient. If that is the case, then a minimum norm solution can be derived using the more accurate SVD described in Section 24.4.5.

24.4.5 LOCAL SINGULARITIES IN ALGEBRAIC EQUATIONS

Under certain conditions, the Jacobian of the constraint equations may be locally singular. This can occur when the orifice areas become very small just before, or after, a transition between modes. In this case, a modified Newton's method is used (see Ref. [15]), where the aim is to determine the minimum norm solution to the equations, that is, the solution where the state vector has the smallest magnitude. The method uses a generalized outer inverse of the Jacobian to update the state vector. The outer inverse is determined using the more robust SVD of the Jacobian. The form of the decomposition is given in Equation 24.40, where $\mathbf{U}$ and $\mathbf{V}$ are orthogonal matrices, $\mathbf{\Sigma}$ is a diagonal matrix containing the singular values, and the rank of the Jacobian (r) is given by the number of nonzero diagonal elements in Σ.

$$\frac{\partial \mathbf{G}}{\partial \mathbf{v}} = \mathbf{U}\Sigma\mathbf{V}^\mathrm{T} \equiv \begin{bmatrix} \mathbf{U}_1 & \mathbf{U}_2 \end{bmatrix} \begin{bmatrix} \Sigma_r & 0 \\ 0 & 0 \end{bmatrix} \begin{bmatrix} \mathbf{V}_1^\mathrm{T} \\ \mathbf{V}_2^\mathrm{T} \end{bmatrix} \tag{24.40}$$

$$U^\mathrm{T}U = V^\mathrm{T}V = I \tag{24.41}$$

$$\Sigma_r = \begin{bmatrix} \sigma_1 & 0 & \cdots & 0 \\ 0 & \sigma_2 & \cdots & 0 \\ \vdots & \vdots & \ddots & \vdots \\ 0 & 0 & \cdots & \sigma_r \end{bmatrix}, \quad \sigma_1 > \sigma_2 > \cdots > \sigma_r \tag{24.42}$$

An outer inverse (denoted by #) of the Jacobian is defined as any matrix where the relationship in Equation 24.43 is satisfied. On the basis of the SVD, an outer inverse can be shown to be given by Equation 24.44, and the modified Newton method is then given by Equation 24.45. The SVD is calculated during the simulation using the FORTRAN **DGESVD** function in Ref. [12], and although it is computationally more expensive to calculate compared to the **QR** decomposition described previously, the fact that it is only required at a few points during the simulation means that it does not adversely affect the efficiency of the simulation approach.

$$\frac{\partial \mathbf{G}^\#}{\partial \mathbf{v}} \times \frac{\partial \mathbf{G}}{\partial \mathbf{v}} \times \frac{\partial \mathbf{G}^\#}{\partial \mathbf{v}} = \frac{\partial \mathbf{G}^\#}{\partial \mathbf{v}} \tag{24.43}$$

$$\frac{\partial \mathbf{G}^{\#}}{\partial \mathbf{v}} = \mathbf{V}_1 \, \Sigma_r^{-1} \, \mathbf{U}_1^{\mathrm{T}} \tag{24.44}$$

$$\tilde{\mathbf{v}}_{\mathrm{a}}^{i+1} = \tilde{\mathbf{v}}_{\mathrm{a}}^{i} - \left(\frac{\partial \mathbf{G}}{\partial \tilde{\mathbf{v}}_{\mathrm{a}}} \right)^{\#} \mathbf{G}_0 \tag{24.45}$$

24.5 EXAMPLE: PRESSURE-CONTROLLED SERVO VALVE

24.5.1 MODEL DEFINITION

This section demonstrates the derivation and implementation in the MathWorks Simulink® [16] product of a hybrid-system model for the hydraulic circuit shown in Figure 24.2. The circuit consists of a two-stage servo valve controlling the pressure in a 6 m long hydraulic pipe that is terminated at one end. The first stage in the servo valve consists of a flapper mechanism driven by a torque motor. The second stage spool is acted upon by a pilot pressure from the first stage as well as a spring force to center the spool when not under load. In addition, there is a small internal volume of fluid within the valve that acts on the spool and provides a feedback mechanism to control the pressure at the output port. A more detailed description of the operation and use of the servo valve is provided in Ref. [17].

The signal flow diagram in Figure 24.3 shows the fluid flow into the four internal junctions (circles) due to either flow through connecting orifices (solid lines) or the motion of the spool itself (dashed lines). The aim is to simulate the dynamics of the servo valve when the fluid within these small connecting volumes is assumed to be incompressible.

Table 24.1 provides a description of each of the parameters within the model, along with representative numerical values. The level of detail within this model is typical of an initial system-level definition before more detailed performance models are developed.

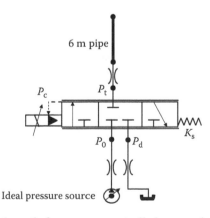

FIGURE 24.2 Circuit schematic for pressure-controlled servo valve.

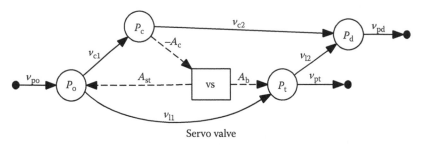

Servo valve

FIGURE 24.3 Signal flow diagram for pressure-controlled servo valve.

TABLE 24.1
Description of Servo Valve Parameters

Parameter	Description	Value
A_b	Spool feedback pressure area	2.10 mm²
A_c	Spool control pressure area	4.19 mm²
A_{c1}, A_{c2}	Flapper nozzle areas	—
A_{l1}, A_{l2}	Flow areas of second-stage orifices	—
A_{smax}	Maximum orifice area	1.64 mm²
A_n	Maximum nozzle area	0.0341 mm²
A_{po}, A_{pt}, A_{pd}	Flow areas at servo ports	230 mm²
A_{st}	Spool stub area	0.261 mm²
R_p	Pipe radius	8.56 mm
β_e	Fluid bulk modulus	580 MPa
ρ	Fluid density	860 kg/m³
g_f	Flapper current gain	0.003 mm/A
K_f	Flapper stiffness constant	23.5 N/mm
K_s	Spool spring constant	14.2 N/mm
P_c	Internal junction pressure	—
P_d	Pressure at return port	—
P_o	Pressure at supply port	—
P_r	Return pressure	5 bar
P_s	Supply pressure	210 bar
P_t	Pressure at output port	—
v_{c1}, v_{c2}	Velocity of fluid through first stage	—
v_{l1}, v_{l2}	Velocity of fluid through second stage	—
v_{po}, v_{pt}, v_{pd}	Velocity of fluid through servo ports	—
v_s	Spool velocity	—
x_d	Flapper displacement	—
x_f	Flapper displacement at nozzle	—
x_l	Lapped region	0.0254 mm
x_{fmax}	Maximum flapper displacement	0.0305 mm
x_s	Spool displacement	—
x_{smax}	Maximum spool displacement	0.381 mm
τ_m	Torque motor time constant	1 ms

24.5.2 MODEL DYNAMICS

Some parts of the model are represented by systems of ODEs that are simulated using standard solvers. The response of the flapper (x_d) in Figure 24.4 to a change in input current (i_v) to the torque motor is assumed to be first order with a time constant equal to τ_m. The transfer function is given in Equation 24.46, where g_f provides the steady-state gain between the current input and the achieved flapper displacement.

$$\frac{X_d(s)}{I_v(s)} = \frac{g_f}{\tau_m s + 1} \tag{24.46}$$

The displacement of the flapper x_f at the nozzle is then given in Equation 24.47, where K_f is the flapper spring constant, A_n is the null position area, and X_{fb} is a bias. The resulting flapper nozzle areas are given by Equation 24.48.

$$x_f = x_{fb} + \frac{P_o A_n}{K_f} + x_d \tag{24.47}$$

$$A_{c1} = \frac{A_n}{2} \frac{x_f}{x_{f\,max}}, \quad A_{c2} = \frac{A_n}{2}\left(1 - \frac{x_f}{x_{f\,max}}\right) \tag{24.48}$$

Although the fluid within the servo is treated as incompressible, the fluid within the connected pipe is treated as compressible as the volume of fluid is significantly greater than that within the servo valve. The compressibility also provides an important memory of the fluid pressure when the valve is centered, and the pipe is effectively isolated from the rest of the system. For this example, the fluid pressure P within the pipe is determined from Equation 24.49, where β_e is the effective bulk modulus, V_p is the volume of fluid within the pipe, and the flow rate Q_t is determined in Equation 24.50. Treating the fluid in this case as a single lumped volume

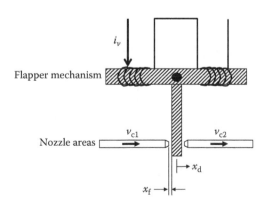

FIGURE 24.4 First-stage flapper displacement and nozzle area.

is a simplistic representation used to demonstrate the methodology. More detailed transmission line dynamic models described in Ref. [17] could equally have been used to investigate pressure peaking and oscillations following step inputs from the valve.

$$\frac{\mathrm{d}P}{\mathrm{d}t} = \frac{\beta_e}{V_p} Q_t \qquad (24.49)$$

$$Q_t = \left(\pi R_p^2\right) v_{pt} \qquad (24.50)$$

24.5.3 MODE TRANSITION (FINITE STATE MACHINE)

The discrete modes of operation must be established before the model dynamics can be derived. The modes can be defined by the relationship between the orifice areas and the position of the flapper or spool. The fluid flows through the second stage of the valve at speeds v_{11} and v_{12}, and the respective orifice areas A_{11} and A_{12} are functions of the spool position shown in Figure 24.5. The figure shows that there are five discrete modes depending on whether the spool has reached its end stops (in which case $v_s = 0$), or whether the orifices are fully closed in which case either $v_{11} = 0$ or $v_{12} = 0$. The flapper mechanism provides three additional modes depending on whether the flapper blocks one of the nozzle areas. In total, there are 15 discrete modes, and each has its own defined set of minimal order dynamic equations.

Control over switching between sets of dynamic equations is governed by the finite state machine implemented using the MathWorks Stateflow® product [18] and shown in Figure 24.6. The flapper and the spool are controlled through two independent parallel states. The transition between modes is governed by events triggered by a zero-crossing signal in the model. Events can occur because of either an orifice opening or closing or contact between a physical object and its end-stop. In the case of contact occurring, the subsequent release of the object is triggered by a change in sign of the calculated reaction force.

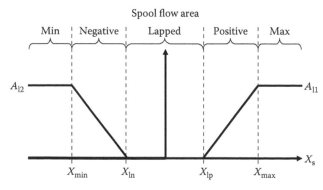

FIGURE 24.5 Second-stage orifice areas versus spool position.

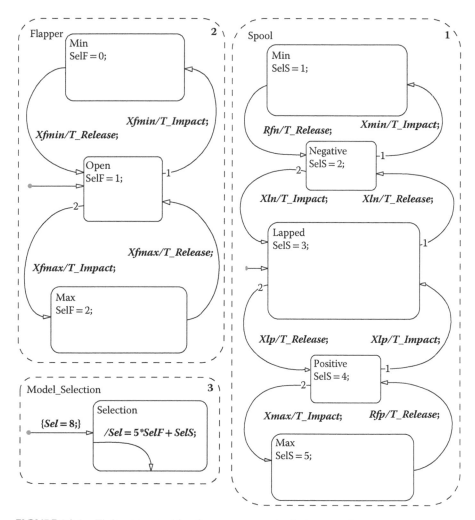

FIGURE 24.6   Finite state machine for pressure-controlled servo valve.

The very nature of the finite state machine representation means that it is ideal for including additional features that are of interest to the control engineer. In particular, the effect on the system behavior of introducing faults into the system is easily investigated by introducing trigger signals that force the model to transition into a failure state. An example where this would be appropriate is the case where a valve becomes stuck open or closed and cannot move until the failure is cleared.

24.5.4   DAE EQUATIONS ("POSITIVE" MODE)

This section defines the system of DAEs when the spool is in the *Positive* mode where the orifice between the return line and the outlet port (A_{12}) is fully closed. The full velocity state vector and the internal junction pressures are defined in

Equation 24.51. Two mass cases are considered for comparison. The first assumes that that there is a small mass of fluid in each line and in the second case all the mass elements are assumed to be negligible and are set to zero. If there is a zero mass matrix, then the dynamics of the flow through the servo valve reduces to a system of algebraic equations that must be solved at each time step for the reduced order state vector.

$$
\mathbf{v} = \begin{bmatrix} v_{11} \\ v_{pd} \\ v_{pt} \\ v_{c1} \\ v_{c2} \\ v_{12} \\ v_s \\ v_{po} \end{bmatrix}, \quad \mathbf{P}_J = \begin{bmatrix} P_o \\ P_d \\ P_t \\ P_c \end{bmatrix}
\tag{24.51}
$$

The flow constraints at the junctions are defined by the matrix $\mathbf{W}_J$ given in Equation 24.52. These constraints are maintained for all modes of operation and can be considered to be configuration constraints. The additional constraint ($\mathbf{W}_C$) introduced because of the closed orifice A_{12} is given in Equation 24.53.

$$
\mathbf{W}_J = \begin{bmatrix}
-A_{11} & 0 & A_{11} & 0 \\
0 & -A_{pd} & 0 & 0 \\
0 & 0 & -A_{pt} & 0 \\
-A_{c1} & 0 & 0 & A_{c1} \\
0 & A_{c2} & 0 & -A_{c2} \\
0 & A_{12} & -A_{12} & 0 \\
A_{st} & 0 & A_b & -A_c \\
A_{po} & 0 & 0 & 0
\end{bmatrix}
\tag{24.52}
$$

$$
\mathbf{W}_C = \begin{bmatrix} 0 & 0 & 0 & 0 & 0 & 1 & 0 & 0 \end{bmatrix}^T
\tag{24.53}
$$

In this mode, there are eight velocity states and five linear constraints. This leads to a minimum order system consisting of three independent states. The selection of the independent states from the full state vector is made by direct inspection of the symbolically derived basis vectors for the null space of the constraint matrix. The derived symbolic matrix is shown in Equation 24.54, with the rows and columns rearranged so that the order of the states is consistent with that in Equation 24.11.

$$
\mathbf{v} =
\begin{bmatrix}
1 & 0 & 0 \\
0 & 1 & 0 \\
0 & 0 & 1 \\
-\dfrac{A_c\,A_{11}}{A_b\,A_{c1}} & \dfrac{A_{pd}}{A_{c1}} & \dfrac{A_c\,A_{pt}}{A_b\,A_{c1}} \\
0 & \dfrac{A_{pd}}{A_{c2}} & 0 \\
0 & 0 & 0 \\
-\dfrac{A_{11}}{A_b} & 0 & \dfrac{A_{pt}}{A_b} \\
\dfrac{A_{11}(A_b + A_{st} - A_c)}{A_b A_{po}} & \dfrac{A_{pd}}{A_{po}} & -\dfrac{A_{pt}(A_{st} - A_c)}{A_b A_{po}}
\end{bmatrix}
\begin{bmatrix}
v_{11} \\
v_{pd} \\
v_{pt}
\end{bmatrix}
\qquad (24.54)
$$

24.5.5 SIMULATION RESULTS

The simulated responses to a series of step changes in demanded output pressure are shown in Figures 24.7 and 24.8. Each figure shows the simulation results from the following four model implementations:

1. An incompressible fluid model with a full mass matrix
2. An incompressible fluid model with all masses within the servo made equal to zero
3. A purely compressible fluid model, simulated using a "stiff" solver
4. An incompressible fluid model with a zero mass matrix running in discrete time (updated at 1000 Hz)

Figure 24.7 shows for each of the above models the pressure at the internal junction within the servo valve and the controlled pressure at the output port. Figure 24.8 shows the spool position and the resulting flow into the pipe. The dynamic response of each model implementation is very close, demonstrating that little accuracy has been lost by making the assumption that small volumes of fluid within the servo valve act instantaneously.

The transition between systems of reduced order dynamic equations is demonstrated in Figure 24.9, which shows the spool position and the identification number for the model that is being executed. The plot demonstrates that all five modes relating to the second-stage spool have been entered during the simulation.

A comparison of the average time taken to run each of the Simulink models is shown in Table 24.2. The models were simulated using MathWorks MATLAB® version R2006b, running on a PC with an Intel® Dual Core2 CPU running at 2.13 GHz. Model (1) clearly takes the longest time to simulate, demonstrating the detrimental effect of introducing small masses of fluid to act as connecting elements. By comparison, the simulation time of Model (3) is significantly

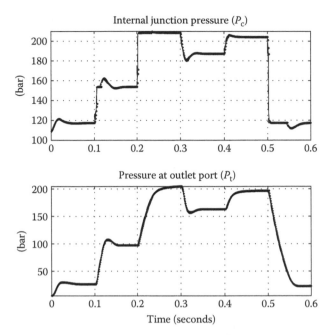

FIGURE 24.7 Simulated junction pressures: incompressible with a full mass matrix (solid line); incompressible with a singular mass matrix (dashdot line); compressible (dashed line); incompressible with a singular mass matrix and discrete time (dotted line).

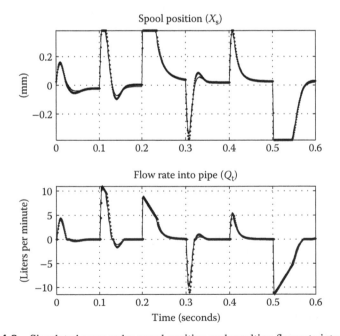

FIGURE 24.8   Simulated servo valve spool position and resulting flow rate into pipe.

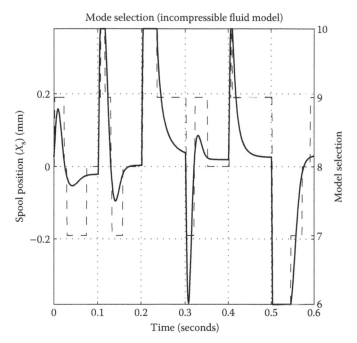

FIGURE 24.9 Simulated switching behavior of the spool position (solid line) and the corresponding model selection (dashed line).

TABLE 24.2
Average Simulink Runtime (Simulated Time = 0.6 s)

Model	Fluid Type	Mass Matrix	Solver	Time (s)
1	Incompressible	Full	ODE45	37.76
2	Incompressible	Singular	ODE45	0.79
3	Compressible	—	ODE23tb	1.90
4	Incompressible	Singular	Discrete	0.31

faster, demonstrating that current "stiff" solvers are very efficient. Simulating this model using the default variable step solver (ODE45) proved to be prohibitively slow by comparison. The simulation time of Model (2) is less than half that of Model (3), which demonstrates the proposed aim of this work, namely that being able to take larger time steps during the simulation because of the removal of the high-frequency modes outweighs the additional computational overhead to implement a DAE solver. The greatest improvement in runtime is achieved with Model (4) where all the continuous time integrators were replaced with discrete time states, and the very fast dynamics of the torque motor were replaced by a constant gain.

24.6 CONCLUSIONS

This chapter has demonstrated a systematic approach to deriving hybrid system models of a hydraulic circuit using a combination of symbolic and numeric techniques. The models are defined in terms of the original circuit layout, but the dynamics of small volumes of fluid are abstracted to instantaneous changes in state. The efficiency of the resulting models, using standard off-the-shelf solvers, has been tested by comparing simulation times with models that use small volumes of compressible fluid as connecting elements and current state-of-the-art "stiff" solvers. The simulation times of the hybrid system models have been shown to be lower than the reference models using a "stiff" solver. The greatest improvement in simulation time is achieved when the hybrid system models are run in discrete time, which is not an option when relying on the "stiff" solvers. The ability to run models in discrete time faster than real time demonstrates the potential to run simulations of hydraulic circuits in true real time, while still maintaining a faithful representation of the inherent nonlinear behavior of the circuit.

24.7 SYMBOL LIST

A	Basis vectors for null space of flow constraint matrix	$\mathbf{W_O}$	Pressure forces across open orifices
B	Basis vectors for null space of reduced order mass matrix	$\mathbf{W_R}$	Constraints due to physical end-stops
$\mathbf{A}$ to $\mathbf{D}$	Matrix coefficients for algebraic constraints	ρ	Fluid density
$\mathbf{F_R}$	Reaction forces	C_d	Discharge coefficient
$\mathbf{F}$	Forcing function	$\Delta \mathbf{P_C}$	Pressure drop across closed orifices
$\mathbf{G_0}$	Algebraic constraints	$\Delta \mathbf{P_O}$	Pressure drop across open orifices
$\mathbf{M}$	Mass matrix	$\hat{\mathbf{v}}$	Dependent states taken from $\mathbf{v}$
$\mathbf{P_J}$	Pressures at internal junctions	$\mathbf{\Pi_a}$	Transformation of $\tilde{\mathbf{v}}_a$ into $\tilde{\mathbf{v}}$
$\mathbf{P}$	Basis vectors for range space of mass matrix	$\mathbf{\Pi_i}$	Transformation of $\tilde{\mathbf{v}}_i$ into $\tilde{\mathbf{v}}$
$\mathbf{V}$	Velocity vector	$\tilde{\mathbf{v}}$	Independent states taken from $\mathbf{v}$
$\mathbf{W_C}$	Flow constraints due to closed orifices	$\tilde{\mathbf{v}}_a$	Algebraic states taken from $\tilde{\mathbf{v}}$
$\mathbf{W_J}$	Flow constraints at junctions	$\tilde{\mathbf{v}}_i$	Integrated states taken from $\tilde{\mathbf{v}}$

ACKNOWLEDGMENT

The authors wish to thank Airbus Operations Ltd for their support during this research.

REFERENCES

1. Pfeiffer, F., and F. Borchsenius. 2004. "New Hydraulic System Modelling." *Journal of Vibration and Control* 10(10): 1493–515.
2. Gillespie, R. B., and J. E. Colgate. 1997. "A Survey of Multibody Dynamics for Virtual Environments." In *Proceedings of the ASME International Mechanical Engineering Conference and Exposition*, Dallas, Texas, Vol. 61, pp. 45–54.
3. Kane, T. R. 1961. "Dynamics of Nonholonomic Systems." *Journal of Applied Mechanics* 28: 574–78.
4. Gillespie, R. Brent. 2003. "Kane's Equations for Haptic Display of Multibody Systems." *The Electronic Journal of Haptics Research* 3: 1–20.
5. Gillespie, R. Brent., V. Patoglu, I. I. Hussein, and E. R. Westervelt. 2005. "On-Line Symbolic Constraint Embedding for Simulation of Hybrid Dynamical Systems." *Multibody System Dynamics* 14(3–4): 387–417.
6. Mosterman, P. J. 1999. "An Overview of Hybrid Simulation Phenomena and Their Support by Simulation Packages." In *Hybrid Systems: Computation and Control: Second International Workshop*. Berlin/Heidelberg: Springer.
7. Merritt, H. E. 1967. *Hydraulic Control Systems*. 2nd ed. New York: John Wiley & Sons, Inc.
8. Wampler, C., K. Buffington, and J. Shu-hui. 1985. "Formulation of Equations of Motion for Systems Subject to Constraints." *Journal of Applied Mechanics* 52: 465–70.
9. Singh, R. P., and P. W. Likens. 1985. "Singular Value Decomposition for Constrained Dynamical Systems." *Journal of Applied Mechanics* 52: 943–48.
10. The MathWorks, Inc. 2006. *Symbolic Math Toolbox User's Guide*. 3.1.5 ed. Natick, MA: The MathWorks, Inc.
11. Rabier, P. J., and W. C. Rheinboldt. 2001. "Theoretical and Numerical Analysis of Differential-Algebraic Equations." *Handbook of Numerical Analysis* 8: 183–542.
12. Anderson, E., Z. Bai, C. Bischof, J. Demmel, J. Dongarra, J. Du Croz, A. Greenbaum, S. Hammarling, A. Mckenney, and D. Sorensen. 1990. "Lapack: A Portable Linear Algebra Library for High-Performance Computers." Technical Report CS-90-105, University of Tennessee, Knoxville, TN.
13. Quintana-Ort, G., X. Sun, and C. Bischof. 1998. A BLAS-3 Version of the QR Factorization with Column Pivoting. *SIAM Journal on Scientific Computing* 19(5): 1486–94.
14. Foster, L. V., and X. Liu. In press. "Comparison of Rank Revealing Algorithms Applied to Matrices with Well Defined Numerical Ranks." *SIAM Journal on Matrix Analysis and Applications*, unpublished manuscript. http://www.math.sjsu.edu/~foster/rank/rank_revealing_s.pdf.
15. Nashed, M. Z., and X. Chen. 1993. "Convergence of Newton-Like Methods for Singular Operator Equations Using Outer Inverses." *Numerische Mathematik* 66: 235–57.
16. The MathWorks, Inc. 2006. *Simulink Reference*. 6.5 ed. Natick, MA: The MathWorks Inc.
17. Tunay, I., E. Y. Rodin, and A. A. Beck. 2001. "Modeling and Robust Control Design for Aircraft Brake Hydraulics." *IEEE Transactions on Control Systems Technology* 9: 319–29.
18. The MathWorks, Inc. 2006. *Stateflow and Stateflow Coder Reference*. 6.5 ed. Natick, MA: The MathWorks Inc.

Index

A

AATC, *see* Advanced Automatic Train Control
Absent-event approach, 268
Abstract buffer, 278
Abstraction levels, 270–271, 286
Abstract semantic space, 305
Abstract specification language, 309–310
Acausal modeling, 557–561
Accelerated simulation, 220, 226
Access Point Event Simulator (APES), 215
Active event set, 392
Active objects, 141
Active–passive interface connections, 265, 268
Active service model interface, 264
Activity-based simulation, 6
Activity diagrams, 131
Actor-oriented programming model, 207
Adams-Bashforth approach, 27
Advanced Automatic Train Control (AATC),
      134–135, 292, 293, 294
Advanced General Aviation Transport
      Experiment (AGATE), 334
Aerodynamic derivatives, 339
Aerospace application, 235
AES, *see* All-electric ship
AGATE, *see* Advanced General Aviation
      Transport Experiment
Aggregate-Level Simulation Protocol (ALSP),
      373–374
Algebraic equations, local singularities in,
      610–611
Algebraic loops, 562
Algorithms Physiologically Derived to Promote
      Learning Efficiency (APPLE), 380
All-electric ship (AES), 233–234
Allocation Modeling (Alloc), 148
ALSP, *see* Aggregate-Level Simulation Protocol
Analog computers, 13
APCI, *see* Application Protocol Control
      Information
API, *see* Application programming interface
APPLE, *see* Algorithms Physiologically Derived
      to Promote Learning Efficiency
Application Layer, 355, 356
Application modeling, 271–272
Application programming interface (API), 248,
      361, 524
Application Protocol Control Information
      (APCI), 356

Application Service Data Unit (ASDU), 356
Application Services, 425–426
Application-specific instruction set processor
      (ASIP), 284–285, 286
Arbitrary event processes
   concurrent simulation approach for
      notation and definitions, 402–404
      observed and constructed sample path
         coupling dynamics, 404–408
      speedup factor, 408–409
      TWA, 401, 409–410
Architecture Analysis and Design Language
      (AADL), 149
Architecture modeling, 272–273, 277
Army Research Institute report, 381–382
Artop project code, 536
ASA, *see* Augmented system analysis
ASDU, *see* Application Service Data Unit
ASIP, *see* Application-specific instruction set
      processor
Assignment statement, 210
Asynchronous activities, TDL, 164
Atomic model, 41, 67, 113
Atomic Rational Time Advance, 70
Atomic RT-DEVS model, 114
Augmented system analysis (ASA), 391,
      399–400
Automatic control (autopilot), 332
Automotive applications, 232–233
Automotive domain, 125
Automotive Open System Architecture
      (AUTOSAR), 523
   behavioral level, 528
   BSW and ECU level, 539–540
   BSW runnable, 545
   deployment tools, 530
   goal of, 526
   methodology, 530–532
   RTE level, 529–530
   simulation and analysis, models for
      in development process, 537–538
      VFB and behavior level, 538
   Simulink model, 540–544
   standard SW platform, 527
   SWCs, 524
   timing model, 532–536
   use and role of, 536–537
   VFB level, 527–528
Automotive RT simulation, modeling
      domains in, 503

Autopilot function in heading control, 339, 340
AUTOSAR, *see* Automotive Open System
    Architecture
Auxiliary propulsion systems, 233

B

Backward differentiation, 13
Balanced hybrid simulation, 22–24
BART system, *see* Bay Area Rapid Transit
    system
Basic software (BSW), 524
Battery model, 552
Bay Area Rapid Transit (BART) system, 134,
    135, 149, 150, 292, 293
Behavior
    class, 141
    diagrams, 131, 132
Behavioral semantics, 140–142
Bifurcated design process, 185–187
Binding-time analysis, 110
Black-box
    approach, 506
    components, 270
Blocking service request, 264, 265
Block-structured languages, 14
Blueprints, 125, 126
Body-worn system, 378
BON, *see* Business Oriented Notation
Boolean functions, 90
Boost.Fusion, 108
Bouncing ball simulation, 19
BSW, *see* Basic software
Buffer
    FIFO, 282, 283
    service model, 279–280
Built-in wind model, 337
Business Oriented Notation (BON), 303

C

C++
    implementation, 367
    template metaprogramming, 106–108
Canadair Regional Jet 200 (CRJ200)
    series, 338
CD++, 66
CDTRP, *see* Controller design-test-redesign
    platform
Centralized simulation, 56
CFD, *see* Computational fluid dynamics
Channel
    model, 193
    usage phase, 191
Charmy tool, 303
Check Train Status sequence diagram, 137
Chrona's Validator, 215

Class diagrams, 131
    domain model as, 138
    UML, 133
Client/server communication ports, 528
Climate control system, 576
Clock, 308, 392
Closed-loop
    real-time simulation, 576
    simulation, architecture of, 204
    testing, 230
CM, *see* Concurrency manager
CO condition, *see* Constructability condition
Code encapsulation concept, 334
Code equivalence problem, 244, 245
Code generation, 170
Cognition algorithm, 190–191
Cognitive modem, 188, 190
Cognitive radio (CR)
    design, 187–188
    network
        design of, 192–194
        design procedure and implementation
            environment, 188–189
        experiment results, 194–196
        single cognitive modem, design of,
            190–192
Combined continuous-discrete benchmark,
    22, 23
Commands Received message, 320–321
Commercial-off-the-shelf (COTS), 185,
    227, 333, 382, 383
Commercial simulation tools, 228
Common Intermediate Language, 110
Common random numbers (CRN), 392
Communicating Sequential Processes
    (CSP), 303
Communication diagrams, 131, 133
Communication-oriented design methods, 37
Complex control systems, 235
Complex electric systems, testing of, 230
Compliant model, 560
Compliant vehicle model, 559, 560
Component diagram, 131
Component Implementation, 532
Component Internal Behavior Description, 532
Compositionality, TDL, 162
Comprehensive methodology, 125
Computational approach, 36
Computational fluid dynamics (CFD), 503–504
Computational parts, 252, 254
Concurrency manager (CM)
    emulation of, 247–248
    real-time, 250–251
Concurrent simulation techniques, 390
Conducting mode, 19–20
CONDUIT, *see* Control Designer's Unified
    Interface

Connect-and-play capability, 334, 335
Connectors, 553, 554, 555
Conservative scheduling, 7
Conservative synchronization, 6
Consistency
    modeling language, 128
    UML, 153
⟨CONSTRAINT⟩, 310, 311
Constraint matrix, 609
Constraints semantics, 305–308
Constructability (CO) condition, 391, 397
Constructed sample path coupling dynamics,
        404–408
Context switch capabilities, 250
Continuous mathematical model, 9
Continuous models, 4–5, 442
    continuous simulations
        example of, 15
        software for, 13–15
        time management for, 9–13
    with discrete element, 19–21
    nature of, 8–9
Continuous simulations
    example of, 15
    software for, 13–15
    time management for, 9–13
Control Designer's Unified Interface
        (CONDUIT), 334
Controller algorithm codes, 334–335
Controller design-test-redesign platform
        (CDTRP)
    benchmark plants of, 473–474
    categorization of modes, 468–470
    experimental setup of developed, 471, 472
    plant emulator card with PIC microcontroller,
        470–471
    structure and functions of, 466–467
    taxonomy of real-time simulation modes,
        463, 464, 467–468
    verification and validation of
        benchmark plants, 475
        controller design-test-redesign procedure,
            475–482, 483
        mixed operating modes, 482, 484
Controller prototype, 224–225
Controller synthesis case, 463
Control station, 350
Control system
    application, 164
    tester simulator, 444
Conventional simulations, 183, 190, 192
Cooperative robotic system, 57
Core elements, 145
Co-simulation, 214, 223
COTS, see Commercial-off-the-shelf
Coupled models, 66, 72, 73, 113, 114
CR, see Cognitive radio

Crankshaft position sensor, 545
CRC, see Cyclic redundancy check
Critica infrastructures (CI), 351
CRJ200 series, see Canadair Regional Jet
        200 series
CRN, see Common random numbers
Cryptography, 352
CSP, see Communicating Sequential Processes
Custom Computer Services C Program Compiler
        Version 4.084 software, 466
Cyber defense technology experimental research
        (DETER), 350, 352
Cycle accurate models, 260
Cyclic redundancy check (CRC), 350, 354

D

DAE, see Differential algebraic equations
DARPA, see Defense Advanced Research
        Projects Agency
DAS, see Digital Analog Simulator
Data aggregator, 350, 359–360
Data dependencies, resolving, 172
Dataflow-oriented design approach, 37
Data generation, 100–101
Data Link Layer, 354, 355–356
DC, see DynamicsController
DDSBE, see Debriefing Distributed Simulation-
        Based ExercisesDeadlock system, 393
Debriefing Distributed Simulation-Based
        Exercises (DDSBE), 381
Debugging, 208–211
Decision-making model, 57
Defense Advanced Research Projects Agency
        (DARPA), 334, 373
Defense Modeling and Simulation Office
        (DMSO), 374, 375
Defense sector, 382–383
Defense training community, 373
Delta-delay
    based representation, 275
    mechanism, 276
Delta event list, 276–277
Department of Defense Architecture
        Framework, 126
Department of Energy (DOE), 351
Deployed execution, 56
Deployment diagrams, 131
DES, see Discrete event systems
Description logic (DL), 298
Design methodology, 246, 252–256
Design space exploration (DSE), 244
DETER, see Cyber defense technology
        experimental research
DEVS, see Discrete Event System Specification
DEVSJAVA models, 47, 50, 51, 53, 189
DieCast, 420–421

Differential algebraic equations (DAE)
    index of, 606–607
    simulating, 600–602
Differential equations in mathematical
        model, 13
Digital Analog Simulator (DAS), 13
Digital signal processors (DSPs), 26,
        182, 188, 227
Digital system models, 6
DIS, *see* Distributed Interactive Simulation
Discrete-based hybrid simulation, 21–22
Discrete event modeling, 274
Discrete event simulation, 204, 205
Discrete event systems (DES), 391, 392–394
Discrete Event System Specification (DEVS),
        53, 183
    atomic model, 65
    cognitive network, 194
    formalism, 6–7, 18
    as formalized aid to system design, 44–45
    formal verification, 68–69
    graph, 67
    model-based system design, 40
    rational time-advance, 69–70
    real-time, 112–114
    real-time distributed systems, design aid and
            verification for
        DEVSJAVA models, 50, 51, 53
        P2P-based distributed real-time
                systems, 55–56
        P2P-based network, 52, 54
        RTI, 50
        UML-RT, 49
    real-time distributed VE, design aid
            for, 45–49
    and RT-DEVS, 40–44
    simulation, metaprogramming application
            to, 115–117
    verification methodology, 71–80
Discrete modeling, 4–8
Discrete optimization approach, 412
Discrete simulation, 7–8
Discrete-time compensation techniques, 222
Discrete-time simulation, 220, 455
DiskSim, 428, 432
Dismounted Infantry Virtual After Action
        Review System (DIVAARS), 381
Distributed AAR, human effort required for,
        380–381
Distributed generation (DG), integration of, 230
Distributed Interactive Simulation (DIS),
        373–374
Distributed real-time
    cooperative robotic systems, 56–59
    simulations, 25
    system semantics, ontology for, 307

Distributed real-time simulation-based training,
        372–373
    challenges
        defense sector, 382–383
        effectiveness assessment, lack of,
                381–382
        fidelity, 377–379
        instructional strategies, 379–380
        instructor workload, 380–381
        interoperability, 375–377
Distributed SBT, 373–375, 382
Distributed simulation, 56, 390
DIVAARS, *see* Dismounted Infantry Virtual
        After Action Review System
DL, *see* Description logic
DML, *see* Domain Modeling Language
DMM, *see* Dynamic metamodeling
DMSO, *see* Defense Modeling and
        Simulation Office
DNP3.0, 350, 351, 354–356
    attacking, 356
    modeling, 357–358
    protocol, 358
    RINSE, 352–354
    TCP/IP, 363
DO-178B standard, 235
Domain architectures, 376
Domain Modeling Language (DML), 353
    and EMS configuration, 365
    network topologies, 364
    PowerWorld, 360
Domain-specific languages (DSLs), 101–105
Domain specific modeling language (DSML),
        421–422
Domain-specific optimization, 111
Domain, TDL, 173–174
DO-254 standard, 235
Drive cycle simulations, 576
Drive design pattern, integrator as, 456
Driver function, 84
DSE, *see* Design space exploration
DSLs, *see* Domain-specific languages
DSML, *see* Domain specific modeling
        language
dSPACE platform, 572
DSPs, *see* Digital signal processors
D-UML, 155
Duration constraints, 147
Dymola, 571
Dynamic allocation scheme *vs.* fixed resource
        allocation vector, 414
Dynamic consistency, 297
Dynamic engine thermal model, 574
Dynamic metamodeling (DMM), 300
Dynamic resource allocation, 412–415
DynamicsController (DC), 201, 203

E

Earliest deadline first (EDF), 249
ECC, *see* Engine cycle calculation
ECD++ toolkit, implementation on, 84–88
E-code, 167, 174
ECS, *see* Engine Controller System
ECU, *see* Electronic control unit
EDF, *see* Earliest deadline first
EDSLs, *see* Embedded domain-specific
    languages
Education application, real-time simulation,
    236–237, 238
Efficient system design, 35
EGR valve, *see* Exhaust gas recirculation valve
Eiffel, 304
Electric train networks, 233–234
Electromagnetic transient (EMT)
    simulators, 223
Electronic control unit (ECU), 124, 530–531
Elementary specification ⟨ELEM-SPEC⟩,
    308, 310, 311
Elevator
    controller model, DEVS graph, 77, 78
    model states, 75
    RTA-DEVS models, 74
    timed automata model, 76
E-machine, 165, 171
Embedded code, 165
Embedded domain-specific languages
    (EDSLs), 103, 104
Embedded DSLs, 103–105
Embedded real-time (RT) software systems, 63
Embedded simulation, 441
Embedded system, typical architecture of, 201
Emergency Brake, 150
    component, 293, 295, 320, 321
    state-machine diagrams for, 152
Emerging applications, real-time
    simulation, 237–239
Energy Management System (EMS),
    365–366, 367
Engine Controller System (ECS), 211, 213
Engine cycle calculation (ECC), 506–507
Engineering systems development process, 333
Environmental Model, 150, 294
Environment inputs, 79
Environment timed automata model, 91
E-puck
    controller DEVS model hierarchy, 82
    robot, 81
    robotic application
        DEVS model specification, 81–84
        ECD++ toolkit, implementation
            on, 84–88
        e-puck models, executing, 88–89

Epuck0 atomic component
    external transition function of, 85–86
    state diagram, 83
ESL, *see* European Simulation Language
Ethernet Capture Effect, 420, 424–425
Euler integration method, 10, 26, 27
Euler method, 15
European Simulation Language (ESL), 17, 23
Event-based models, 442
Event-based simulation, 6
Event Matching algorithm, 400
Event routing, 117
Event-triggered activities, 164
Event Triggering Constraint, 534
Execution Order Constraint, 534
Exhaust gas recirculation (EGR) valve system
    actuation method of, 513, 514
    Newton's Second Law, 514–515
Expanding fidelity, 378
Experimental frames (EFs), 56, 58, 59
Expert analysis, 364
Explicit algorithms, 590
    vs. implicit algorithms, 12
Explicit solvers, 590
Extension mechanisms, UML, 139
External transition functions, 67, 73, 75, 85, 117
Extract ECU-Specific Information, 531
ExtractHistory() function, 316

F

Factory acceptance test (FAT) simulation,
    444–445
Failures Divergence Refinement (FDR), 303
Fair play concept, 377
Fast Fourier Transform (FFT), 195
Fast-switching power electronic devices,
    simulation of, 231
FAT simulation, *see* Factory acceptance test
    simulation
FCS, *see* Flight control system development
    process
FD-DEVS, *see* Finite and Deterministic DEVS
FDR, *see* Failures Divergence Refinement
Feasibility assumption, 401
Federal Communications Commission, 187
Federated Architecture, 524
Field-programmable gate arrays
    (FPGAs), 26, 227
FIFO, *see* First in first out
Filterbank sensing module, 190
Finite and Deterministic DEVS (FD-DEVS), 68
Finite state machine (FSM), 162, 442, 614–615
First column modes, 464, 467
First in first out (FIFO), 248, 282, 283
First-order method, 11

FITE, *see* Future Immersive Training
    Environment
Fixed-cost simulation, 591
Fixed-cost solver, 586–587
Fixed-step implicit solver, 590
Fixed-step solver, 584
Fixed time-step simulation, 220
Fixed *vs.* variable step routines, 11–12
Flat fading wireless channel, 194
Flight control system (FCS) development
    process, 331–332
    challenges of, 341
    code encapsulation concept, 334
    design case study, 338–341
Flight training device (FTD), 337
Flow dependency, UML, 142
FlowNet, 46
FMI, *see* Functional Mockup Interface
Formal composition techniques, 421–422
Formal design methods, 36
Formalized system design, 37, 39
Formal system specification language, 36
Formal techniques, 64
Forward Euler method, 505
4-speed automatic gearbox, 572
Fourth-order Runge-Kutta methods, 11
FPGAs, *see* Field-programmable gate arrays
Freeze mode, TDL module, 173
FSM, *see* Finite state machine
FTD, *see* Flight training device
Functional languages, 106
Functional Mockup Interface (FMI), 571
Function Call Creation, 543–544
Function-Call Subsystem, 170, 171, 172
Fundamental design method, 37
Future Immersive Training Environment
    (FITE), 378

G

Game-based learning (GBL), 443
Generic Component Model (GCM), 149
Generic Quantitative Analysis Modeling
    (GQAM), 149
Generic Resource Modeling (GRM), 148
Global Purpose Simulation System (GPSS), 102
gnu debugger plugin, 208, 209, 210
GNU GCC compiler, 247, 250
Governing equations, 20
Gradient-based optimization approach, 412
Graphical user interface for platform, 467

H

Handwritten simulator, 117
Hardware co-simulation, 214
Hardware description languages (HDL), 244

Hardware-in-the-loop (HIL), 182, 200, 225–226,
    228, 518
    methods, 502, 570
        combustion engines, advantages of,
            512–513
    and RCP, combining, 229
    simulation, 56, 463–464, 468
        motor drive, 236
    testing, 583
Hardware-in-the-loop (HITL) model, 333
Hardware Resource Modeling (HRM), 149
Hardware/software (HW/SW), 244
HDL, *see* Hardware description languages
Heterogeneous embedded systems, 261
Heterogeneous models of computation/
    communication, 155
HEV, *see* Hybrid electric vehicle
Hierarchical coupled model, 42
High-Level Application Modeling (HLAM), 148
High-level architecture (HLA)
    protocol, 374
    simulations, 25
High-speed real-time (HSRT)
    hybrid simulations, 25–26
    multirate simulation, 28–30
    numerical integration for, 26–28
HIL, *see* Hardware-in-the-loop
HITL model, *see* Hardware-in-the-loop model
HLA, *see* High-level architecture
HMI, *see* Human-machine interface
HSCB, *see* Human Social Culture Behavior
HSRT, *see* High-speed real-time
Human-machine interface (HMI), 443
Human Social Culture Behavior (HSCB), 378
HW/SW, *see* Hardware/software
Hybrid
    DSLs, 103
    model, 4, 20
Hybrid electric vehicle (HEV), 574
Hybrid simulations
    examples of, 19–24
    real-time, 24–30
    software for, 16–19
    time management for, 16
Hydraulic cylinder drives model, 454
Hydraulic models, 588
Hydraulic simulation models, 600
Hydraulics models, 607
Hydraulic steering system, 575
Hydraulic system models, 603
HYPERSIM™, 227

I

ICE simulation, *see* Internal combustion engine
    simulation
IEDs, *see* Intelligent electronic devices

if statements, 17, 18
Implicit algorithms *vs.* explicit algorithms, 12
Implicit solvers, 590
Incompressible fluid models, network equations
    for, 602–603
In-cylinder pressure
    closed-loop control, 518
    feedback control, 518–519
Infinite-state systems, 68
Informal methods, 64
Inline- integration method, 512
Input events filtering, 117
Instructional Systems Design (ISD), 374
Instruction set simulators (ISSs), 200,
    214, 244, 246
Instructions generation, 101
Integrated Architecture, 524
Intelligent electronic devices (IEDs), 350
InteractionFragment, 317
Interactive flight control system development
    test bed, 334–338, 346
Interdisciplinary domains, 128, 154
Intermodel communication, 264, 265, 268
Internal combustion engine (ICE)
    simulation, 504
Internal constraint forces, 603
Internal DSLs, *see* Embedded domain-specific
    languages (EDSLs)
Internal transition function, 67, 86
Interoperability, 363, 375–377
Inter-runnable communication, 543
Interrupt management, 251–252
Interrupt service routine (ISR), 206, 207, 251
Invariability assumption, 401
Inverse models, 566–568
Invertible systems, 403
Isaac scenario, 430, 431
ISD, *see* Instructional Systems Design
ISR, *see* Interrupt service routine
ISSs, *see* Instruction set simulators
Issue New Commands, 138
ITEA2 EUROSYSLIB projects, 570
ITEA2 MODELISAR projects, 570, 571

J

Jacobian matrix, 609
Java, 277, 278
JModelica.org, 557
Junction pressures, 605–606

L

Lab setup, 361–363
Language syntax, 528
Large-scale systems, 426
    cooperative robotic, 58

formal composition techniques, 421–422
    simulation techniques, 422–423
Legacy code, modeling and simulating, 214–215
Linear quadratic regulation (LQR) approach, 341
Linear simulation result, 343
Linear undamped pendulum, 490–491, 493
Linear velocity constraints, simulation
    with, 604–606
Link Trainer, 373
Linux RTAI operating system, 574
LISP Code, 106
Live, virtual, and constructive (LVC), 375
Live Virtual Constructive Architecture
    Roadmap (LVCAR), 376
Lock-free synchronization approach, 164
Logical architecture models, 126, 133
Logical execution time (LET), 162, 163
Logical models, 127
LOOP operator, 136
LQR approach, *see* Linear quadratic
    regulation approach
LVC, *see* Live, virtual, and constructive
LVCAR, *see* Live Virtual Constructive
    Architecture Roadmap

M

MAC layer module, 190
Main mode, TDL module, 173
Markov chains, 398
Markov processes, 400
MARTE systems, *see* Modeling and Analysis of
    Real-Time and Embedded systems
Mass matrix, 602
Mathematical models, 8, 9, 443
MathWorks MATLAB®, 617
MathWorks®, 581
MathWorks Simulink®, 611
MathWorks Symbolic Math Toolbox™, 605
MATLAB®, 189
    code, 193–194
    7.04 code, 466
    commands, 586
    version 7.6 platform, 337
Matrix polynomial equations, 608–609
MBE, *see* Model-based engineering
MC, *see* MotorController
MDA®, *see* Model-Driven Architecture®
Mean value engine models (MVEMs), 507
Mechatronics applications, 236, 237
Memory emulation, 434–435
MessageOccurrenceSpecification, 316
Message Origination Time Tag (MOTT), 137
Metaclass, 139
Meta-metaclass Class, 130
MetaML language, 106
Metamodeling architecture, 129–131

Meta Object Facility (MOF™), 130
Metaprogram, 100, 101, 116
    C++ template, 107
    definition of, 99
    to DEVS simulation, application of,
        115–117
    techniques, comparison of, 111–112
    text generation, 100–101
MetaProgramming Library (MPL), 108
Metasimulator, 115
Metropolis, 215
MIDAS, *see* Modified Integration DAS
Minimum order dynamic equations, 607–608
Minimum order momentum equations, 605
Missing events, 403
Mixed-mode integration, 512, 571
Mixed modes, 467
Mobile audio processing platform
    accuracy, 286
    ASIP, 284–285
    overview of, 284
Modbus protocol, 350, 351, 364
Model
    composability, 377
    configuration, 569–570
    continuity, 40, 184, 185
        methodology, 57
    dynamics, 613–614
    validation/verification, 510
Model-based design (MBD), 36, 227–230, 583
Model-based diagnostics, EGR valve
        system, 517
Model-based engineering (MBE), 289–290
    in embedded systems, 124–125
    multiview models and consistency
        challenges, 290–292
    process, 125–127
Model-based system design, 40
Model-checking algorithm, 68
Model-Driven Architecture® (MDA®), 291
Modelica
    acausal modeling, 557–561
    applications from, 572–576
    compiler, 567
    example of, 552–556
    history of, 551–552
    inverse models, 566–568
    language, 551
    model configuration, 569–570
    open platform, 556–557
    real-time language extensions and
        interfaces, 570–571
    symbolic manipulation, 561–566
    tool-specific features, 571–572
    website, 552
Modelica Association, 551

Modelica standard library (MSL), 552, 555, 567
Modeling and Analysis of Real-Time and
        Embedded (MARTE) systems,
        139–140, 290, 294, 319
    basics, 145–149
    example of, 150–152
    nonfunctional properties annotations in, 147
    semantics, 149–150
    SysML and, 144–145
    TimedConstraints in, 148
Modeling and simulation (M&S), 36, 42,
        56, 64, 373
Modeling and Simulation Coordination Office
        (MSCO), 374
Modeling languages, requirements for, 127–129
Models of computation (MoC), 172
    clocked synchronous, 275
    discrete-time, 270
    process states, 268
    service model concept, 269
Modern simulator, 237
Mode transition, 614–615
Modified Integration DAS (MIDAS), 13
Modules, 162
Module Sender, 165, 173, 174
Monte Carlo simulations, 226
MotorController (MC), 201, 203
Motor timed automata model, 90
MPSoCs, *see* Multiprocessor systems-on-chip
M&S, *see* Modeling and simulation
MSCO, *see* Modeling and Simulation
        Coordination Office
MSL, *see* Modelica standard library
MSPLs, *see* Multistage programming languages
Multidisciplinary approach, 236
Multidomain physical system modeling, 576
Multiparadigm language, 106
Multiple simulation tools, 223
Multiprocessor systems-on-chip (MPSoCs), 243
Multirate benchmark simulation, 29
Multirate simulation, HSRT, 28–30
Multistage programming concept, 99
Multistage programming languages
        (MSPLs), 105–108
MVEMs, *see* Mean value engine models

N

Next-generation Expeditionary Warfare
        Intelligent Training (NEW-IT), 380
NFP, *see* Nonfunctional Properties Modeling
Node-i-thread-i models, 53
Node model, 186
NoEvent operator, 563
Noise, vibration, and harshness (NVH), 558
Nonconducting mode, 20–21

Non-domain-specific optimization, 111
Nonfunctional properties, 128, 154
Nonfunctional Properties Modeling (NFP),
    145–146
Noninterruption condition, 392
Nonlinear aircraft model, 339, 340
Nonlinear model for simulation, 341, 342
Nonlinear resistor, 564, 565
Nonlinear simulation, 344
Non-real-time applications, 16
Notion of service, 261
    service model, 263
n-stage executions, 99
Numerical integration algorithm, 5, 12
    types of, 9–11
NVH, *see* Noise, vibration, and harshness

O

OB condition, *see* Observability condition
OBD, *see* On-Board Diagnostics
Object Management Group® (OMG®), 124,
    130, 140, 143, 144, 290
Object orientation concept, 456
Object-orient design approach, 38
Object-oriented formal system specification
    language, 39
Object-oriented modeling language, 18
Observability (OB) condition, 396
Observed sample path coupling dynamics,
    404–408
OCP, *see* Open control platform
ODE, *see* Ordinary differential equations
Offline simulation, 231–232
OMG®, *see* Object Management Group®
On-Board Diagnostics (OBD), 513
Ontology, core elements of, 305, 306
Open control platform (OCP), 334
OpenModelica, 557
Open platform, 556–557
OpenVPN client, 362
Operating system (OS), 243
    emulation layer, 253
Optimistic synchronization, 6
Optimizing real-time simulation, code generation
        and metaprogramming
    concepts and definitions, 99–100
    DSLs, 101–105
    MSPLs, 105–108
    partial evaluation, 108–111
    text generation, 100–101
Ordinary differential equations (ODE), 551,
    561, 562, 600
Original equipment manufacturers (OEM),
    124, 571
OS, *see* Operating system

OutMotor output port, driver interface function
    for, 87–88
Output mapping, DEVS, 84
Overspecialization, 110

P

PADS, *see* Parallel and distributed simulation
PanFS system, 431, 432
Pantelides algorithm, 562
Parallel and distributed simulation (PADS), 371
Parallelism, 101, 188
Parameter abstraction, of small masses, 602
Parametric diagrams, 144
ParkingController (PC), 203
Partial evaluation, 108–111
Passive objects, 141
Passive service model interface, 264
Patient visual object, 48
PC, *see* ParkingController
Perfect controller, 568
    Modelica code, 569
Perfect controller model variant, 569
Perturbation effects, disturbance and
        parameter, 478
Perturbed fluid-aerodynamic force
        (moment), 339
Petri net-based approaches, 37, 39
Petri net programs, 101
Phantom event sources, 409
Physical DC motor plant, 477, 480
Physical simulators, adaptation of, 4
Physical system components, 185
PIC microcontroller 18F452, 466, 470
    driver card, 475, 476
PID, *see* Proportional-integral-derivative
PIL, *see* Processor-in-the-loop
Pilot-in-the-loop (PITL) model, 333
PIM, *see* Platform-independent model
PITL model, *see* Pilot-in-the-loop model
Plant emulator card, hardware of, 470, 471, 472
Platform-based design approach, 215
Platform-independent model (PIM), 291
Platform mapping, 170
Platform model, 262, 272–273
    ASIPs, 285
    mapping, 281
Platform specific model (PSM), 291
Plug-and-play capability, 334, 335
Pneumatic system, 589
Poisson process, 188
Polled report-by-exception, 368
Polynomial equations, solving systems of,
    609–610
Portability, 161
Ports, TDL, 162

Port syntax, 528
Port-type incompatibility, 116
Positive mode, 615
POSIX-Thread, *see* Pthreads
PowerDEVS, 19
Power electronic converters, 224
Power generation applications, 230–232
Power grid, 350–351
Power network creation, 364–365
Power plant simulation, 229
PowerWorld, 350
    EMS, 367
    sample design, 364–365
    State Server, 360
pow function, 106
P2P-based distributed real-time systems, 55–56
P2P-based network, 52, 54
P-Port, 527
Pressure-controlled servo valve, 611–612, 615
    DAE equations, 615–617
    model dynamics, 613–614
    mode transition, 614–615
    simulation results, 617–619
Primary users (PUs), 187, 193, 196
Process-based models, 443
Process-based simulation, 6
Processor-in-the-loop (PIL), 512
Process trainer simulations, 444
Process Wait, 274–275
Producer–consumer application
    abstract buffer, 278
    body of producer service model, 278–279
    buffer service model, 279–280
    framework, 277
    processing element, 281, 282, 283
Profile diagrams, 140
Profile mechanism, 139
    UML, 143–145
Program generator, definition of, 99
Programming language, 125, 126
Progressive simulation-based design
        (PSBD), 182
    CR network
        design of, 192–194
        design procedure and implementation
            environment, 188–189
        experiment results, 194–196
        single cognitive modem, design of,
            190–192
    for networked real-time embedded systems
        bifurcated design process for, 185–187
        conventional simulation, 183
        model continuity, 184, 185
        physical system experiment, 184
        virtual environment simulation,
            183–184
Promela model, 303

Proportional-integral-derivative (PID),
        566, 567
    control structure, 341
Proportional-integral-derivative-based closed-
        loop control, 495
ProtocolMessage class, 353, 363
ProtocolSession class, 353, 363
Prototype Verification System (PVS), 304
Prototyping modes, 468
PSBD, *see* Progressive simulation-based design
Pseudo-Transport Layer, 355, 356
PSM, *see* Platform specific model
Psychological fidelity, 377
Pthreads, 244, 247
    as real-time concurrency model, 248–250
Ptolemy, 207
Ptolemy II integration, 172–174
PVS, *see* Prototype Verification System
Python, 249, 250
    integration, 251

Q

QoS-aware system, 52
Quality of service (QoS), 144
Quantitative performance estimation, 262
Quasi-Boolean logic, 301
Queries semantics, 305–308
⟨QUERY⟩, 311
Queuing models, 5–6
Queuing Network Analysis Package 2
        (QNAP2), 102

R

RACER tool, 298
RACOoN tool, 298
Random sensor inputs, environment
        model, 92
Rapid control prototyping (RCP), 224–225
Rapid prototyping
    model, 333
    of motor controller, 235
Rational Rose tool, 300
Rational time-advance DEVS (RTA-DEVS),
        64, 72, 73
    component, 74
    external transition, 77
R&D, technological and instructional, 374
Reaction forces, 605–606
Realizability, 128, 153
Real-time applications, 16
Real-time atomic model, 44
Real-time behavior, in validator tool, 201–203
Real-time concurrency manager, 250–251
Real-time concurrency model, Pthreads as,
        248–250

Real-time DEVS (RT-DEVS), 40–44, 68, 112–114
Real-time distributed systems, 35
  design aid and verification for
    DEVSJAVA models, 50, 51, 53
    node-i-thread-i models, 53
    P2P-based distributed real-time systems, 55–56
    P2P-based network, 52, 54
    QoS-aware system, 52
    RTI, 50
    service queue AM, 51
    task generation model, 53
    UML-RT, 49
  formal approaches to, 37–39
Real-time distributed VE, design aid for, 45–49
Real-time embedded systems, 184
Real-time hybrid simulations
  high-speed, 25–26
  HSRT
    multirate simulation, 28–30
    numerical integration for, 26–28
    timing issues in, 24–25
Real time immersive network simulation environment (RINSE), 350, 352, 365–366
  C++ implementation, 367
  OpenVPN server, 362
  State Server, 360–361
  view, 366
Real-time interface (RTI), 50
Real-time language extensions and interfaces, 570–571
Real-time OS (RTOS), 243–245
Real-time parts, 252, 254
Real-time scheduler, 250
Real-time simulation, 24, 26, 97–98, 181, 443, 572
  adjusting models to, 596–597
  aerospace application, 235
  all-electric ships and electric train networks, 233–234
  automotive applications, 232–233
  benefits of, 583–584
  challenges of, 584
  code generation and metaprogramming
    concepts and definitions, 99–100
    DSLs, 101–105
    MSPLs, 105–108
    partial evaluation, 108–111
    text generation, 100–101
  discrete-time simulation, 220
  education application, 236–237, 238
  electric drive and motor development and testing, 235–236
  emerging applications, 237–239
  examples of, 581–583

  HIL, 225–226
  MBD using, 227–230
  mechatronics applications, 236, 237
  modes, 463
  moving from desktop to, 585–586
  power generation applications, 230–232
  rapid control prototyping, 224–225
  SIL, 226
  simulator bandwidth requirements, analysis of, 223–224
  technology, 226–227
  timing and constraints, 221–223
  tool, 389, 390
Real-time (RT) system, 139–140, 244, 249, 315
Real-time systems simulator (RTSS), 336–337
Real-time systems simulator and a flight training device (RTSS-FTD), 336, 337
Real-time (RT) techniques, 502
  for automotive systems, 503–505
  in engine simulation
    data preparation, 508–510
    executable simulator, 511–512
    modeling combustion engines approach, 506–508
    model verification and validation, 510–511
  implementations in automotive industry, 512–513
    component diagnostics, 513
    EGR valve, 513–517
    in-cylinder pressure feedback control, 518–519
Real-time units, 148
Real-Time Workshop® Embedded Coder™, 168
Reduced order state vector, 604–605
Reengineer legacy systems, 211–213
REF operator, 137
Register transfer level, 260
Regression
  techniques, 509
  testing, validator, 211, 212
Regular event list, 276
Relative velocity time dilation, 420
Relay, 350
Remote terminal units (RTUs), 350, 351
Renewable energy sources (RES), 230
Report-by-exception, 368
  polling, 368
Reporting phase, 190
Requirement R11, 297
Requirement R12, 297
Requirement validation and verification (RV&V), 332–333
R-E-R mode, implementation of, 487
Reset Timer to Wait Commands, 322
Residual function, 109, 110
Residual program, 108, 117

Resource models, 128, 154
Reverse debugging, 209, 211
Rigid drivetrain, 559
Rigid vehicle model, 559, 560
RINSE, *see* Real time immersive network
    simulation environment
Robot controller timed automata model, 89
Robotic systems, 57
Robot-in-the-loop simulation, 58, 193
Round-robin, 248
R-Port, 527–528
R-R-R operating mode, 468
RTA-DEVS, *see* Rational time-advance DEVS
RT-DEVS, *see* Real-time DEVS
RTE, *see* Run-time environment
RTL model, 285, 286
RTOS, *see* Real-time OS
RTSS, *see* Real-time systems simulator
RTSS-FTD, *see* Real-time systems simulator
    and a flight training device
RT system, *see* Real-time (RT) system
RT techniques, *see* Real-time techniques
RTUs, *see* Remote terminal units
RUBiS, 430, 431
Rule-based approaches, 297–298
Run, definition of, 307
Runge-Kutta methods, 11, 27
Run-time environment (RTE)
    events, 525, 542
    level for AUTOSAR process, 529–530
    simulation at, 538
RV&V, *see* Requirement validation and
    verification

S

Safety control component, 150
Sample path constructability, 395
    ASA, 399–400
    CO condition, 397
    G/G/1/3 and G/G/1/2, state transition
        diagrams, 397, 398
    OB condition, 396
    problem, 391
    SC approach, 398–399
Sample simulation, 554
Sargent Circle, 510, 511
SC, *see* Standard Clock; System calls
SCADA system, *see* Supervisory control and
    data acquisition system
Scalable Simulation Framework (SSF), 352
Schweitzer Engineering Laboratories (SEL)
    devices, 361
Scicos, 557
SDK, *see* Software development kit
SEC, *see* Software-Enabled Control
Secondary Base (SB) station model, 189

Secondary users (SUs), 187–188, 193, 196
S-E-E mode, 475
Select function, 117
SemiLinear operator, 563
Sender/receiver communication ports, 528
Sensing phase, 190
Sensitive server, 412
SEP, *see* Systems engineering process
Sequence diagram, 136
    for Check Train Status, 137
    for computing and delivering train
        commands, 151
Sequential function chart, 456
S-E-R mode, 475
    noise disturbances in, 480–482
Service models, 263–264
    implementations
        composition, 270–271
        states, 268
    interfaces, 263, 264–265
Service queue AM, 51
Service requests, 266–267
S-function, 171, 172, 337, 338
SiE-WG, *see* Simulation in Europe Basic
    Research Working Group
Signatures detection algorithms, 432
SIL, *see* Software-in-the-loop
SimHydraulics models, 582, 588
Similarity assumption, 401
SIMNET, *see* Simulation Network
Simple Time package, 134
Simplified Communication diagram, 138
Simplified linearized dynamics model, 336
Simplified model-based engineering process, 126
Simplified vehicle models
    compliant and, 559
    with rigid drivetrain, 559
Simscape™, 581
Simscape physical network, 590, 593
Simula language, 8
Simulated motor drive, 235
Simulation
    depth of, 454–455
    differential algebraic equations, 600–602
    GBL, 443–444
    languages for, 445–446
    modeling process, 441–443
    for operator training, 440–441
    of plant and machinery, 444–445
    real time, 443
    recent developments and trends, 446
    techniques, 422–423
    time step, 224
    tools, 215
    tool selection and program coding, 455–457
    training, 457–458
    VE, 57

Simulation engine
    discrete event simulation, 274–275
    event lists, 276–277
    time representation, 275–276
Simulation in Europe Basic Research Working
        Group (SiE-WG), 551
Simulation Language for Alternative Modeling
        (SLAM), 102
Simulation Network (SIMNET), 373
Simulation-specific instructional
        strategies, 379
Simulation speed
    ASIPx1-system model, 286
    investigated system models, 287
Simulator
    technology, real-time, 226–227
    training
        control program, reduction of states, 453
        hardware configuration of, 448
        hardware expense, 452
        machine states reduction, 452–453
        requirement analysis of, 447–448
        system design of, 448–451
Simulator bandwidth requirements, analysis of,
        223–224
Simulink®, 590
    and MATLAB®, integration with, 168–172
    model, 540–544
    (version 7.1) platform, 337
    and Ptolemy II integration, 174
    S-function, 171
Simulink extension Stateflow®, 168
Simultaneous internal events scheduling, 117
Single cognitive modem, 191, 197
    design of, 190–192
    simulation model of, 190
Single-step vs. multistep methods, 12
Singular value decomposition (SVD), 604
6-degree-of-freedom (DOF), 573
Sketching, 125
SLERT, see Suse Linux Enterprise Real-Time
Slot selection, 163
Smart grid, 350
Smooth operator, semantics of, 563
Soft real-time systems, 584
Software
    modules, 335–336
    protocols, 375–376
Software components (SWCs), 524
Software co-simulation, 214
Software-defined radio (SDR), 182
Software development kit (SDK), 556
Software-Enabled Control (SEC), 334
Software-in-the-loop (SIL), 200, 201, 226, 512,
        570
    simulation, 464, 468
Software platform architecture, 524

Software Resource Modeling (SRM), 149
Sol, 572
Solid-state drive (SSD) flash drives, 435
Solver Configuration block, 590, 591, 593
Solver settings, tuning, 586–596
Solver technology, 582
Source code, 552, 553
S parallel servers, queueing system with, 410
⟨SPEC⟩, 312, 313–314, 321
Speedup factor, 408–409
SPIN model checker, 303
Spring-damper element, 559
SR models, see Synchronous reactive models
SSF, see Scalable Simulation Framework
S-S-S modes, 468, 469
Stability derivatives, 339
STA modeling framework, see Stochastic Timed
        Automaton modeling framework
Standard Clock (SC)
    approach, 398–399
    method, 391
State event, 16
State-free approach, 507
State invariability assumption, 401, 402
State-machine diagrams
    for Emergency Brake, 152
    for Train Controller, 138
Statement-structured languages, 14
State-of-the-art real-time simulators, 222
State-of-the-art technology, 232
StateSelect language element, 564
State Server, 360–361
State-Space Nodal, 220
State-transition approach, 27–28
Static consistency, 297
Static polymorphism, 117
Station AATC system, 294
Stepwise model refinement approach, 183
Stereotypes, 139
Stiff systems, 13, 454
Stochastic clock structure, 394
Stochastic Timed Automaton (STA) modeling
        framework, 392
Subassembly, 449–450
Substation, 350
Superfluous operation, removal of, 117
Supervisory control and data acquisition
        (SCADA) system, 350, 351, 352, 361,
        363, 364
SUs, see Secondary users
Suse Linux Enterprise Real-Time (SLERT), 575
SVD, see Singular value decomposition
Swapping features, 335, 336
SWCs, see Software components
Symbolic analysis, 562
Symbolic manipulation, 561–566
Synchronization Timing Constraint, 535

Synchronized Lorenz chaotic systems
    analog hardware implementation of,
        482, 484
    master-slave synchronization of, 484,
        485–488
Synchronous generators, 230–231
Synchronous reactive (SR) models, 525
SysML, 144
System call emulation, 246–248
System calls (SC), 246
    input/output, 248
System Configuration
    Description, 531
    Input, 530–531
System-level performance estimation,
        260–261
System model, 262, 273–274
Systems engineering process (SEP), 331–332
    RV&V, 332–333
System simulation concept, 333

T

Table lookup method, 358
Tag definition, 140
Tagged values, 140
Task activation, 162, 163
Task *clientTask*, periodic execution of, 167
Task execution profile plots, 212
Task generation model, 53
Tasks *inc*, periodic execution of, 167
Task source code annotation, 207
Task splitting, 163–164
Taylor series, 10, 11
TCIP, *see* Trustworthy Cyber Infrastructure for
        the power Grid
TCP/IP, *see* Transmission Control protocol/
        Internet protocol
TCP Nagle algorithm problem, 432
TDL, *see* Timing definition language
TDL-Machine, 212, 213
TDLModule actor, 173
TDL toolchain, 174
TDL:VisualCreatortool, 169, 170
TE, *see* Trap emulator
Tearing algorithm, 562
Technical architecture models, 126, 127, 133
Technology integration, 367
Template instantiation, 106
Text generation, 100–101
Thermal network model, 552
Thread-context switching, 278
Thyristor converter, 222
Time-based models, 442
Time coherence, 377
Time-constrained DEVS (TC-DEVS), 69
Time constraint, 147

Timed automata (TA), 38, 64, 65, 71, 75, 89,
        90, 215, 392
    environment model, 91
    UPPAAL, controller model in, 77, 78
TimedConstraints, in MARTE, 148
TimedDescriptionEvent, 532
Time determinism, 161
Timed event, 16
Timed-functional model, 203
Time dilation, 420–421
    with DieCast
        CPU scheduler and network
            emulation, 429
        Ethernet Capture Effect, 430
        paravirtualized vs. fully virtualized
            VMs, 428–429
        TDF and SF, 426–427
    motivation for, 423–424
        Application Services, 425–426
        Ethernet Capture Effect, 424–425
        large-scale system, 426
    in physical memory, 432–435
TimedInstantObservation, 151
TimedObservation, 147
Timed Petri nets, 215
Time management
    for continuous simulations, 9–13
    for discrete simulation, 7–8
    for hybrid simulations, 16
Time model, UML, 154
Time, real-time systems, 128–129
Time safety, TDL, 161
Time synchronization/model composability, 377
Time-triggered activities, 164
Time-triggered system, 161
Time Warp, 8
Time Warping Algorithm (TWA), 391, 401,
        405–408
    extensions of, 409–410
    for real-time optimization, 410–415
Timing definition language (TDL), 160, 211
    language constructs, 162–164
    MATLAB® and Simulink®, integration,
        168–172
    modules, example of, 164–166
    properties, 161–162
    Ptolemy II, integration, 172–174
    toolchain, 166–168
        for MATLAB® and Simulink®, 168–169
Timing diagrams, 131, 134
toMSG(), 316
Touch sensor model, 48
Traceability, 128, 153
Trade-off involving solver choice, 594
Traditional offline simulation software, 230
Train Controller, 150
Training simulator

control program, reduction of states, 453
hardware configuration of, 448
hardware expense, 452
machine states reduction, 452–453
requirement analysis of, 447–448
system design of, 448–451
Transaction-level modeling, 265
Transformation methodology, 71
Transition, 318
Translation-based approaches, 298, 301
Transmission Control protocol/Internet protocol (TCP/IP), 350, 354
Transmitter emulator, 493, 494
Transparent distribution, 161
Transport Canada Authority, 337
Trap emulator (TE), 246
structure of, 247
for thread, 250
Trapezoidal rule, 592
Truncation error, 11
Trustworthy Cyber Infrastructure for the power Grid (TCIP), 352
Tuning parameters, 364
TWA, see Time Warping Algorithm

U

Ultra flexible reversing mill, 440
UML, see Unified Modeling Language
UML-RT, 143–144
UML4SysML, 144
UML 2.x, 149
Unified Modeling Language (UML), 38, 124, 290
AATC system, 134–135
BART system, 134, 135
behavioral semantics, 140–142
consistency
checking, 297–301
management, example of, 315–322
notion of, 312–315
requirements, 296–297
extension mechanism, 139–140
Issue New Commands, 138
metamodeling architecture, 129–131
modeling logical and technical architectures with, 133–134
real time, extensions for, 142–145
REF operator, 137
requirements modeling with, 131–133
semantics, 301–304
abstract specification language, 309–310
notational preliminaries and system formalization, 308–309
queries and constraints, 305–308
specification language semantics, 310–312

sequence diagram, 136
for Check Train Status, 137
Unmanned underwater vehicle (UUV), 29, 30
Unsolicited report-by-exception, 368
Untimed automaton, 392
UPPAAL, 79
deadlock, verification for, 91
elevator verification results in, 80
model checker, 77
TA model, 69, 77, 78
tool, 92
Utility function, 190
UUV, see Unmanned underwater vehicle

V

Validator library, 207
Validator tool suite
basic features of, 205
co-simulation, 214
embedded system validation and verification with, 208–213
legacy code, modeling and simulating, 214–215
real-time behavior in, 201–203
simulation
architecture of, 203–205
setup of, 206–207
solid system verification and validation, 200–201
Value determinism, 161
Value Specification Language (VSL), 145–146, 320
Variable monitoring, 205
Variable-order algorithms, 13
Variable-step solver, 586, 590
Variable vs. fixed step routines, 11–12
Vehicle dynamics module, 336
Vehicle model architecture, 570
Verification, validation, and accreditation (VV&A), 375
Verilog, 285
Vertical consistency, 297, 313, 324–325
VFB, see Virtual Functional Bus
Virtual calls dispatch, 116–117
Virtual data aggregators, 363
control flow, 359–360
Virtual DNP3 protocol, 358
Virtual environment simulation, 183–184, 187, 192, 193, 195
Virtual Functional Bus (VFB), 526
level for AUTOSAR system, 527–528
view, 533
Virtualizing hosts, 359–361
Virtual Power System TestBed (VPST), 352
Virtual relays, 358–359, 364
Virtual robots, 58

Virtual startup simulation, 444
Virtual Test Bed (VTB), 30
Virtual-time simulation mode, 88
Visual objects, behavior validation of, 47
VPST, *see* Virtual Power System TestBed
V-shape diagram, 332
VSL, *see* Value Specification Language
VTB, *see* Virtual Test Bed
VV&A, *see* Verification, validation, and
            accreditation

W

waitFor(event), 275
waitFor(service request, event type), 275
waitFor(time), 274
WaitPoint, 529
WARPED libraries, 423
when clause, 17
Wide open throttle (WOT), 560, 561

X

Xchek model, 301
X3D Scripts, 49
Xen, 426
xlinkit, 300
XML Metadata Interchange (XMI®), 300

Y

Y-chart approach, 261

Z

Zero-crossing events, 587
Ziegler-Nichols method, 479, 483
Zonal electric distribution systems (ZEDS), 234